MILITARY JUSTICE HANDBOOK

Military Justice Handbook

ANTHONY PAPHITI

Second Edition

Howgate Publishing Limited

First published in 2021 by
Howgate Publishing Limited
Station House
50 North Street
Havant
Hampshire
PO9 1QU
Email: info@howgatepublishing.com
Web: www.howgatepublishing.com

British Library Cataloguing-in-Publication Data
A catalogue record for this book is available from the British Library

ISBN 978-1-912440-23-8 (pbk)
ISBN 978-1-912440-24-5 (ebk - PDF)
ISBN 978-1-912440-25-2 (ebk - ePUB)

CONTENTS

CASE LAW

STATUTORY REFERENCES

LIST OF TABLES AND FIGURES

Tables

Figures

FOREWORD

It is an absolute delight to write the foreword to the second edition of what has already become an essential part of the court martial practitioners toolkit. The second edition represents a significant expansion on the first. It covers every aspect of the service justice system and administrative action. The practitioner can now turn to the Handbook for guidance on any topic from the intricacies of service police powers, to how to apply for legal aid or even to decipher the many military acronyms to be found in every case.

Regardless of the topic, Anthony Paphiti is a thoroughly reliable guide to the sometimes difficult terrain of the military justice system. He has a wealth of experience upon which to draw. He was the first legal adviser to the NATO Allied Command Europe Rapid Reaction Corps Headquarters (HQ ARRC) and was involved in the negotiations to establish the Memorandum of Understanding between all participating nations. He was a government adviser in the case of Grieves v United Kingdom [2003], which was heard before the Grand Chamber of the European Court of Human Rights (ECtHR), concerning a challenge to the structure of the UK court martial system as a violation of Article 6 § 1 of the Convention on the grounds of a lack of independence and impartiality and fairness. He also worked with the Armed Forces Bill Team in the development of the Armed Forces Act 2006, which harmonised the disciplinary systems of the three services. As Brigadier Prosecutions he was the day-to-day head of the Army Prosecuting Authority. A former co-vice chairman of the Employed Barristers Committee of the Bar Council, he retired from the army in 2006 and is currently a military law consultant with the Aspals Consultancy. He has given evidence to the Response Systems To Adult Sexual Assault Crimes Panel in the United States, and has submitted advisory papers to the United Kingdom Defence Select Committee of the House of Commons, in respect of combat immunity and hybrid warfare. He is a Visiting Research Fellow in Conflict, Rule of Law and Society at Bournemouth University.

The last few years have, once again, seen great changes, not only in the civilian justice system but also in its service counterpart. The Armed Forces Act 2016 saw new powers for the Director of Service Prosecutions in relation to offenders who assist investigations and the CMAC has again looked at the issue of bare majority verdicts. Prosecutors must now face applications to dismiss the charges. Everywhere, HM forces are asked to do more and more with less and less. There are fewer service personnel, who are now more likely to be stationed in the UK. There are consequently fewer judges and fewer courts. Increasingly, in-house civilian prosecutors are to be seen representing the Crown. The military justice system is under intense scrutiny and has been the subject of criticism, some of which is fair, some decidedly not. At the time of writing 'fresh consideration' is being given to the MoD's decision not to adopt the recommendations of the Service Justice System Review carried out by HH Shaun Lyons. It remains to be seen whether change will be imposed upon the system or not. The court martial must however remain a portable asset able to deploy anywhere in the world so as to support the operational effectiveness of HM forces.

Whether novice practitioner working out the *mens rea* necessary to contravene a standing order, or seasoned courtroom veteran, Brigadier Paphiti's handbook will be an invaluable *vade mecum*.

Peter Glenser QC
August 2020

PREFACE

In 2006, a major systemic reform was introduced in the Armed Forces. The service disciplinary acts of each service (the Army Act 1955, the Air Force Act 1955 and the Naval Discipline Act 1957) were repealed and replaced by the Armed Forces Act 2006 (known as the 'Act') which harmonised the provisions for the three services, creating common offences. The three separate prosecuting authorities were abolished and replaced by a tri-service organisation, the Service Prosecuting Authority (SPA).

Bearing in mind the volume of law and regulation, this book is principally designed to give some useful background information about the service disciplinary system and provide an insight into the main offences charged at summary hearing and court martial. It is a sort of 'road map' of the military justice system. Its size will make it easy to carry around for use as a quick reference tool. The digital version also serves as a quick reference guide, providing the full possibilities of electronic search. It is not an in-depth analysis of the legislation, case law or procedures. To achieve that would require a work of considerable size and would defeat the object of this book as a concise reference work.

I have retained the title of the original edition, although some colleagues from the other services have indicated that it sounds too 'army oriented' and should be renamed as the Service Justice Handbook. I have certainly intended no slight to the other services, for whom I have the greatest respect. I retained the title because all three services are professions of arms and I feel that 'military' represents a true collective description of the profession of those who defend our country. This is encapsulated in the *Your Dictionary* definition: 'The definition of military is the armed forces responsible for securing and defending a country.'[1] Indeed, the Court Administration Officer is head of the Military Court Service, which serves all three services.

Since the first edition of this book was published at the end of 2013, there have been changes introduced by two subsequent Armed Forces Acts of 2011 and 2016, together with their respective subordinate legislation. Consequently, it was considered necessary to update the original text to reflect them. In this edition, I have attempted to break down further some of the complexity of the legislation, which, at times, is difficult to follow, to render the system more comprehensible. Suffice to say, it has still been necessary to include many statutory references for readers to understand better the sources of the propositions set out, so that they may refer to them for further reading.

The body of law relating to the military is huge and it would be an extraordinary task to cite it all in a manageable book. As the size of the text has grown, I looked for any parts of the first edition that might be safely removed. To this end, I decided to remove the chapter on operational cases and rules of engagement. The relevance of self-defence to rules of engagement is amply covered by the criminal law in section 76 of the Criminal Justice and Immigration Act 2008 and by Article 2 of the European Convention on Human Rights. Responding to an order to do something – or not – may fall to be evaluated in the context of individual self-defence. This produces an uneasy tension with the traditional view that orders must be obeyed, as self-defence is a legal defence of the individual that overrides an order, whether legal or illegal. There are some military lawyers who, for understandable reasons, disagree with my view on this matter on the basis, inter alia, that it is for leaders to lead and to require their lawful orders to be obeyed, as they are the ones that direct the battle plan and see each part of it in context of the overall plan. In any event, this is very much a topic more suited to operational law rather than military justice and, for that reason, it has been omitted in this edition.

I did consider removing the chapter on civilians in the Service Justice System but decided against it as I was advised that dependants and other civilians subject to service discipline remain overseas, in places like Germany and Cyprus, despite the significant drawdown in our military personnel there. With the government expressing a desire to redefine its role as a global power prepared to intervene against countries that 'flout international law', and, according to Gavin Williamson, the former UK defence secretary, there is an intention to open two new military bases overseas in the Caribbean and in south-east Asia, so nothing is certain.

Some parts of the text, for example, those referring to powers of entry, stop, and search, dealt with in Chapter 5, are repeated, but this is because the provisions themselves, which cater for actions by the service police and those authorised by the Commanding Officer, are very similar and are repeated in the respective statutory references. These situations have been dealt with separately, so that the reader can refer directly to the relevant section of the book. I have tried to minimise the need to cross reference, but it is still necessary.

The changes in the 2006 Act ensure the Commanding Officer remains at the centre of the disciplinary process but, as discussed in Chapter 8 on Summary Hearings, they no longer have the power to prevent serious offences from going to trial.

In relation to court martial proceedings and appeals, a brand new power was introduced by section 321 and Schedule 11 of the Armed Forces Act 2006, which permits the Criminal Cases Review Commission to make a reference to the Court Martial Appeal Court (CMAC) for appeal of a conviction, verdict, finding or sentence with or without an application being made by the person to whom it relates. This is discussed in Chapter 13, Appeals.

The end of a period of service does not always mean the end of association with the Armed Forces. Chapter 3, Disciplinary System, discussed how some individuals have what is known as a reserve liability or are members of the Reserve Forces and subject to service law in a more limited way, governed by the Reserve Forces Act 1996 and the amendments made by the Armed Forces Act 2006.

At the end of 2013, the government created a new Joint Cyber Reserve unit, part of the Joint Forces Cyber Group, which targets recruits from three sectors: regular personnel leaving the Armed Forces, current and former reservists with the necessary skills, and individuals with no previous military experience, but with the technical knowledge, skills, experience and aptitude to work in this highly specialised area. All personnel applying to join are subject to a security clearance process. They work alongside regular forces.

In order to avoid duplication, where the jurisdiction covers both civilians and service personnel, the relevant provisions will appear under the most appropriate heading. For example, the section concerned with trial and sentencing of civilians in the court martial is to be found in Chapter 15, Civilians in the Service Disciplinary System, and does necessitate some dipping back and forth, but it was felt unavoidable.

In this second edition I have included the updated repeals and activations brought about by the AFA2011(C4)O 2013, effective from 1 November 2013, and the new Chapter 3A inserted into the AFA 2006, as modified by the AFA 2016, concerning Testing For Alcohol and Drugs. This is included at Appendix N.

For those unfamiliar with the three services, their structures and language, entering the world of military law can be a daunting prospect. Military dialogue often relies upon abbreviations and in-service 'speak' which can sound like a foreign language to those from civilian life. My wife was rather bemused when, on joining me at my new posting in Germany, which was her first within a service community, she was introduced at a social occasion to the ADC to the CinC. She politely replied, 'how nice', clueless as to who the individual was or what he did. It took several weeks of immersion in the military environment for her to become acclimatised to the new language. Sadly, unlike the owners of the *Military Justice Handbook*, she did not have a helpful *vade mecum* at her side to guide her through many of the linguistic mysteries that confronted her. The list of commonly used abbreviations at the beginning of this book will assist the reader to better understand documents and references within them which may be contained in case papers for disciplinary purposes or correspondence from assisting officers.

The other major problem for anyone new to the service environment is that they are confronted with a huge range of titles and ranks which are completely alien to them. In chambers, the silks walk on water, as do the partners in a solicitors' office. But everyone else is a professional equal. However, in the services it would be unfortunate, for example, to confuse a corporal with a colonel, no disrespect intended to either. Both will have earned their rank. The service defendant will show their legal adviser both courtesy and respect so, when advising a service defendant, it therefore helps to demonstrate an understanding of their world and show them respect according to their rank. Usefully, the rank structures for the three services are set out in Appendix A.

There is often an additional feature which it is important for civilian legal advisers to recognise when they enter the service environment: service personnel may be holders of a medal and/or a military honour for gallant or distinguished service. While they may be modest about their achievements, being able to recognise these awards will give the legal adviser a better insight into the character of the person they represent. So being able to identify these awards is truly important. A selection of medals and awards are explained in Chapter 17.

Readers are asked to understand that internet references, although correct at the time of publication, are subject to change at short notice. I would also mention that the use of personal pronouns has been made as non-gender specific as possible but, where legislation has been quoted or referred to, I have tried not to alter the text. Of course, a reference to any particular gender is to be taken as all inclusive, unless the context makes it otherwise clear.

Finally, I would like to extend my thanks to my family for supporting me during the revision process, particularly my granddaughter, Lea, who is never slow to engage me in debate and challenge me, and my grandson, Alexander, who was very eager to assist with the typing. I would also like to thank, once more, my colleague Rueben Scott, an experienced former military policeman and now a solicitor practising in the military justice field. I am also deeply indebted to Peter Glenser QC, an experienced court martial advocate who has appeared in some of the most serious court martial cases, appearing as both prosecutor and defence counsel; to Matthew Bolt, Secretary of the Association of Military Advocates (AMCA) and a member of the criminal bar, and Allan Steele of Armed Forces Legal Action (AFLA), who took time to review parts of the book and make some very helpful suggestions. Lord Thomas, the Honorary Chairman of the Association of Military Court Advocates, Reuben and Matthew, along with other members of AMCA, have been very supportive of this project from the beginning. Finally, I would like to thank my publishers, Howgate Publishing, in particular I would mention Kirstin Howgate, who was most enthusiastic and encouraging about my draft manuscript, and Colonel Grant Davies, L/AGC(ALS), a fellow lawyer, who spent long hours reviewing the text and to whom I am deeply grateful for his insightful and constructive comments and observations, and Janet MacMillan who painstakingly sought to forge the final draft into shape. Nevertheless, any errors or omissions are my responsibility alone.

Anthony Paphiti
December 2020

ABBREVIATIONS AND GLOSSARY

2*	An officer holding the rank of rear admiral, major-general or air vice-marshal
2IC	Second in Command
2Lt	Second Lieutenant (army), Midshipman (RN), Pilot Officer (RAF)
AA 55	Army Act 1955
AAO	Accused's Assisting Officer
AB	Army Book; Army Board; Able Seaman, Able Rate (equivalent to private in the army and airman in the RAF)
ACOS	Assistant Chief of Staff
Acting Rank	A temporary rank, senior to the substantive rank held, which is paid at rates attributable to that higher rank. Pension also accrues at the rate attributable to the higher rank. Officers holding acting rank take precedence over all those holding local rank of the same grade.
ADC	Aide-de-Camp
Adjt	Adjutant
AF(ALPSCD)R 2013	Armed Forces (Alcohol Limits for Prescribed Safety-Critical Duties) Regulations 2013
AF(CCDSO)(MCJA2003)R 2009	Armed Forces (Civilian Courts Dealing with Service Offences) (Modification of the Criminal Justice Act 2003) Regulations 2009
AF(CM)(A)R 2013	Armed Forces (Court Martial) (Amendment) Rules 2013
AF(CM)R 2009	Armed Forces (Court Martial) Rules 2009
AF(CP)R 2009	Armed Forces (Custody Proceedings) Rules 2009

AF(CRFC)O 2009	Armed Forces (Conditional Release from Custody) Order 2009
AF(CSSD)O 2009	Armed Forces (Civilians Subject to Service Discipline) Order 2009
AF(CWC)R 09	Armed Forces (Custody Without Charge) Regulations 2009
AF(DP)R 2009	Armed Forces (Disposal of Property) Regulations 2009
AF(DT)R 2009	Armed Forces (Drug Testing) Regulations 2009
AF(DTRF)(2)R 2009	Armed Forces (Discharge and Transfer to the Reserve Forces) (No. 2) Regulations 2009
AF(EIATSC)R 2009	Armed Forces (Evidence of Illegal Absence and Transfer to Service Custody) Regulations 2009
AF(EPCC)R 2009	Armed Forces (Evidence in Proceedings before Civilian Courts) Regulations 2009
AF(FD)R 2009	Armed Forces (Forfeitures and Deductions) Regulations 2009
AF(FPEO)(A)R 2013	Armed Forces (Financial Penalty Enforcement Orders) (Amendment) Regulations 2013
AF(FPEO)R 2009	Armed Forces (Financial Penalty Enforcement Orders) Regulations 2009
AF(FS)(2)R 2009	Armed Forces (Forfeiture of Service) (No 2) Regulations 2009
AF(FS)R 2009	Armed Forces (Forfeiture of Service) Regulations 2009
AF(ITADT)R 2013	Armed Forces (Interpretation, Translation and Alcohol and Drug Tests) Rules 2013
AF(MA)R 2019	The Armed Forces (Miscellaneous Amendments) Regulations 2019
AF(MCO)R 2009	Armed Forces (Meaning of "Commanding Officer") Regulations 2009
AF(MPLPRR)R 2009	Armed Forces (Minor Punishments and Limitation on Power to Reduce in Rank) Regulations 2009
AF(P)(C)R 2009	Armed Forces (Proceedings) (Costs) Regulations 2009
AF(P5AFA2006)R 2009	Armed Forces (Part 5 of the Armed Forces Act 2006) Regulations 2009

AF(PANOO)O2019	Armed Forces (Prescribed Air Navigation Order Offences) Order 2009
AF(PSSSSR)(A)0 2012	Armed Forces (Powers of Stop and Search, Search, Seizure and Retention) (Amendment) Order 2012
AF(PSSSSR)(A)O 2014	Armed Forces (Powers of Stop and Search, Search, Seizure and Retention) (Amendment) Order 2014
AF(PSSSSR)O 2009	Armed Forces (Powers of Stop and Search, Search, Seizure and Retention) Order 2009
AF(RF)O 2013	Armed Forces (Remission of Fines) Order 2013
AF(ROCMS)(SP) R 2009	Armed Forces (Review of Court Martial Sentence) (Supplementary Provision) Regulations 2009
AF(ROCMS)O 2009	Armed Forces (Review of Court Martial Sentence) Order 2009
AF(RSO)O 2013	Armed Forces (Retrial for Serious Offences) Order 2013
AF(SAC) R 2009	Armed Forces (Summary Appeal Court) Rules 2009
AF(SCC) R 2009	Armed Forces (Service Civilian Court) Rules 2009
AF(SCFA)A 2015	Armed Forces (Service Complaints and Financial Assistance) Act 2015
AF(SCMP)R 2015	Armed Forces (Service Complaints Miscellaneous Provisions) Regulations 2015
AF(SCR)(A)R 2015	Armed Forces (Service Courts Rules) (Amendment) Rules 2015
AF(SHASSSD)R 2009	Armed Forces (Summary Hearing and Activation of Suspended Sentences of Service Detention) Rules 2009
AF(SI)R 2008	Armed Forces (Service Inquiries) Regulations 2008
AF(SSPO)R 2009	Armed Forces (Service Supervision and Punishment Orders) Regulations 2009
AF(WASO)R 2009	Armed Forces (Warrants of Arrest for Service Offences) Rules 2009
AFA	Armed Forces Act (followed by year)
AFA 2006	Armed Forces Act 2006
AFA 2016(CN3)R 2019	Armed Forces Act 2016 (Commencement No. 3) Regulations 2019

AFA2006(AOS2)O 2018	Armed Forces Act 2006 (Amendment of Schedule 2) Order 2018
AFA2006(TP)O 2009	Armed Forces Act 2006 (Transitional Provisions, etc.) Order 2009
AFA2011(C1TTP) O 2012	Armed Forces Act 2011 (Commencement No. 1, Transitional and Transitory Provisions) Order 2012
AFA2011(C2)O 2012	Armed Forces Act 2011 (Commencement No. 2) Order 2012
AFA2011(C3)O 2013	Armed Forces Act 2011 (Commencement No. 3) Order 2013
AFA2011(C4)O 2013	Armed Forces Act 2011 (Commencement No. 4) Order 2013
AFA2016(C3)R 2019	Armed Forces Act 2016 (Commencement No. 3) Regulations 2019
AFA2016(CN2)R 2018	Armed Forces Act 2016 (Commencement No. 2) Regulations 2018
AFDA 2000	Armed Forces Discipline Act 2000
AGAI	Army General and Administrative Instructions AGAI 67 covers policy and procedure for administrative punishments.
ALS	Army Legal Service
ANO 2005	Air Navigation Order 2005
AOC	Air Officer Commanding
ARRC	Allied Command Europe Rapid Reaction Corps (ACE Rapid Reaction Corps)
ASBCPA 2014	Anti-social Behaviour, Crime and Policing Act 2014
ASBCPA2014 (C8STP)O 2015	Anti-social Behaviour, Crime and Policing Act 2014 (Commencement No. 8, Saving and Transitional Provisions) Order 2015
AVM	Air Vice Marshall (RAF), equivalent to VAdm (RN), Maj Gen (army)
AWOL	Absent without leave
BCM(CM)	Better Case Management in the Court Martial
BM Force	11 Group (RAF) Battlespace Management

Bn	Battalion
Bombardier	A person holding the rank of Corporal, serving in the Royal Regiment of Artillery (Bdr). Lance Bombardier (LBdr) equates to Lance Corporal.
BPS	Building Performance Standards
BRd2	Book of Rules Volume 2 (Queens Regulations for the Royal Navy)
Brig	Brigadier
Brigade	A Brigade will comprise between 3,000 and 5,000 soldiers. It may consist of between three and six battalions/regiments.
BSA	Boarding School Allowance
BSM	Battery Sergeant Major
CAO	Court Administration Officer
CB	Commander of the Order of the Bath
CBE	Commander of the Order of the British Empire
CCA 2013	Crime and Courts Act 2013
Cdr	Commander (RN); equivalent to Lt Col (Army and Wg Cdr (RAF)
Cdre	Commodore RN, equivalent to a brigadier (army), Air Commodore (RAF)
CDT	Compulsory drug testing
Cfn	Craftsman (Private in the Royal Electrical and Mechanical Engineers)
CFR-EU	Charter of Fundamental Rights of the European Union
CGC	Conspicuous Gallantry Cross
Chf Tech	Chief Technician
CinC	Commander in Chief
CJ(AFCPVC)R 2015	Criminal Justice (Armed Forces Code of Practice for Victims of Crime) Regulations 2015
CJA	Criminal Justice Act (followed by the relevant year)
CJCA2015 (CNo1STP)O2015	Criminal Justice and Courts Act 2015 (Commencement No. 1, Saving and Transitional Provisions) Order 2015
CJCSA 2000	Criminal Justice and Court Services Act 2000

CJIA 2008	Criminal Justice and Immigration Act 2008
CJPOA 1994	Criminal Justice and Public Order Act 1994
CJPOA1994 (AAF)O 2009	Criminal Justice and Public Order Act 1994 (Application to the Armed Forces) Order 2009
CLA 1977	Criminal Law Act 1977
CM(A)A 1951	Courts-Martial (Appeals) Act 1951
CM(A)A 1968	Court-Martial (Appeals) Act 1968
CM(A)R 1997	Court Martial (Army) Rules 1997
CM(PA)O 2006	Courts-Martial (Prosecution Appeals) Order 2006
CM(PA)O 2009	Court Martial (Prosecution Appeals) Order 2009
CMAC	Court Martial Appeal Court
CMAC(B)O 2009	The Court Martial Appeal Court (Bail) Order 2009
CMAC(E)O 2009	Court Martial Appeal Court (Evidence) Order 2009
CMAC(IT)R 2013	Court Martial Appeal Court (Interpretation and Translation) Rules 2013
CMACR 2009	Court Martial Appeal Court Rules 2009
CMAO	Court Martial Administration Office
CMASCC (YJCEA1999)(A) R2015	Court Martial and Service Civilian Court (Youth Justice and Criminal Evidence Act 1999) (Amendment) Rules 2015
CMASCC (YJCEA1999)R 2009	Court Martial and Service Civilian Court (Youth Justice and Criminal Evidence Act 1999) Rules 2009
CMCHA 2007	Corporate Manslaughter and Corporate Homicide Act 2007
CMPCM	Practice in the Service Courts Collected Memoranda
CM-SACG	The Court Martial and the Summary Appeal Court Guidance
CO	Commanding Officer
CO Regs	Armed Forces (Meaning of "Commanding Officer") Regulations 2009
CoJA 2009	Coroners and Justice Act 2009
CoJA(13)O 2013	Coroners and Justice Act 2009 (Commencement No. 13) Order 2013

Col	Colonel
COS	Chief of Staff who often acts as a second-in-command of a headquarters.
Coy Comd	Company Commander
CP(AF)A 1987	Crown Proceedings (Armed Forces) Act 1987
CPA	Coalition Provisional Authority
CPA 1947	Crown Proceedings Act 1947
CPIA1996(AAF)O 2009	The Criminal Procedure and Investigations Act 1996 (Application to Armed Forces) Order 2009
CPIA1996(CP)(AF) O 2009	The Criminal Procedure and Investigations Act 1996 (Code of Practice) (Armed Forces) Order 2009
Cpl	Corporal
CPO	Chief Petty Officer (RN), equivalent to SSgt (Army), Flt Sgt (RAF)
CPSG	Crown Prosecution Service Guidance
CQMS	Company Quartermaster Sergeant
CSM	Company Sergeant Major
C-SPT	Commander for Support
D(TF)A 1964	Defence (Transfer of Functions) Act 1964
DAO	Defendant's Assisting Officer
DBS	Disclosure and Barring Service
DC	Defence Council
DCM	District Court Martial
DCOS	Deputy Chief of Staff
DFC	Distinguished Flying Cross
Director (The)	The Director Service Prosecutions (same as DSPA)
DIU	Defence Inquests Unit
Div	Division eg 1 (UK) Armd Div. A body of approximately 15,000 personnel.
DJAG	Deputy Judge Advocate General
DO	Deciding Officer
DPA	Data Protection Act
DPP	Director of Public Prosecutions

DSC	Distinguished Service Cross
DSO	Distinguished Service Order
DSPA (or DSP)	Director, Service Prosecuting Authority
DTO 2009	Drug Testing Regulations 2009
DVCVA	Domestic Violence, Crime and Victims Act 2004
EAW	Expeditionary Air Wings
ECAB	Executive Committee of the Army Board
ECHR	European Convention on Human Rights and Fundamental Freedoms
ECtHR	European Court of Human Rights
EN	Explanatory Notes to legislation (subject to Crown Copyright)
ESC	European Social Charter
FLAGO	Fleet General and Administrative Orders
Flight	RAF sub-unit of a squadron, usually divided into two sections, each containing two to three aircraft, which share ground staff with the other section, and are usually commanded by a Flight Lieutenant.
Flt Lt	Flight Lieutenant (RAF), equivalent to Captain (Army/RM), Lt (RN)
FO	Flying Officer (RAF), equivalent to Lt (Army/RM), Sub Lt (RN)
FOSF	Flag Officer Surface Fleet
FOSM	Flag Officer Submarines
FOST	Flag Officer Sea Training
FSgt	Flight Sergeant (addressed as 'Flight')
Fus	Fusilier (equivalent rank of Private in the Fusiliers)
FW	Formal Warning
G1	Staff branch responsible for personnel matters including manning, discipline and personnel services
G2 (J2)	Staff branch responsible for intelligence and security. In a joint headquarters with another Service or in a multinational HQ the branch title is prefixed by a 'J' to signify 'Joint'.

G3 (J3)	Staff branch for military operations, planning and exercise planning
G4 (J4)	Staff branch responsible for Logistics and quartering
G5 (J5)	Staff branch responsible for strategic planning and policy
G6 (J6)	Branch Communications and IT
G7	Training branch
G8 (J8)	Branch Resource Management (Finance and Contracts)
G9 (J9)	Branch responsible for CIMIC (Civil and military cooperation)
Gdsm	Guardsman (rank equating to Private)
Gnr	Gunner (Rank equating to Private in Royal Artillery)
GOC	General Officer Commanding
GPMG	General Purpose Machine Gun
GSM	General Service Medal
GSMLD	Garrison Sergeant Major, London District
HCR	Household Cavalry Regiment
Higher Authority	Any officer in the commanding officer's disciplinary chain of command who is superior in that chain of command to the commanding officer
Host Nation	The country within whose borders the force of a Sending State and civilian component is based
HQ	Headquarters
I. GDS.	Irish Guards
ICC	International Criminal Court created pursuant to the Rome Statute
IFF	Identification Friend or Foe
IFOR	Implementation Force
IH	Initial Hearing
IHAT	Iraq Historic Allegations Team
IHL	International Humanitarian Law (namely, the law of armed conflict and the Geneva Conventions and Protocols)

INT CORPS	Intelligence Corps
INTSUM	Intelligence Summary
ISAF	International Security Assistance Force
ISTAR	Intelligence Surveillance Target Acquisition and Reconnaissance
JA	Judge Advocate (see also JAG and VJAG)
JA Memoranda	Practice in the Court Martial: Collected Memoranda
JAG	Judge Advocate General
JFACHQ	Joint Force Air Component Headquarters
JHQ	Joint Headquarters
JNCO	Junior NCO
Jnr Tech	Junior Technician
JOCS	Joint Operational Command Systems
JSP	Joint Service Publication
jus ad bellum	The law justifying going to war
jus in bello	The law governing hostilities (i.e. the law of armed conflict)
KFOR	Kosovo Force
Kgn	Kingsman (private in the Kings Regiment)
KRH	King's Royal Hussars
L/Sgt	Lance Sergeant – in the Foot Guards and Honourable Artillery Company, being a rank which equates to Corporal (although they wear three chevrons)
LAC	Leading Aircraftman
LASPOA 2012	Legal Aid, Sentencing and Punishment of Offenders Act 2012
LASPOA2012 (C4SP)O 2012	Legal Aid, Sentencing and Punishment of Offenders Act 2012 (Commencement No. 4 and Saving Provisions) Order 2012
LASPOA2012 (C9SPSCD)O 2014	Legal Aid, Sentencing and Punishment of Offenders Act 2012 (Commencement No. 9, Saving Provision and Specification of Commencement Date) Order 2014
LBdr	Lance Bombardier (LCpl in the Artillery)
LCpl	Lance Corporal

LD	Light Dragoons
LH	Leading Hand / Leading Rate (eg Leading Seaman, Leading Regulator), can be colloquially referred to as 'killick'.
LOAC	The law of armed conflict
Local Rank	A temporary rank higher than the substantive rank held, which does not confer financial benefits unless specifically allowed by regulations. It is to be relinquished on the date that the holder ceases to fulfil the function for which local rank was granted.
LoU	Letter of Understanding
LS&GC	Long Service and Good Conduct Medal
Lt	Lieutenant (army), equivalent to Sub Lieutenant (RN) and Flight Lieutenant (RAF)
Lt Col	Lieutenant Colonel (army), Cdr (RN), Wg Cdr (RAF)
MAA	Minor Administrative Action
MAcr	Master Aircrew
Maj	Major, equivalent to Lt Cdr (RN) and Sqn Ldr (RAF)
Maj Gen	Major General (the rank of a Divisional Commander)
MC	Military Cross (gallantry award)
MCA	Magistrates Courts Act 1980
MCS	Military Court Service
MCTC	Military Corrective Training Centre, Colchester
MDP	Ministry of Defence Police
MDR	Minimum Drawing Rate
MiD	Mention in Despatches (gallantry award) – no longer awarded
Mid	Midshipman (RN), 2Lt (army), Pilot Officer (RAF)
MJS	Military Justice System
Mne	Marine
MoU	Memorandum of Understanding
MSL	Manual of Service law
NAAFI	Navy, Army, and Air Force Institutes

NATO	North Atlantic Treaty Organisation
NDA 1957	Naval Discipline Act 1957
OAPA 1861	Offences Against the Person Act 1861
OC	Officer Commanding a sub unit, such as a Troop or Squadron
OJAR	Officer's Joint Appraisal Report
ORA 2014	Offender Rehabilitation Act 2014
ORA2014(C2)O 2015	Offender Rehabilitation Act 2014 (Commencement No. 2) Order 2015
PA 1997	Police Act 1997
PA1997(CRCRM)(A)(EW)O 2013	Police Act 1997 (Criminal Record Certificates: Relevant Matters) (Amendment) (England and Wales) Order 2013
Pam(Air)	Pamphlet (Air)
PCC(S)A 2000	Powers of Criminal Courts (Sentencing) Act 2000
PFA2012	Protection of Freedoms Act 2012
PFA2012(CA)O 2013	Protection of Freedoms Act 2012 (Consequential Amendments) Order 2013
Phase 1 Training	Training which provides basic military skills
Phase 2 Training	This is Trade Training, specific to the role or unit or branch to be joined.
Phase 3 Training	Further specialist trade training
PIT	Public interest test
PJHQ	Permanent Joint Headquarters
PLAGO	Personnel, Legal, Administrative and General Orders
Pln Comd	Platoon Commander
PO	Petty Officer (RN), equivalent to Sgt (Army), Sgt (RAF)
PO	Pilot Officer (RAF), equivalent to 2nd Lt (Army / RM), Midshipman (RN)
PTHP	Plea and Trial Preparation Hearing
PVR	Premature Voluntary Release (from service in the Armed Forces)
QCB	Queen's Commendation for Bravery (gallantry award)

QCVS	Queen's Commendation for Valuable Service
QDG	Queen's Dragoon Guards
QMSI	Quarter Master Sergeant Instructor
QOY	Queen's Own Yeomanry
QRH	Queen's Royal Hussars
QRs	Queen's Regulations (issued in respect of each of the three services under Royal Prerogative by the Defence Council, setting out the policy and procedure to be observed in command and administration)
RAdm	Rear Admiral (RN), equivalent to Lt General(army), Air Marshall (RAF)
RCM	Regimental Corporal Major
RDG	Royal Dragoon Guards
Receiving State	The state in whose territory the force or civilian component of a Sending State is located, whether it be stationed there or passing in transit.
Regulator	A term in the Royal Navy for a policeman
Rfn	Rifleman
RFO 1996	Reserve Forces Act 1996
RFR	Royal Fleet Reserve
RL	Royal Lancers
RMAS	Royal Military Academy Sandhurst
RMP	Royal Military Police
ROA 1974	Rehabilitation Of Offenders Act 1974
ROA1974(E)O 1975	Rehabilitation of Offenders Act 1974 (Exceptions) Order 1975
RQMS	Regimental Quarter Master Sergeant
RSM	Regimental Sergeant Major
RTR	Royal Tank Regiment
RWxY	Royal Wessex Yeomanry
RY	Royal Yeomanry
SAC	Senior Aircraftman

SAC(YJCEA1999) R 2015	Summary Appeal Court (Youth Justice and Criminal Evidence Act 1999) Rules 2015
SCC	Service Civilian Court
SCC Rules	Armed Forces (Service Civilian Court) Rules 2009
SCO	Service compensation order
SCOP	Service police codes of Practice
SCSRSR 2009	Service Custody and Service of Relevant Sentences Rules 2009
SDAs	Service Discipline Acts (i.e. The Army Act 1955, or the Air Force Act 1955 or the Naval Discipline Act 1957)
Sending State	The State to which the force belongs when based overseas
Service Civilian	Shorthand term used to refer to a person subject to service discipline
SHAPE	Supreme Headquarters Allied Powers Europe
SHPO	Sexual Harm Prevention Orders
SIB	Special Investigation Branch of the RMP
Sig	Signaller (Private in the Royal Signals)
SJAR	Soldier's Joint Appraisal Report
SMCW(SO)R 2013	The Special Measures For Child Witnesses (Sexual Offences) Regulations 2013
SNCO	Senior non-commissioned officer
SNIY	Scottish and North Irish Yeomanry
SO	Staff Officer (see G1, G2, etc.)
SO2	Senior Staff Officer, usually at OF3 level (Major, etc.)
SO3 G1	Staff officer responsible in a headquarters, usually at OF3 (Captain) level, reporting to the ACOS. In disciplinary matters, it is the SO3 G1(Disc)
SOA	Sexual Offences Act 2003
SOCPA 2005	Serious Organised Crime and Police Act 2005
SOFA	Status of Forces Agreement
SOPO	Sexual offences prevention orders
SoS	Secretary of State (for Defence)

Sp Coy	Support Company
SPA	Service Prosecuting Authority
Spr	Sapper (Private in the Royal Engineers)
Sqn (Army)	Sub-unit of a Regimental/battalion, of about 130 personnel
Sqn (RAF)	The basic fighting unit of the Royal Air Force composed of 2 or 3 flights. Manning numbers will range from about 100 personnel upwards
Sqn Ldr	Squadron Leader (RAF) equivalent to Major (Army/RM), Lt Cdr (RN)
SSAFA	Soldiers, Sailors, Airmen and Families Association
SSM	Squadron Sergeant Major
SSPO	Service supervision and punishment order
Sub Lt / S Lt	Sub Lieutenant (RN), equivalent to Lt (army), Flying Officer (RAF)
Subordinate Commander	Officer commanding a military squadron, company or equivalent sub-unit
Tpr	Trooper (Private in Household Cavalry, Royal Armoured Corps, Special Air Service and Honourable Artillery Company)
UKBC	UK-based civilian
UKSC	UK Supreme Court
VAdm	Vice Admiral (RN), equivalent to Major General (army), Air Vice Marshall (RAF)
VC	Victoria Cross (highest gallantry award)
VFA 1952	Visiting Forces Act 1952
VJAG	Vice Judge Advocate General
WG CDR	Wing Commander (RAF), equivalent to Lt Col (Army) and Cdr (RN)
Wing	An organizational unit which controls squadrons and subunits of the RAF.
WO1	Warrant Officer Class 1
XO	The Executive Officer who is Second in Command.
YJCEA 1999	Youth Justice and Criminal Evidence Act 1999

YJCEA 1999(ASC) (A)(2)O 2015	Youth Justice and Criminal Evidence Act 1999 (Application to Service Courts) (Amendment) (No. 2) Order 2015
YJCEA1999 (ACMAC)O 2006	Youth Justice and Criminal Evidence Act 1999 (Application to the Courts-Martial Appeal Court) Order 2006
YJCEA1999(ASC) O 2009	Youth Justice and Criminal Evidence Act 1999 (Application to Service Courts) Order 2009

INTRODUCTION

For many defence practitioners engaged in court martial defence on behalf of a service defendant, finding source materials is a difficult task. The body of statute law applicable, from the main Armed Forces Acts of 2006, 2011 and 2016 to the many supplementary Statutory Instruments and orders, together with quinquennial statutes, show the somewhat overly complex and technical nature of the military justice system, with the requirement to hunt down subordinate legislation, and cross-reference other statutory provisions, with much of the detail in the small print. It does not present the service justice system as completely transparent and easily understood. Many materials are not even available over the internet. That said, the *Manual of service law* is a fine publication that contains much detail, but is bulky. Nevertheless, it is the authoritative text for how the military dispenses discipline. It is available online and is a vast improvement in transparency from earlier years. Administrative policy documents are much harder to find, even though they are based on Joint Service Publications, which are replicated in each single Service internal guidance. For example, in the Army, these are Army General Administrative Instructions (AGAIs).

Justification for a Military System

The military justice system (MJS) – or service justice system (SJS) – comprises the body of law that governs the disciplinary processes within the three Services. It is important to have a separate justice system to ensure that, wherever in the world a service person is, whether on land, sea or in the air, they are in the unique position that, if they commit a crime or a disciplinary infraction, they know that they will be dealt with according to military law, which also reflects domestic law.[1] Moreover, should they be convicted by court martial of an offence involving loss of liberty, they know they will serve any sentence in the Military Corrective Training Centre or a UK penal establishment rather than languish in a prison where human rights standards may be different to our own. When based in the UK, the service person is subject also to domestic jurisdiction, which has primacy.

Well-trained and disciplined service people are key to an effective and efficient fighting force. In spite of challenges in the European Court of Human Rights, Service law has stood the test of time, has adapted and has provided a defined body of law that the service person knows and understands as shaping their living and working environment, in peacetime and in war, or operations other than war, at home or overseas. It is not static. It is modified or changed as necessary by means of an Armed Forces Act (AFA). Today's professional soldier is well trained and well motivated and is a far cry from those about whom the Duke of Wellington, prior to Waterloo, reputedly said '[my] army is composed of the scum of the earth – the mere scum of the earth … I don't know what effect these men will have upon the enemy, but by God they terrify me!'[2]

Some outside of the military may regard the disciplinary system in a stereotypical way. As imperfect as the summary system is, a desire for fairness lies at its heart. No commanding officer (CO) wishes to be regarded as anything other than a fair tribunal who treats their personnel justly. To do otherwise would sap morale and undermine respect not only for the CO but for the entire summary system. They take their responsibilities very seriously.

1 *Cox v Army Council* [1963] AC 48, cited with approval in *R v Boyd & Ors* [2002] UKHL 31: 'anyone who is guilty of an act or omission that would be punishable by the law of England if committed in England is guilty of an offence under [Service law] wherever he commits it, whether in some other part of the United Kingdom or elsewhere in the world'.

2 *Oxford Dictionary of Quotations*, 1999, at p. 809 §20 (Philip Henry Stanhope Notes of Conversations with the Duke of Wellington (1888) 4 November 1831) and §14 ('As Lord Chesterfield said of the generals of his day, "I only hope that when the enemy reads the list of their names, he trembles as I do"', letter, 29 August 1810, in Supplementary Despatches … (1860) vol. 6).

Armed Forces Act 2016[3]

The AFA 2016 made more changes to the system, to bring it into closer alignment with civilian practice. The provisions of section 14 of the 2016 Act concern repeals relating to discharge for homosexual acts (amending ss. 146(4) and 147(3) of the Criminal Justice and Public Order Act 1994).

The substantive provisions of the 2016 Act make the following amendments:

a. An extension of the commanding officer's power to require preliminary alcohol and drugs tests under section 93A to 93I of AFA 2006 where members of the armed forces and civilians subject to service discipline are suspected of committing, or of having committed, certain offences [s. 2]. This provision was brought into force by the Armed Forces Act 2016 (Commencement No. 2) Regulations 2018, effective from 1 January 2019. Detailed regulations to govern the procedures to be followed will be produced, as intimated in Reg. 2. The amended Chapter 3A is set out in full at Appendix M Alcohol and Drugs Testing.
b. Direct referral to the Director, Service Prosecuting Authority (DSPA), in non-Schedule 2 cases that the CO cannot deal with. The duty of service police following investigation, set out in section 116 of the AFA 2006, is amended to provide that, in all cases which cannot be dealt with by the CO, the decision on the charge and prosecution are the responsibility of the Director.
c. An Amendment to section 117(3) so that service police will be able to refer a case to the Director, DSPA, if, after consulting the Director, they consider it appropriate to do so because of a connection with another case which is referred to the Director under section 116 [s. 3(5)].
d. Changes to the power of the CO to charge where a case is a linked case under section 117 [s.(4)], that is, where the investigation is of multiple offences or offenders. Where an officer has referred a case, the officer must also refer to the Director of Service Prosecutions any other case where there is, or may be, a connection.

3 This summary is based upon the Explanatory Notes to the AFA 2016.

e. Extends the power in section 121 AFA 2006 for the DSPA to charge in respect of cases referred to them by the service police or CO and also provides for charges brought by the Director to be allocated to the appropriate service court on the same basis as where charges are brought at their direction [s. 5].
f. Section 190 of AFA 2006, concerning the period for which a sentence of service detention may be suspended (the 'operational period'), is amended by increasing the operational period the court martial may impose from 12 to 24 months [s. 6].
g. New sections 304A to 304H are inserted into AFA 2006. As at 1 September 2020, these provisions had not been brought into force. By way of background, these sections will create a statutory framework for immunity from prosecution, restrictions on the use of evidence and sentence reductions for offenders who cooperate in investigations and prosecutions. This will allow the Director of Service Prosecutions to give a person immunity from prosecution for any service offence in return for assistance given by that person to the investigation or prosecution of a service offence. This operates in conjunction with the provisions in section 8.
h. A new section 304B is inserted into the 2006 Act allows the Director to give a person an undertaking that information will not be used against them in service proceedings in return for assistance given to the investigation or prosecution of a service offence. However, the undertaking may be withdrawn in the event of a failure to comply with any conditions specified in a written notice to that person [s. 8].
i. A new section 304C is inserted, concerning reduction in sentence of a person convicted of a service offence following a plea of guilty in a court martial, in return for assistance that they have given, or offered to give, to an investigator or prosecutor pursuant to an agreement with the Director [s. 9].[4]

4 New s. 304C makes provision equivalent to the provision in s. 73 of the Serious Organised Crime and Police Act 2005.

j. A new section 304D is inserted into AFA 2006 makes provision equivalent to that in section 74 of the Serious Organised Crime and Police Act 2005 for sentences to be reduced on review following an offer of assistance. However, where there is a subsequent failure to cooperate, the court may increase the sentence under section 11 which inserts new a section 304E into AFA 2006 [s. 11].
k. A new section 304F is added providing that where a court reduces a person's sentence under new sections 304C or 304D it is required to make a statement in open court confirming this and identifying the sentence it would have passed had that reduction not been given, unless it would not be in the public interest to disclose that the sentence has been discounted [s. 12].

In some areas of service justice, most notably that relating to alcohol and testing, the complexity and volume of the law and rules, through the various updating subordinate legislation, would justify a separate volume, as testing covers unfitness for duty, unfitness for specified safety critical duties and unfitness for non-specified safety critical duties, together with the use of random drug testing for units. Perhaps that is a project for another day.

Overview of Content

For those interested in history, Chapter 1 provides a brief summary of the evolution of the current system, starting with the *Findlay* reforms. It is an important window into how the prosecution and court martial systems in the three Services moved away from structures controlled by the chain of command and emerged as wholly independent. Quite an epic journey. It is very useful for understanding how acutely aware the Services are of the need to observe independence and impartiality.

Chapter 2 examines the matter of jurisdiction, both at home and abroad, for the investigation and prosecution of criminal offences. This includes a look at Status of Forces Agreements under which military forces may operate.

An examination of the disciplinary system in overview is in Chapter 3, looking at sources of law and its applicability to those who are subject to service law or subject to service discipline – an important distinction revealed therein.

The three Services operate what is called the Chain of Command (CoC), a hierarchical structure where orders and instructions are passed from superiors to juniors, which are to be followed (unless manifestly illegal). A detailed definition is given of the COC for the three Services in Chapter 4, including the disciplinary COC, which is the structure to be followed when a service person is to face a disciplinary charge. There is also an abbreviated form of the structures to be found at Appendix B.

Chapters 5 and 6 discuss the applicable law concerning arrest and investigation, setting out the authorities of the service police and the powers of COs.

Chapter 7 sets out many of the most commonly used service offences, including a number that carry severe penalties.

The CO's power to deal administratively with minor matters, rather than impose a disciplinary sanction, is covered in Chapter 8. This alternative process was devised in 2005.

In Chapter 9, the summary hearing process is explained, what it is, what offences may be dealt with and what the effect of the process is, bearing in mind many offences dealt with are recordable.

A brief look at the legal aid scheme applicable to service personnel in Chapter 10 covers what circumstances are eligible, how it operates and the availability of financial assistance pre-charge when overseas.

Chapter 11 is an overview of fitness to stand trial, what are the qualifying criteria, findings of insanity and the powers of the court to make orders in the event of either plea succeeding.

A very detailed examination of the court martial process, commencing with the role of the Director Service Prosecutions, their powers and approach to prosecutions by means of the prosecution test, including the concept of the 'service interest', something that is often misunderstood is covered in Chapters 12 and 13. The chapters also explain the requirement for 'Service Papers/Advance Information' and what they must contain. Disclosure, defence statements, skeleton arguments and the Plea and Case Management Hearing are among the topics discussed. The procedure at trial by court martial is also explained, along with the sentences available.

Chapter 14 explains the sentences that are available to the Court Martial and the restrictions where a person has elected trial.

Chapter 15 discusses civilians in the service disciplinary system, jurisdiction over them, what powers are available to try them before service courts and the sentences available.

The various appeal routes available for anyone found guilty by the CO or by a service court, including the eligibility for bail can be found in Chapter 16.

Chapter 17 is dedicated to young offenders and examines the prohibition on imprisonment, dealing with dangerous offenders and other available sentences.

Chapter 18 looks at the plight of victims of, and witnesses, to crime and the great strides made by the Services to address this important issue.

Chapter 19 contains a number of other important provisions in the service system, from special types of punishments or orders, including recognizances, rehabilitation periods for offenders, forfeitures from pay by way of compensation, minor punishments that are available, admissibility of evidence before a civilian court and a selection of caselaw showing the evolution of Court Martial Appeal Court (CMAC) Sentencing Appeals.

Chapter 20 explains service inquiries and coroners inquests, the circumstances when they are required and when legal aid is available.

Finally, a detailed index will assist in quickly accessing the required topic or point required.

Sources and Copyright

The main sources used in this book include the Armed Forces Acts of 2006, 2011 and 2016, the Explanatory Notes to the Acts, the Service rules and regulations published as subordinate legislation, Joint Service Publication 838 (the *Manual of service law*), the Guide to Sentencing in the Court Martial and The Practice Memoranda for judge advocates. Crown copyrighted materials are used in accordance with the terms of the Open Government Licence.[5] Copyright in respect of all other external materials referred to

5 Accessible here: http://www.nationalarchives.gov.uk/doc/open-government-licence/. See, also, *Funke Medien NRW GmbH v Federal Republic of Germany:* Opinion Of Advocate General Szpunar, 25 October 2018 (1) CJEU, Case C-469/17, 'Copyright has two main objectives: (1) to protect the personal relationship between the author and his work as his intellectual creation; (2) to enable authors to exploit their works economically and thus earn an income from their creative endeavours. This involves the area of property rights. Article 11 of the Charter, read in

remains with those bodies and no assumption is made of those rights. The size of this text requires that the reader augments it with a copy of the relevant legislation. References to all original works have been cited, but any omission is not deliberate and should not be interpreted as an assumption of authorship.

conjunction with Article 52(1) thereof, must be interpreted as precluding a Member State from invoking copyright.'

1 HISTORY The *Findlay* Reforms

Pre-1996

Until the European Court of Human Rights (ECtHR) decision in *Findlay v UK* [1997] ECHR 8 (25 February 1997), courts martial were very much under the control of the chain of command. Although military lawyers would advise on the strength of the evidence, it was the chain of command that made the decision whether or not a matter should proceed to trial, and at what level: District Court Martial (for which a Brigadier in command would be the Convening Officer) or General Court Martial (for which the commanding General would be the Convening Officer). The Convening Officer could also decide whether to discontinue the proceedings and made all decisions over charges to be brought and the acceptance of pleas. More controversial still was the fact that all court martial proceedings were subject to confirmation[1] by the Confirming Officer who, in fact, was usually the same person as the officer who convened the court. The Confirming Officer received advice from the Judge Advocate's Office, but was not bound to follow it. It is no understatement to say that many legal officers were unhappy with the way in which the disciplinary system operated and sought changes to bring it more into line with the civilian system.

It was possible – and sometimes did happen – that the Confirming Officer took a different view of the evidence to that taken by the court that tried the defendant, even though the Confirming Officer only had the written trial record to refer to. When decisions were overturned due to non-confirmation, it often left victims feeling that they had been robbed of justice. Victims in sexual cases were particularly aggrieved. A completely non-judicial process had intervened to quash the findings of a lawfully constituted court. It also placed the service person accused in a more advantageous position to their civilian counterpart whose only recourse, quite properly, lay in an appeal to the Court of Appeal.

It was a welcome relief, therefore, that the *Findlay* reforms, enacted in the Armed Forces Act 1996 (AFA 1996), abolished the role of Convening Officer/Confirming Officer,[2] established an independent Prosecuting Authority for each of the three Services[3] and set up an independent office for the convening of courts martial.[4] Since then, prosecuting officers have no involvement in the provision of advice to the chain of command.[5] Once a case is referred to the Prosecuting Authority, it is the prosecuting officer alone who decides the charges and directs the mode of trial of the accused. Unfortunately, the much-criticised process of post-trial confirmation was not quite consigned to the darker pages of legal history. That was yet to come. There was a process of 'one stop review', carried out by the Reviewing Officer on advice from the Office of the Judge Advocate General, which could still overturn a conviction without any chance for the prosecution to challenge the basis before a tribunal such as the Court Martial Appeal Court.

The Prosecutor

The next target for attack was the role of the prosecutor. Under the Armed Forces Act 1996, the head of each Service Prosecuting Authority was also head of the respective service legal branch. This was criticised as lacking independence from the chain of command. The specific criticism was of the role of the Director Army Legal Services (DALS) who was 'dual hatted' in the sense that he was head of the Army Legal Services and also appointed as the Army Prosecuting Authority. The ECtHR decided, in the case of *Morris v United Kingdom* [2002] ECHR 162 that:

2 *Ibid.*, s. 15 & para 69 of Sch 1.

3 *Ibid.*, paras. 14–16 of Sch 1.

4 Courts-Martial (Army) Rules 1997, r. 16(1).

5 Although there ought not to be any objection to so doing. See: *Wright (Ex parte); R v RAF GCM and Anr*, [1999] EWHC Admin 465, where it was held there was no abuse of process when the same officer of the RAF Directorate of Legal Services who had advised a commanding officer on whether to refer charges to a higher authority or to dismiss them, later made the decision in the same case whether to prosecute.

a. The Court's own case law illustrated that a military court can, in principle, constitute an 'independent and impartial tribunal' for the purposes of Article 6(1) of the Convention.
b. Although the DALS is also the Prosecuting Authority, the Court is of the view that sufficient safeguards of independence exist in that, in his advisory role, he does not deal with disciplinary matters and, in any event, he is in that role answerable to the Adjutant General, while as Prosecuting Authority he is answerable to the Attorney General.
c. The presence of the Permanent President did not call into question the independence of the court martial. Rather, he was a significant guarantee of independence on an otherwise ad hoc tribunal.
d. The two serving officers who sat on the applicant's court martial were not appointed for any fixed period of time. Rather, they were appointed on a purely ad hoc basis, in the knowledge that they would return to their ordinary military duties at the end of the proceedings. The ad hoc nature of their appointment made the need for the presence of safeguards against outside pressures all the more important in this case. In particular, that those officers had no legal training, remained subject to army discipline and reports and there was no statutory or other bar to their being made subject to external army influence when sitting on the case.
e. There was no violation of Article 6(3)(c) where an individual was required to pay a contribution to the cost of providing legal assistance and had sufficient means to pay.
f. The reviewing authority was empowered to quash the applicant's conviction and the sentence imposed by the court martial. The very fact that the review was conducted by such a non-judicial authority is contrary to the principle that the power to give a binding decision that may not be altered by a non-judicial authority is inherent in the very notion of 'tribunal'.

The process of Review was finally laid to rest for courts martial by section 16 of the AFA 1996, although it remains for Summary Hearing by the Commanding Officer (CO).[6]

The AFA 1996 also introduced the power to appeal against sentence, as well as conviction, to the Court Martial Appeal Court. This anomaly was long overdue for reform as it meant that where a soldier and a civilian were jointly tried before and convicted by a court martial, the civilian could appeal their sentence, but the soldier could not.

The Legislation

In spite of *Findlay*, there was yet more reform needed. The Prosecutor still did not have the power to refer perverse decisions of law to the Court Martial Appeal Court,[7] or to challenge an unduly lenient sentence.[8] Both of these powers were available in the civilian criminal justice system. One of these powers came to the military justice system with the passing of the Armed Forces Act 2001 (AFA 2001), amending the three Service discipline acts. Section 21 permitted a review of courts martial sentences by the Courts Martial Appeal Court[9] where the defendant had been convicted of a criminal offence (under section 70 of both the Army Act 1955 and Air Force Act 1955, section 42 of the Naval Discipline Act 1957 (NDA 1957)), being an offence tried by court martial that could be tried by a civilian court only on indictment, and any offences specified by order. This is now superseded by section 273 of the Armed Forces Act 2006 (AFA 2006) and supplemented by the Armed Forces (Review of Court Martial Sentence) Order 2009[10] and the amendments in the Armed Forces (Miscellaneous Amendments) Regulations 2019 that govern those cases where the Attorney considers a sentence imposed by the

6 See s. 152, Review of summary findings and punishments. This includes a power for the reviewing officer to refer a finding or punishment to the Summary Appeal Court. In such a case, 'reference to the Summary Appeal Court under subsection (4) or (7) shall be treated for the purposes of this Act as an appeal under section 141 brought by the person to whom the finding or punishment relates against the finding or punishment'.

7 S. 36 Criminal Justice Act 1972 (Reference of point of law following acquittal on indictment).

8 S. 36 Criminal Justice Act 1988 (unduly lenient sentence).

9 By inserting a new s. 113C into the AA 1955 and AFA 1955, and after s. 71A of the NDA 1957.

10 The process is contained in the Armed Forces (Review of Court Martial Sentence) (Supplementary Provision) Regulations 2009, which details applicable time limits and which, as from 13 May 2013. includes s. 59A and s. 61 of the Sexual Offences Act 2003 (as amended).

court martial was unduly lenient[11] and the corresponding offence under the law of England and Wales would be indictable only, or is of a description specified for the purposes of section 273.[12]

From July 2019, sentences in respect of repealed offences under the Sexual Offences Act 1956,[13] namely, under section 6 (intercourse with a girl between 13 and 16) and section 14 or 15 (indecent assault on a woman or a man), and an offence under section 1 of the Indecency with Children Act 1960 (indecent conduct towards young child), are also included as cases specified[14] for the purposes of section 273(3) of the AFA 2006.

After the Court Martial Appeal Court has concluded its review of a case referred to it under section 273(1), concerning an unduly lenient sentence, either the Attorney General or the offender may refer to the Supreme Court a point of law involved in any sentence passed on the offender in the proceedings. Section 274(1) permits the Attorney General or the offender to refer to the Court of Appeal a point of law involved in any sentence passed on the offender in the proceedings. The AFA 2006 makes no provision for a general reference for the purpose of clarifying the law, for example, after an acquittal. That is, there is no similar provision to section 36 of the Criminal Justice Act 1972 enabling a reference of a point of law following acquittal on indictment. Acquittals by court martial are excluded. However, section 34 of the Court Martial (Appeals) Act 1968[15] and Rule 51 of the Court Martial Appeal Court Rules 2009 permit a reference by the Criminal Cases Review Commission, the Judge Advocate General or the Secretary of State for Defence. So, the anomaly exists that an independent prosecuting authority is unable to refer a matter, even via the Attorney General, for clarification of a point of law,[16] yet the Secretary of State for Defence, from whom the Director of the Service Prosecuting Authority (SPA) is independent, can.

---

11 Armed Forces (Review of Court Martial Sentence) Order 2009, s. 2. See also page 556.

12 Such as, threats to kill, various sexual offences, improper importation of goods, cruelty to persons under 16, production or supply of a controlled drug, and certain attempts. The Schedule should be consulted for the full list.

13 These offences were repealed by Schedule 7 of the Sexual Offences Act 2003.

14 Armed Forces (Miscellaneous Amendments) Regulations 2019.

15 The title was amended by Schedule 8 of the AFA 2006, from Courts-Martial (Appeals) Act 1968.

16 This may be corrected in a future Armed Forces Bill.

The AFA 2006, in section 321 and Schedule 11, also provides a power for the Criminal Cases Review Commission to refer a case to the Court Martial Appeal Court.

As noted above, until the passing of the AFA 2006, each of the three Services had its own Service discipline Act: the Army Act 1955 (AA 1955), the Air Force Act 1955 (AF 1955), and the NDA 1957. The Army Act and Air Force Act were very similar in layout and content, even insofar as the section numbers of the main offences. The NDA 1957, on the other hand, was different. Its summary proceedings were different (its punishments for summary offences more severe), as were the section numbers of some main offences. The main legislation was supported by key subordinate legislation for each Service, which dealt with custody and summary dealing by the CO, procedures before Summary Appeal Courts (SACs), and Court Martial Rules.

Importantly, the Court Martial Rules for the Army and Royal Air Force contained a provision in rule 5 for holding a formal preliminary examination (FPE). This was a process whereby the prosecutor could test the evidence of key witnesses to help determine whether the prosecution test[17] was satisfied, thereby weeding out weak cases or those where witnesses were not credible. FPEs were generally conducted in the presence of the accused and their legal representative, although they did not have to be there if they did not wish. The process was endorsed by Senior Treasury Counsel who considered it to be absolutely Article 6 compliant. Significantly, the Crown Prosecution Service subsequently adopted[18] a similar, though less comprehensive, system where key witnesses are examined by a prosecutor, without the presence of the defendant or their legal representative. The proceedings are recorded on video. It is used by the Crown Prosecution Service particularly in rape and serious sexual cases. Unfortunately, the AFA 2006 omitted this valuable facility, which had existed for many years

17 Explained on page 340.

18 In the aftermath of the Damilola Taylor case in 2002, when there was severe criticism of the Crown's failure to ascertain unreliable witnesses prior to trial, in particular, a girl of 12, codenamed 'Bromley'. In his report on the case, Bishop Sentamu's team said: 'The absence of any process to test the veracity of Bromley's testimony before trial left the prosecution of the case vulnerable to problems.' In spite of reservations about the Bar conducting pre-trial interviews, published in the *Law Society Gazette*, by Polly Botsford, which appeared on 28 March 2008, a system of pre-trial interviews was introduced by the CPS and is working successfully.

in the Army and Air Force systems, for reasons never explained properly.[19] It is difficult to understand what possible objections there might be to such a useful facility for prosecutors, which was widely accepted as actually assisting the interests of justice. Civilian lawyers were generally very much supportive of it, as it let them, too, assess the demeanour of key witnesses, without having to lay out their defence or give evidence. It is only to be hoped that future legislation will reintroduce the FPE.[20] The power for the Secretary of State to include such provision in a statutory instrument exists in section 128 of the AFA 2006.[21]

The AFA 2006 is now the legislation common to the three Services. It is a substantial piece of legislation, with 386 sections and 17 schedules,[22] and is supplemented by an array of statutory instruments and rules, making it a complex and sometimes confusing system to navigate.

The 2006 Act was a major consolidating piece of legislation which Ministry of Defence lawyers described as 'harmonising' the three separate Service disciplinary systems. This was complemented by a tri-service manual and subordinate legislation harmonising the processes for arrest, custody, summary disposal, summary appeals and court martial. Instead of having the disparity in layout and offences between the Army and Air Force on the one hand and the Navy on the other, all three Services use the same legislation in the disciplining of their service personnel.

---

19 In the 1955 Acts it was known as the Summary of Evidence, being a process created by the Court Martial Rules. The FPE was created by the Courts-Martial (Army) Rules 1997 (CM(A)R 1997), continuing up to the Courts-Martial (Army) Rules 2007.

20 It is submitted that this procedure does not require any legislation to be introduced, as it is a prosecution process to inform the decision to prosecute. It could therefore be introduced by the SPA as a matter of policy for assessing complainants and victims in serious cases. The CPS introduced this procedure in the aftermath of the notorious Damilola Taylor case. In the foreword to the report entitled *Pre-Trial Witness Interviews by Prosecutors*, published in December 2004, the then Attorney General wrote that he had asked the Director of Public Prosecutions (DPP) 'particularly to consider whether the time had come to introduce a system where, before trial, key witnesses could be interviewed by the prosecutor'. A pilot scheme was successfully implemented between January 2006 and February 2007, eventually becoming a national non-statutory scheme in early 2008. An Aide Memoire was produced for CPS prosecutors, setting out what should be included during the introductory section of the pre-trial witness interview.

21 This section covers the power of the Director, Service Prosecuting Authority (DSP) to direct the bringing of a charge after referral of the case to him. Therefore, a power to permit him to test key prosecution witness evidence to inform the prosecutor's test would be within its scope.

22 The printed text runs to 344 pages.

Summary Dealing

While the court martial system underwent many changes, so did the system of summary dealing by the CO. That this part of the system has survived unscathed from legal challenge is surprising, as it is clearly non-compliant with the requirements of Article 6 of the European Convention on Human Rights. The proceedings lack independence, the service person is unrepresented and only allowed a unit assistant, there is the potential for hearsay and inadmissible evidence to be adduced and there is no right to legal advice or assistance. It is considered to be saved by the right to appeal to a SAC.

The approach of the Court Martial Appeal Court gave rise to a quite interesting decision in the case of *Baines v Army Prosecuting Authority*.[23] Baines appeared before his CO on 5 July 2004, charged with battery contrary to section 39 of the Criminal Justice Act 1988. The case was originally referred by the Commanding Officer to the Army Prosecuting Authority (APA) for a decision on trial on a charge of assault occasioning actual bodily harm, contrary to section 47 of the Offences Against the Person Act (OAPA) 1861 as the victim's arm was thought to have been broken. It later transpired that it had not. As the case was already listed for trial, had it not been for the approach of the defence solicitor, the APA would have allowed the matter to go before a District Court Martial on the charge of battery. However, after representations by the defence to the APA, the case was remitted to the CO on a lesser charge of common assault, which was within the CO's jurisdiction. At the Summary Hearing, Baines was offered the right to elect trial, but he did not do so. The charge was found proved.

The day after the Summary Hearing, the accused lodged an appeal to the Army SAC, against both finding and sentence. At that hearing, he argued that Article 6 European Convention on Human Rights and Fundamental Freedoms (ECHR) had been breached, on the grounds that the CO was not an independent tribunal; the appellant's choice to be tried by the CO was not a free and unambiguous choice; and the appellant had not had the right to legal advice and representation at the hearing. Strong reliance was placed on the decision of the ECtHR in *Thompson v UK*.[24] The Judge Advocate ruled

23 [2005] EWHC 1399 (Admin) QBD.

24 (2005) 40 EHRR 11. The Court found a breach of Art. 6 ECHR. A waiver of a right guaranteed by the Convention, in so far as it is permissible, must not run counter to any important public

against him and he appealed to the Divisional Court, by way of case stated under section 83ZH(2) of the AA 1955. That court held that an accused soldier can be tried *de novo* by a court martial or by the SAC and they have a free and unrestrained choice to elect the former rather than summary trial and the latter if a finding is made against them on a summary dealing. It followed that an accused soldier who is dealt with summarily for an offence under Part II of the Army Act did enjoy the rights conferred by Article 6 ECHR. Accordingly, the summary dealing conducted by the CO did not breach the appellant's rights under Article 6(1) and (3). The court held that, as a matter of general principle, a person can waive the Article 6(1) and (3) rights to be tried by an independent and impartial tribunal and to legal assistance if they make a voluntary, informed and unequivocal election not to claim these rights or to raise no objection to the lack of independence and impartiality of the tribunal and the lack of legal assistance.

What may have been quite significant in this case is the involvement of legal representation during the successful negotiation with the APA. Baines would have been granted legal aid to defend the case before court martial and prior to the Summary Hearing. So, his decisions were informed by that legal advice. For this important reason, it is suggested the key points raised by the defence remain open to argument before the ECtHR as, ordinarily, a service person is not entitled to legal advice or assistance prior to, or representation at, the Summary Hearing.

Subsequently, an interesting decision of the Court of Appeal, in the civilian case of *Saunders*,[25] concerning the point whether there was an informed, voluntary and unequivocal waiver of the applicant's right to legal advice, may have ramifications for the summary dealing process and the question of informed consent by the accused service person, which is distinguishable from the situation confronting the applicant in that case. The Court of Appeal cited the dicta of Lord Hope that, to be valid a waiver must be 'voluntary, informed and unequivocal'.[26]

The Armed Forces Discipline Act 2000 (AFDA 2000) introduced among its provisions the right to appeal the summary award of the CO, even if the

---

interest, must be established in an unequivocal manner and requires minimum guarantees commensurate to the waiver's importance.

25 *R v Saunders* [2012] EWCA Crim 1380.

26 *Ibid.*, s. 9.

finding was one that did not result in a financial penalty or loss of liberty. Such appeals come before a new SAC, comprising a specially appointed judge advocate and two military officers. The legislation provided for an appeal against a finding of guilty by way of a rehearing (so that the whole case will be heard again). Appeals against sentence are also by way of rehearing but only the evidence relevant to sentencing is reheard. The SAC is unable to award any punishment more severe than that which CO awarded. The AFA 2006 retained this court.

The AFDA 2000 also confirmed the accused's right to elect trial by court martial. Where they did so, the court could not award any punishment which could not have been awarded by the CO or appropriate superior authority had the election for trial not been made (that is, up to 60 days with permission from a higher authority).[27]

An appeal from the SAC is to the High Court by case stated on a point of law. These provisions are largely preserved in the AFA 2006, which is dealt with in Chapter 16 on page 539.

27 See Election for Trial by Court Martial, in Chapter 9, page 287.

2 JURISDICTION

United Kingdom

As was aptly pointed out in the former Manual of Military Law,[1] although the provisions apply *mutatis mutandis* to the other two Services:

> A man who joins the Army … [d]oes not cease to be a citizen. With a few exceptions, his position under the ordinary law of the land remains unaffected. If he commits an offence against the civil law he can be tried and punished for it in the civil courts. In respect of civil rights, duties and liabilities the ordinary law in general also applies to him, although a few privileges are granted to him and a few restrictions are imposed upon him for the purpose of enabling him the better to fulfil his military duties.
>
> Whilst officers, warrant officers, non-commissioned officers and soldiers remain subject (with the above qualifications) to the ordinary law of the United Kingdom, they are subject to an entirely distinct legal code known as 'military law'.

In the United Kingdom, the civil authorities have primacy of jurisdiction. Therefore, when a service person commits a criminal offence a decision must be reached between the military and civil authorities about who investigates and who prosecutes. Prosecution decisions are determined in accordance with the Prosecutors' Convention[2] which,

> sets out the responsibilities of prosecutors where a suspect's conduct could be dealt with by criminal or civil/regulatory

---

1 Part 1, Amendment 20, page 1, paras 2 and 3.

2 See: http://www.cps.gov.uk/legal/p_to_r/prosecutors_convention/#content.

sanctions and/or where more than one prosecuting authority and investigating body share the power to take some action.

In such cases, depending on the extent to which more than one authority is pursuing or wishes to pursue an investigation, decisions will have to be taken about whether and how investigations should proceed and as to the choice of criminal charges or civil or regulatory sanction.

The aim is to ensure that cases are conducted in a way that is just and best serves the overall public interest, so that the public can have confidence in the outcome. Through the Convention, Prosecuting Authorities commit themselves to

- Actively managing the interface with other prosecutors and investigators from the earliest possible stage;
- Working together in a co-ordinated manner to ensure that decisions in individual cases are taken promptly and that they serve the overall public interest, having regard to the nature and seriousness of the harm and offences alleged; and
- Enforcing the law in a fair and effective way.

The Service Prosecuting Authority is a party to the Convention and is named in Schedule 1 thereto. There have been some recommendations made to propose that, in the Armed Forces Act 2021 (AFA 2021),

> Court Martial jurisdiction to no longer include the most serious offences (murder, rape and manslaughter) when committed in the UK (except where the consent of the Attorney General is given), in future the SP [Service Prosecutor] should no longer investigate those offences in the UK. Such investigations should revert to the civilian police who should enter into a formal protocol to conduct joint civilian led police/SP engaged investigations.[3]

Section 13 of the Armed Forces Act 2016 (AFA 2016) applies the provisions of the Armed Forces Act 2006 (AFA 2006) to the Isle of Man and British overseas territories except Gibraltar (where it has since expired, under section 382).

---

3 Recommendation 10, *Service Justice System Policing Review (Part 1)*, by Professor Sir Jon Murphy QPM DL LLB, 2018.

Nevertheless, our service personnel do not cease to be citizens, with commensurate rights and obligations. However, there are limitations placed upon what service people can and cannot do, which their civilian counterparts are not subject to. For example, while a service person may take part in the democratic process by casting a vote in an election, Queen's Regulations stipulate it is impermissible to actively participate in party politics[4] or correspond with the press without authority. Similarly, with minor exceptions, trade union membership is not allowed, unless it is a professional association or necessary in order to enhance their trade skills and professional knowledge and as an aid to resettlement into civilian life. They are not to participate in industrial action or in any form of political activity organised by civilian trade unions or professional association.[5] Whether these limitations are all still necessary in a democratic society has not yet been tested by a British service person before the European Court of Human Rights (ECtHR), although some countries, such as France, Germany and Holland,[6] do not have quite the same restrictive approach.

Two cases in the ECtHR examined the ban in the French Armed Forces on membership of a federation. In *Matelly v France*,[7] the applicant, a gendarme, complained that an order was given that, as a policeman, he was not to join an association, thereby giving rise to a violation of Article 11 of the Convention on Human Rights and Fundamental Freedoms (ECHR).

---

4 Although they may attend political meetings, provided that uniform is not worn.

5 Musicians may be members of the Musicians Union, barristers belong to the Bar Council, solicitors to the Law Society. The right to association for members of the professional staff of the armed forces was examined by the Standing Committee on behalf of the Council of Europe's Parliamentary Assembly, 3 Sept. 2002, which adopted Council of Europe Recommendation 1572 (2002) 11. It recalled Resolution 903 (1988) on the right to association for members of the professional staff of the armed forces, in which it called on all Member States of the Council of Europe to grant professional members of the armed forces, under normal circumstances, the right to association, with an interdiction of the right to strike.

6 Dutch Army soldiers are represented by a union, the Algemene Federatie Militair Personeel General (AFMP) – Federation of Military Personnel – which was recognised by the Dutch government in 1966 and represents both current and retired soldiers. The AFMP is a member of the Dutch Federation of Trade Unions. In countries such as Sweden or Germany, independent military associations exist that are financed by members' fees and that employ their own advisory staff. In other countries, such as Bulgaria, finance comes from the Ministry of Defence. The arguments for and against union membership for service personnel and Convention rights are discussed in detail in Chapter 9 of the *Handbook on Human Rights and Fundamental Freedoms of Armed Forces Personnel*, OSCE/ODIHR, DCAF.

7 [2014], Court (Fifth Section), 02/10/2014 Application no. 10609/10.

Article 5 of the European Social Charter (revised) provides the right to organise:

> With a view to ensuring or promoting the freedom of workers and employers to form local, national or international organisations for the protection of their economic and social interests and to join those organisations, the Parties undertake that national law shall not be such as to impair, nor shall it be so applied as to impair, this freedom. The extent to which the guarantees provided for in this article shall apply to the police shall be determined by national laws or regulations. The principle governing the application to the members of the armed forces of these guarantees and the extent to which they shall apply to persons in this category shall equally be determined by national laws or regulations.

Article 12(1) of the Charter of Fundamental Rights of the European Union (CFR-EU) reads:

> Everyone has the right to freedom of peaceful assembly and to freedom of association at all levels, in particular in political, trade union and civic matters, which implies the right of everyone to form and join trade unions and join in the defence of its interests.

Relying on Article 11 of the ECHR (Freedom of expression and information), the applicant complained of an interference with the exercise of his freedom of association. He challenged both the legality and lack of sufficient predictability of the law.

The ECtHR held there had been a violation of Article 11 of the Convention. The reasons given by the authorities to justify the interference with the applicant's rights were neither relevant nor sufficient, since their decision was an absolute prohibition for the military to join a trade association formed to protect their professional and moral interests. If freedom of association of the military may be subject to lawful restrictions, outright prohibition to form a union or become a member goes to the heart of this freedom, being a restriction prohibited by the Convention.

The decision coincided with *ADEFDROMIL v France*[8] decided on the same day, where the Court unanimously concluded that a blanket ban prohibiting military personnel from forming or joining a trade union violated Article 11.

Since these decisions, there has been little, if anything, in the United Kingdom press or media, nor has there been any debate by serving personnel. It is clearly not a matter of pressing importance for the UK armed forces.[9]

Outside the United Kingdom

When based overseas, outside of armed conflict, the law of the state is not affected by the presence of the UK military who are, by default, subject to the host nation's laws *in toto* unless altered by agreement between the two sovereign powers. Jurisdiction is usually determined according to Status of Forces Agreements (SOFAs)[10] or Memoranda of Understanding (MoU) – or Letters of Understanding (LoU).

NATO Member States

Main Provision: NATO SOFA (see Appendix I); Agreement Supplementary to the SOFA (Germany)

Within NATO member states the relationship between the visiting force and the host nation (both NATO members) is governed by the NATO Status of Forces Agreement (NATO SOFA), of 19 June 1951.[11] Jurisdiction is resolved through the provisions of Article VII of the NATO SOFA.[12] There are three basic types of jurisdiction exercised under this provision:

8 ADEFDROMIL is the Association de Défense des Droits des Militaires (Association for the Protection of the Rights of Military Personnel). See: [2014] Court (Fifth Section), 02/10/2014 (Application no. 32191/09).

9 In UK, the British Armed Forces Federation (BAFF) is an independent representative all-ranks staff association open to serving members of any branch of the Services. It was launched in late 2006 by serving and former members of HM Forces.

10 The status of our forces based in NATO countries is governed by the NATO SOFA. See in particular, Article VII on the matter of jurisdiction.

11 The NATO SOFA (or London Agreement) was signed on 19 June 1951 and ratified by UK on 13 May 1954. See Appendix J.

12 The US military authorities relied upon Article VII to claim jurisdiction to try Captain Ashby, a US Air Force pilot accused of reckless flying in the Italian Alps, when his aircraft severed a

a. The authorities of the sending state (for example, the UK) have the right to exercise exclusive jurisdiction over offences, including offences relating to its security, punishable by the law of the sending state, but not by the law of the receiving state.
b. The authorities of the receiving state have the right to exercise exclusive jurisdiction over members of a force or civilian component and their dependents with respect to offences, including offences relating to the security of that state, punishable by its law but not by the law of the sending state.
c. Offences common to both jurisdictions (concurrent jurisdiction). These comprise the vast majority of offences.

An example of the first category is where the conduct or act is not an offence in the host nation. For example, it is not an offence to purchase and possess a flick knife in certain jurisdictions. However, it is an offence contrary to English law to possess an offensive weapon in public, without a lawful authority or reasonable excuse. So a service person buying a flick knife could be committing an offence contrary to the Prevention of Crime Act 1953, or section 139 Criminal Justice Act 1988 (bladed and pointed articles) even though under the domestic law of the host nation/receiving

ski-lift cable. Twenty tourists were killed when the ski-gondola in which they were travelling crashed to the ground. The pilot was tried by US court martial and acquitted. A news report of the reaction appeared in *The Times*, 5 March 1999. The court's decision caused outrage in Italy. In contrast *Dunn v Secretary Of State For Foreign And Commonwealth Affairs* [2020] EWHC 3185 (Admin), judicial review was sought, unsuccessfully, in relation to matters arising out of a road traffic accident on 27 August 2019, when Harry Dunn (then aged 19 years) was killed following a collision between his motorbike and a vehicle driven on the wrong side of the road by Mrs Anne Sacoolas, the wife of Jonathan Sacoolas, a member of the US Government's Administrative and Technical Staff at RAF Croughton. Following the accident there was dialogue between the US Embassy and the Foreign and Commonwealth Office concerning the potential diplomatic immunity of Mrs Sacoolas from UK criminal jurisdiction. The NATO SOFA did not apply, as the US personnel at RAF Croughton were State Department personnel rather than Defence Department personnel and they came within the provisions relating to diplomatic immunity under the 1961 Vienna Convention on Diplomatic Relations, brought into UK law by the Diplomatic Privileges 1964 Act, s. 2(1). Upon the conclusion of that dialogue, on 15 September 2019, Mrs Sacoolas left England, the FCO having accepted that Mrs Sacoolas was entitled to immunity from criminal proceedings in the United Kingdom. On 22 December 2019, the CPS began extradition proceedings against Mrs Sacoolas in respect of a charge of causing Harry's death by dangerous driving. The US State Department declined to progress the extradition, asserting Mrs Saccolas had immunity at the time of the accident. Harry's parents have been granted permission to appeal against the High Court ruling.

state, there is no offence. In this case the UK, as sending state, would have exclusive jurisdiction.

The vast majority of offences fall within the third category and are usually the subject of negotiation between the receiving and sending states.

Germany

When service personnel are serving outside the UK, the AFA 2006 extends jurisdiction to cover civilians who are dependants of serving service personnel, or who are Crown servants or who otherwise fall within its provisions. Through the medium of section 42,[13] English civilian criminal offences are triable by the military authorities in these circumstances. See the discussion below, concerning the case of *R v Martin*,[14] which involved a conviction for murder by a 17-year-old dependant of a serviceman in Germany, where the defence unsuccessfully argued inter alia that trial by court-martial was an abuse of process in itself and that the civilian accused should not have been tried for murder before a court martial in Germany but, rather, by a Crown Court in England, as the English courts would have had jurisdiction under section 9 of the Offences Against the Person Act 1861.

Sometimes the NATO SOFA will be supplemented by other agreements that clarify in detail the specific rights and how they operate within that territory. However, such supplementary agreements cannot remove rights granted by the SOFA. In respect of our forces based in Germany, it should be pointed out that, pursuant to Articles 17 and 19 of the Bonn Agreement,[15] often referred to as the 'Supplementary Agreement', there is a standing waiver of jurisdiction in favour of the British military authorities in respect of the matters covered by paragraph 3[16] of Article VII of the SOFA (concurrent

---

13 (1) A person subject to Service law, or a civilian subject to service discipline, commits an offence under this section if he does any act that— (a) is punishable by the law of England and Wales; or (b) if done in England or Wales, would be so punishable. For the meaning of 'a civilian subject to service discipline', see Chapter 3 below, on page 36.

14 House of Lords, 16 December 1997. See Chapter 3 below, on page 27.

15 An agreement to supplement the NATO SOFA provisions in respect of foreign forces stationed on Germany territory (the Bonn agreement), signed on 3 August 1959. More informally referred to as 'The Bonn Agreement,' subsequently revised and available in Volume 3, Chapter 40 of the Manual of service law (MSL).

16 'With the consent of the German authorities, the military authorities of a sending State which has requested the waiver under paragraph 1 of this Article may transfer to the German courts or authorities for investigation, trial and decision, particular criminal cases in which jurisdiction rests with that State. (b) With the consent of the military authorities of a sending

jurisdiction). Murder, rape, robbery, theft, etc. are all offences of concurrent jurisdiction under the NATO SOFA. Nevertheless, the German authorities may recall that waiver within a period of 21 days after receipt of notification by the British authorities that a particular case involves a member of the UK force, civilian component or a dependant. This, in practice, is exercised very rarely by the German authorities with whom, over the many years since the agreements have been operating in Germany, a very good working relationship has developed. In Germany, under the Supplementary Agreement to the SOFA, a specific liaison organisation deals with resolving all concurrent jurisdiction issues with the German authorities.

There is one other provision that defence advocates will encounter from time to time whenever a case involves a German witness or, rather, a witness living in Germany who is not subject to service discipline (for example, an expatriate or local civilian living and working in Germany). Witness summonsing is now the responsibility of the Court Martial Administration Office (CMAO). In the case of German witnesses, the summons is actually issued by the German authorities, at the request of, and in liaison, with the Joint Services Liaison Office and local Service Liaison Officers of the British force (who, in turn, are acting on the instruction of the CMAO), in accordance with treaty provisions relating to mutual cooperation. In view of the time it takes to complete the procedures, there is a realistic requirement for approximately six weeks' notice. Any summons issued by the German court will then only be for the specified days for which it is anticipated the witness will be required, as explained below. Defence advocates requiring the attendance of a German witness should therefore contact the CMAO and provide them with all relevant details to enable the witness to be traced and summonsed.

It is axiomatic that German legal procedures are quite different to ours. It is often the case that a witness will attend before a German court on a specific day at a specific time in order to give their evidence. If it is not completed then, they may be asked to return on another date and time. German witnesses are unaccustomed to being kept hanging around at court indefinitely to give evidence. If the specified date passes and their evidence

---

State which has requested the waiver under paragraph 1 of this Article, the German authorities may transfer to the military authorities of that State for investigation, trial and decision, particular criminal cases in which jurisdiction rests with the Federal Republic' [19(5)].

has not been called, they are not bound to attend on any subsequent day without another summons being issued. As mentioned above, within the context of courts martial, this can take up to six weeks to process. However, to cater for unforeseen circumstances, such witnesses are often subpoenaed for more than one day. Advocates should therefore plan carefully their strategy for calling German witnesses and ensure that they are subpoenaed for more than one day, especially where they give crucial and/or lengthy evidence. If a German witness's evidence is incomplete by the end of the day for with they are summoned, and there is no subpoena for their attendance on the following day, advocates may have serious difficulty in securing their attendance to complete their evidence.

In addition to the commonly accepted grounds of illness and infirmity, it is a reasonable excuse for non-attendance at court, under German law, if a witness has booked a holiday over the period of the trial and, therefore, cannot attend on the date required. Defence advocates who may wish to call such witnesses would do well to bear these difficulties in mind. As indicated above, any requirement for a German witness should be communicated without delay to the CMAO at Upavon, so that steps may be taken to procure the witness as soon as possible. Leaving it until the start of a trial will not be sufficient, and there is no guarantee that the judge advocate will accede to an application for an adjournment without very good reason.

If a German witness deliberately fails to answer a summons issued by the German authorities, that will be a matter for investigation by those authorities and, should the witness fail to provide a reasonable excuse, for prosecution by them. Additionally, any German witness who behaves in a contemptuous manner before a court martial can only be dealt with by the German authorities, to whom the matter may be reported for such further action as those authorities deem appropriate. A court martial could not itself deal with that contempt, as it clearly lacks jurisdiction.

A challenge to the court martial system was mounted in the case of *Martin*,[17] where it was alleged that it was an abuse of process to try by court martial in Germany a 17-year-old son of a serviceman charged with

---

17 16 December 1997, House of Lords. Alan Martin murdered Vanessa Chapple, a British school teacher at the local forces school, in a wood adjacent to the military headquarters at Rheindahlen, Germany. He was the son of a serviceman. See above, on page 25.

the murder of a British school teacher. It was argued that that the civilian accused should have been tried for murder by a Crown court in England, as the English courts would have had jurisdiction. This was not accepted by the House of Lords. Although their Lordships expressed some initial concern at the prospect of a civilian being tried by a military court, they recognised the clear difficulties presented, not least of which related to the inability to compel German witnesses to attend any trial in the UK. Lord Slynn held, at page 195:

> It is, however, clear, that parliament has provided for a court-martial to be held in the circumstances of this case and, after full argument on both sides, I do not consider that there has been any breach of the rules required to be followed or that there has been any abuse of process.

Lord Hope, at page 212, said:

> There is no sound basis for thinking that, at the time when a decision about this would have had to have been taken, a prosecution in the English courts within a reasonable time would have been seen to be practicable. The alternatives lay between proceedings by way of court-martial in Germany, leaving the matter in the hands of the German public prosecutor or taking no proceedings at all … The only way to secure the ends of justice other than returning the case to the German prosecutor was to proceed by way of a court-martial. There is no ground for saying that the appellant did not receive fair trial.

The case proceeded to the ECtHR,[18] where the Court held that, while it cannot be contended that the Convention absolutely excludes the jurisdiction of military courts to try cases in which civilians are implicated, as there is nothing in the provisions of Article 6 to exclude the determination by service tribunals of criminal charges against service personnel, the existence of such jurisdiction should be subjected to particularly careful scrutiny, since only in very exceptional circumstances could the determination of criminal charges against civilians in such courts be held to be compatible with Article 6. The question to be answered in each case is whether the individual's doubts

18 [2006] Application no. 40426/98.

about the independence and impartiality of a particular court martial can be considered to be objectively justified and, in particular, whether there were sufficient guarantees to exclude any such legitimate doubts. The power of military criminal justice should not extend to civilians unless there are compelling reasons justifying such a situation, and if so, only on a clear and foreseeable legal basis. The existence of such reasons must be substantiated in each specific case. It is not sufficient for the national legislation to allocate certain categories of offence to military courts *in abstracto*. The Court did hold, at paragraph 45 of its judgment, that the power to try the applicant by court martial had a clear and foreseeable legal basis, namely section 209(2) of the AFA 1955 (now, s. 370 and Sch. 15 of the AFA 2006).

This was a pre-*Findlay* case. The ECtHR considered that the distinctions between the court martial in the *Martin* case and that in *Findlay* were not sufficiently material to persuade it to reach a different conclusion. The essential safeguards that were lacking in *Findlay* were also absent in the present case and, as in *Findlay*, the Judge Advocate did not provide the same guarantees of independence and impartiality as in *Cooper*.[19] The Court considered the applicant's concerns about the independence and impartiality of his tribunal to be objectively justified and there had been a violation of Article 6(1).

It is still an area of uncertainty whether the concerns raised in *Martin* are completely addressed in the AFA 2006, especially when it comes to the matter of trying serious offences that are not covered by UK jurisdiction. Of course, the option is to leave the matter to the local courts. If those courts are ECHR compliant, it is less of a worry. However, a defendant might not consider the *Martin* case worth relying on if they were to be left to the mercies of a jurisdiction that was not a party to the ECHR.

The subsequent ECtHR decision in *Mustafa v Bulgaria*[20] confirmed that there must be 'compelling reasons … demonstrated in each case' that the power of military criminal justice should extend to civilians.

In those cases where English extraterritorial jurisdiction applies, trying a case in the UK, where the witnesses are mainly based abroad, may

19 *Cooper v United Kingdom* [2003] ECtHR Grand Chamber.

20 Application no 1230/17, 5th Section, 28 November 2019, at s. 45: *'la Cour rappelle le principe établi dans sa jurisprudence selon lequel le pouvoir de la justice pénale militaire ne devrait s'étendre aux civils que s'il existe des raisons impérieuses justifiant une telle situation.'*

be easier these days, with the ability to use video links. But, when a foreign witness gives their account by video link, there is at present no ability to punish any contempt of court as they are outside the jurisdiction.

Article VII(4) of SOFA provides that no right is implied for the military authorities of the United Kingdom to exercise jurisdiction over persons who are nationals of or ordinarily resident in the receiving state, unless they are members of the force of the sending state. So, for example, a German wife of a British serviceman stationed in Germany who commits an offence in Germany, even on service property, cannot be tried by the British military authorities. This is also reflected in the provisions relating to civilians in Schedule 15 to AFA 2006, as amended. See below, Chapter 15 at page 497.

Non-NATO Member States

In non-NATO countries, jurisdiction over British troops will be determined either in accordance with the United Nations SOFA,[21] for UN mandated operations, or through a separate SOFA or MoU or LoU between the British and host nation governments.[22] It is a key provision of such agreements that jurisdiction for criminal and disciplinary offences remains with the British military authorities. It would be a serious failure on the part of Her Majesty's Government to deploy our forces anywhere without ensuring beforehand that there were proper jurisdictional safeguards in place to protect the rights of personnel subject to service law in relation to the investigation, trial and punishment of alleged offences. The relevant MoU provisions are normally based upon the wording of the NATO SOFA, but it is always worth checking that is the case. All the same, it is very doubtful that the government would deploy service personnel to any foreign country absent such an agreement. If the agreement is not formally completed by the time of a deployment, there will usually be a LoU and a diplomatic exchange of correspondence to put in place the requisite guarantees about jurisdiction pending the completion of the MoU. In the absence of an agreement for exclusive jurisdiction, the government has a duty to secure the best arrangements possible.

In the NATO mission to the Former Yugoslavia, codenamed Operation Joint Endeavour, the Implementation Force operated on the basis of a SOFA agreed at the Dayton Conference, which took place at Wright Air Force

---

21 UN Doc A/45/594, 9 October 1990.

22 See the explanation on Chapter 5, page 101 below.

Base, from 1 to 21 November 1995. This document set out the broad heads of agreement for how the force would operate within the whole of the Former Yugoslavia. The military legal adviser there was a very experienced US military lawyer, Colonel James Burger, US Army. It was left to the deploying headquarters to resolve the detail and draft subsidiary agreements to make the SOFA's objectives achievable in theatre.

The author arrived in HQ Commander for Support (C-SPT), in Zagreb on the evening of 9 December 1995. The following day he set to work on drafting the Technical Arrangement that, at that time, consisted of an overarching agreement – the Technical Arrangement – and 14 annexes, each one dealing with a specific issue, for example, de-mining, medical care, customs and border crossings, tax exemptions, jurisdiction, airports, ports. These drafts were prepared that day for use at a meeting the following day with the Croatian government. Thereafter, they were honed into the final documents that successfully governed the *modus vivendi* of the force in its interface with the host government. The drafts for Croatia were used as the basis for similar agreements in Bosnia. In respect of the deployment to Afghanistan, a military technical agreement was drawn up between the Afghan Interim Administration and the International Security Assistance Force, signed on 4 January 2002.[23]

The occupation of Iraq, under the governance of the Coalition Provisional Authority (CPA) in 2003 during a transition to Iraqi civilian government, and headed by a senior American diplomat (L. Paul Bremer III), was based on a comprehensive body of law made under the authority of UN Security Council Resolution 1483 (2003) and the international legal provisions relating to occupation. The CPA issued Orders, which were binding instructions or directives to the Iraqi people, that created penal consequences or had a direct bearing on the way Iraqis were regulated, including changes to Iraqi law.

Since 2003, the UK government may have MOUs or LOUs with countries such as Iraq, from where it is conducting operations ostensibly against ISIS in Syria and Iraq. Its presence in Syria for any purpose, *a fortiori* military operations, would be unlawful unless with the agreement of the Syrian government or the authority of the United Nations under Chapter VII of the Charter. Neither exists.

---

23 UN S/2002/117, 25 Jan. 2002.

3 THE DISCIPLINARY SYSTEM IN OVERVIEW

Sources

Service personnel are subject to a wide body of law and policy. The primary legal provisions are to be found in the Armed Forces Act 2006 (AFA 2006), Armed Forces Act 2011, and the numerous Statutory Instruments and Orders made therefrom. The Armed Forces Act 2016 made some small amendments to prosecution processes, but these do not affect the range of offences set out in the parent Act of 2006. The next major price of legislation is expected to be the Armed Forces Act of 2021, but the date of implementation is unknown.

In addition, each Service has its own Queen's Regulations (QRs), made under the prerogative, which lay down the policy and procedure to be observed in the command and administration of each Service. They also provide Commanding Officers with direction on the command and administration of their units; the effectiveness of their Service as a whole depends on the efficiency of those units. QRs provide for the delegation of authority for the management of each Service. As pointed out in Joint QRs:

> **J1.001** The government and command of each of the fighting Services is vested in Her Majesty The Queen, who has charged the Secretary of State with general responsibility for the defence of the Realm and established a Defence Council having command and administration over Her armed forces.
>
> **J1.002.** The Defence (Transfer of Functions) Act 1964 transferred to the Secretary of State the statutory functions of the previous Secretaries of State for War and Air and (except for certain specified functions relating to command and administration) of the former Board of Admiralty. The Act transferred to the Defence Council the statutory functions of the former Army

> Council and Air Council and corresponding statutory functions (including the excepted functions referred to above) of the former Board of Admiralty.
>
> **J1.003.** The Letters Patent of the Defence Council are produced in Annex A(J) to this Chapter. The Council are given the prerogative functions of the former Board of Admiralty, Army Council, and Air Council, are given administrative functions, are given command of all members of the forces and are directed to set up an Admiralty Board, Army Board, and an Air Force Board.
>
> **J1.005.** The Defence (Transfer of Functions) Act 1964 empowers the Army Board to discharge the statutory functions of the Defence Council, subject to any Directions of the Council. The effect of the above Directions is to confine this power, in the main, to statutory functions relating to the military forces. Certain prerogative functions under Orders in Council relating to the Royal Navy and Royal Marines were transferred to the Secretary of State who in practice is advised on such matters by the Admiralty Board. The Army and Air Force Boards have been empowered by Royal Warrant to exercise concurrently certain prerogative functions.

Defence Council Instructions are also issued to the three Services. The Defence Council is the senior departmental committee. It is chaired by the Secretary of State, and comprises the other ministers, the Permanent Under Secretary, the Chief of Defence Staff and senior service officers and senior officials who head the armed Services and the department's major corporate functions. It provides the formal legal basis for the conduct of defence in the UK through a range of powers vested in it by statute and Letters patent.[1] The role of the Defence Council is

> to administer such matters pertaining to Our Naval Military and Air Forces as We shall through Our Principal Secretary of State for Defence direct them to execute and to have command under

1 https://www.gov.uk/government/organisations/ministry-of-defence/about/our-governance.

> Us of all Officers and Ratings Soldiers and Airmen of Our Naval Military and Air Forces.[2]

In addition to primary and secondary legislation and QRs, each Service produces administrative instructions to set down processes for dealing with the day-to-day management of the Service. In the Army such instructions are called Army General and Administrative Instructions (AGAIs). In the Royal Air Force these instructions are referred to as Air Publications (APs). AP 1 is the Ethos, Core Values and Standards of the Royal Air Force. In the Royal Navy, the administrative guidance is to be found in RN Books. BRd2 is The Queen's Regulations for the Royal Navy (April 2017 edition).[3]

AGAI 67, for example, defines the Army's procedures for administrative action taken against personnel. This is part of the Army's employment mechanism that deals with unacceptable behaviours and is complimentary to the formal disciplinary process. The procedures for administrative action fall into four categories.

The first is minor 'admin' action (used to deal with minor infractions of discipline, for example, being late for work which, prior to the introduction of the administrative option in the early 2000s, would have led to a charge being laid for failing to attend a duty, etc.). The standard of proof is the civil standard: the deciding officer must be satisfied that it is more likely than not that the alleged failure in performance or standards of personal behaviour occurred.

The second category of admin action under AGAI 67 is major admin action, used for more serious misconduct, where the Service test is breached. The conduct must breach the Service test, which is: 'Have the actions or behaviour of an individual adversely impacted or are they likely to impact on the efficiency or operational effectiveness of the Service?' A failure to abide by Values and Standards[4] – breaching the Service test – may follow disciplinary action, or civil conviction (for example, for a drink driving offence).

---

2 Queen's Regulations for The Royal Air Force, Appendix 1 and http://webarchive.nationalarchives.gov.uk/20121026065214/http://www.mod.uk/NR/rdonlyres/079FF8D3-E131-47D1-BFF0-AE8EE0C65591/0/20090713DCLettersPatentU.pdf.

3 https://www.royalnavy.mod.uk/reference-library/brd2.

4 https://www.army.mod.uk/who-we-are/our-people/a-soldiers-values-and-standards/

The remaining two categories are career management actions, namely, Formal Warning and Removal from Appointment (where the person's position within the unit becomes untenable).

Joint Service Publications (JSPs) are also issued from time to time and cover matters common to the three Services. JSP 397 is the Service police codes of practice. JSP 838 deals with the procedure for the grant of legal aid.

It should be understood that, while the vast majority of employment legislation does not apply to the Services, the revised Service complaints procedure in Part 14A of AFA 2006[5] provides that a service person has the right to complain if they think themself wronged in any matter relating to their service. This is a wide-ranging right and provides those hearing such complaints a broad discretion in how they redress the complaint, if they find it proved. Such a complaint would be dealt with in accordance with the procedures set out in JSP 831 and the administrative instructions for the Service to which the individual belongs. Therefore the combination of the rights under JSP 831 and the administrative instructions really provide what is akin to a form of Service employment law. Indeed, the power to grant redress open to the Service Board is arguably wider than the ambit of remedies available in civilian employment law. Ultimately, the person seeking redress may raise their complaint with the Service Complaints Ombudsman.

Applicability

Throughout the legislation there are references to persons 'subject to Service law' and persons 'subject to Service discipline'. The distinction is important.

A person is 'subject to Service law' if they fall within sections 367–369 of AFA 2006, which means every member of the regular forces, every member of the reserve forces on call-out or training or serving on the permanent staff of a regular force, reserve forces and members of British overseas territories' forces serving with UK forces. If a service person based, for example, in Germany, goes on holiday to Italy, they are still subject to Service law at that time, even though they are outside the area commanded by a British military commander, as section 367 AFA 2006 makes it clear that every 'member of the regular forces is subject to Service law at all times'.

Section 370 defines a person 'subject to Service discipline' as a civilian who –

a. is not subject to Service law; and
b. is within any paragraph of Part 1 of Schedule 15, which lists the categories.[6]

Part 2 of Schedule 15 (exclusion and definitions relating to Part 1) has effect and provides that a person who is not a United Kingdom national is not within any of paragraphs 4 to 10 at any time when he is in a country –

a. of which he is a national; or
b. in which he is ordinarily resident.

The effect is that a spouse of a service person, who is a national of the country in which the service spouse is serving, is not a person 'subject to Service discipline'. The local laws will apply to them. The Armed Forces (Civilians Subject to Service Discipline) Order 2009, which came into force on 31 October 2009 determine whether a person is a civilian subject to Service discipline in those cases where this status depends in some cases partly on whether they are a member or employee of a specified organisation. Article 2 and Schedule 1 provide that certain members or employees are civilians of a military organisation (the North Atlantic Treaty Organisation, as specified in Sch 1) subject to Service discipline when outside the British Islands.

Article 3 and Schedule 2 provide for the specification of those other organisations whose employees are civilians subject to service discipline when in a designated area.

Some categories of person are civilians subject to service discipline if they are in a designated area. Article 4 and Schedule 3 provide for the designation of areas (referred to in respect of persons within paragraphs 4, 6 and 9 of Schedule 15 to the Act) for the different categories.[7]

---

6 See Appendix J.
7 See the End Notes to the 2009 Order.

4 CHAIN OF COMMAND

The chain of command of each Service runs from the Sovereign,[1] through the Defence Council. Examples of the chain of command for each Service, depicting the hierarchy in brief overview, can be found in Appendix B, Unit Chain of Command. However, it is the Service Boards that issue policy and direction for their respective Service. However, those wielding command over service personnel are responsible for maintaining discipline within their ship, regiment or squadron. The command structure is hierarchical. Using the Army as an example, divisions and brigades are responsible for administering groupings of smaller units. Major units are regiment or battalion sized with minor units being smaller, either company sized sub-units or platoons.

Defence Council

The Defence Council, is

> a formal body, with almost all of its work conducted by the Defence Board. The Secretary of State is the most senior individual in the Defence Council, which comprises of other Defence ministers, the Permanent Secretary, the Chief of Defence Staff, senior service officers and senior officials who head the armed services and the department's major corporate functions. It provides the formal legal basis for the conduct of defence in the UK through a range of powers vested in it by statute and Letters Patent. In addition, the three Service boards (the Admiralty

---

1 'As Sovereign, The Queen is Head of the Armed Forces.' *The Queen and the Armed Forces*: https://www.royal.uk/queen-and-armed-forces-0.

> Board, the Army Board and the Air Force Board), which are sub-committees of the Defence Council, meet annually for each service chief to report to the Secretary of State on the health of their respective services.[2]

Each Service Board,[3] that is the Admiralty Board, Army Board and Air Force Board, is actually, responsible for the day-to-day management of its Service and development of policy for its respective Service, as permitted by its delegated powers. The statutory authority is to be found in the Defence (Transfer of Functions) Act 1964 (D(FT)A), which provides for the delegation of functions to Service Boards. The Act makes for interesting reading in other respects, in that it explains the authority held by the Secretary of State for Defence as a corporation sole (with a corporate seal) for all purposes relating to the acquisition, holding, management or disposal of property. As clarified in section 4 of the Act, nothing in the D(TF)A shall prejudice any power exercisable by virtue of the prerogative of the Crown.

Admiralty Board

The Admiralty Board has command over the officers, ratings and marines of Her Majesty's Naval and Marine forces.[4] It is also charged with the administration of all such matters relating to the Royal Navy and Royal Marines as may be administered by the Defence Council subject to Defence Council directions. In practice, it delegates these responsibilities to the Navy Board, which is the body responsible for running the Royal Navy. Members of the Navy Board are responsible for assisting the First Sea Lord in discharging his responsibilities as Head of Service and Royal Navy Top Level Budget Holder. Principal among these is to ensure that the Navy is fit to fight and can deliver the military capability required to defend the

---

2 'Our governance', https://www.gov.uk/government/organisations/ministry-of-defence/about/our-governance. The Act transferred to the Defence Council (DC) the statutory functions of the former Army Council and Air Council and corresponding statutory functions (including the excepted functions referred to above) of the former Board of Admiralty. QRs J1.002. above on Chapter 3, page 33. The Letters Patent of the Defence Council are at QRS Annex A(J) which lists the members of the DC.

3 Information obtained from each respective Service website.

4 See https://www.royalnavy.mod.uk/our-people/senior-naval-staff/admiralty-board.

UK and its interests. The Navy Board meets monthly, or more often as required. Meetings of the Admiralty Board normally take place once a year in January.

Army Board

In the case of the Army, the Army Board is the senior management directorate whose responsibility is the routine management of the Army. The Executive Committee of the Army Board (ECAB), attended by senior UK Army commanders, dictates the policy required for the Army to function efficiently and meet the aims required by the Defence Council and government. The Chief of the General Staff is the chairman of the Executive Committee of the Army Board. The military staff at the various headquarters worldwide act upon Army Board and ECAB decisions. Whereas the Service Boards are normally constituted with very senior commissioned officers of the relevant Service, in 2015 the Army departed from precedent and appointed the Army Sergeant Major as the first enlisted soldier to be a member of the ECAB.

Air Force Board

The routine management of the Royal Air Force is the responsibility of the Air Force Board. The Air Force Board Standing Committee (AFBSC) dictates the policy required for the Royal Air Force to function efficiently and meet the aims required by the Defence Council and government. The Chief of the Air Staff is the chairman of the Air Force Board Standing Committee. Air staff at various headquarters worldwide implement decisions made by the Defence Council or the Air Force Board. The Chief of the Air Staff is the officer ultimately responsible for the Royal Air Force's contribution to the national defence effort. They maintain control through the AOC (Air Officer Commanding) and the staff branches of the various Royal Air Force headquarters.

Disciplinary Chain of Command

Main Provisions: Appendix A Rank Structure; Appendix B Unit Chain of Command; CO's Powers

Every service person has a commanding officer (CO), even though the immediate CO may not be the same officer as that person's CO for a particular purpose under the Act.[5] The CO is central to the service discipline system, and is responsible for the effective administration of discipline at unit level.

A very loose outline of the disciplinary chain of command of each of the three Services responsible for delivering the direction and policy of the Service Boards under the Defence Council is below. It is to be read in conjunction with the discussion on the statutory provisions, also below. The chain of command is hierarchical, which requires that lawful orders of superiors be obeyed. The full rank structure is in Appendix A below and the examples of unit structures are in Appendix B.

In a disciplinary context the chain of command is the hierarchical arrangement that leads from the most junior to the most senior officers in positions of command. For disciplinary purposes within a unit, it starts with the CO who may delegate to subordinate officers, under section 128 and Reg 16 of the Armed Forces (Part 5 of the Armed Forces Act 2006) Regulations 2009 AF(P5AFA2006)R 2009, certain functions. For example, the Captain of a ship, the CO of the regiment or battalion, or a Royal Air Force (RAF) Station Commander, may delegate powers for disciplinary purposes to subordinate officers[6] who are in charge of a sub-unit, such as a department on a ship, or a company, or a flight. At more senior level, a senior officer may be specifically appointed to act as CO of a less senior officer charged with an offence.

A CO has power to delegate some of their powers of investigation, charging and determining mode of trial to officers who are qualified.[7]

5 For example, an Army Legal Services Officer will not ordinarily be a CO for disciplinary purposes for any legal officer under their command, even though they professionally control and direct the legal work undertaken under their overall supervision. So, in the event of a disciplinary offence, a CO external to the legal branch will be appointed to deal with the legal officer in question for the disciplinary offence alleged.

6 See 'Subordinate Commander' below.

7 Officers not below the rank of naval lieutenant, military or Royal Marine captain or flight lieutenant. See Part 5 of the 2006 Act and regulation 16 of the Armed Forces (Part 5 of the Armed Forces Act 2006) Regulations 2009.

Disciplinary Chain Within the Royal Navy

In general terms, powers of summary discipline vest in officers who are COs of units (i.e. ships or establishments) or appointed as 'Commanding Officer' of a body of service personnel. The Armed Forces Act 2006 (AFA 2006) also empowers 'Subordinate Commanders' who may exercise powers of summary discipline under the authority of the CO. Naval service policy requires that those who exercise powers of summary discipline are authorised in advance (either as COs of ships or establishments, or as otherwise authorised) – and that the relationship between subordinate commander, CO and Higher Authority is approved.[8]

The Naval Service comprises the Royal Navy, Royal Marines, the Reserve Naval and Marine forces and the Naval Careers Service. A range of organisations (including external agencies and the Royal Fleet Auxiliary), largely staffed by civilians, support the Naval Service. Those organisations do not form part of the Naval Service, but where they contain Royal Navy personnel, those personnel individually are part of the Naval Service. The Queen Alexandra's Royal Naval Nursing Service (QARNNS) officers and ratings form part of the Royal Navy. The Admiralty Board of the Defence Council governs the Naval Service.

Some of the main groupings can be further subdivided into:

a. **Branch.** This is a professional functional sub-division, with Arms (some or all of General Service, Fleet Air Arm and Submarine), which can be further divided into specialisations. Examples in the Naval Service are Warfare, Engineering, Logistics and Royal Marines. All personnel, except for those in the Careers Service, belong to a branch.

b. **Specialisation.** This is a professional sub-division of a branch and likely to reflect a particular field of expertise. Some personnel will have more than one specialisation and others may have none.

c. **Sub-specialisation.** This is a division of Specialisation, and is based on professional grounds. It is the lowest recognised level within the tri-Service structure.[9]

8 Naval Personnel Management, Chapter 20, Discipline Policy And Processes.

9 Personnel Management Organisation & Strategy, Chapter 1, ANNEX 1A

The Naval Service chain of command is set out in (BRd 2) (April 2017 Version 2), the Queens Regulations for the Royal Navy. The disciplinary chain of command is explained in BRd 3. This is the single authority under which Naval Service Commanding Officers may exercise powers of discipline. It is a detailed arrangement for each ship or naval station and shows which appointment is the Subordinate Commander, who exercises powers of the CO, who is the 1st Higher Authority (at 1 star rank/Commodore),[10] who is the 2nd Higher Authority (at 2 star rank/Vice Admiral) and contains contact details of the Fleet Legal Advisers.

The BRd 3(1) specifies:

20045. Delegation of Disciplinary Powers to a Subordinate Commander

a. Commanding Officers may delegate (in writing) their disciplinary functions to one or more Subordinate Commanders (Sub Cdr), for example the XO or OC. The Sub Cdr must be of Lieutenant RN/Captain RM rank or higher (see also Fleet General and Administrtive Orders Chapter 16).

b. When the Subordinate Commander is temporarily unavailable to exercise his or her responsibilities for the administration of justice, the CO of a ship, unit or establishment may delegate another officer to assume such responsibilities. This delegation must be in writing, is to be promulgated within the ship, unit or establishment and is to specify the period during which the named officer is to fulfil these duties.

c. Only one Subordinate Commander may hold the CO's delegation for a particular charge. In effect this means that the CO may delegate disciplinary powers to the XO/OC, but if the XO's/OC's powers are insufficient to deal with a particular charge then the CO would have to revoke the delegation in respect of that charge and deal with the matter personally. Such revocation need not be in writing but may be executed by the action of bringing the charge or hearing the charge personally. However, ideally revocation should be in writing.

10 See Appendix A Rank Structure (Further Notes).

The Executive Officer is the Second in Command.[11] They are responsible to the CO for the general organization of the ship and for the discipline, morale and welfare of the whole ship's Company together with the coordination of their training both for war and peace, at sea and in the harbour. They are to be prepared to assume command if the CO becomes incapacitated for any reason, in accordance with 0335 and 0336. They should also be prepared for duties in a command Watch Bill opposite the CO; to which end they may be required to assume greater responsibility and be delegated greater authority in certain circumstances as required by the CO and as authorised by the CO.

Disciplinary Chain and Structure Within the Army

A very basic outline[12] of the structure/chain of command for disciplinary[13] matters is:

Sub-Unit/Subordinate Commander (usually a Major).
Commanding Officer (usually a Lieutenant Colonel).
Higher Authority (usually a Brigadier).

Regiments and Battalions

The backbone of the Army is the regimental system. The Regiment – or Battalion – is a body of about 650 personnel, commanded by a Lieutenant Colonel (usually referred to as the Colonel). As will be seen, the unit terminology and structure are more complicated than the other two Services. In his evidence to the Defence Select Committee on 'Future Army 2020', General Sir Peter Wall said,

> A unit is the generic term for a military organisation that is the basic building block of a specific operational capability. It includes service personnel predominately from a single cap-badge but with small detachments of other cap-badged personnel that provide specialist capability. A unit may be called a Regiment or a Battalion, dependent on the capability it provides and will generally consist of a Headquarters, a support sub-unit and three other sub-units called companies, squadrons

11 BRd 2, Chapter 9, §0901.

12 See also Appendix B Unit Chain of Command.

13 As opposed to administrative.

> or batteries, dependent on the unit type. The size and structure of each unit varies considerably (408 up to 729) and is dependent on its role and specialisation.
> Within the Infantry the standard unit is the battalion and consists of a battalion headquarters, a headquarter company, a support company and three rifle companies; the size is dependent on role. The Reaction Force infantry battalions are 729 or 709 strong, dependent on capability, each rifle company consisting of three rifle platoons. Within the Adaptable Force, infantry battalions are slightly smaller at 581 or 561 personnel.[14]

The size of the regiment will also affect the size of its sub-units.[15] As General Wall mentioned, terminology differs according to the type of unit. For example, the Parachute Regiment, which is an organisation of airborne troops, is divided into four battalions (Bn) (the 4th Bn being a reserve parachute battalion with its headquarters in Pudsey, West Yorkshire). Each battalion is commanded by a Lieutenant Colonel. Similar arrangements subsist in other units with more than one regimental-sized body of soldiers. Another example would be The Guards regiments (Grenadier, Coldstream, Scots, Irish, and Welsh), where the command unit is known as a battalion. Artillery units may, on the other hand, be called regiments.

Sub-units

The sub-units comprise, usually, three companies, or squadrons, each commanded usually by a Major (or a Captain). The company or squadron may have between 80 and 250 personnel.

Below each company/squadron level the sub-unit is a troop or platoon, commanded by a Captain or Lieutenant. This sub-unit may comprise about 20 to 30 personnel. Under the troop/platoon is a section,

---

14 Question 38, Session 2013–2014, published 15 May 2013. See http://www.publications.parliament.uk/pa/cm201213/cmselect/cmdfence/writev/army2020/m01.htm.

15 Unit names differ for historical reasons. An infantry regiment is an administrative and ceremonial organisation only and may include several battalions. An infantry battalion is equivalent to a cavalry regiment. For operational tasks a battle group will be formed around a combat unit, supported by units or sub-units from other areas. An example would be a squadron of tanks attached to an armoured infantry battle group, together with a reconnaissance troop, artillery battery and engineering support. See MoD website, 'British Army Structure – Naming Conventions'.

or squad, commanded by a non-commissioned officer, usually a corporal. These sub-units consist of about 12 personnel, which may then be divided into five teams of four. Otherwise, a Lance Corporal is the Section Second in Command (2IC).

There are special appointments within the regimental system. The Adjutant is an officer normally of the rank of Captain (or sometimes Major), who assists the CO with disciplinary matters and the issuing of orders. They are staff officer, in that they serve on the staff supporting the CO, rather than as a sub-unit commander.

Each regiment or battalion will have a Regimental Sergeant Major (RSM). This is an appointment held by a Warrant Officer Class 1 (WO1), who is the most senior non-commissioned officer in the unit and who is responsible for discipline. The RSM is a very experienced soldier and commands a great deal of respect, from officers and soldiers alike.

Other appointments are Regimental Quartermaster Sergeant (RQMS) and Quarter Master Sergeant Instructor (QMSI), usually a WO2, who is the second most senior Warrant Officer (WO) after the RSM.

Below these ranks are the Platoon or Troop Staff/Colour Sergeant (SSgt/CSgt), and the Company Quartermaster Sergeant (CQMS). The appointment of CQMS is held by a Staff Sergeant.

The main senior non-commissioned regimental appointments are:[16]

Warrant Officer First Class (WO1) It is more usual for a WO1 to be referred to by their appointment, for example Academy Sergeant Major, rather than by their rank, some which are:

- Army Sergeant Major
- Conductor (RLC)
- Academy Sergeant Major (AcSM) Royal Military Academy Sandhurst (RMAS)
- Garrison Sergeant Major, London District (GSMLD)
- Royal Artillery Sergeant Major
- Master Gunner RA
- Garrison Sergeant Major, except London District

16 This information was obtained from Alan Harding's very helpful website, http://www.alanharding.com/index.html. The links have been disabled.

- Regimental Corporal Major (RCM) – Household Cavalry
- Regimental Sergeant Major (RSM)
- Bandmaster
- Staff Sergeant Major RLC or AGC(SPS)
- Foreman of Signals Sergeant Major (RS)
- Clerk of Works Sergeant Major (RE)
- Artificer Sergeant Major (ASM)
- Farrier Corporal Major – Household Cavalry.

The four most senior warrant officer appointments in the British Army are generally considered to be, in descending order of seniority, Army Sergeant Major, the Conductor RLC, Academy Sergeant Major, and the Garrison Sergeant Major, London District, although there is some debate about their precedence within the Army. WOs with appointments have laurel wreathes around their insignia.

Warrant Officer Second Class (WO2)

Regimental duty appointments:

- Company Sergeant Major (CSM)
- Squadron Sergeant Major (SSM)
- Battery Sergeant Major (BSM)
- Squadron Corporal Major (SCM) in Household Cavalry.

Appointments as Quarter Master and Technical Grades:

- Regimental QMS (RQMS) – RQMC in Household Cavalry Regiment (HCR)
- Artificer QMS (AQMS)
- Foreman of Signals Quartermaster Sergeant
- Band Sergeant Major (BSM) – BCM in HCR
- Quartermaster Sergeant Instructor (QMSI) APTC, RE.

Staff Sergeant:

- Colour Sergeant (CSgt, C/Sgt) – Infantry, RM
- Staff Corporal – HCR
- Company/Squadron/Battery Quartermaster Sergeant (CQMS, SQMS, BQMS)

Higher Command

Above regimental level, the next level of command is the brigade, commanded by a Brigadier (a 1 star officer who, in the British Army, no longer holds the title of General). A brigade may comprise anything from three to six regiments or battalions. The brigade is a part of a Division, which will have two or three brigades and is commanded by a Major General (2 star officer).

Corps are larger sized formations and are generally divided into regiments. The Army Air Corps is a good example. It has seven regiments, each commanded by a Lieutenant Colonel.[17] Another is the Royal Armoured Corps, comprising Household Cavalry (HCR, 1st The Queen's Dragoon Guards (QDG), Royal Scots Dragoon Guards (Carabiniers and Greys) (SCOTS DG), Royal Dragoon Guards (RDG), The Queen's Royal Hussars (The Queen's Own and Royal Irish) (QRH), The Royal Lancers (Queen Elizabeths' Own) (RL), King's Royal Hussars (KRH), Light Dragoons (LD), Royal Lancers ('Queen Elizabeths' Own') (RL), Royal Tank Regiment (RTR). These are augmented by Army Reserve – Yeomanry, namely, Royal Yeomanry (RY),[18] Royal Wessex Yeomanry (RWxY), Scottish and North Irish Yeomanry (SNIY), Queen's Own Yeomanry (QOY).

Professional organisations are also organised into Corps. For example, the Royal Medical Corps, the Royal Army Veterinary Corps, the Royal Army Dental Corps, Royal Army Nursing Corps, the Adjutant General's Corps (AGC),[19] the Intelligence Corps, the Corps of Royal Engineers, and the Corps of Royal Electrical & Mechanical Engineers.

Disciplinary Chain Within the Royal Air Force

Within the Royal Air Force (RAF), the structure is loosely similar to that of the Army. A CO is responsible to the air officer commanding for the maintenance of discipline and efficiency in, and for the proper administration of, the station and units or unit under their command. QR (RAF) 994 provides the definition of 'CO' for Disciplinary Purposes consonant with

17 1 Regt, 2 Regt, 3 Regt, 4 Regt, 5 Regt, 6 Regt (Volunteers), and 7 Regt.

18 The senior cavalry regiment in the Territorial Army.

19 Made up of Provost Branch (RMP/SIB, Military Provost Staff, and Military Provost Guard Service), Army Legal Services, Staff & Support Services, and Educational & Training Services. Prior to the formation of the AGC, the Army Legal Service (ALS) was known as the Army Legal Corps.

the provisions of the AFA 2006, which places the CO at the centre of the Service Justice System and confers a range of powers on them, being at the apex of a unit's command and control structure. Consonant with the other two Services, in order to deal summarily with an officer the CO must be two ranks higher. Where the CO is not two ranks higher, an officer will be specifically appointed to be CO for the purpose of hearing the charge. The officer next higher in the CO's chain of command would usually be specifically appointed as CO in these circumstances. The CO may delegate some or all of their disciplinary functions to subordinate commanders and make conditions on such delegations as they consider appropriate. The delegation will continue until the CO revokes it. A very basic outline of the structure/chain of command for disciplinary matters is as follows:

Sub-Unit/Subordinate Commander (usually a Flight Lt/Squadron Ldr).

CO (Usually a Wing Commander but could be a Group Captain if the RAF Station is a large unit).

Higher Authority (usually an Air Commodore).

As to the structure of the RAF, the HQ Air Command is based at RAF High Wycombe, Buckinghamshire.[20] All officers ranked Air Commodore and above are referred to as holding Air rank, or Air Officers.

Unit command encompasses RAF Group Command.[21] Currently there are six group, each of which is commanded by an Air Vice Marshal (see Appendix A, Rank Structure).

No 1 Group's mission is 'To understand, control, and exploit the battle space in order to deliver battle winning Combat Air Power and to support the generation, development and sustainment of the RAF's Support Helicopter Force so that they can best meet the Joint Helicopter Command Mission.' No 1 Group coordinates all of the RAF's frontline, fast-jet force elements and the Intelligence Surveillance, Targeting and Reconnaissance (ISTAR) work.

No 2 Group has three distinct roles and thus there are three distinct elements within No 2 Group – Air Transport and Air-to-Air Refuelling (AT/AAR); Intelligence Surveillance, Targeting and Reconnaissance (ISTAR);

20 RAF website, http://www.raf.mod.uk/.

21 See https://www.raf.mod.uk/our-organisation/groups/.

and Force Protection (FP). Its role is to generate and employ military capability.

11 Group combines the capabilities of the Chief of Staff Ops AO with the Air Battle Staff, and comprises the deployable Joint Force Air Component (JFAC), the National Air & Space Operations Centre (NASOC) and the Executive Team. It also incorporates the BM Force, providing the building blocks for the generation of a powerful non-kinetic effects capability.

No 22 (Training) Group, which exists to recruit and train RAF personnel and provide trained specialist personnel to the RAF and the other two Services.

38 Group HQ is located at RAF High Wycombe. The Group comprises engineering and logistics, communications and Medical Operations Force Elements, in addition to units such as RAF Music Services. Air Officer Commanding Number 38 Group has responsibility for UK-based United States Visiting Forces (USVF) units and for RAF personnel detached to other armed forces around the world. Overall, the Group employs around 2,600 Service personnel and 310 Ministry of Defence civilians across multiple defence sites in the United Kingdom and deployed locations worldwide.

No 83 Expeditionary Air Force Group, commanded by an Air Commodore, is the RAF's operational headquarters in the Middle East. The Air Commodore commands the four Expeditionary Air Wings (EAW) – each in turn commanded by a Group Captain – created in early 2006 to improve the ability of the Royal Air Force to support Permanent Joint Headquarters' (PJHQ) aim of successful overseas operations. Below the EAW are EAW flights, each commanded by a Squadron Leader. 83 Expeditionary Air Group is the RAF's operational headquarters in the Middle East.

RAF Station Commanders (usually a Group Captain) are responsible for an RAF base, which houses other units, service personnel and families. Various branches with specific responsibilities are with the Station headquarters, e.g. Personnel Support, which deals with personnel and disciplinary matters. A Wing Commander may command a subdivision of the RAF station.

The RAF Regiment is the ground fighting force of the RAF. The Regiment is a diverse organisation that provides first-rate worldwide force protection to UK air assets. Its members are officers and gunners. There are

10 regular and 4 reserve squadrons, usually commanded by a Squadron Leader.

Commanding Officer

The CO is broadly akin to a managing director / chief executive officer of a civilian organisation, being the leader and director of corporate (regimental) output for day-to-day operations. Strategic guidance will be issued by their higher chain of command (akin to a board of directors).

Every service person must have a CO, who is the person responsible for exercising lawful authority and direction over those under their command, together with disciplinary powers defined by statute. Even civilians accompanying a force are subject to service discipline and will have a CO for disciplinary matters, if they commit an offence. As the Manual of service law points out: '[e]very officer appointed as a CO should be clear for whom he has disciplinary responsibility. This includes sub-units, attached personnel and relevant civilians. These disciplinary relationships should be published routinely in unit orders.' It goes on to explain that, '[e]very Service person and relevant civilian should have a CO for disciplinary purposes who, in the case of a Service person, will normally be the CO of the unit of which the Service person is a member.'[22]

Just who is a person's CO for disciplinary purposes is defined in the Armed Forces (Meaning of Commanding Officer) Regulations 2009,[23] made under powers conferred by section 360 of the AFA 2006 and section 127(3) of the Reserve Forces Act 1996. The following is an extract.

The Commanding Officer is defined in accordance with the following provisions:

> **Specific Appointments**
>
> 3. An officer may be appointed by or on behalf of the Defence Council, or by an officer authorised by the Defence Council, to be the Commanding Officer of a person for the purpose of any relevant provision either generally or in relation to a particular matter.

---

22 JSP 830 MSL Version 2.0 1-2-3

23 http://webarchive.nationalarchives.gov.uk/20121026065214/http://www.mod.uk/NR/rdonlyres/B111C908-F810-4707-A4B9-C19C9B0C6D90/0/AF_MeaningofCommandingOfficer_Regs2009.pdf.

Members of the Regular Forces

4. (1) This regulation is subject to regulations 3 and 8 and applies to determine the commanding officer of a person who is a member of the regular forces.

(2) The person's Commanding Officer for general purposes is –

(a) subject to paragraphs (b) to (d), the officer in command of the unit of which he is a member;

(b) subject to paragraphs (c) and (d), if he is attached to another unit, the officer in command of the unit to which he is attached;

(c) if he is in a service hospital as a patient, the officer in command of the service hospital, except insofar as the officer relinquishes authority over him for any purposes of any provision made by or under the 2006 Act;

(d) if he is at the MCTC[24] in Service custody, the officer in command of the MCTC.

(3) For the purposes of section 124[25] or any provision of Part 6[26] of the 2006 Act, or any rule made under section 153(1)[27] of that Act, the person's commanding officer is

(a) his commanding officer for general purposes, if that officer is his superior by more than one rank;

(b) if his commanding officer for general purposes is his superior by one rank, that officer's immediate superior in the disciplinary chain of command;

(c) if his commanding officer for general purposes is of the same rank as the person, that officer's superior by two ranks in the disciplinary chain of command.

24 Military Corrective Training Centre, Colchester.

25 CO to hear charge allocated for summary hearing.

26 Summary Hearing and Appeals and Review.

27 The power of the Secretary of State to make provision with respect to summary hearing of charges by COs and hearings as regards the making of orders under section 193 (activation of suspended sentence of service detention).

(4) For the purposes of section 251(5)(a)[28] of the 2006 Act, the person's commanding officer is the officer who imposed on the person the fine or service compensation order in respect of which the power in section 251(3)[29] or (4)[30] is exercisable.

(5) For the purposes of paragraph 2(c), the officer in command of a service hospital may relinquish authority over the person for the purposes of any provision made by or under the 2006 Act by informing

(a) subject to paragraph (b), the officer in command of the unit of which the person is a member,

(b) if the person is attached to another unit, the officer in command of the unit to which the person is attached,

That he relinquishes authority for that purpose.

(6) For the purposes of this regulation, a naval chaplain, other than a principal chaplain or the Chaplain of the Fleet, shall be treated as if he held the rank of commander.

Members of the Reserve Forces

5. (1) This regulation is subject to regulations 3 and 8.

(2) This regulation applies to determine the Commanding Officer of a person who is a member of the reserve forces, except for the purposes of any provision made by or under the 2006 Act which applies in relation to him or her by virtue of his being or having been a civilian subject to service discipline.[31]

(3) Except in a case within paragraph (4), the person's Commanding Officer for general purposes and for the purpose of any provision made under the 1996 Act[32] is

(a) subject to paragraphs (b) to (d), the officer in command of the unit of which he is a member;

---

28 S 251(5)(a) if the fine or service compensation order was imposed by an officer and subsection (6) applies, the CO of the relevant person;
S 251(6) This subsection applies if the relevant person is for the time being—
(a) subject to Service law;
(b) a member of a volunteer reserve force; or
(c) a member of an ex-regular reserve force who is subject to an additional duties commitment.

29 Power of a court to impose a fine or service compensation order by instalments.

30 Power of the court to vary the fine or service compensation order.

31 See above, on page 35.

32 Reserve Forces Act 1996.

(b) subject to paragraphs (c) and (d), if he is attached to another unit, the officer in command of the unit to which he is attached;

(c) if he is in a service hospital as a patient, the officer in command of the service hospital except in so far as that officer relinquishes authority over him for the purposes of any relevant provision;

(d) if he is at the MCTC in Service custody, the officer in command of the MCTC.

(4) If the person is not a member of a unit, his commanding officer for general purposes and for the purposes of any provision made by the 1996 Act is the officer in command of the last unit, if any, of which he was a member.

(5) For the purpose of section 124[33] or any provision of Part 6 of the 2006 Act, or any rule made under section 153(1) of that Act, the person's commanding officer is-

(a) his commanding officer for general purposes, if that officer is his superior by more than one rank;

(b) if his commanding officer for general purposes is his superior by one rank, that officer's immediate superior in the disciplinary chain of command;

(c) if his commanding officer for general purposes is of the same rank as the person, that officer's superior by two ranks in the disciplinary chain of command.

(6) For the purposes of section 251(5)(a) of the 2006 Act, the person's commanding officer is the officer who imposed on the person the fine or service compensation order in respect of which the power in section 251(3) or (4) is exercisable.

(7) For the purposes of paragraph 3(c), the officer in command of a service hospital may relinquish authority over the person for the purposes of any provision made by or under the 2006 Act by informing –

(a) subject to paragraph (b), the officer in command of the unit of which the person is a member;

---

33 See page 52, note 25, above.

(b) if the person is attached to another unit, the officer in command of the unit to which the person is attached.

That he relinquishes authority for that purpose.

Former Members of Her Majesty's Forces

6. (1) This regulation is subject to regulations 3 and 8.

(2) Subject to paragraph 3, if a person has ceased to be a member of Her Majesty's forces, for the purpose of any relevant provision which applies in relation to him by reference to the fact that he or she was previously Subject to Service Law, his commanding officer is the officer in command of the last unit, if any, of which he was a member.

(3) If a person who has ceased to be a member of Her Majesty's forces is at the MCTC in Service custody, his or her Commanding Officer for the purposes of any relevant provision is the officer in command of the MCTC.

Certain Civilians Subject to Service Law

7. (1) This regulation is subject to regulations 3 to 6 and 8.

(2) If a person is within paragraph 3[34] of Schedule 15 and is at the MCTC, his commanding officer for the purposes of any provision made by or under the 2006 Act is the officer in command of the MCTC,

(3) Subject to paragraph (2), paragraphs (4) to (11) apply to determine who is the Commanding Officer for the purposes of any provision made by or under the 2006 Act of a person (A) who is within either

(a) only one paragraph of Schedule 15; or

(b) paragraph 3 of Schedule 15 and only one other paragraph of that Schedule.

(4) If A is within paragraph 1[35] of Schedule 15, and a person Subject to Service Law is in command of the aircraft which A is

34 Persons in service custody by virtue of any provision of or made under the 2006 Act or being arrested, or of having an attempted arrest made of him, by a person who has a duty under Service law to apprehend him.

35 Persons in one of Her Majesty's aircraft in flight.

in, A's commanding officer is that person's commanding officer for general purposes.

(5) If A is within paragraph 2[36] of Schedule 15, his commanding officer is the officer in command of the ship which A is in.

(6) If A is within only paragraph 3 of Schedule 15 and has previously been within another paragraph of that Schedule, his commanding officer is the officer who was his Commanding Officer when he was last within a paragraph of that Schedule, other than paragraph 3.

(7) If A is within paragraph 4 of Schedule 15,[37] and is working in support of one unit, his Commanding Officer is the officer in command of that unit.

(8) If A is within paragraph 7[38] of Schedule 15, and is working in support of one unit, his Commanding Officer is the officer in command of that unit.

(9) If A is within paragraph 8[39] of Schedule 15, his Commanding Officer is the officer in command of the unit of which the person Subject to Service Law with whom A resides or is staying is a member.

(10) If A is within paragraph 9[40] of Schedule 15, his Commanding Officer is the commanding officer for general purposes of the relevant person (within the meaning of that paragraph) with whom A resides or is staying.

(11) If A is within paragraph 10[41] of Schedule 15, his Commanding Officer is the Commanding Officer for general purposes of the relevant person (within the meaning of that paragraph) with whom A resides or is staying.

(12) In this regulation 'Schedule 15' means Schedule 15 to the 2006 Act.

---

36 Persons in one of Her Majesty's ships afloat.

37 'Crown servants whose normal place of work in that role is in that designated area; or otherwise is in a designated area and he came there wholly or partly for the purposes of his work in that role' – as amended by AFA 2011, s. 22.

38 Persons outside the British Islands designated by or on behalf of the DC.

39 Persons residing or staying with person subject to service law in a designated area.

40 Persons residing or staying with Crown servants, or members or employees of other specified organisations, in designated area.

41 Persons residing or staying with persons working for specified military organisations.

These provisions look exceptionally complex, but they cater for the myriad circumstances of service life, from the obvious relationship between a service member of a unit and the person commanding that unit, to those attached to other units (including a service hospital or the Military Corrective Training Centre, Colchester (MCTC)) and includes civilians who are dependants of service personnel, or family living with or staying with them in their service married quarter,[42] or are Crown servants working with the military or the family of Crown servants residing with them, or otherwise are persons working for the Armed Forces (e.g. contractors) and their families residing with them. It is also important to remember that persons who are no longer subject to service law, but who committed a service offence when they were so subject, can be brought back for disciplinary purposes, as described below on page 66. They, too, will need a CO. Usually, this is the officer in command of the last unit of which they were a member, unless a bespoke appointment is made. However, policy may dictate that, for those no longer serving, it will be a bespoke appointment of a CO who is close geographically to the accused's location for the sake of convenience. This is particularly so for ex-soldiers whose former units are overseas, or a long distance away from their home, or no longer exist, as it makes no sense having a dislocated unit dealing with discipline and serving necessary papers.

Where there are two or more service personnel, subject to service law, who are involved in the commission of an offence, it is usual for one CO to be appointed for both. This will be agreed between the COs concerned and then an official appointment order will be issued by the relevant personnel branch in accordance with regulation 3 of the Armed Forces (Meaning of 'Commanding Officer') Regulations 2009 (AF(MCO)R 2009). If it is decided that each CO shall deal with their own offender, they should ensure there is parity of dealing.

Where a person does not have a readily identifiable CO, for example, they are a senior officer of Lieutenant Colonel/Wing Commander/

42 The Regulations do not limit the scope of the Act in this way, but it is submitted it would be difficult to exercise discipline over the visiting family of a service person or a spouse, when the service person was living in private accommodation e.g. they purchased a private flat or house. There is no service connection between the military and the visiting family in such circumstances. The matter becomes more problematic if tax-free facilities are granted to visiting members of the service person's family.

Commander rank, a specific appointment has to be made. Specific appointments can only be made by or on behalf of the Defence Council or by an officer authorised by the Defence Council.[43]

If a civilian falls within more than one paragraph of Part 1 of Schedule 15 to the Act,[44] which concerns civilians subject to service discipline, and neither of those paragraphs is paragraph 3,[45] the Defence Council regulations do not automatically identify a CO. It is therefore necessary for a CO to be specifically appointed for them.

For a more detailed examination of the command arrangements, see Chapter 2 of the Manual of service law.

The disciplinary powers of the CO are addressed in Chapter 9.

Subordinate Commander

In relation to the definition of and delegation of powers to a subordinate commander, see Chapter 9, page 321 *et seq*.

43 See regulation 3 of The Armed Forces (Meaning of 'Commanding Officer') Regulations 2009.

44 As amended by s 22 Armed Forces Act 2011.

45 Persons in service custody.

5 ARREST

Main Provisions: sections 50, 70–83, AFA 2006; Sch 4(3)(2) AFA 2011; AFA2011(C4)O 2013; MSL Ch 4; EN

There are two principal ways that an investigation will commence: information will come to the attention of the Commanding Officer (CO), or it will be imparted to the service police.

Investigations under service law are conducted by the service police, who hold statutory powers under the Act, although they are not police constables under the various Police Acts. They do not have civil police powers of arrest and the Service police codes of practice (SCOP) are similar but not identical to the Police and Criminal Evidence Act (PACE). A person who is reasonably suspected of being engaged in committing, or of having committed, a service offence may be arrested by a person subject to service law. Most cases will be concerned with an arrest under section 67 of AFA. There is an additional authority for service police to arrest a person whom they reasonably suspect of being about to commit a service offence (s. 69).

In terms of search, seven situations are covered by the legislation, namely,

a. search on arrest by service police (s. 70)
b. search on arrest by authorised persons (s. 71)
c. stop and search by service police (s. 75)
d. stop and search by authorised persons (s. 76)
e. authorised entry and search of premises by service police (s. 87)
f. authorised entry and search of premises by other persons (s. 88)
g. entry and search authorised by a judge advocate (s. 83).

A service offence is defined in section 50(2) AFA 2006[1] as

a. any offence under Part 1;
b. an offence under section 93A, 93E or 93G (testing for alcohol or drugs on suspicion);[2]
c. an offence under section 107 (breach of requirement imposed on release from custody);
d. an offence under section 229 (breach of service restraining order);
e. an offence under section 266 (failure to comply with financial statement order);[3]
f. an offence under section 305 (random drug testing);[4]
g. any offence under regulations under section 328 (false answer during enlistment in a regular force) or section 343 (service inquiries) that the regulations provide is a service offence;
h. an offence under paragraph 2, 3 or 4 of Schedule 2A (offences committed by a lay member of the Court Martial, such as research by lay members, and engaging in other prohibited conduct, such as conduct from which it may reasonably be concluded that the person intends to make a finding on a charge or a decision about a sentence otherwise than on the basis of the evidence presented in the proceedings);
i. an offence under paragraph 5 of Schedule 2A (disclosing information about members' deliberations, etc.) committed by a person who was a member of the Court Martial for the proceedings, or at the time the offence was committed, was a person subject to Service law or a civilian subject to Service discipline [sub-paragraph (2) of that paragraph];
j. an offence under section 18 or 20 of the Armed Forces Act 1991 (c. 62) (orders for the protection of children);

---

1 As amended by s. 82 of the Anti-social Behaviour, Crime and Policing Act 2014, which repealed s. 17 and Sch 4(3)(3) (breach of a service sexual offences prevention order or extended prohibitions order).

2 Inserted by Sch 4(3)(2) of the AFA 2011. See also, AFA2011(C4)O 2013.

3 This includes assets. See CCA 2013, Sch 16(38).

4 Amended by Sch 4(3)(4) of AFA 2011. See also, AFA2011(C4)O 2013. See consequential amendments that remove from s. 305(2)(b) the wording from 'or an investigation' onwards. This is pursuant to the repeal of s. 306.

k. an offence under any of sections 95 to 97 of the Reserve Forces Act 1996 (c. 14) (reserve forces offences); or
l. an offence under paragraph 5(1) of Schedule 1 to that Act (false answer during enlistment in a reserve force) committed by a person within paragraph 5(3) of that Schedule.

Powers of stop and search are subject to some restrictions, set out in section 79 AFA 2006, concerning persons found in a garden or yard, or on other land, occupied with and used for the purposes of a dwelling or any service living accommodation. In such a case the individual may not be searched unless the service policeman/woman or authorised individual has reasonable grounds for believing the person does not reside in the dwelling or service accommodation, and that they are not there with the express or implied permission of a person who does reside in the dwelling or service living accommodation. Section 79(3) and (4) contain relevant provisions relating to the search of any vehicle situated in a garden or yard connected to the above-mentioned properties.

Where powers of search are exercised under either section 87 or section 88 and any property has been seized and retained, both the powers of search and the property seized are subject to review by a judge advocate as soon as reasonably practicable (s. 89). The review is conducted in accordance with Article 34 of Armed Forces (Powers of Stop and Search, Search, Seizure and Retention) Order 2009 (AF(PSSSSR)O 2009). This allows, inter alia, for representations to be made by the owner of the seized property for its return and takes into account any claim to legal privilege in respect to any part of, or the whole of, the seized property. There is a loophole here when compared to civil police powers of destruction, in particular with regard to indecent images, etc. Service police have no powers to destroy other than through judicial order. The Armed Forces Act 2021 (AFA 2021) may address this when it comes into force.

Arrest

Main Provisions: section 61 and Part 3 of the AFA 2006, sections 67–97; sections 110–1111 AF(PSSSSR)O 2009; AF(PSSSSR)O 2014; Chapter 3 AFA 2006 (ss. 313–320); Code G

Powers of arrest, search and entry are contained in Part 3 of the AFA 2006, sections 67–97. Accompanying Regulations are made under the AF(PSSSSR) O 2009. These provisions cover the power of arrest for a service offence, under sections 67–69, search on arrest (ss. 70–74), stop and search (ss. 75–82), powers of entry, search and seizure (ss. 83–91), and additional powers of entry, search and seizure (ss. 92–93) with supplementary provisions for the remaining sections (ss. 96–97).

The Service police will usually carry out arrest and search. However, where this is not possible, or it is a case of urgency, the CO may authorise the arrest. Arrest may be exercised personally; or by giving orders for the arrest of the person who is to be arrested; or where that person is subject to service law, by ordering them into arrest.[5] A person who is reasonably suspected of being engaged in committing, or of having committed, a service offence may be arrested in accordance with section 67, as follows:

- (a) An Officer may be arrested[6] by
  - (1) an officer of superior rank or, if engaged in a mutiny, quarrel or disorder, by an officer of any rank;
  - (2) a Service policeman; or
  - (3) a person who is lawfully exercising authority on behalf of a provost officer (defined in s. 374 as a Service policeman), and who
    - (i) is an officer; or
    - (ii) is acting on the order of an officer.
- (b) A person of or below the rank or rate of warrant officer may be arrested by
  - (1) an officer;
  - (2) a warrant officer or non-commissioned officer of superior rank or rate;
  - (3) by a person who is lawfully exercising authority on behalf of a provost officer, and who
    - (i) is an officer; or
    - (ii) is acting on the order of an officer.[7]

5 *Ibid.*, s.67(6).

6 As amended by AFA 2011, s. 29 and Sch 3, in force on 2 April 2012: AFA2011(C1TTP)O 2012.

7 As amended by Armed Forces Act 2011, Sch 3, brought into force by the AFA2011(C1TTP) O 2012.

(4) a person who is lawfully exercising authority on behalf of a provost officer; or

(5) if a member of a ship's company or an embarked force, by a person exercising authority as a member of the staff of the officer of the day.

(c) A civilian subject to Service discipline may be arrested under subsection (1) by

(1) an officer;

(2) a Service policeman; or

(3) a person who is lawfully exercising authority on behalf of a provost officer.

In addition, service police may arrest in anticipation of the commission of a service offence where they reasonably suspect the offender is about to commit a service offence (s. 69), and in other specifically authorised circumstances (s. 61(2)).

The effect of the above is that the authority to arrest in the situations referred to above relate to service offences committed by service personnel /civilians, whereas the circumstances under section 61(2) additionally includes the arrest of civilians no longer subject to service law (a specially authorised circumstance). This would arise in the event of the arrest of a former service person or civilian against whom the Attorney General has consented to prosecute in accordance with section 61(2), in which case a 'person who is reasonably suspected of being engaged in committing, or of having committed, a service offence may be arrested by a Service policeman'.[8] See below under Time limits for offences. Section 68 clarifies that a person who is suspected of having committed a service offence while a member of Her Majesty's forces and who is not still a member of Her Majesty's forces is to be treated in relation to the offence as being of the rank or rate which they held when they were last a member of Her Majesty's forces. Where the person suspected was a civilian subject to service discipline at the time of the offence, they are to be treated in relation to the offence as if they were a civilian subject to service discipline.

Civilian police may also effect an arrest where a judge advocate has issued a warrant for the arrest of anyone who is reasonably suspected of

---

8 Modifications of s. 67 by s. 68(4).

having committed a service offence (s. 313). The warrant is addressed to an officer or officers of a UK police force or British Overseas Territory[9] police force and must specify the name of the person for whose arrest it is issued and the offence which they alleged to have committed. See Chapter 3 of AFA 2006 (ss. 313–320), which covers, inter alia, arrest under warrant of the judge advocate, arrest by civilian police of deserters and absentees without leave, deserters and absentees without leave surrendering to civilian police, proceedings before a civilian court where a person is suspected of illegal absence and certificates in connection with transfer of individuals to service custody. It is reassuring to note that civilian police may use reasonable force, if necessary, to make an arrest.[10] Why this needs statutory reinforcement is unclear, as any arrest where excessive force is used would be an assault.

Under section 67(6), 'the power of arrest conferred on any person under that section may be exercised—

- (a) personally;
- (b) by giving orders for the arrest of the person who is to be arrested; or
- (c) where that person is Subject to Service law, by ordering him into arrest.'

For example, if the CO authorised an officer to arrest another officer, of inferior rank, that officer may delegate the duty to another officer, so long as the officer actually carrying out the arrest is at least one rank higher than the person to be arrested. However, if there is a 'mutiny, quarrel or disorder', the arrest can be made by an officer of any rank. Mutiny is defined as 'acts with the intention of overthrowing or resisting authority' and include a

9 There are 14 British Overseas Territories, subject to British sovereignty. In the cases of the Falkland Islands and Gibraltar, British sovereignty is disputed by Argentina and Spain, respectively. According to a confidential signal dated 5 January 2009, disclosed to the *Daily Telegraph* by WikiLeaks, '[T]he BOTs are not constitutionally part of the UK. Each has a distinct constitution and a unique legal relationship to the UK. HMG guarantees the defence of all BOTs and handles their foreign relations.' https://www.telegraph.co.uk/news/wikileaks-files/london-wikileaks/8305236/A-GUIDE-TO-THE-BRITISH-OVERSEAS-TERRITORIES.html.

10 *Ibid.*, s. 320: 'Power to use reasonable force
Where a power of arrest is conferred on any person by virtue of this Chapter, he may use reasonable force, if necessary, in the exercise of the power.'

disobedience to authority 'in such circumstances as to subvert discipline' (s. 6(2)). The terms 'quarrel or disorder' are not defined and should, it is submitted, be given their ordinary meaning, bearing in mind that, as the Values and Standards of each service emphasise, it is the responsibility of the chain of command to maintain good order and discipline. An arrest may be effected for a service offence that is within section 50 (see earlier). For example, if a Flying Officer (OF1) witnessed a brawl (disorder) between a Squadron Leader (OF3) and another person, the Flying Officer would be justified in arresting the superior officer, if necessary. A brawl falls within section 21 (fighting or threatening behaviour, or affray, etc.).

Where a service person has been charged with, or is awaiting sentence for, a service offence, the CO also has the power, under section 110 AFA 2006, to order their arrest if satisfied that taking that person into service custody is justified on the basis they will fail to attend any hearing; or they will commit an offence; or they will interfere with witnesses or otherwise obstruct the course of justice. This is known as post-charge custody and is the most common form of custody encountered within the Service Justice System.

Section 110 also applies to any case where the CO, having dealt with the accused summarily, determines to await further evidence before sentencing or where the service person is awaiting sentence by service court, and information comes to the CO's attention that the individual is imminently likely to commit one of the offences mentioned above or there are reasonable grounds for believing that they have failed or will fail to comply with one of the conditions of their release (see the Explanatory Notes to legislation (EN), §§ 239–240). The circumstances envisaged in section 110 to justify taking a person into service custody are those where there are reasonable grounds for suspecting that, if not taken into service custody, they would

(a) fail to attend any hearing in the proceedings against him;
(b) commit an offence; or
(c) interfere with witnesses or otherwise obstruct the course of justice, whether in relation to himself or any other person.

Taking a person into service custody is also justified for the purposes of this section if

(a) he has failed to attend any hearing in the proceedings against him;
(b) there are reasonable grounds for suspecting that he should be taken into service custody for his own protection or, if he is aged under 17, for his own welfare or in his own interests; or
(c) there are reasonable grounds for suspecting that
    (i) if not taken into service custody, he *would fail* to comply with a requirement imposed under section 107(3), namely a condition imposed by a judge advocate on his release from custody; or
    (ii) he has failed to comply with such a requirement.

This power might be used, for example, where the accused at a summary hearing did not elect to serve their sentence and was granted 14 days to decide whether to appeal, where the CO forms a reasonable suspicion that there are one or more of the above risk factors. A CO is unlikely to risk an individual absconding where there is ample evidence to support this concern.

If the person is arrested and kept in service custody, they must be brought before a judge advocate as soon as is practicable for a review of whether they should continue to be kept in service custody. This is further explained in the Bail/Custody section in Chapter 6.

Arrest at the Direction of the Court

Arrest can be directed by the judge advocate[11] where a person has been arraigned and the proceedings are still ongoing, in the sense that they have not finally concluded, for example because the court is awaiting reports, and the judge advocate is satisfied that taking that person into service custody is justified (on the same grounds in section 110 that a CO must consider). Review of the requirement for custody is in accordance with sections 108 and 109. The judge advocate must be satisfied that taking that

11 *Ibid.*, s. 111. '(1) Where—
(a) a person has been arraigned before the Court Martial or the Service Civilian Court, and
(b) proceedings before the court have not concluded,
a judge advocate, if satisfied that taking him into service custody is justified, may direct the arrest of that person.'

person into service custody is justified on the basis they will fail to attend any hearing; or they will commit an offence; or they will interfere with witnesses or otherwise obstruct the course of justice.

Service Police

Service police have the additional power provided by section 69 to arrest a person whom they reasonably suspect of being about to commit a service offence. After arrest, the fact of the arrest must be reported as soon as practicable to the arrestee's CO. The arrestee may be kept in service custody until such time as a Service policeman is satisfied that the risk of them committing the service offence concerned has passed. Code G of the Service police codes of practice contains the code of practice for the statutory power of arrest by Service police, setting out the criteria they have to consider when exercising powers of arrest under sections 67, 69, 110, 111 and 303[12] of the AFA 2006.

Persons No Longer Subject to Service Law or to Service Discipline

Main Provisions: sections 61(2), 67, 68, 367 to 369, 370 and 374 of the AFA 2006

Where an authority is given to charge a former service person or service civilian, defence advocates should consider what powers of arrest, entry search and seizure exist. Supplementary powers of arrest have been conferred on the Service police by section 68(2) and (3) AFA 2006 in respect of anyone suspected of having committed a service offence while subject to service law or subject to service discipline, and who is not now a member of HM forces, or a civilian subject to service discipline, respectively. In other words, they have ceased any connection with the services. Such person is to be treated in relation to the offence as if they were still subject to service law in the rank or rate that they held when last a member of HM forces or, if a civilian, as if they were still subject to service discipline.[13] Subject to what is said below, a former service person may be arrested by anyone listed in section 67(3), namely, by an officer, warrant officer or non-commissioned officer of superior rank or rate, by a service policeman or woman, by a

12 This relates to the power of service police to arrest and detain a person unlawfully at large.
13 This is also discussed below.

person who is lawfully exercising authority on behalf of a provost officer; or if a member of a ship's company or an embarked force, by a person exercising authority as a member of the staff of the officer of the day. A civilian formerly subject to service discipline may be arrested by an officer, a service policeman or woman or a person lawfully exercising authority on behalf of a Provost officer (s. 67(4)).

As noted in Chapter 6 under Time Limits for Offences and in Chapter 12, offences outside the six-month time limit[14] can only be proceeded with if the Attorney General gives consent. This does not prevent an arrest being made before the Attorney has actually given consent.[15] Nevertheless, where a person may be charged (within the meaning of s. 61(1)) with an offence only with the consent of the Attorney General, section 68(4) modifies the wording of section 67(1), so that it confers a power of arrest only upon Service police and nobody else. These provisions are important in the prosecution of those suspected of historic and or serious crime committed during their period of service or as service civilians.

The 2006 Act does, for some unknown reason, give rise to a difference in wording in section 68, where it attempts to explain how those who were formerly subject to service law or discipline are to be treated.

Pre-AFA 2006

Under the Service Discipline Acts (SDA), there were clear provisions concerning the trial and punishment of offences under military law. As with the 2006 Act, the SDAs distinguished between civilians accompanying the force and those serving. Section 209 and Schedule 5 of the Army Act 1955 and the Air Force Act 1955 applied their provisions (with modifications to matters such as trial and punishment) to service civilians so that they too became liable to prosecution for offences under Part II of the relevant SDA. This provision had the equivalent effect to that of section 370 of the 2006 Act.

---

14 Ss 57 (formerly subject to service law) and 58 (civilian formerly subject to service discipline).

15 The civilian case of *R v CW and MM* [2015] EWCA Crim 906 was an unsuccessful interlocutory appeal under s. 35(1) of the Criminal Procedure and Investigations Act 1996, from a judge's ruling that the indictment and proceedings for conspiracy were null and void because consent from the Attorney General to prosecution had not been obtained in time. The appeal was rejected. http://www.bailii.org/ew/cases/EWCA/Crim/2015/906.html.

For example, section 131 of the Army Act 1955 and Air Force Act 1955 very clearly explained that,

> (1) … where an offence under this Act triable by court-martial has been committed, or is reasonably suspected of having been committed, by any person while subject to military [Air Force] law, then in relation to that offence he *shall be treated*, for the purposes of the provisions of this Act relating to arrest, keeping in custody, investigation of charges, summary dealing with charges (including appeals against findings recorded, or punishments awarded, on summary dealing) trial and punishment by court-martial (including review and suspension) and execution of sentences *as continuing subject to* military [air force] law notwithstanding his ceasing at any time to be subject thereto. [emphasis added][16]

The section went on to explain that where such a person was in military or air force custody, they were to be treated in accordance with the rank they formerly held when serving. Section 132 dealt with applications to the Attorney General for leave to prosecute offences that would otherwise be out of time. Thus, there was a specific reference to the former service person as someone who was treated as if they *continued to be* subject to service law.

Post AFA 2006

The AFA 2006 is not quite so clear on the matter. Section 374, in conjunction with sections 367 to 369, defines who is subject to service law. Little help is to be gleaned from the EN as they do not clarify this matter in relation to those formerly subject to service law.

Section 367 states that every member of the regular forces is subject to service law at all times. Every member of the reserve forces is subject to service law while undertaking duty or training, etc. In reality, this provision for reserve forces causes considerable difficulties, particularly regarding applicability during travel to and from duty, 'drill night' time-scales, etc.

16 In the case of civilians, the references in s. 131 to 'to being, continuing, or ceasing to be subject to military law' were replaced by 'being, continuing to be or ceasing to be in such circumstances that the said Part II applies…'. Part II covered the provisions relating to discipline and trial and punishment of military offences.

There is no discernible case law for discipline of reservists. Anecdotally, many are dealt with summarily and/or administratively. Where the offence is serious, however, court martial proceedings may be instituted (*R v Knock*).[17]

Section 368 provides that references to 'members of the regular forces' include a person recalled to service under any provision of the Reserve Forces Act 1980 or the Reserve Forces Act 1996, or any other recall obligation of an officer. They are to be regarded as being a member of the regular forces from acceptance into service to release or discharge. An officer who is not on the active list, and is not a member of the reserve, is not to be regarded as being a member of the regular forces.

Section 67(3) confers a power of arrest on service police. For the purposes of section 67(3), a person who is suspected of having committed a service offence while a member of Her Majesty's forces, and is not a member of Her Majesty's forces or a civilian subject to service discipline at the time of arrest, is to be treated in relation to the offence as being of the rank or rate which they held when they were last a member of Her Majesty's forces. *Treating* someone according to a certain status is not the same as subjecting them once more to the regulations and processes that adhered when serving. Unlike the clarity of the former legislative provisions, this does not expressly state that the person is to be treated as *continuing to be* subject to Service law, 'for the purposes ... relating to arrest, keeping in custody, investigation of charges, summary dealing with charges .... trial and punishment by court-martial ... and execution of sentences as continuing subject to military law', although that is the implication.

However, when one reads the similar provision applicable in section 68(3) in relation to civilians formerly subject to service discipline, it is to the effect that they are to be treated in relation to the offence 'as if he were a civilian subject to Service discipline'; in other words, as if they continued to be a civilian subject to Service discipline.

---

17 [2014] EWCA Crim 1986. The appellant served as a member of the Territorial Army. Under the Reserve Forces (Call Out and Recall) Financial Assistance Regulations 2005, she was potentially eligible for a range of allowances, including childcare and gardening expenses incurred as a result of deployment. The necessary process under the regulations required the submission to the Adjudication Officer of verifying documents. She made a number of false claims for expenses. She pleaded guilty to five charges of forgery and five charges of using a false instrument.

Reading the provisions holistically, the meaning of section 61(2) and section 68(2) is that former service personnel suspected of having committed a service offence while a member of Her Majesty's forces become subject to service law once more in that they are required to be treated in accordance with the rank they held during their previous service. This is consonant with the purposes covered by the powers of arrest authorisation and charge. It follows, therefore, that the Service police may also enter, search and arrest a former service person for an offence against service law committed during their service. Such arrests were made in relation to allegations of misconduct in Iraq.[18]

Remedies

Aside from raising the matter during the course of the proceedings, where the service arresting authorities, be they service police or those authorised by the CO, act in excess of their power, or use unreasonable (excessive) force, the service person is not without a remedy. The service authorities do not exercise absolute power. An unreasonable use of force may amount to an assault and consequently be dealt with as a disciplinary matter against the person responsible. If dealt with by the CO or a service court, compensation may be awarded.[19]

In more serious cases a civil remedy may also be available through the Crown Proceedings Act 1947, which provides an avenue to sue the Crown. Section 2 of the Act covers the liability of the Crown for all those liabilities in tort to which, if it were a private person of full age and capacity, it would be

18 For example, former Captain Rachel Webster was arrested in a dawn raid in 2014 over comments made about an ex-partner a decade previously. Investigators from the Iraq Historic Allegations Team (IHAT) (who had no power of arrest) are described as having arrived at her home in the early hours of 10 January 2014, and arrested her on suspicion of misconduct in a public office, a common law offence. This, in itself, is an unusual offence. Its nearest equivalent in service law is s. 19 of the Conduct Prejudicial to Good Order and Discipline. The arrests were actually effected by Naval Regulators, in accordance with the AFA 2006, who then transported her to Portsmouth for interview. MoD paid £5,000 in damages for her rough treatment.

19 See Chapter 9, pages 301 *et seq* and Chapter 15, page 521.

subject.[20] This includes torts committed by its servants or agents.[21] Further provisions relate to any breach of those duties that a person owes to their servants or agents at common law by reason of being their employer.

A further option for the service person would be to submit a redress of grievance under the Armed Forces (Service Complaints and Financial Assistance) Act 2015 (AF(SCFA)A 2015)[22] to the Service Complaints Ombudsman. Essentially, if a person subject to service law thinks themself wronged in any matter relating to their service, the person may make a complaint about the matter in accordance with the Armed Forces (Service Complaints Miscellaneous Provisions) Regulations 2015. Certain matters are excluded from this process, as detailed in the schedule to the rules, but they do not apply to the situation under discussion. The right to submit a complaint is discussed in more detail in Chapter 8.

Civilian Arrest Provisions

Main Provisions: sections 313–320 AFA 2006; AF(EIATSC)R 2009; AF(WASO)R 2009

Military personnel who commit crimes in the UK are liable, as any UK citizen is, to be arrested and tried by the civilian authorities. The civil authorities may also effect an arrest for those service personnel who are reasonably suspected of having committed a service offence, or who are deserters or absentees, or who are unlawfully at large. The provisions in the AFA 2006 that cover the arrest and detention by the civil authorities are in sections 313–320.

Where a judge advocate is satisfied by evidence given under oath or affirmation that there are reasonable grounds for doing so, they may issue a warrant for the arrest of a person who is reasonably suspected of having committed a service offence. The warrant will be addressed to an officer

20 There was a provision in s. 10 of the 1947 Act that excluded the Crown from liability in tort in cases involving the Armed Forces, but this was repealed by the Crown Proceedings (Armed Forces) Act 1987.

21 See also the dictum of Lord Mance in *Smith and others (FC) (Appellants) v The Ministry of Defence* (Respondent) [2013] UKSC, at s. 111. Lord Mance drew a distinction with situations involving combat.

22 The office of Service Complaints Commissioner was abolished by s. 1, which inserted s. 365B into the AFA 2006. S. 2 inserted Part 14A headed, 'Redress of service complaints'.

or officers of a UK police force or British overseas territory police force, in respect of a person who is reasonably suspected of having committed a service offence (s. 313); or for the arrest of deserters and absentees without leave (s. 314), and deserters and absentees without leave who surrender to the civilian police (s. 315) other than at a police station, in a relevant territory (defined as England and Wales, Scotland, Northern Ireland, the Isle of Man, or a British overseas territory other than Gibraltar).

Where a person who is unlawfully absent surrenders to the civilian police, including surrender at a police station, the person in charge of the police station, or a person authorised by them, must consider the case and arrange for the person to be transferred to service custody; or arrange for them to be brought, as soon as practicable, before a court of summary jurisdiction in the relevant territory in which the police station is situated; or release them subject to reporting conditions that they report as and when specified, to a specified police station, for the purpose of enabling them to be taken into service custody (s. 315).

Where a person suspected of illegal absence is brought before a civilian court of summary jurisdiction, the procedures are explained in section 316. This covers matters such as the arrangement for transfer of the absentee into service custody. The police may arrange for the absentee to be brought, as soon as practicable, before a court of summary jurisdiction in the relevant territory in which the police station is situated.

Admission of Illegal Absence

If the person admits to the court that they are a person subject to service law who has deserted or is absent without leave, or the court has in its possession evidence of a description prescribed by regulations made by the Secretary of State,[23] the court must either arrange for them to be transferred to service custody; or release them subject to a condition that they report, at or by such time as may be specified in the condition, to such place or person as may be so specified for the purpose of enabling them to be taken into service custody. If the person is in custody for some other cause, the court may arrange for them to be transferred to service custody. Where the person admits illegal absence, or the court has evidence of it, and there

---

23 AF(EIATSC)R 2009, reg 3.

is likely to be a delay before they can be transferred to service custody under subsection (3)(a)(i) or (b), the court may commit them to be held in custody pending their transfer in a prison, in a police station, or in any other place provided for the confinement of persons in custody which is situated in the relevant territory in which the court is situated.

No Admission

If the person brought before a civilian court does not admit illegal absence, and is not in custody for any other matter, the court must release them. Even where they are in custody for some other cause, section 136(4)(b) permits the court to release them. But this does not prevent the person from being subsequently arrested under section 67, or under a warrant issued under section 313, as a person who is reasonably suspected of having committed an offence under section 8 (desertion) or 9 (absence without leave).

Subsequent Arrest

Where a person is released under section 315(4)(c)[24] or 316(3)(a)(ii)[25] and they fail to comply with the condition subject to which they were released, an arrest warrant may be issued under section 317 by a judge advocate or by a person who is authorised to issue the warrant by the court that imposed the condition.

Civilian police may arrest without warrant a person who has been sentenced to service detention and who is unlawfully at large (s. 318). A 'civilian policeman' is defined in section 314(4A) as an officer of a UK police force or British overseas territory police force, other than a force or body constituted in Gibraltar.

Further detail about the relevant certificates and information they must contain can be found in the Armed Forces (Evidence of Illegal Absence and Transfer to Service Custody) Regulations 2009 (AF(EIATSC) R 2009). Essentially, these rules cover the issue of certificates issued by COs

---

24 Where the person surrendered to the authorities and is made subject to a condition that 'he reports, at or by such time as may be specified in the condition, to such place or person as may be so specified for the purpose of enabling him to be taken into service custody'.

25 Where person admits illegal absence to the court and the court released 'him subject to a condition that he reports, at or by such time as may be specified in the condition, to such place or person as may be so specified for the purpose of enabling him to be taken into service custody'.

or service authorities for illegal absence, and certificates issued by civilian police or courts relating to transfer to service custody, release subject to conditions, and certificates to be signed and authorised persons.

The Armed Forces (Warrants of Arrest for Service Offences) Rules 2009 (AF(WASO)R 2009) allow a judge advocate to issue a warrant for the arrest of a person subject to service law or service discipline, who is reasonably suspected of having committed a service offence and to issue a warrant for the arrest of a person subject to service law, who is reasonably believed to be illegally absent and who has previously been released by a civilian policeman or court of summary jurisdiction under condition that they report to a specified person or place at a specified time, and who has failed to comply with those conditions.

Search on Arrest

Main Provisions: sections 67, 70, 71 of the AFA 2006; r. 4 AF(PSSSSR) 2009; r. 5 AF(PSSSR)O 2009; AF(PSSSR)O 2014

The AFA 2006 search provisions cover a number of situations, which can be broadly grouped as:

a. search on arrest by the Service police (s .70)
b. search on arrest by authorised persons (s. 71)
c. service police powers of stop and search (s. 75)
d. stop and search authorised by the CO (s. 76).

There are additional provisions relating to entry and search of premises by the service police (s. 87) and entry and search of premises by authorised persons (s. 88). There are concomitant powers of seizure and retention of items amounting to stolen or prohibited property, etc. Service police may also enter and search premises in order to arrest a person (s. 90). A judge advocate may also issue a warrant permitting entry, search and seizure.

Search by Service Police

When a person is arrested by service police under the authority of section 67, or section 69 (a person being about to commit a service offence), or section 110 (after charge or during proceedings by order of commanding officer), or section 111 (during proceedings at direction of court) or section 303 (arrest of a person unlawfully at large), they may lawfully be searched when there are reasonable grounds for believing that they may present a danger to themself or others, or if there are reasonable grounds for believing that they may have concealed on them any items that might aid an escape or which might constitute evidence relating to a service offence (s. 70). The policeman or woman is required by AF(PSSSSR)O 2009, r. 4, to make a record of the search containing the specified information relating to

(a) the object of the search;
(b) the grounds for making it;
(c) the date and time when it was made;
(d) the place where it was made;
(e) whether anything, and if so what, was found;
(f) whether any, and if so what, injury to a person or damage to property appears to the Service police to have resulted from the search; and
(g) the identity of the Service policeman/woman making it.

Search by Other Persons

Under section 71 AFA 2006, where a person (other than service police) is exercising a power of arrest under section 67 (see above), they may search the arrested person if they have reasonable grounds for believing that the arrested person may present a danger to themself or others. Similar provisions apply to those relating service police powers, discussed above, as section 71(8) stipulates that search may be authorised after an arrest under section 67, section 110 or section 111.

Subsection (4) of section 71 grants a power to search the arrested person for things subject to search. On the direction of the arrested person's CO, if they have reasonable grounds for believing that the arrested person may have concealed on them anything that is subject to search, a person other than service police is permitted to search an arrested person for anything that they might use to help them escape or which is evidence relating to an offence that they might conceal, damage, alter or destroy (s. 71(2)). The CO can only direct such a search if the assistance of service police or civilian police cannot be obtained, and the CO has reasonable grounds to believe that if the search is delayed the arrested person would escape or conceal or damage evidence (s. 71(3)). Even though there is a requirement for the CO to have reasonable grounds for authorising the search in the first place, it is also incumbent upon the person authorised under subsection (4) to carry out the search only if they reasonable grounds for believing that the arrested person may have concealed on them anything that is subject to search (s. 71(6)). In other words, it is not a carte blanche to search. There must exist reasonable grounds relating to concealment of evidence subject to search.

Persons who exercise lawfully a power to search on the grounds that the arrested person may be concealing anything which might help them to escape or that might be evidence relating to an offence, may seize and retain things they have reasonable grounds to believe might be used by the arrested person to escape from custody, or which are evidence of an offence or which have been obtained in consequence of the commission of an offence, unless those items are subject to legal privilege within the meaning of section 10 of PACE.[26]

Stop and Search

Service Police

Main Provisions: sections 75–82 AFA 2006; AF(PSSSSR)O 2009; AF(PSSSSR)O 2014; JSP 397 – Service police codes of practice – Code A, and Code B

Where service police have reasonable grounds for suspecting that a search of a place permitted by section 78 will reveal stolen or prohibited articles, or will reveal Her Majesty's stores that have been unlawfully obtained, they have the power of stop and search of persons, vehicles, etc., provided they have reasonable grounds for suspecting the search will reveal stolen or prohibited articles, or HM stores, or drugs. This exclusive service police power arises under section 75 AFA 2006 and authorises search of any of the following for stolen or prohibited articles, controlled drugs, psychoactive substances or Her Majesty's stores:

26 EN s. 268 and s. 73 AFA 2006.

(a) any person who is, or whom the Service policeman has reasonable grounds for believing to be, a person subject to Service law or a civilian subject to Service discipline;
(b) a service vehicle which is in the charge of any person;[27]
(c) any vehicle which is, or which the Service policeman has reasonable grounds for believing to be, in the charge of a person subject to Service law or a civilian subject to Service discipline;
(d) anything which is in or on a service vehicle or a vehicle within paragraph (c).

The person, property or vehicle may be detained for the purposes of the search (s. 75(3)).

Under section 78, the power of stop and search may be exercised in,

a. any place to which (at the time of exercise of the power) the public or any section of the public has access, on payment or otherwise;
b. any other place to which people have ready access (at the time of exercise of the power) but which is not a dwelling or Service living accommodation;
c. any premises which (at the time of exercise of the power) are permanently or temporarily occupied or controlled for the purposes of any of Her Majesty's forces but are not Service living accommodation.

It will be seen that stop and search powers are not exercisable in a dwelling or service living accommodation.[28] Nor are they exercisable to search a person or vehicle on other land occupied with and used for the purposes of a dwelling unless the Service police have reasonable grounds for believing that the person, or the person in charge of the vehicle, does not reside there, *and* the person or vehicle is there without the express or implied permission

27 Defined in Code A s. 1.5g as belonging 'to any of Her Majesty's forces, or (2) is in use for the purposes of any of those forces'. As will be noted, this definition covers a private vehicle owned by a member of HM forces.

28 Namely, 'any building or part of a building which is occupied for the purposes of any of Her Majesty's forces but is provided for the exclusive use of a person subject to Service law or civilian subject to service discipline, or of such a person and members of his family, as living accommodation or as a garage' – Code A, s. 1.5f(1). This also exempts married quarters.

of a person who lives there (s. 79).[29] An example would be where the driver drives onto someone's private driveway in an attempt to evade the service police. However, even if the driver actually does have permission from the property owner, it will not invalidate the stop and search provided the service police have reasonable grounds for believing that the person, or the person in charge of the vehicle, does not reside there. If this situation were to arise outside the UK, although the service police would be lawfully exercising the power of stop and search under the AFA 2006, they may be, nevertheless, committing an offence under domestic law against the property of the actual owner, or the tort of trespass.

After searching someone stopped on reasonable suspicion, service police may not only detain that person and or vehicle, for the purpose of the search, but may seize any stolen or prohibited article, stores or evidence (of misuse of drugs) that they discover in the course of a search (s. 75(3) and (4)).

Guidance on the handling of any item seized by police can be found in the SCOP – Code A, which is concerned with 'Statutory Powers of Stop and Search and the Recording of Encounters with the Service Community'.[30]

Code A points out that

> Powers to stop and search must be used fairly, responsibly, with respect for people being searched and without unlawful discrimination .... The intrusion on the liberty of the person stopped or searched must be brief and detention for the purpose of the search must take place at or near the location of the stop.

A person who has not been arrested but is searched during a search of premises should be searched in accordance with Code A (Code B §2.4).

Section 75 authorises service police to stop and search service vehicles driven by personnel who are not subject to service law or a civilian subject to service discipline, provided they have reasonable grounds for believing that person to be a person subject to service law or a civilian subject to service discipline. Code A clarifies that the driver may be detained for only as long as is necessary to conduct the search of their vehicle but they may

29 This places a limitation on searching persons or vehicles in certain gardens, etc.

30 Introduced on 30 September 2003, updated 10 July 2017. Available online from the MoD: https://assets.publishing.service.gov.uk/government/uploads/system/uploads/attachment_data/file/626986/20170314-SPCoP_amended_2017-FINAL.pdf.

not be physically searched under this Code. Anything found within the vehicle, such as a bag or box, may be searched unless it is on the person of an individual who is not subject to service law or civilian subject to service discipline. Where it transpires that a vehicle stopped is a civilian vehicle driven by a person who is not subject to service law or is not a civilian subject to service discipline, neither the vehicle nor the driver may be searched by the service police.

In every case of stop and search by service police, the provisions of Articles 3 and 4 of AF(PSSSSR)O 2009 must be complied with. These cover procedural requirements (Art. 3) and the compilation of a record of search (Art. 4), which must be made on the spot, in writing, unless it is not reasonably practicable to do so, in which case it must as soon as is practicable after the completion of the search.

Authorised by Commanding Officer

Main Provisions: sections 1, 76, 78 of the AFA 2006; Part 4 of the AF(PSSSSR) O 2009; AF(PSSSSR)O 2014; section 11 of PACE 84

There is some overlap between section 71, which governs the procedure on search by persons other than service police upon arrest (see earlier) and section 76, which concerns the authorisation by an officer to order or authorise a person subject to service law (other than service police) to conduct a stop and search on arrest only in relation to a particular person or vehicle (s. 76(1)). Powers of stop and search under the AFA Chapter 2 apply to ships and aircraft as in the same way they apply to vehicles (s. 82). The authorising officer, for the purpose of section 76 is the CO (see s. 76(2)).

Under section 71 (search on arrest), the CO can authorise a person (other than service police) who is exercising a power of arrest to search an arrested person who they have reasonable grounds to believe may present a danger to themself or others. It further allows a person other than service police to search an arrested person for anything that the arrested person might use to help them escape or which might be evidence relating to an offence on the direction of the arrested person's CO. The CO can only direct such a search if the assistance of service police or civilian police cannot be obtained, and the CO has reasonable grounds to believe that if the search is delayed the arrested person would escape or conceal or damage or destroy evidence.

The CO's power under section 76 to authorise a person subject to service law (who is not a member of the service police) to conduct stop and search can be exercised if it would not be practicable to obtain the assistance of either the police or a UK police force in time and the CO has reasonable grounds for believing that it is likely that an offence under section 42 would be committed (criminal conduct), or a person who has committed such an offence would avoid apprehension, in which cases the CO may order or authorise a person subject to service law, who is not service police, to search the following persons for stolen or prohibited articles, controlled drugs, psychoactive substances or Her Majesty's stores, in a place permitted by section 78:

a. a person subject to service law whose commanding officer is the authorising officer; or
b. a civilian subject to service discipline whose commanding officer is that officer; or
c. a person whom the CO or the authorised person has reasonable grounds for believing to be a person to be subject to service law or service discipline;
d. a vehicle in the charge of such a person;
e. anything which is in or on such a vehicle.

The CO may only give authorisation in relation to a particular person or vehicle. The person authorised by the CO to conduct the search may exercise the power conferred only in the circumstances mentioned *mutatis mutandis* in section 75(2),[31] namely, where the circumstances are that the authorised person themselves has reasonable grounds for suspecting that the search will reveal stolen or prohibited articles or that the search will reveal Her Majesty's stores that have been unlawfully obtained.

The CO's authorisation may include the power to detain (s. 76(1)(b)) for the purposes of the search, and to seize items found (s. 76(1)(c)). The search must be for stolen or prohibited articles, controlled drugs or Her

31 See s. 76(5): 'A person authorised under subsection (1) may exercise the power of search conferred by virtue of that subsection only in the circumstances mentioned in section 75(2) (references to the service policeman being read as references to the authorised person).'

Majesty's stores.[32] The manner of the search must comply with section 71 (see above) and Article 5 of AF(PSSSSR)O 2009.[33]

The power to detain under section 76(6) stop and search arises in those cases where service police or UK civilian police are not available and there is an urgency to act on the basis of the above criteria before the assistance of service or UK civilian police could be obtained.[34] The range of persons who may be searched extends to those who, on reasonable grounds, are believed to be a civilian subject to service discipline or a person subject to service law.[35]

Section 78 specifies the places where powers under sections 75 and 76 may be exercised. Both a dwelling and service living accommodation are expressly excluded from this search authority.[36] See below for the CO's powers under section 87 to authorise a search of premises by service police and for authorised entry and search by persons other than service police (s. 88).

For those exercising the power of search under either section 70 (Search by service policeman upon arrest) or section 71 (Search by other persons upon arrest),[37] nothing in section 70 or 71 authorises anyone to require an arrested person to remove any of their clothing in public other than an outer coat, jacket, headgear or gloves, neither is there any authority to require removal in public of headgear worn for religious reasons.

Items discovered on a lawful search may be seized and retained, under section 73, if there are reasonable grounds to believe they might be used by the arrested person to escape from custody, or which are evidence of an offence, or which have been obtained in consequence of the commission

32 The definitions of matters relevant to stop and search are in s. 77.

33 This inter alia requires the authorised person, referred to as 'the relevant person', to provide specified information to 'the appropriate person', who is defined in Art. 2 as '(a) the person to be searched; or (b) the person in charge of a vehicle that it is proposed to search'.

34 EN, s. 172; s. 76(6) AFA 2006.

35 Service police have the additional power to stop and search *any person* who is, or whom the service policeman or woman has reasonable grounds for believing to be, a person subject to service law or a civilian subject to service discipline (s. 75(1)(a)).

36 'Any place to which (at the time of exercise of the power) the public or any section of the public has access, on payment or otherwise; any other place to which people have ready access, but which is not a dwelling or service living accommodation; and any premises which are permanently or temporarily occupied or controlled for the purposes of any of HM forces *but are not service living accommodation*' (s. 78, emphasis added).

37 *Ibid.*, s. 72.

of an offence, unless those items are subject to legal privilege. The time for which a person or vehicle may be detained for the purposes of a search under the section 75 or 76 stop and search provisions is such time as is reasonably required to permit a search to be carried out either at the place where the person or vehicle was first detained or nearby.

Persons authorised by an officer must comply with the procedural requirements of Article 5 AF(PSSSSR)O 2009.[38] This requires, for example, that the authorised person identifies themself by producing their identity card to the 'appropriate person' (the person to be searched or the person in charge of a vehicle to be searched), and provides their name, rank or rate and the name of their unit, together with their authority to conduct the search, the object of the proposed search, and tells the appropriate person that they are entitled to a copy of the record of search, if requested. The request may be made within 12 months of the date on which the search was made (Art. 5(13)). The appropriate person means the person to be searched; or the person in charge of a vehicle that it is proposed to search.

Entry and Search of Premises by Service Police

Main Provisions: Code B; sections 83, 84, 87, 90 AFA 2006; AF(PSSSSR)O 2009; AF(PSSSSR)O 2014; s.30(1) and Sch 4(6) AFA 2011; and s.3 AFA 2011 (No. 2)O 2012

Hovering above the statutory powers under the AFA 2006 is the European Convention on Human Rights,[39] in particular Article 8:

> (1) everyone has the right to respect for his private and family life, his home and his correspondence. (2) There shall be no interference by a public authority with the exercise of this right except such as in accordance with the law and is necessary

38 Art. 5 of AF(PSSSSR)O 2009 places a duty on the authorised person conducting the search to inter alia, take reasonable steps before they commence the search to notify the person being searched of their name, rank, unit and authority for the search, and what the objective is and their entitlement to a copy of the relevant record. See the Article for its more detailed provisions. The AF(PSSSSR)O 2009 was amended by the AF(PSSSSR)(A)O 2012.

39 S. 1(2) of the Human Rights Act 1998 states that the specified Articles (which include Art. 8) are to have effect for the purposes of the Act subject to any designated derogation or reservation.

> in a democratic society in the interests of national security, public safety or the economic well-being of the country, for the prevention of disorder or crime, for the protection of health or morals, or for the protection of the rights and freedoms of others.

The statutory provisions have attempted to strike a true balance between the interests of justice, the interests of discipline and the protection of individual service persons and their families from interference with their private lives.

While there is a general right under section 95 for the CO and service police to search premises that are not service living accommodation, the entry to a person's service living accommodation or residence and search can only be made for a lawful reason and, generally speaking, on the authority of a judge advocate or, in a case of urgency, by the CO. Section 90 specifies two circumstances when such authority is not required by service police for the purposes of a search of service living accommodation etc., which are examined below. Bearing in mind that the process for applying for a warrant under section 83 is relatively straightforward, and can be made by live link, or even telephone, if it is not reasonably practicable to make the application in person,[40] the circumstances when no such application is made should be narrow.

Premises

The following definitions apply to the subsequent headings on entry and search, including the authorisation by warrant by a judge advocate. Within the context of entry and search, the term 'premises' is defined in section 87(3) as:

> (a) Service living accommodation of a person whose Commanding Officer is the officer [authorising the search];
> (b) premises occupied as a residence (alone or with other persons) by
>   (i) a person subject to Service law whose Commanding Officer is that officer; or

40 Art. 7, AF(PSSSSR)O 2009.

(ii) a civilian subject to Service discipline whose Commanding Officer is that officer;

(iii) premises which that officer has reasonable grounds for believing to be within paragraph (b).

In addition, section 96(1A)(3) provides that a tent or movable structure, as well as any vehicle, ship or aircraft can be classed as premises.

Service Police Without Warrant

In specific circumstances, service police may enter and search premises[41] without a warrant, for the purpose of making an arrest, where they have reasonable grounds for believing that the person to be arrested is on the premises (s. 90). Subsection (4) gives power, in an emergency, where service police may, for the purpose of saving life or limb or preventing serious damage to property, enter and search any service living accommodation or residence.

Service police also have powers to search without a search warrant certain premises and vehicles in which the arrested person was present immediately prior to arrest or where arrest takes place for a serious service offence (s. 74 and Art. 12 AF(PSSSSR)O 2009). There is some overlap with the powers of stop and search, discussed above.

Where, after an arrest under section 67, a search is conducted of a person being held in custody without charge for a serious service offence (Art. 13), there is an important restriction on entry and search powers: the service police must first obtain written authority of an authorising service policeman or woman (see below). Nevertheless, they may conduct a search following an arrest under section 67 for a serious service offence without obtaining the authorisation required, if they have reasonable grounds for believing that there is evidence for which a search of relevant residential premises (defined below) is permitted and if they have reasonable grounds for believing that if no search was carried out before the earliest time by which it would be practicable to obtain an authorisation, the purpose of the search would be frustrated or seriously prejudiced. In such a case they

41 Defined as 'service living accommodation and premises occupied as a residence by (i) a person subject to Service law; (ii) a civilian subject to service discipline; or (iii) the person to be arrested' (s. 90(2)).

must inform an authorising service policeman or woman that they have made the search as soon as practicable after the completion of the search.[42]

It is axiomatic that nothing prevents a search where the person entitled to grant entry to the premises does so. If practicable, that consent must be given in writing on the Notice of Powers and Rights (MOD Form 1755) before the search takes place.

Commanding Officer Authorisation

In all other circumstances, where a relevant offence within the meaning of section 84 has been committed and there is insufficient time to obtain a warrant from a judge advocate, CO authorisation under section 87 will be provided for service police entry and search of premises where the purpose is searching for material of substantial value that would be likely to be admissible in evidence at a trial for the offence and are not items subject to legal privilege, excluded material or special procedure material. The CO authority may only be exercised where the conditions for obtaining a warrant from a judge advocate under section 83 exist (in other words, the grounds that would satisfy a judge advocate) and the CO has reasonable grounds for believing that it is likely that the time needed to get a warrant would result in the search being frustrated or seriously prejudiced.

Section 87(4) (as qualified by s. 89) provides for the service police to seize and retain anything for which the search was authorised, subject to review by a judge advocate as soon as practicable. Section 89 requires a mandatory review by a judge advocate where any property is seized and retained during a search authorised by the CO, under either section 87 or 88 (where the CO authorises other persons).

The statutory provisions of section 87(1),[43] by which an officer (the CO) may authorise service police to enter and search premises (service living accommodation or premises occupied as a residence (alone or with other persons)) are summarised below, and require that the officer has reasonable grounds for believing

a. that a relevant offence within the meaning of section 84 has been committed (see below);

42 JSP 397 Code B, s. 5.15 to s. 5.17.

43 As amended by s. 30(1) and Sch 4(6) of the AFA 2011, brought into force on 14 December 2012 by s. 3 of the Armed Forces Act 2011 (Commencement No. 2) Order 2012.

b. that material which is likely to be of substantial value (whether by itself or together with other material) to the investigation of the offence is on the premises;
c. that the material would be likely to be admissible in evidence at a trial for the offence;
d. that it does not consist of or include items subject to legal privilege, excluded material or special procedure material (within the meaning given by section 84);
e. that at least one of the conditions specified in section 83(4) applies. These are
    1) that it is not practicable to communicate with any person entitled to grant entry to the premises;
    2) that it is practicable to communicate with a person entitled to grant entry to the premises, but it is not practicable to communicate with any person entitled to grant access to the evidence;
    3) that entry to the premises will not be granted unless a warrant is produced;
    4) in the case of Service living accommodation within section 96(1)(b) or (c), it is not practicable to communicate with the person or (as the case may be) any of the persons for whom the accommodation is provided; or that there is no such person with whom it is practicable to communicate who will agree to grant access to the accommodation without the production of a warrant; and
    5) that the purpose of a search may be frustrated or seriously prejudiced unless Service police arriving at the premises can secure immediate entry to them.
    6) that it is likely that the purpose of the search would be frustrated or seriously prejudiced if no search could be carried out before the time mentioned in subsection (2).

Section 87(2) provides:

That time is the earliest time by which it would be practicable (a) for Service police to obtain and execute a warrant under section 83 authorising the entry and search of the premises; or (b) in a case where a member of a UK police force could obtain a warrant under section 8 of PACE (authorisation by a Justice of the Peace) or any other enactment authorising the entry and search of the premises, for a member of such a force to obtain and execute such a warrant.

Relevant Offence A 'relevant offence' is defined in section 84, which should be read alongside Article 33 and Schedule 3 of the AF(PSSSSR)O 2009, as any of the following:

a. an offence under section 42 as respects which the corresponding offence under the law of England and Wales is an indictable offence (i.e. a criminal offence);
b. a Service offence specified for the purposes of this subsection in article 33 and Schedule 3 of the AF(PSSSSR)O 2009, namely,
    (1) Assisting an enemy
    (2) Misconduct on operations
    (3) Obstructing operations
    (4) Looting.
    (5) Mutiny
    (6) Failure to suppress mutiny
    (7) Desertion, with the intention of avoiding a period of active service within the meaning of section 8 of the Act
    (8) Hazarding a ship
    (9) Dangerous flying
    (10) An attempt to commit any of the above offences
    (11) Encouraging or assisting the commission of any of the above offences other than an attempt.
c. a Service offence whose commission has led to, or is intended or is likely to lead to, any of the consequences mentioned in subsection (5), namely,
    (1) Serious harm to the security of the State or to public order;

(2) Serious interference with the administration of justice or with the investigation of offences or of a particular offence;
(3) The death of any person;
(4) Serious injury to any person;
(5) Substantial financial gain to any person;
(6) Serious financial loss to any person;
(7) The undermining of discipline or morale among members of any of HM forces.

Service police have powers to search certain premises without a search warrant. The main ones provided by the AFA 2006 include powers to search premises:

a. to make an arrest, under s. 90 Armed Forces Act 2006;
b. to enter and search premises authorised under s. 87 by the Commanding Officer;
c. after an arrest where the person is being held in custody without charge for certain offences known as 'serious service offences'.[44]

Code B deals with the search of premises by service police and the seizure of property found. Before an application is made to a judge advocate for a search warrant or production order, authority must be granted by an authorising service policeman or woman.

An authorising service policeman or woman, for the purposes of Code B, must be a service policeman or woman at or above the rank of:

a. Lieutenant (Royal Navy); or
b. Captain (Army); or
c. Flight Lieutenant (Royal Air Force).

44 Art. 12 AF(PSSSSR)O 2012, which amends Art. 8 of the 2009 Order by, inter alia, substituting a new paragraph (3)(b) to provide that 'if the application is for an unspecified premises warrant, the application is to specify: (i) as many sets of premises which it is desired to enter and search as it is reasonably practicable to specify; (ii) the person who is in occupation or control of those premises and any others which it is desired to enter and search; (iii) why it is necessary to search more premises than those specified under sub-paragraph (i); and (iv) why it is not reasonably practicable to specify all the premises which it is desired to enter and search'.

If it is not practicable to comply with this requirement, any service policeman or woman senior in rank to the service policeman or woman seeking authorisation may act as an authorising service policeman or woman.

The AF(PSSSSR)O 2009 govern the procedural requirements to be complied with for entry search and seizure under the parent Act. For example, the application for a warrant may be made by live television or telephone link.[45]

Article 2 (interpretation) of the 2009 order is amended by the AF(PSSSSR)(A)O 2014 so that the words 'within the meaning of PACE' are inserted after the words 'an indictable offence' in paragraph (a) of the definition of 'serious service offence'.[46] The effect is to provide that any reference to an 'indictable offence' has effect as if it included a reference to low-value shoplifting.

Serious Service Offence is defined in Article 2 as:

(1) an offence under section 42 of the Armed Forces Act 2006 for which the corresponding offence under the law of England and Wales is an indictable offence;
(2) any other Service Offence which may not be dealt with at a summary hearing by a Commanding Officer;
(3) an offence under section 11(1) of the Armed Forces Act 2006 (using violence towards a superior officer);
(4) an offence under section 18(3) or (4) of the Armed Forces Act 2006 (offences in relation to official documents and records with intent to deceive);

---

45 See below, Judge Advocate Authorisation of Search by Service Police.

46 AF(PSSSSR)(A)O 2014 entered force on 13 May 2014. S. 176(6) of the Anti-social Behaviour, Crime and Policing Act 2014 inserts in the Magistrates' Courts Act 1980 (c. 43) a new section 22A, which provides that low-value shoplifting is a summary offence. Any reference in PACE to an 'indictable offence' has effect as if it included a reference to low-value shoplifting (as defined in section 22A(3) of the Magistrates' Courts Act 1980). Section 176(6) of the 2014 Act provides that certain powers, including powers of entry and search, conferred by the Police and Criminal Evidence Act 1984 on the civilian police in respect of indictable offences, remain available in respect of low-value shoplifting, notwithstanding that it is reclassified as summary. S. 176(6) does so by providing that in the 1984 Act all references to an 'indictable offence' include a reference to low-value shoplifting. The 2014 Order makes equivalent provision, subject to modifications, in respect of certain powers of entry and search used for the investigation of service offences under the AFA 2006.

(5) an offence under section 24(1) of the Armed Forces Act 2006 (Intentional or reckless damage to or loss of public or service property (6) an offence under section 39 of the Armed Forces Act 2006 of attempting to commit an offence within sub-paragraphs (3), (4) or (5));

(6) an offence under section 39 of the Armed Forces Act 2006 of attempting to commit an offence within sub-paragraphs (3), (4) or (5);

(7) an offence under section 40 of the Armed Forces Act 2006 of encouraging or assisting the commission of an offence within sub-paragraphs (3), (4) or (5). [47]

Articles 8 to 10 set out the safeguards governing the issue, and execution by service police under section 83 of the Act or under Schedule 1 of the Order,[48] of warrants to enter and search premises.

Article 12 AF(PSSSSR)O 2009[49] (search of premises where person arrested) applies to service police in any case where a person is arrested under section 67 of the Act for a serious service offence, as defined in Article 2 (see above, page 89). To the extent that is reasonably required for the purpose of discovering any relevant evidence relating to the offence for which a person has been arrested, it permits a service policeman or woman to enter and search any relevant residential premises, or any vehicle, in which the arrested person was when arrested or immediately before they were arrested, and to open and search any locker falling within section 96(1)(c) of the Act which the arrested person had open when or immediately before they were arrested.

47 '(3) … an act that would have constituted an offence under Part 2 of the Serious Crime Act 2007 if the offence encouraged or assisted had been an offence under the law of England and Wales; (4) In determining whether an act would have constituted an offence under that Part, section 49(4) of that Act has effect as if for "offences under this Part and listed offences" it read "offences under sections 39 and 40 of the Armed Forces Act 2006". (5) Any requirement in that Part to specify matters in an indictment applies for the purposes of this section as it applies for the purposes of that Part, but with references to the indictment being read as references to the charge sheet.'

48 The making of orders by judge advocates concerning excluded and special procedure material.

49 The Order was made by the Secretary of State in exercise of the powers conferred by ss 74, 81, 84(2)(b), 85, 86, 89(2), 92, 93, 323 and 380 of the AFA 2006.

Article 13 applies to any premises occupied or controlled by a person who *has been arrested* under section 67 of the Act for a serious service offence and is being held in service custody without being charged. If they have reasonable grounds for suspecting that there is relevant evidence on the premises, service police may search for such evidence and seize and retain anything that is related to the offence for which the arrested person has been arrested, or if it relates to some other serious service offence that is connected with or similar to that offence. Evidence is not 'relevant evidence' if it is subject to legal privilege. There is an obligation to return items subject to legal privilege and special procedure material,[50] and any person with a relevant interest in the seized property can make an application to a judge advocate (see Schedule 1, Excluded and Special Procedure Material). 'Excluded material' is defined in section 11 of PACE 84.[51] (See page 102 below about access to excluded material.) Service police may conduct a search for relevant evidence without obtaining a written authorisation from an authorising service policeman or woman if they have reasonable grounds for believing that it is likely that, if the search is not carried out before it would be practicable for them to obtain authorisation, the purpose of the search would be frustrated or seriously prejudiced; they then must inform an authorising service policeman that they have made the search as soon as practicable after it takes place.

Under Article 14, a general power of seizure is exercisable by service police who, in connection with the investigation of a service offence, are lawfully on premises that are searchable by virtue of Part 3 of the Act (Powers of Arrest, Search and Entry). In those circumstances, service police may seize anything that is on the premises if they have reasonable grounds for believing that it has been obtained in consequence of the commission

---

50 As defined under s. 14 PACE 84, being material, other than items subject to legal privilege and excluded material, in the possession of a person who acquired or created it in the course of any trade, business, profession or other occupation or for the purpose of any paid or unpaid office; and holds it subject to an express or implied undertaking to hold it in confidence; or to a restriction on disclosure or an obligation of secrecy contained in any enactment. Special Procedure material includes journalistic material other than excluded material.

51 Personal records acquired or created in the course of any trade, business, profession or other occupation or paid or unpaid office and held in confidence; human tissue or tissue fluid taken for the purposes of diagnosis or medical treatment and held in confidence; and journalistic material which a person holds in confidence and which consists of documents; or records other than documents.

of a service offence, and that it is necessary to seize it to prevent it being concealed, lost, damaged, altered or destroyed. Service police may also seize anything that is on the premises if they have reasonable grounds for believing that it is evidence in relation to an offence being investigating or any other service offence and that it is necessary to seize it to prevent the evidence being concealed, lost, altered or destroyed.

Part 4 of the AF(PSSSSR)O 2009 (as amended) deals with the additional powers of seizure from premises and persons, including the arrangements for the examination and return of property seized, including the obligation under Article 24 to return items subject to legal privilege. It is open to any person with a relevant interest in the seized property to apply to a judge advocate, on one or more of the grounds mentioned in Article 28(3), for the return of the whole or a part of the seized property, namely,

- (a) that there was no power to make the seizure;
- (b) that the seized property is or contains an item subject to legal privilege that is not comprised in property falling within article 23(3);[52]
- (c) that the seized property is or contains any excluded material or special procedure material which
  - (i) has been seized under a power to which article 24 applies [authorised retention of property];
  - (ii) is not comprised in property falling within article 24(3)[53] or (4);[54] and

52 S. 23(3) '(a) the whole or a part of the rest of the property is property falling within paragraph (4) or property the retention of which is authorised by article 25; and (b) in all the circumstances, it is not reasonably practicable for that item to be separated from the rest of that property (or, as the case may be, from that part of it) without prejudicing the use of the rest of that property, or that part of it, for purposes for which (disregarding that item) its use, if retained, would be lawful.'

53 'Property in which any excluded material or special procedure material is comprised falls within this paragraph if— (a) the whole or a part of the rest of the property is property for which the person seizing it had power to search when he made the seizure but is not property the return of which is required by this article or article 23; and

(b) in all the circumstances, it is not reasonably practicable for that material to be separated from the rest of that property (or, as the case may be, from that part of it) without prejudicing the use of the rest of that property, or that part of it, for purposes for which (disregarding that material) its use, if retained, would be lawful.'

54 'Property in which any excluded material or special procedure material is comprised falls within this paragraph if— (a) the whole or a part of the rest of the property is property the retention of which is authorised by article 25; and (b) in all the circumstances, it is not

(iii) is not property the retention of which is authorised by article 25;[55] and

(d) that the seized property is or contains something seized under article 19 or 20 which does not fall within article 22(3).

The provisions of Code B cover all searches and seizure of property by service police.

Entry and Search Other than by Service Police

Main Provisions: sections 83, 84, 87, 88, 89, 91 and 95 AFA 2006; Sch 3 AFA 2011; Art 34 AF(PSSSSR)O 2009; Code B SCOP; Home Office Code of Practice, Powers of Entry

Nothing in the Act affects the general power of service police or COs to enter and search, or order the entry and search of, premises that are occupied for the purposes of any of Her Majesty's forces, to the extent that the premises *do not constitute* service living accommodation. Moreover, the CO, otherwise than in connection with the investigation of a service offence or the exercise of any power of arrest, may enter and search, or order the entry and search of, service living accommodation or to search, or order the search of, a person or to stop and search, or order the stop and search of, a service vehicle. Additionally, service police or COs may search, or order the search of, a service vehicle which is not in the charge of any person (s. 95).

The principles set out in the Home Office Powers of Entry Code of Practice 2014[56] should be followed where the CO has authorised entry and

reasonably practicable for that material to be separated from the rest of that property (or, as the case may be, from that part of it) without prejudicing the use of the rest of that property, or that part of it, for purposes for which (disregarding that material) its use, if retained, would be lawful.'

55 Contains the provisions relating to the authorisation of the retention of property.

56 Made under s. 47 of the Protection of Freedoms Act 2012. 'This Code applies to powers of entry and associated powers as specified under any Act or Statutory Instrument other than devolved powers of entry and devolved associated powers', Protection of Freedoms Act 2012, s. 1.5. The Code does not apply to the entry to premises for purposes and in circumstances that are subject to a separate statutory Code of Practice. As there is no separate statutory code of practice for the CO. The Home Office Powers of Entry Code of Practice is applicable to a CO's authorisation under sections 87, 88 and 91 of the Act and to all powers of entry and associated powers exercised by a service person other than a service policeman in accordance with the CO's authorisation. See JSP 830 MSL Version 2.0, s. 91A at 1-4-24

search of premises under this general authority. This Code points out, at section 2(3), that section 67(9) of PACE 1984 requires persons *other than police officers* who are charged with the duty of investigating offences or charging offenders to have regard to PACE Code B. However, in circumstances where the Code *does not* apply,[57] and if *no other relevant Code* applies, then relevant persons should have regard to the Home Office Code. PACE does not apply to COs or service police.[58]

Where any premises have been entered by force, before they leave, the authorised person in charge should make sure the property is left as secure as possible and in as close to the condition before entry as practicable.

Commanding Officer Authorisation

Section 88 gives 'an officer' (the CO) a similar power to that in section 87 to authorise a member of HM forces who is not a service policeman or woman to conduct the search, where the CO has reasonable grounds for believing that it is likely that the time needed to obtain the assistance of a service policeman or woman or, where a member of a UK police force could obtain a warrant under section 8 of PACE, would frustrate or seriously prejudice the search. This power can only be used to search service living accommodation, as defined in section 96 (see below on page 100).

In addition to any building or part of a building occupied for the purposes of any of Her Majesty's forces and provided for the exclusive use of a person subject to service law, or of such a person and members of their family, as living accommodation or as a garage, living accommodation might also include shared temporary accommodation on operations overseas or on exercise. A locker connected with sleeping accommodation used by a person subject to service law is also classed as 'Service living accommodation'.

---

57 Paragraph 2.5 of PACE Code B states that it does not apply if the exercise of a power is not dependent on the existence of grounds for suspecting that an offence may have been committed and the person exercising the power has no reasonable grounds for such suspicion.
58 Service Codes of Practice issued by the Secretary of State under s. 113(3) and (5) of the Police and Criminal Evidence Act 1984 are only applicable to service police. The Codes deal with the contact between the service police and members of the service community in the exercise of their powers. The Codes regulate service police powers and procedures in the investigation of offences and set down safeguards and protections for members of the service community.

Under the power in section 88, as amended by Schedule 3 of the AFA 2011, an officer may authorise a person subject to service law (other than service police) to enter and search service living accommodation within subsection (3) if the officer has reasonable grounds for believing:

a. that a relevant offence within the meaning of section 84 has been committed [see the definition above, on page 87];
b. that material which is likely to be of substantial value (whether by itself or together with other material) to the investigation of the offence is on the premises;
c. that the material would be likely to be admissible in evidence at a trial for the offence;
d. that it does not consist of or include items subject to legal privilege, excluded material or special procedure material (within the meaning given by s. 84);
e. that at least one of the conditions specified in section 83(4) applies[59] (the reference in s. 83(4)(e) to a Service policeman being read as a reference to a person authorised under this subsection); and
f. that it is likely that the purpose of the search would be frustrated or seriously prejudiced if no search could be carried out before the time mentioned in subsection (2), namely, the earliest time by which it would be practicable for a Service or UK policeman to obtain and execute a warrant.

As mentioned above, searches conducted under section 87 or section 88 must be reviewed by a judge advocate as soon as practicable where anything is seized and retained during a search.[60]

---

59 These conditions are: 1. It is not practicable to communicate with any person entitled to grant entry to the premises; 2. Even if so practicable, it is not practicable to communicate with any person entitled to grant access to the evidence; 3. Entry to the premises will not be granted unless a warrant is produced; 4. In the case of service living accommodation (a) it is not practicable to communicate with the person or (as the case may be) any of the persons for whom the accommodation is provided; or (b) there is no such person with whom it is practicable to communicate who will agree to grant access to the accommodation without the production of a warrant; 5. the purpose of a search may be frustrated or seriously prejudiced unless a Service policeman arriving at the premises can secure immediate entry to them.

60 AFA 2006, s. 89 and Art. 34 AF(PSSSSR)O 2009.

Section 91 confers an additional power on the CO to authorise a person subject to service law (other than service police) to enter and search premises[61] for the purpose of making an arrest, only if they have reasonable grounds for believing that the person to be arrested is on the premises. This, as in other situations mentioned above, is premised upon the urgency of the situation, where the arrest could not be made before the earliest time by which it would be practicable to obtain the assistance Service police, or UK police. However, the exercise of this power is subject to the conditions in subsection (3), namely,

(a) the arrest is to be made under section 67;
(b) the offence in respect of which the arrest is to be made is a relevant offence [as defined by s. 84[62]]; and
(c) the officer has reasonable grounds for believing that, if the arrest could not be made before the earliest time by which it would be practicable to obtain the assistance mentioned in subsection (4)
    (i) the person to be arrested might evade arrest, conceal, damage, alter or destroy evidence, or present a danger to himself or others; or
    (ii) discipline or morale among members of any of Her Majesty's forces might be undermined.

There is the further provision under subsection (6) that the CO may authorise a person subject to service law in relation to premises only if it is not practicable to obtain the assistance of a service policeman or woman in time to take the necessary action to save life or limb or prevent serious damage to property.

As mentioned above, on page 92, Part 4 of the AF(PSSSSR)O 2009 (as amended) deals with the additional powers of seizure from premises and persons, including the obligation to return items subject to legal privilege

61 Defined in s. 91(2) as '(a) service living accommodation of a person whose commanding officer is the officer mentioned in subsection (1); (b) premises occupied as a residence (alone or with other persons) by (i) a person subject to Service law whose commanding officer is that officer; or (ii) a civilian subject to service discipline whose commanding officer is that officer; (c) premises which that officer has reasonable grounds for believing to be within paragraph (b).'
62 See page 87.

and special procedure material,[63] and the ability of any person with a relevant interest in the seized property to make application to a judge advocate.

Judge Advocate Authorisation of Search by Service Police

Main Provisions: sections 83–89 AFA 2006; section 7 AFA 2011; section 8 PACE 1984; Code B; SOCPA 2005; paragraph 4 of Schedule 3 AFA 2011; AF(PSSSSR)O 2009; AF(PSSSSR)(A)O 2012; AF(PSSSSR)(A)O 2014

Section 83 of AFA 2006 was modified by section 7 of the AFA 2011, which substituted a completely new section 83. The authorised search provisions in sections 83–89 need to be read in conjunction with AF(PSSSSR)O 2009, AF(PSSSSR)(A)O 2012 and AF(PSSSSR)(A)O 2014.

In spite of the exceptional statutory provisions permitting a search without a warrant, where a relevant offence has been committed the provisions of section 83 empower a judge advocate, in specified circumstances, to issue a warrant authorising service police to enter and search one or more sets of premises specified in the application, or any *relevant residential premises* occupied or controlled by a person specified in the application, for purposes of obtaining evidence. These provisions are based on section 8 of PACE 1984, which empowers a justice of the peace to issue a warrant upon the application of a constable. PACE was amended by the Serious Organised Crime and Police Act 2005 (SOCPA 2005) to enable a constable to apply for an 'all premises warrant' if it is necessary to search all premises occupied or controlled by a particular person, but it is not practicable to identify all such premises at the time of the application. AFA 2011 has now updated section 83 of the AFA 2006 to bring it into line with the amended section 8 of PACE.

63 As defined under s. 14 PACE 1984, being material, other than items subject to legal privilege and excluded material, in the possession of a person who acquired or created it in the course of any trade, business, profession or other occupation or for the purpose of any paid or unpaid office; and holds it subject to an express or implied undertaking to hold it in confidence; or to a restriction on disclosure or an obligation of secrecy contained in any enactment.

It should be remembered that section 83 judicial authorisation does not apply to other premises occupied by the services. This is because COs and the service police need no statutory power to enter those other areas (s. 95).

Requirements for Issue of a Warrant

The relevant requirements that must be met to justify the issue of a warrant are in section 83 and are based on reasonable grounds for believing that a relevant offence (above, page 87) has been committed. These requirements are that the judge advocate is:

> satisfied that each set of premises specified in the application is relevant residential premises and that there are reasonable grounds for believing
>
> (a) that a relevant offence has been committed;
> (b) that material which is likely to be of substantial value (whether by itself or together with other material) to the investigation of the offence is
>   (i) in the case of a warrant authorising entry and search of specified premises, on those premises;
>   (ii) in the case of a warrant authorising entry and search of any relevant residential premises occupied or controlled by a specified person, on one or more sets of such premises occupied or controlled by that person;
> (c) that the material would be likely to be admissible in evidence at a trial for the offence;
> (d) that it does not consist of or include items subject to legal privilege, excluded material or special procedure material; and
> (e) that at least one of the conditions specified in subsection (4) applies in relation to each set of premises specified in the application.

For the purposes of powers of entry, search and seizure, the term 'premises' includes any place and, in particular, includes any vehicle, ship or aircraft and any tent or movable structure.

Unspecified Premises

An 'unspecified premises' warrant authorises service police to enter and search all relevant residential premises occupied or controlled by the person named in the application for the warrant, including premises that are specified (see Art. 3 PSSSSR(A)(O) 2012). Premises not specified in the warrant may not be entered or searched unless there is written authority for entry; that authority would be given by the authorising service police (as defined above on page 88).

Where the service police apply for an unspecified premises warrant, the application must specify:

(i) as many sets of premises which it is desired to enter and search as it is reasonably practicable to specify;

(ii) the person who is in occupation or control of those premises and any others which it is desired to enter and search;

(iii) why it is necessary to search more premises than those specified under sub-paragraph (i); and

(iv) why it is not reasonably practicable to specify all the premises which it is desired to enter and search; and

(v) it is in the public interest, having regard—

(vi) to the benefit likely to accrue to the investigation if the material is obtained; and

  (a) to the circumstances under which the person in possession of the material holds it,

  (b) that the material should be produced or that access to it should be given.

Multiple Entry Warrant[64]

Section 8 of PACE also makes provision for authorising entry to and search of premises more than once (a 'multiple entry warrant'). The judge advocate must be satisfied that it is necessary to authorise multiple entries to achieve the purpose for which the warrant is issued. If the warrant authorises multiple entries, the number of entries authorised may be unlimited, or limited to a maximum (s. 83(6)).

64 S. 83(5). Again these provisions were not reflected in s. 83 of the AFA 2006 as originally enacted.

Relevant Residential Premises

Residential premises are those premises in which a person resides or resided. It is wide enough to encompass barrack room accommodation, service married quarters and private accommodation.

Such premises are defined in section 84(3) as:

(a) Service living accommodation; or
(b) premises occupied as a residence (alone or with other persons) by
  (i) a person subject to Service law;
  (ii) a civilian subject to Service discipline; or
  (iii) a person who is suspected of having committed an offence in relation to which the warrant is sought.

Service Living Accommodation

Section 96 defines service living accommodation as: any building or part of a building which is occupied for the purposes of any of HM forces but is provided for the *exclusive use* of a person subject to Service law or Service discipline, or of such a person and members of his/her family, as living accommodation or as a garage. In addition, any other room, structure or area (whether on land or on a ship) which is occupied for the purposes of any of HM forces and which is used for the provision of sleeping accommodation for one or more persons subject to Service law or Service discipline; or any locker provided by any of HM forces for personal use by a person subject to Service law in connection with his sleeping accommodation, but is not in a room, structure or area occupied as sleeping accommodation. However, the definition excludes premises being used for keeping persons in service custody, such as unit guardrooms or detention facilities (s. 96(1A)(2)).

The amendments in Sch 3 to the 2011 Act, which inserted section 96(1A), make it clear that service living accommodation includes accommodation occupied by a civilian subject to service discipline.

Practicalities

Complications may arise where a service person lives in their own or their spouse's accommodation or the spouse's parents' house. In such a case, the person occupies [relevant residential premises] as a residence (alone or with other persons). In the UK, a civilian spouse is not subject to service discipline. While private accommodation does not fall within the definition of service living accommodation, it falls within the section 83(3) definition of 'relevant residential premises'.

Although it would appear that a warrant might be issued to search such premises, as the premises are 'occupied by a person subject to Service law' or (outside the UK) service discipline, it is not clear that a search could be always lawfully effected by the service authorities and may well constitute a trespass, or even a criminal offence in some circumstances. For example, the service authorities have no jurisdiction to enter private accommodation overseas. Any attempt would require compliance with the domestic laws of the host nation and would usually be executed by host nation authorities, if in conformity with their law, and any relevant Status of Forces Agreement (SOFA) or other jurisdictional agreement. Moreover, while a local police officer may enter such premises and carry out the arrest, if authorised under host nation law, there must be a provision in host nation law that authorises the handover of the arrested person to the service authorities. While it is clear that this provision exists under the NATO SOFA, a local agreement by way of a Memorandum or Letter of Understanding may not be so obvious.[65] It is possible, of course, that local laws provide a citizen's power of arrest, such as that found in section 24A of PACE in respect of indictable offences. Service police authorities based overseas should apprise themselves as to whether these powers exist and to what extent. For example, where a local police officer is not available in time to prevent the escape of an offender, or before the officer can assume responsibility for the offender, is there a local power to apprehend in the form of a citizen's arrest?

Were the power of entry into private accommodation owned by a non-service member to be exercised, it would quite likely be the subject of much public and press interest.

Code B states, at paragraph 3.5:

---

65 These are agreements between a sending state and receiving state/host nation and are usually drafted for short-term deployment of forces to non-NATO nations. The MoU and LoU are less formal than a treaty as they are intended to be temporal in nature. For more information about these documents, see *Treaties and Memoranda of Understanding (MoUs), Guidance On Practice And Procedures* (Foreign And Commonwealth Office, March 2014).

Except in a case of urgency, if there is reason to believe that a search might have an adverse effect on relations between the Service or Civil Police and the community, then the local Service or United Kingdom Police community liaison officer shall be consulted before it takes place. In urgent cases, the local police/community liaison officer should be informed of the search as soon as practicable after it has been made. Furthermore, if there is reason to believe that a search might have an adverse effect on a local police operation, an appropriate officer of that police force is to be consulted before the search commences.

Access to Excluded Material

Main Provisions: sections 83, 86 AFA 2006; section 8 AFA 2011; PACE 1984; Part 2 and Part 6 and Sch 1 AF(PSSSSR)O 2009; AF(PSSSSR)(A)O 2014; r. 5 and r. 151A AF(CM)(A)R 2013

Under PACE and AFA 2006, certain material is subject to special safeguards in relation to the grant of search warrants. That material may be 'items subject to legal privilege', 'excluded material' or 'special procedure material'. These expressions are defined in PACE and have essentially the same meanings in the AFA 2006.[66] A consolidated summary of terms is set out below.

Legal Privilege

'Items subject to legal privilege' means, per PACE, section 10:

(a) communications between a professional legal adviser and his client or any person representing his client made in connection with the giving of legal advice to the client;

(b) communications between a professional legal adviser and his client or any person representing his client or between such an adviser or his client or any such representative and any other person made in connection with or in

66 S. 84(4).

contemplation of legal proceedings and for the purposes of such proceedings; and

(c) items enclosed with or referred to in such communications and made
  (i) in connection with the giving of legal advice; or
  (ii) in connection with or in contemplation of legal proceedings and for the purposes of such proceedings,

when they are in the possession of a person who is entitled to possession of them.
Items held with the intention of furthering a criminal purpose are not items subject to legal privilege.

Excluded Material
This is defined in section 11 PACE as:

(a) personal records which a person has acquired or created in the course of any trade, business, profession or other occupation or for the purposes of any paid or unpaid office and which he holds in confidence;
(b) human tissue or tissue fluid which has been taken for the purposes of diagnosis or medical treatment and which a person holds in confidence;
(c) journalistic material which a person holds in confidence and which consists
  (i) of documents; or
  (ii) of records other than documents.

Personal Records
PACE section 12 defines this as meaning documentary and other records concerning an individual (whether living or dead) who can be identified from them and relating

(a) to his physical or mental health;
(b) to spiritual counselling or assistance given or to be given to him; or

(c) to counselling or assistance given or to be given to him, for the purposes of his personal welfare, by any voluntary organisation or by any individual who—
   (i) by reason of his office or occupation has responsibilities for his personal welfare; or
   (ii) by reason of an order of a court has responsibilities for his supervision.

Journalistic Material

According to PACE section 13, this means material acquired or created for the purposes of journalism:

> (a) Material is only journalistic material for the purposes of this Act if it is in the possession of a person who acquired or created it for the purposes of journalism.
> (b) A person who receives material from someone who intends that the recipient shall use it for the purposes of journalism is to be taken to have acquired it for those purposes.

Special Procedure Material

PACE section 14 defines this as paraphrased:

> (a) ... material other than items subject to legal privilege and excluded material, in the possession of a person who acquired or created it in the course of any trade, business, profession or other occupation or for the purpose of any paid or unpaid office; and who holds it subject
> (i) to an express or implied undertaking to hold it in confidence; or
> (ii) to a restriction or obligation such as is mentioned in section 11(2)(b), namely, a restriction on disclosure or an obligation of secrecy contained in any enactment, including an enactment contained in an Act passed after PACE 1984; and
> (b) journalistic material, other than excluded material.

These provisions demonstrate that, although the power to authorise entry, search and seizure arises under section 83 of the AFA 2006, the section does not permit the issue of a warrant to search for items subject to legal privilege, excluded material or special procedure material.

Production Order

A provision permitting seizure of excluded material, etc. exists and is in section 86, as amended by section 8 of the AFA 2011, and augmented by Schedule 1 to the AF(PSSSSR)O 2009. These provisions set out the requirements to be met before a judge advocate may grant an authority to service police to obtain access to excluded material or special procedure material on relevant residential premises. Section 86 makes provision equivalent to that of Schedule 1 to PACE, enabling service police to obtain access to excluded material, or special procedure material on *relevant residential premises* by making an application to a judge advocate for a production order, requiring the person apparently in possession of the material to produce it to be taken away by service police or to give service police access to it.[67] This includes material that consists of information stored in any electronic form, for example, computer disks, drives and files. USB drives, which are very popular for their convenience and portability, fall within this category. In limited circumstances a judge advocate may issue a warrant authorising service police to enter and search the premises. This authority does not extend to access to items subject to legal privilege.

An application by service police to a judge advocate, in the case of excluded and special procedure material on relevant residential premises, or to material (other than items subject to legal privilege) on premises other than relevant residential premises, will only be successful if the application is made under Schedule 1 to AF(PSSSSR)O 2009 and in accordance with that Schedule. The judge advocate to whom the application is made must be satisfied that prior to making a production order, one or other of the sets of access conditions set out in Schedule 1 is fulfilled.[68]

Where a person fails to comply with a production order, this essentially amounts to a contempt of court under section 86(2)(c).[69] Rule 151A AF(CM(A)

---

67 An order under paragraph 5 of Schedule 1 to the Armed Forces (Powers of Stop and Search, Search, Seizure and Retention) Order 2009. See r. 5 of AF(CM)(A)R 2013.

68 These are specified in Arts 2 and 3 of the Schedule.

69 An order under s. 86, for access to excluded material, etc., may 'make provision equivalent to section 311(2) and (3) of this Act (certification to civil court of offences akin to contempt) in

R 2009[70] provides the power to certify a person as failing to comply with a judge advocate's order under paragraph 5 of Schedule 1 to the AF(PSSSSR) O 2009 by which a person was ordered to produce material to service police or to give service police access to it. It is a matter for a judge advocate to direct a hearing under this rule, specifically for the purpose of dealing with the contempt.

Neither section 83 nor section 86 allows access to items subject to legal privilege that are held on premises *other than* relevant residential premises. This makes Schedule 1 to the 2009 Order largely ineffective because relevant material that qualifies as excluded material or special procedure material (such as bank records or social workers' files) is, by its nature, unlikely to be held on relevant residential premises (EN 35).

Paragraph 3.14 of Code B points out that

> In certain circumstances it may not be appropriate to access excluded or special procedure material on relevant residential premises through a Production Order. In such circumstances and where certain conditions are met, an application can be made for a search warrant under paragraph 12, Schedule 1 to the AF(PSSSSR)O 2009 authorising a Service Policeman to enter and search relevant residential premises in order to access excluded and special procedure material. Such an application shall if appropriate indicate why it is believed service of notice of an application for a Production Order may seriously prejudice an investigation. (Additional guidance at paragraph 7.16–7.20)

As previously mentioned, it is unclear how any warrant issued under these provisions may be executed by the military authorities in residential premises that are not under the authority of the services, for example private accommodation in the host nation overseas. While such relevant residential premises might be occupied by a person specified in the application, they may be owned or rented by a national of the host nation, for example a spouse, in-law or other relative, or a girlfriend, etc., who falls outside the definition of a person subject to service discipline. Refusal of entry may be

---

relation to a failure by a person within section 309(6) to comply with an order made by a judge advocate under the enabling order'.

70 Inserted by AF(CM)A)R 2013.

made by the owner in such a case and cannot be truly regarded as a refusal by the service person named in the warrant. The warrant would have to be effected by the relevant authorities of the host nation, under the terms of any agreement with them and in accordance with their own domestic law, so as not to infringe the legal rights of their own citizens.

6 INVESTIGATION

Main Provisions: sections 113, 116(2)(b), 120 and Schedule 2 AFA 2006; ss 3–5 of the AFA 2016; AF(SHASSSD)R 2009; AF(P5AFA2006)R 2009; AGAI 67; Queen's Regulations; AF(MA)R 2019; BRd 3(1) (RN); AF(SCFA)A 2015; JSP 831; JSP 833; AFA2016(C3)R 2019

Commanding Officer

The Armed Forces Act 2006 (AFA 2006) left the Commanding Officer (CO) at the centre of discipline, although it cut back some of the extensive powers that could be exercised in respect of offences that could not be dealt with summarily, yet could still be dismissed under what was section 76(5) of the Army Act/Air Force Act 1955. The present position is that, after investigation, the CO must first determine whether the circumstances indicate that the matter is a minor one that would be more appropriately dealt with administratively, or whether formal disciplinary action is required. A discussion of minor administrative action appears is in Chapter 8, Administrative Action by CO. The changes brought about by the 2006 Act, and amendments, mean that the system is more streamlined, with serious cases going directly from the service police to the Service Prosecuting Authority (SPA), without any direct involvement of the CO.

The procedure is dealt with in Chapter 9, Summary Hearing.

Who Investigates?

This will usually depend upon the seriousness of the offence. In the majority of cases, the service police will conduct an investigation on the basis of information received from the unit or from other sources and, if the matter is suitable for summary dealing, present a report to the CO, or the SPA if it is not. Where the CO becomes aware of an allegation that does not require them to notify the service police under section 113, and is not an allegation where there are 'prescribed circumstances' under section 114,[1] the officer must either ensure that the matter is investigated in such way and to such extent as is appropriate or ensure, as soon as is reasonably practicable, that a service police force is aware of the matter. The army reporting policy is set out in Army General and Administrative Instructions (AGAI) 62, but this document is not publicly released.

Decision Whether to Charge

Once the CO's investigation is complete, they can make the decision whether to deal with the matter administratively, or by summary dealing, or by making a referral to the SPA (in the more serious non-Schedule 2 cases), or by not proceeding at all. If it is decided to charge, the CO will apply a two-stage test, different to that applied by the SPA and Crown Prosecution Service. First, an evidential test. MSL 1-6-29 §93 states: 'A CO should not charge a person with a service offence if it is plain that either there is no basic evidence against the suspect or the evidence is clearly outweighed by evidence in the suspect's favour.' This places the likelihood of a prosecution as quite high, as it is only where *the evidence is clearly outweighed by evidence in the suspect's favour* that no prosecution will result. It is quite different to the 'more likely than not'/balance of probabilities test applied by the SPA. This is important when one considers the jeopardy a service person may face through prosecution in a tribunal that is not compliant with the European Convention on Human Rights (ECHR), for example for assault occasioning actual bodily harm or, indeed, any recordable offence (see Chapter 9, page 277).

The *Manual of service law* (MSL) describes this second-stage test[2] as being whether it is in the *service interest* to prosecute.

> Where there is evidence to support the bringing of a charge, a CO should also consider whether there are any factors that would make disciplinary action *inappropriate*. (emphasis added)

1 For example, a case of AWOL, where the start and end dates are clear.
2 JSP 830 MSL Version 2.0, 1-6-29 at §§ 94–95.

This must be a typographical error, as it is the wrong way round. There should not be the appearance of a presumption in favour of prosecution. The consideration must be whether disciplinary action is *appropriate*, in much the same way that the public interest is predicated upon whether it is in the public interest *to prosecute*. The public interest is paramount, albeit that in the Armed Forces, it may be influenced by factors unique to service life. This is redeemed, perhaps, by the MSL text that follows:

> He should, however, be careful to ensure that any exercise of his discretion not to take disciplinary action is consistent with justice and with the fair and efficient maintenance of discipline and operational effectiveness.

In spite of the misleading language of the first sentence, what follows seems to confirm the view that the service interest is, indeed, part of, but not superior to the public interest.

When one considers that a matter as serious as an assault occasioning actual bodily harm may be dealt with summarily by a CO, and there is sufficient evidence to meet the evidential test, it cannot be right that a summary dealing would not go ahead because the accused is, for example, an international sports person and is required elsewhere, for a sports tournament abroad and, thus, it was deemed not in the service interest to prosecute. Such a decision would be judicially reviewable by the complainant as an improper application of the public interest.[3] The public interest in seeing justice delivered must remain paramount to any service interest, although it may be informed by service factors.

The service interest is discussed further in Chapter 12 on page 340 *et seq*.

The options on the method of dealing are dealt with below in this chapter.

The definitions of terms used in the Act and Regulations are set out on page 118 *et seq*.

3 See also, JSP 839, and Chapter 18, Victims of and Witnesses to Crime.

Non-Schedule 2 Offence

The offences that may be dealt with at a summary hearing are to be found in section 53 of the AFA 2006, which covers a broad range of conduct contrary to service discipline. Certain criminal offences found in Schedule 1 may also be dealt with by the CO, although those listed in Part 2 to the Schedule require permission from higher authority. As noted above, where the service offence is not one that requires the CO to notify the service police under section 113 or 114, following which the service police will usually conduct the investigation, the CO has a choice and must either ensure that the matter is investigated in such a way and to such an extent as is appropriate, or ensure, as soon as is reasonably practicable, that a service police force is aware of the matter (s. 115). In other words, the CO can order a unit investigation or the matter can still be sent to the service police for investigation. This might happen, for example, in the case of a charge under section 21 AFA 2006 of fighting in a unit bar where, owing to the involvement of persons from different units and / or services, it is beyond the abilities or resources of the CO of the person reported to investigate.

In addition to the duty under section 113, the CO must ensure the service police are aware of 'certain circumstances of a prescribed description'. These are defined in section 114 as circumstances prescribed by regulations, which are the Armed Forces (Part 5 of the Armed Forces Act 2006) Regulations 2009 (AF(P5AFA2006)R 2009). These circumstances, in summary, are cases of serial conduct, harassment, assault causing serious injury by a superior while on duty, and cases involving death where no service offence has been committed (e.g. suicide or accident). Some of this conduct may, in isolation, be suitable for summary dealing, but the cumulative conduct makes it a graver offence unsuitable for summary dealing. Reg 3 defines 'circumstances of a prescribed description'. See more detail on page 113.

There remain a few non-Schedule 2 offences that are not within the power of the CO to deal with summarily under section 53 (but which used to go to the CO directly), for example, a section 20 wounding.[4] Where the service police consider there is enough evidence to charge a person with a service offence, the case must now be sent directly to the SPA.

4 The offence under s. 20 of the Offences against the Person Act 1861, of inflicting grievous bodily harm, is a Schedule 2 offence.

This amended approach, brought about by sections 3–5 of the AFA 2016, makes the referral procedure more streamlined and efficient, and saves time by removing the CO from the process, as the charge was not capable of summary dealing anyway – a remnant of the fallacy that it was an exercise of the CO's disciplinary powers, even though it added to the unit administrative burden and effectively meant that the CO acted as nothing more than a postbox. All the same, it does remove any temptation for COs to reduce charges to ones that can be dealt with summarily, for example, drunkenness or conduct prejudicial to good order, or fighting.

Schedule 2 Offences

Where the CO becomes aware of an allegation or circumstances that would indicate to a reasonable person that a Schedule 2 offence (serious specified offences) has or may have been committed by a *relevant person*, they must as soon as is reasonably practicable ensure that a *service police force* is aware of the matter.[5] Serious offences, listed in Schedule 2 of AFA 2006 (see Appendix E) must be investigated by the service police who, if they believe that there is sufficient evidence to charge a person with a Schedule 2 offence, will directly refer the case to the SPA.

A 'relevant person' is defined as a person whose CO is the officer who becomes aware of the allegation of a Schedule 2 offence, etc. The 'circumstances' adverted to in section 113 are those that 'would indicate to a reasonable person that a Schedule 2 offence has or may have been committed by a relevant person'. This is described in detail below. The reference to a 'reasonable person' shows that the judgment to be made does not require expert knowledge, but the application of a modicum of common sense.

Disciplinary Action

Unlike civilian practice, where all cases are dealt with by the Crown Prosecution Service, which decides whether to prosecute by applying the Prosecutor's Test, there is a two-tier approach in the services. Some, mostly minor, cases will never be seen by the SPA, but will be dealt with by the CO themself, who acts as prosecutor, judge, jury and sentencer. The accused has no right to legal representation at the summary hearing before the CO,

5 *Ibid.*, s. 113.

although may they be assisted by an Accused's Assisting Officer (AAO) who may be nominated by the accused under rule 10 of the Armed Forces (Summary Hearing and Activation of Suspended Sentences of Service Detention) Rules 2009 (AF(SHASSSD)R 2009) to represent them at the hearing. To practitioners, representation normally connotes professional representation. In fact, a person may only be nominated as an AAO if they are subject to service law and are of at least the rank or rate of petty officer, military or marine sergeant, or air force sergeant, and consent to be nominated.[6] Once appointed, the AAO may, on the accused's behalf, exercise any right of the accused to question witnesses or make submissions at the hearing.

In order to assist a CO with any charge that is within their summary power, they will receive legal advice from the appropriate service staff legal office. If the accused wishes to have legal advice, they must pay for it privately.

Prescribed Circumstances

This provision needs to be contrasted with that relating to the obligations on the service police. Where certain prescribed circumstances exist, the CO's obligation under section 114 is 'as soon as is reasonably practicable [to] ensure that a Service police force is aware of the matter'.[7] What amounts to 'prescribed circumstances' in relation to matters before the CO is determined in accordance with Reg 3, AF(P5AFA2006)R 2009. Essentially, these occur when

> (a) an allegation has been made which would indicate to a reasonable person, or there are other circumstances which would indicate to a reasonable person, that a person Subject to Service Law has or may have been the victim of
>
> (i) a course of conduct[8] by a person Subject to Service Law, involving on at least two occasions an assault in which that individual participated as a principal offender or as a secondary party; or

6 AF(SHASSSD)R 2009, r. 10.

7 S. 114(1).

8 This is defined in Reg 5 of the AF(P5AFA2006)R 2009. See on page 126 below.

(ii) an offence under section 42 as respects which the corresponding offence under the law of England and Wales is the offence under section 4 of the Protection from Harassment Act 1997 (putting people in fear of violence), committed by a person Subject to Service Law;

(b) an allegation has been made which would indicate to a reasonable person, or there are other circumstances which would indicate to a reasonable person, that a person Subject to Service Law has or may have been the victim of an assault causing serious injury, inflicted by a person of superior rank or rate while the assailant was otherwise carrying out his duties;

(c) there are what appear to the prescribed officer to be reasonable grounds to believe that the death of any person, or serious injury to **a relevant person**, has occurred in a **relevant place**, unless the prescribed officer is satisfied that there is no allegation which would indicate to a reasonable person, or circumstances which would indicate to a reasonable person, that the death or injury was, or may have been, the result of a Service offence committed by a person of whom he is the commanding officer;

(d) the death of a person has occurred and

(i) it appears to the **prescribed officer** that the person had at any time been held in a relevant place in Service custody; and

(ii) there are reasonable grounds to believe that the misconduct, during the period that person was in such custody, of a person Subject to Service Law or a civilian subject to service discipline may have caused (directly or indirectly), or may have contributed to, the death.

The references to 'relevant person', 'relevant place' and 'prescribed officer' (defined below) are emphasised in bold in the above quoted text to indicate they have a special meaning for the purposes of these provisions, set out in the AF(P5AFA2006)R 2009.

CO's Power to Charge

The CO's power to charge arises under section 120 of the AFA 2006, and the procedure is governed by AF(P5AFA2006)R 2009, as amended by The

Armed Forces (Miscellaneous Amendments) Regulations 2019 (AF(MA)R 2019).[9] In any case that, even though not listed in Schedule 2, falls outside the CO's power to deal summarily, the amended section 116 of AFA 2006[10] provides for the service police to refer it straight to the Director, Service Prosecuting Authority (DSP) where there is sufficient evidence to charge an offence. The DSP then has control of the charging process. 'The procedure for bringing a charge against a person ("A") under section 120(2), 121(1A)(3) or 122(1) is as follows (r. 11, as amended)':

(a) a charge sheet must be prepared in accordance with the Schedule to AF(P5AFA2006)R 2009;
(b) the person bringing the charge must sign the charge sheet; and
(c) a copy of the signed charge sheet must be served on A.

The Schedule to the Regulations requires the charge sheet to state:

(a) the full name of the person who is to be charged;
(b) if he is subject to Service law –
    (i) his service number, rank or rate; and
    (ii) the name of the ship, establishment or unit, of which he is a member; and
(c) the charge brought against him, or amended, or substituted, or added.

The list does not include the necessity to aver the place where the offence is alleged to have taken place. This is because the Act states that service personnel are subject to service law at all times, wherever they are. Thus, despite the practice to the contrary, there is no specific requirement for a charge to aver the place of an alleged commission of an offence, unless it is an essential element of the offence, crucial to the defence, or the defence is misled by the particularised time, such as where an accused raises an alibi, or where the accused was required to hold a valid certificate during

9 The Regulations also govern the charging procedures to be followed by the DSP.
10 Brought fully into force on 1 July 2019, by AFA2016(C3)R 2019.

a particular period of time. Place only becomes material for purposes of ensuring the court has jurisdiction.[11]

A charge sheet may contain more than one charge if all the offences charged are founded on the same facts or form or are a part of a series of offences of the same or a similar character.

The interesting case of *LSA v The Queen*[12] considered the circumstances where two charges arising out of the same facts are preferred. The defendant was charged with two offences contrary to the Station Standing Orders of her military base: (i) driving when unfit to do so through drink and (ii) dangerous driving. The offences were said to have been committed on the same evening. The Judge Advocate ruled (first) that they were in effect offences charged on the same facts and went on to direct acquittal on the charge of unfitness to drive. Later, in a second ruling, he stayed the dangerous driving charge as an abuse of the process of the court. His reasoning for both rulings was based on a note in *Archbold Criminal Pleading and Practice* ('Where a defendant is facing two charges, one of dangerous driving and the other of driving with excess alcohol, based upon the same facts, the prosecution should choose either to proceed with the charge of dangerous driving and consider bringing the lesser charge if the defendant is acquitted, or to proceed with the excess alcohol charge alone. The invariable rule is that where a person is tried on a lesser offence, he is not to be tried again on the same facts for a more serious offence' paras 29–32), which in turn says that it is based upon *R v Forest of Dean Justices, ex p Farley*.[13] The Crown submitted, inter alia, on appeal that both rulings were wrong.

The Court Martial Appeal Court (CMAC) said, *obiter*, it had the benefit of careful submissions on both sides as to the effect of *ex p Farley* and the accuracy of the passage in *Archbold* and that it was a great pity that the report in *ex p Farley* was not obtained (see *LSA* para. 33). It is in no sense wrong if two charges arising out of the same facts are put before the same court on the same occasion.

---

11 See the Canadian CMAC case of *R v Edwards* 2019 CMAC 4, *Coram*: Chief Justice Bell, Mosley J.A., Brown J.A. Place does become important to aver, however, when a charge is laid against a civilian subject to service jurisdiction, as such persons are only liable to prosecution by a service court for any service offence committed outside the British Islands.

12 [2008] EWCA Crim 1034.

13 [1990] RTR 228.

> To the contrary, sometimes the presentation of different charges in the same court, arising from the same facts, is positively the right thing to do since it may enable the tribunal of fact to determine the several issues which arise – for example whether driving was dangerous because of the manoeuvre undertaken or also because of the driver was unfit through drink, or, in an allegation of homicide, which party was at the outset carrying a firearm which was used … We are not to be taken, by silence, to endorse the decision that the correct way to give effect to the first ruling in this case, even if it were right, was to direct acquittal. The more usual way of dealing with the matter would be to amend the indictment/charge sheet by deletion of the offending charge, or (more often) to discharge the court from giving a verdict upon it. [14]

Where the charge sheet is signed by the DSP, it must be served by hand on the accused by their appointed CO, together with the case papers listed in rule 8 of the Armed Forces (Summary Hearing and Activation of Suspended Sentences of Service Detention) Rules 2009 (AF(SHASSSD)R 2009) comprising, inter alia, the case summary, charge sheet, evidence and unused material.

The amended procedure under Regulation 2 AF(AM)R 2019 (amending Reg. 11 of AF(SHASSSD)R 2009) for bringing a charge enables a CO to serve a charge sheet by post or by leaving it at a person's address if that person has ceased to be subject to service law or a civilian subject to service discipline. If a person in that position is charged by the DSP, the Director may also serve the charge by post, etc. A CO or the DSP may authorise a person to serve a copy of a signed charge sheet on their behalf.

If the accused has ceased to be subject to service law or is a civilian subject to service discipline, a copy of the signed charge sheet must be served (r. 11(3))

(a) by hand on the accused by the commanding officer of the accused; or

14 *Ibid.*, paras 35 and 38.

(b) by being left at, or sent by post to, the accused's usual or last known address by—
    (i) the commanding officer; or
    (ii) the DSP if the charge is brought under section 121(1A).

Definitions

A 'relevant person' is defined in Regulation 6 AF(P5AFA2006)R 2009 as someone

a. who is not a member of the regular or reserve forces; or
b. a person who is a member of the regular or reserve forces and—
    (1) is under 18 years old;
    (2) has enlisted in the regular or reserve forces and has not completed Phase 1 and Phase 2 Training;
    (3) is an officer or officer cadet and has not completed Phase 1 Training; or
    (4) is in Service custody.

Regulation 6 defines 'relevant place' in two ways:

a. In regulation 3(c)[15] and 5(c)[16] relevant place means:
    (1) any premises or other place which at the time of the **death or serious injury** was permanently or temporarily occupied or controlled for the purposes of Her Majesty's forces; and
    (2) any vehicle, ship or aircraft which at the time of the **death or serious injury** was in use for the purposes of Her Majesty's forces.

15 Where the CO is to ensure the service police are aware of certain circumstances: Reasonable grounds to believe that the death of any person, or serious injury to a relevant person has occurred in a relevant place.

16 Where there is a referral of a case following investigation by service or civilian police and the evidence referred to in section 116(2)(b) is evidence of inflicting serious injury on a relevant person in a relevant place; or when under a duty to safeguard a relevant person in a relevant place failed to prevent the infliction of serious harm; or when under a duty to safeguard a relevant person in a relevant place failed to prevent the death of that person in that place.

b. In regulations 3(d)[17] and 5(d)[18] AF(P5AFA2006)R 2009 'relevant place' within the context of service custody, means:
   (1) any premises or other place which at the time of the suspected misconduct was permanently or temporarily occupied for the purposes of Her Majesty's forces; and
   (2) any vehicle, ship or aircraft which at the time of the suspected misconduct was in use for the purposes of Her Majesty's forces. (emphasis added)

A 'prescribed officer' is defined in Regulation 2 as 'the officer of a prescribed description for the purposes of section 114'. Nevertheless, it is in r.4 where a definition is to be found in relation to a circumstance prescribed in regulation 3 (above, on page 113):

a. In relation to r. 3(a),[19] the prescribed officer is the **commanding officer of a person subject to service law against whom there is an allegation** which would indicate to a reasonable person, or in relation to whom there are circumstances which would indicate to a reasonable person, that he has, or may have
   (1) carried out a course of conduct within regulation 3(a)(i) of two or more offences of assault; or
   (2) committed an offence within regulation 3(a)(ii) of harassment.
b. In relation to **regulation 3(b)**, the prescribed officer is the **commanding officer of a person against whom there is an allegation** which would indicate to a reasonable person, or in

17 Where the CO is to ensure the service police are aware of certain circumstances, where death occurs of a person who appears to the prescribed officer as having at any time been held in a relevant place in service custody and there are reasonable grounds to believe that the misconduct, during the period that person was in such custody, of a person subject to service law or a civilian subject to service discipline may have caused (directly or indirectly), or may have contributed to, the death.

18 Referral following investigation by service police: Where a policeman considers there is sufficient evidence to charge and is aware of circumstances of a prescribed description (s. 116(2)(b)) that the death of a person was caused (directly or indirectly), or contributed to, by the misconduct of a person subject to service law or a civilian subject to service discipline, and that the misconduct occurred while the deceased was being held in a relevant place while in service custody.

19 This stipulates the CO is to make the service police aware of an allegation made of a course of conduct of at least two occasions of assault or an offence of harassment under s. 4 of the Protection from Harassment Act 1997.

relation to whom there are circumstances which would indicate to a reasonable person, that he was, or may have been, the assailant.[20]

c. In relation to regulation 3(c), where there are what appear to the prescribed officer to be reasonable grounds to believe that the death of any person, or serious injury to a relevant person, has occurred in a relevant place, the prescribed officer is **any officer who is a commanding officer**.

d. In relation to **regulation 3(d)**, where the death of a person has occurred who at any time has been held in a relevant place in service custody and there are reasonable grounds to believe that the misconduct of a person subject to service law or a civilian subject to service discipline may have caused (directly or indirectly), or may have contributed to, the death, the prescribed officer is **the commanding officer of a person in relation to whom it appears to that officer that there are reasonable grounds** within regulation 3(d)(ii). (emphasis added)

Service Police

Main provisions: sections 113, 114, 116–118 AFA 2006; AF(P5AFA2006) R 2009; Sch 3 AFA 2011; section 3 AFA 2016; AF(MA)R 2019;

The service police are divided into a two branches. First, a general duties branch that provides patrols and general policing to garrison/station areas, and they may have operational duties in ensuring force protection. The general duties police will also investigate service offences. Second, each service police force has a detective branch or Special Investigations

20 This regulation is concerned with the situation where an allegation is made that a person subject to service law has or may have been the victim of an assault causing serious injury, inflicted by a person of superior rank or rate while the assailant was otherwise carrying out their duties.

Branch (SIB).[21] The Secretary of State[22] may at any time require Her Majesty's Inspectorate of Constabulary to inspect, and report to the Secretary of State on, any or all of the following

a. the independence of investigations carried out by a particular service police force;
b. the effectiveness of such investigations;
c. a particular aspect of, or matter related to, such investigations.

The authority for these inspections is section 321A. Each service SIB is inspected by Her Majesty's Inspectorate of Constabulary (HMIC) whose reports are accessible publicly via the internet. The HMIC reports must be laid before Parliament, although the Secretary of State may exclude from a report laid before Parliament any material the publication of which, in the Secretary of State's opinion, would be against the interests of national security or might jeopardise the safety of any person (s. 321B).

Service detectives will investigate the most serious offences. Overseas, for example, where jurisdiction under the NATO Status of Forces Agreement (SOFA) applies, they will have the authority to investigate murder, manslaughter, sexual offences, armed robbery and espionage, to name a few. In the UK, investigations into serious crime will be conducted in accordance with the Home Office Protocol.[23]

Members of the SIB wear plain clothes, although each member holds military rank. They are trained in many areas of advanced policing, including forensics and scene of crime work. Many service police go on to join civilian police forces after their military service. The Armed Forces Act 2011 (AFA 2011) inserted section 115A into AFA 2006, which places a

21 The SIB deal with offences or incidents reported to the Royal Military Police (RMP) that by their nature, gravity or complexity, require investigation by more specially trained personnel. These investigations usually fall into the category of serious or complex crime but may also include certain minor crimes afforded 'Special Investigation' status because of their sensitivity or other surrounding circumstances. The role of the SIB is equally as applicable on operations overseas as it is here in the UK. (See the SIB web page https://www.army.mod.uk/who-we-are/corps-regiments-and-units/adjutant-generals-corps/provost/royal-military-police/specialistroles/.)

22 The Armed Forces Act 2011 s. 4 places a statutory duty on the appointed inspector or inspectors of HMIC to inspect and report to the Secretary of State (Defence) on the independence and effectiveness of investigations carried out by each service police force (the RMP, Royal Navy Police and the Royal Air Force Police).

23 See below, FN 6, page 632 in Chapter 20, Service Inquiries and Coroners Inquests.

statutory responsibility upon the Provost Marshal of each service police force to seek to ensure that all investigations carried out by the force are free from improper interference, which is to be understood as including, in particular, 'any attempt by a person who is not service police to direct an investigation which is being carried out by the force'. In other words, this provision aims to stop interference by the chain of command where, in any case of reputational sensitivity, there may be an attempt to urge the investigators that it is not in the interests of that service for the investigation to proceed. Not being in the interests of the service is quite different to the service interest, which is explained at Chapter 12, page 345 *et seq* as a reason for prosecuting an offence.

In *R (Ali Zaki Mousa And Others) v Secretary of State for Defence*,[24] the complainants brought judicial review proceedings in February 2010 claiming that the investigation established by the defendant, the Secretary of State, was neither independent nor in adequate compliance with the investigative duties under Articles 2 and 3 of the European Convention on Human Rights (ECHR). In those proceedings they succeeded in establishing that the investigation was not sufficiently independent. The Secretary of State reconstituted the investigation and excluded members of the Royal Military Police from the Iraq Historic Allegations Team (IHAT) investigation, as that Provost Branch had been involved in Iraq. In the second set of proceedings, the complainants unsuccessfully contended that, as reconstituted, IHAT was still not independent and they sought a more far-reaching inquiry.

Legal advice to the service police is provided by the staff legal branches and, in the more serious non-CO and Schedule 2 cases, by the SPA. This ensures there is no duplication of work and that the prosecutor is seized of the matter soonest, to advise what other aspects of the investigation may need to be pursued.

Investigation and Referral

Section 116 applies where the service police have investigated a possible service offence or where a civilian police force has investigated a matter and referred it to the service police. AF(P5AFA2006)R 2009 Regulation 5

24 [2013] EWHC 1412 (Admin).

prescribes the circumstances, for the purposes of section 116, in which the service police are required to refer certain cases to the DSP. Section 116(2) provides that service police must refer the case to the Director of Service Prosecutions (for a decision on whether to charge, etc.) where:

(a) a service policeman considers that there is sufficient evidence to charge a person with a Service Offence that is *not* a CO offence, or
(b) a service policeman considers that there is sufficient evidence to charge a person with a Service Offence that *is* a CO offence, and is aware of circumstances of a description prescribed by regulations under section 128 (the AF(P5AFA2006)R 2009). (emphasis added)

A service offence committed by a person is a 'CO offence' if a charge against the person in respect of the offence is capable of being heard summarily by the person's CO (see s. 52).[25] The duty to refer to the DSP relates to the most serious cases and to a number of other cases in which it is considered especially important to ensure that the key decisions on prosecution are decided by the Director (the 'prescribed circumstances cases').

Section 116(3) makes it clear that there is no duty of referral to the SPA where service police consider the matter to be a CO offence, without prescribed circumstances, and where the case is not linked under section 117(5) to a case already referred, or to be referred, to the SPA. Instead, in such circumstances, the service police must refer the case to the suspect's CO who will then consider it under their statutory powers.[26]

In all cases that *cannot* be dealt with by the CO (i.e. not a CO offence), the decision on the charge and whether to prosecute are the responsibility of the Director. Section 3(1) AFA 2016 amends section 116 of AFA 2006 to provide for the service police to refer any case straight to the Director where there is sufficient evidence to charge an offence and the CO cannot deal with the case. It ensures potentially serious cases do not slip below the DSP's radar.

Just because an offence is not contained in Schedule 2 it does not mean it is one that can be dealt with summarily by the CO. As previously

25 S. 116(5)(b), as amended by s. 3(4)(b) of AFA 2016.
26 *Ibid.*, Regulation 5 and s. 116(2) of the AFA 206 – amended by Sch 3 AFA 2011.

mentioned, an offence of section 20 wounding is not contained in Schedule 2, neither is it one the CO can deal with under section 53. In these circumstances, where service police are required to, under section 116(4), they must consult the DSP before referring a case. The subsection provides:

> s. 116 (4) Subsection (4A) applies if
> (a) the allegation or circumstances would indicate to a reasonable person that a person has committed, or might have committed, a Service Offence which is not one that may be dealt with at a summary hearing (see section 53), or
> (b) any circumstances investigated are circumstances of a description prescribed by regulations under section 128[27] for the purposes of section 114, and a Service policeman proposes not to refer the case to the Director under subsection (2) or section 117(5).
> (4A) If this subsection applies, the Service policeman must consult the Director as soon as is reasonably practicable (and before any referral of the case under subsection (3)).

A referral under section 116(3), adverted to in section 116(4A), is a referral to the CO where, although there is insufficient evidence to charge a non-CO offence, there is evidence to support a CO offence (i.e. for summary hearing), and no prescribed circumstances[28] exist. For example, in a section 20 wounding allegation, there is no evidence to prove a wound, that is, that all layers of the skin were broken, yet the evidence is sufficient to prove actual bodily harm or common assault, both within the CO's summary powers in Schedule 1. There is the additional requirement that the matter is not connected to any case that has been or must be referred to the DSP.

These provisions of section 116(4A) essentially prevent the service police from referring the matter to the CO (perhaps on a lesser charge) without first consulting the DSP where, despite the allegation of a non-CO offence, the service police consider that the evidence is insufficient to prove it or that prescribed circumstances do not exist in an offence that *is* a CO offence. It also ensures that linked cases are not missed.

27 R. 5 of AF(P5AFA200)R 2009 and Armed Forces (Miscellaneous Amendments) Regulations 2019.

28 As defined under r. 3 of AF(P5AFA2006)R 2009.

Regulations stipulate[29] that the following are 'prescribed circumstances' for the purpose of section 116(2)(b) requiring referral by the service police:

(a) the evidence referred to in section 116(2)(b) is evidence that a person Subject to Service Law has been the victim of
   (i) a **course of conduct** by a person Subject to Service Law, involving on at least two occasions an assault in which that individual participated as a principal offender or as a secondary party; or
   (ii) an offence under section 42[30] as respects which the corresponding offence under the law of England and Wales is the offence under section 4 of the Protection from Harassment Act 1997, committed by a person Subject to Service Law;
(b) the evidence referred to in section 116(2)(b) is evidence that an assault causing serious injury was inflicted on a person Subject to Service Law by a person of superior rank or rate while the assailant was otherwise carrying out his duties;
(c) the evidence referred to in section 116(2)(b) is evidence
   (a) that a person ('A') participated (as a principal offender or as a secondary party) in the inflicting of serious injury on a relevant person[31] in a relevant place;
   (b) that A was under a duty to safeguard a relevant person ('B') while B was in a relevant place[32] and that A failed to prevent an assault inflicting serious injury on B in that place; or
   (c) that A was under a duty to safeguard a person ('B') while B was in a relevant place and that A failed to prevent B's death being caused while B was in that place;

29 The Armed Forces (Part 5 of the Armed Forces Act 2006) Regulations 2009. *Ibid.*, Regulation 5.
30 An offence punishable by the law of England and Wales; or if done in England or Wales, would be so punishable.
31 See the definition above, on page 118.
32 Defined above, on page 118.

(d) the evidence referred to in section 116(2)(b) is evidence that the death of a person was caused (directly or indirectly), or contributed to, by the misconduct of a person Subject to Service Law or a civilian subject to service discipline, and that the misconduct occurred while the deceased was being held in a **relevant place** while in Service custody. (emphasis added)

Each person's conduct in relation to each incident is to be regarded as giving rise to a separate case. Sometimes, where more than one person is involved, the accused may belong to different units, or the part played by one individual is more or less serious than that of another. However, where a case that is to be sent to the CO is closely connected with another case that must be sent to the Director, section 117(3) of AFA 2006 now provides that the service police will be able to refer a case to the Director if, after consulting the Director, they consider it appropriate to do so because of a connection with another case that is already referred under section 116. Section 3(6) of the AFA 2016 amends section 118 of the AFA 2006 to require the service police to inform the CO where a case is referred to the DSP under section 116[33] or section 117(5).[34]

Whenever service police consider that there is sufficient evidence to charge a person with a service offence and refer the case under section 116(2) or 117(5) to the Director they must, as soon as reasonably practicable after referring the case, notify the accused's CO of the referral and provide prescribed documents to the accused's CO (s. 118). The 'prescribed documents' are a copy of all reports relating to the case prepared by a service police force or provided to a service police force by a UK police force or an overseas police force (r. 8(1) AF(P5AFA2006)R 2009).

Course of Conduct

The MSL describes a course of conduct as a legal concept on which there is considerable case law. In some cases a course of conduct is obvious, as where there is evidence that one person has undertaken a campaign of misconduct against another. But there may be a course of conduct where

33 Where the offence is either a non-CO offence or, even if it is, the service police are aware of prescribed circumstances; s. 118(1) and s. 116(2).

34 Connection offences, described in s. 117(3)(d), s. 118(1).

the connection between incidents is less obvious. For example, the mere fact of two assaults cannot of itself establish a course of conduct; the sort of factors that may be sufficient to establish a course of conduct would be if there was also evidence that the assailant intended to or had threatened that they would carry out further assaults or the two assaults came after a sustained campaign of threats and intimidation in the past. In all but the most obvious cases where the CO suspects that there may be a course of conduct, they should take staff legal advice.[35]

It is important to note that the phrase 'circumstances of a prescribed description' in regulation 3, AF(P5AFA2006)R 2009 applies to a 'course of conduct' mentioned in both section 114, placing an onus upon the CO to ensure the service police are aware of an allegation relating to 'of certain circumstances', and section 116(2)(b), which places the onus on the service police to refer such cases to the DSP.

Property Acquired During an Investigation – Disposal

Main Provisions: sections 142(3), 155(5) and 278(2) AFA 2006; AF(DP)R 2009

Where property comes into the possession of the service police or a person's CO during the investigation of an offence, regulations exist to determine how it should be dealt with. These are to be found in the Armed Forces (Disposal of Property) Regulations 2009. The Regulations enable a judge advocate, specified for proceedings under sections 142(3), 155(5) and 278(2), 'to order the return of property to the person appearing to be the owner of it or, if the owner cannot be found, to order its disposal as appropriate (including retention by the police, its sale or destruction)'.[36]

After a summary hearing, the CO of a person charged with a service offence may determine that property is to be returned to the person appearing to be the owner or, if the owner cannot be found, determine that it is to be delivered to the service police for disposal as appropriate.

In other circumstances, for example where no one has been charged with a service offence, a CO may determine that property is to be returned

35 See JSP 830 MSL Version 2.0 1-6-E-2.

36 This text is taken from the EN to the Regulations.

to the person claiming to be the owner of it or to the person in whose possession the property was before it was seized.

Where the owner cannot be ascertained, it is possible to retain the property. In such a case, the judge advocate may order, or CO may determine, that the property is to be retained by the service police and the property shall vest in them on the making of the order or determination under Reg 6. Where it is proposed, in these circumstances, to make an order or determination for sale or destruction of property a judge advocate or CO, as appropriate, must invite representations from interested persons concerning the retention of the property and any order or determination shall not take effect until the time limit for an appeal under these Regulations has expired with no appeal being made. Time limits are established after which a decision on disposal by a judge advocate cannot be challenged in civil proceedings but such limits are not applied where a decision on disposal is made by a CO.

A decision made by a CO under these Regulations may be appealed, within 14 days of the original decision, to a judge advocate who can make an appropriate order in respect of the property. If the determination is made by a judge advocate, an appeal is first of all considered by the Judge Advocate General (JAG) who may uphold it. If the JAG is minded to dismiss the appeal without a hearing, the court administration officer shall notify the appellant in writing of that fact. Where the JAG does not determine the appeal either way, or the appellant requires a hearing, they may direct that the matter be heard by a different judge advocate. This procedure is outlined in Reg 8. No further appeal lies against the order of a judge advocate made under Reg 8.

Time Limits for Offences

Main Provisions: sections 55–62 AFA 2006; sections 22, 30 and Sch 4 AFA 2011; RFA 1996; AFA2011(C4)O 2013

Provisions in Chapter 2 of the AFA 2006, in sections 55–62, cover time limits for commencing proceedings. When a person is serving as a member of HM forces, the time limit for an offence to be charged under section 42 of the Act (a criminal offence under the law of England and Wales) is that specified for the offence. For military offences, there is no time bar and a service person will continue to be liable to disciplinary action and/or prosecution for as long as they continue to serve.

Former Service Persons

In relation to those persons no longer 'subject to service law' (i.e. former service personnel), the time bar descends at the end of six months beginning with the date the person ceased to be 'subject to service law'. This is subject to section 61(2) of the AFA 2006, which provides for consent to prosecution being obtained from the Attorney General in the case of those who left the services or ceased to be subject to service discipline more than six months previously. It should be noted that the Attorney cannot consent to the addition of a charge in proceedings that have already been commenced. In the case of a former member of the regular forces, the time bar is still relevant, even if they rejoin the force within those six months.[37] In practice, the consent of the Attorney to prosecution will usually only be sought if the offence is a serious one, for example indictable. Under the former Service Discipline Acts (SDAs), consent could only be given if the offence were a civilian criminal offence contrary to section 70 of the 1955 SDAs or section 42 of the Naval Discipline Act 1957 (NDA 1957). This caused problems when an individual was responsible for a disciplinary matter with others involved in a course of criminal conduct but, by virtue of having left HM forces more than six months previously and not having committed a *criminal* offence themself, could not be charged, as the Attorney could not give their consent. This is no longer the case. See above on page 60 for the powers of arrest. Section 61(2) of the AFA 2006 provides that, where any of sections 55 to 58 prohibits the charging of a person with an offence, they may nevertheless be charged with it if the Attorney General consents. This is discussed below, in the context of the *Shuttlewood* case.

Civilians

The position of civilians ceasing to be subject to service discipline is an interesting one.

37 See s. 55(3) AFA 2006.

Pre-1986

As the law stood then, any trial by court martial had to commence within six months of an accused soldier ceasing to be subject to military law.[38] In the case of a civilian accompanying the force abroad, the same time limit applied to trial by both court martial and the Standing Civilian Court.[39] It used to be the case that civilians in Germany could escape trial on jurisdictional grounds simply by crossing the international German border very shortly after committing an offence. That used to start the jurisdictional clock running. If the trial did not take place within six months of that date, the prosecution was out of time and would have needed to apply to the Attorney for permission to prosecute outside the six-month limit. With minor instances of shoplifting or disorderly behaviour, no such application would ordinarily be made, with the result that the individual would escape trial.

Post-AFA 1986

The lacuna was closed in section 209(3A) of the Armed Forces Act 1986, by the introduction of a new provision into the Army Act 1955 and Air Force Act 1955. This provision dealt with two situations. First, anyone whose ordinary residence, service or employment was within the limits of a military commander would not cease to be subject to military law merely because they left the limits of that command. This caught the day tripper to Holland or Belgium. Second, anyone who interrupted their residence with a family of persons whose place of residence continued to be their home would continue to be subject to military law. This latter provision was designed to catch children away at boarding school.

Post-AFA 2006

The position is now governed by section 58 of the AFA 2006, as amended by the AFA 2011.[40] In essence, this follows the AFA 1986 amendment

---

38 See, e.g., s. 132(3)(b) AA 55.

39 See s. 209(3)(g) AA 55.

40 *Ibid.*, s. 30 and Sch. 4 – this filled some gaps relating to UK representatives working in international naval, military and air-force organisations specified by the Secretary of State, while they are outside the British Islands, in which case 'country' is to be read in accordance with para. 14 of Sch. 15 (inserted by AFA 2011, s. 22 and brought into force on 14 December 2012 by AFA 2011 (Commencement No2) Order 2012. See also, Art. 3 of AFA2011(C4)O 2013.

and continues to treat as 'subject to service discipline' a civilian within Schedule 15 to the Act who continues to reside or stay in a qualifying place. Hence, temporary absences such as visits back to the UK or on holiday to another country will not trigger the six-month time limit. This is dealt with in more detail below, in the discussion on page 487 in Chapter 15.

Certain other time limits are specified in relation to section 107 offences (breach of requirement imposed on release from custody),[41] section 266 offences (failure to comply with financial statement order),[42] and for Reserve Forces Act 1996 offences (see s. 62 of the AFA 2006). In the case of a charge under the Reserve Forces Act 1996, however, the time limit is the point at which all the periods specified in section 62(1) AFA 2006 have expired, namely, the end of whichever of the following periods ends last

a. six months beginning with the date of commission of the offence;
b. two months beginning with the date the offence becomes known to the person's commanding officer;
c. two months beginning with the date the person is apprehended;
d. if the offence was committed when the person was a relevant reservist, six months beginning with the date he ceases to be a relevant reservist.

In *The Queen on the application of Daniel Shuttlewood v Ministry Of Defence*,[43] judicial review was sought by the claimant who had been arrested and tried, by way of court martial in 2015 for a number of sexual offences, some involving children, committed in 2003/2004 and 2009. He was convicted and sentenced to 10 years' imprisonment. At the time the offences were committed, he was a civilian living in Germany and subject to service jurisdiction. His application for leave to appeal his convictions to the CMAC had been refused. He unsuccessfully advanced two grounds of challenge at the judicial review. First, that the Solicitor General's fiat, dated 17 November 2014, was ultra vires section 61(1) of the Armed Forces Act 2006 because it was not signed in the name of the Attorney General with sufficient reasons. Second, there existed a procedural irregularity in that the Attorney General's consent was never disclosed to the claimant until

41 *Ibid.*, s. 59.
42 *Ibid.*, s. 60. But see s. 66, which excludes the SCC as having power to make such an order.
43 [2019] EWHC 1209 (Admin).

recently, when the decision was communicated to him. The court held that the terms of section 1 of the Law Officers Act 1997 provide that any function of the Attorney General may be exercised by the Solicitor General. Anything done by the Solicitor General, in the exercise of or in connection with the function of the Attorney General, has effect as if done by or in relation to the Attorney General. The Attorney General may give consent to prosecute outside the standard time limit. Section 61(2) does not require that consent is given 'in the name of the Attorney General, rather that a person may be charged if the Attorney General consents'. This function was exercised by the Solicitor General and takes effect as if done by the Attorney General. Further, there is no requirement that the Solicitor General give reasons for his fiat, nor is there a requirement that the decision to prosecute outside the time limits be communicated to a defendant in court martial proceedings.

Bail/Custody

Main Provisions: See also, bail on appeal to court martial, on page 437, bail during proceedings in the SCC, on page 409, and bail pending appeal to the CMAC, on page 443; sections 29, 67, 69, Part 4 Chapter 1 (ss. 98–109), 297 of the AFA 2006; AF(CP)R 2009; r 3 of AF(CWC)R 2009; Article 3 and Schedule to AF(CRFC)R 2009

In the context of pre-charge custody the term 'bail' is slightly misleading, because the mechanisms within the services are different to police bail. Importantly, the service police have no power to grant bail. Although the CO has a limited power to detain without charge, the pre-charge judicial process (custody without charge) before the CO only exists to charge or release. Indeed, in custody proceedings the prosecutor is the one who must justify retention in custody. This changes in post-conviction custody, where the service person is awaiting reports prior to sentencing or is intending to appeal. The Armed Forces (Conditional Release from Custody) Order 2009 makes specific reference to 'bail applications' and places the proceedings under the complete control of the judge advocate. There is no longer a requirement in such cases for the prosecutor to initiate the application to justify retention in custody: the court has already determined the custodial issue by way of its sentence, so the initiative lies with the appellant to make the application for bail. This is dealt with in more detail in relation to appeals from the Summary Appeal Court (SAC) to the High Court and appeals from the Service Civilian Court (SCC) to court martial, on page 542 and on page 437 respectively.

The term 'bail' is used here as a form of convenient label to bring the reader to the appropriate section of the text by using a term familiar to practitioners in the civilian courts.

Except in accordance with sections 99 to 102,[44] a person arrested under section 67 may not be kept in service custody without being charged with a service offence. In dealing with those under their command who have been arrested, a CO may delegate any or all of their functions under sections 98 to 102,[45] subject to such restrictions considered appropriate, taking into account the rank or rate and experience of the delegated officer.

Custody Hearing

There are two principal situations where custody hearings may occur:

a. prior to charge, governed by Part 4 Chapter 1 of the AFA 2006 (ss. 98–104) and the Armed Forces (Custody Without Charge) Regulations 2009;
b. after charge, governed by Part 4 Chapter 2 of AFA 2006 (ss. 105–109) and the Armed Forces (Custody Proceedings) Rules 2009.

Service Custody

This is not specifically defined in the AFA 2006. Considering the various references throughout the legislation may help the reader to understand what the phrase means. Some examples are:

Section 29 of the AFA, concerning offences in relation to service custody. Under section 29(2) a person subject to service law, or a civilian subject to service discipline, commits an offence if they use violence against

---

44 Custody without charge authorised by the CO, review of custody by CO, and extension of custody authorised by judge advocate.

45 AF(CWC)R2009, r. 3(1) allows a CO to delegate their functions to an officer under their command or a service policeman or woman not below the rank of naval lieutenant, military or marine captain or flight lieutenant, or, exceptionally, to a service policeman or woman of any rank or rate, except that the delegation cannot be made to the arresting officer [EN].

a person *in whose lawful custody they are*, or their behaviour towards such a person is threatening; (3) For the purposes of this section, (a) references to custody are to 'service custody'.

Section 69 (Power of arrest in anticipation of commission of Service offence) permits service police to *keep* the arrestee *in service custody* until such time as the service police are satisfied that the risk of their committing the service offence concerned has passed.

Section 297 provides for detention in service custody *following the passing of a custodial sentence until* the time they are *committed to the appropriate establishment* for serving their sentence; subsection (3) provides that a person in service custody under subsection (1) is deemed to be in legal custody.

'Service custody premises' is defined in section 300(7) AFA 2006 as 'premises *under the control of* the Secretary of State for the *keeping of persons in Service custody*'.

The Service Custody and Service of Relevant Sentences Rules 2009 rather blandly define 'service custody' in r.1 as 'service custody for the purposes of the [AFA]'.

Finally, rule 2(2) of The Armed Forces (Custody Proceedings) Rules 2009 states that 'custody' means service custody and shall be construed in accordance with sections 98 to 111.[46]

Thus while the AFA 2006 does not expressly say so, it seems from the various provisions in which it is mentioned, 'service custody' means lawful detention under the provisions of the AFA 2000 of a person subject to service law, or a civilian subject to service discipline, by an authorised person subject to service law. This understanding seems consonant with section 319, which requires a certificate in connection with the transfer of a person into 'service custody' by the civilian police.

When considering the conditions a defendant is held in, advocates may find the services accommodation code – volume 1 (JSP 315)[47] – of some assistance. Scale 4 relates to custody facilities. Section 4(9) states that 'Alcohol recovery rooms are to be positioned in the administration area, in

46 Provisions relating to custody without charge, custody, etc. after charge, and arrest after charge.
47 https://www.gov.uk/government/publications/jsp-315-services-accommodation-code-volume-1.

clear view of the staff office to enable constant supervision. The number of rooms will be determined at project brief stage.'

The most recent iteration of JSP 315 omits any reference to custody scales. Where published, Building Performance Standards (BPS) now replace JSP 315 Accommodation Scales and the Interim Guidance issued in 2017. Accommodation scales for custody facilities now appear in BPS 6.2 (see the Index of replaced accommodation scales, published on the Government website[48]).

Custody Without Charge[49]

Unless authorised by the CO under section 99, a person arrested under section 67 may not be kept in service custody without being charged with a service offence. The 'relevant time'[50] which starts the service custody clock, in relation to a person arrested under section 67 or section 69(1), or arrested by a civilian police officer and subsequently transferred into custody,[51] is the time of the arrest, or, in relation to a deserter or unlawful absentee delivered into custody following surrender,[52] the time of the surrender.

When a person is arrested and placed in service custody under section 67, that person's CO must be notified as soon as reasonably practicable of the fact and the grounds for keeping that person in custody without charge.[53] Under section 99(2), until the person making the arrest has made an arrest report to the CO, the arrested person may be kept in service custody without being charged with a service offence, but only if the person who made the arrest has reasonable grounds for believing that keeping them in service custody without being charged is necessary

(a) to secure or preserve evidence relating to a Service Offence for which he is under arrest; or
(b) to obtain such evidence by questioning him or her.

---

48 https://www.gov.uk/government/publications/mod-armed-forces-standards-of-accommodation-scales/transitional-index-of-replaced-scales.
49 When arrest is by service police in anticipation of the commission of a service offence under s. 69, see above, page 66.
50 Reg. 2 AF(CWC)R 2009.
51 Under ss 313(4), 316(3) or 317(4).
52 Under s. 315.
53 AFA 2006, s. 99(1).

Once notified, the CO must determine whether there are grounds to authorise continued custody.[54] If the CO has reasonable grounds for believing (a) that keeping the person in service custody without being charged with a service offence is necessary to secure or preserve evidence relating to a service offence for which they are under arrest or to obtain such evidence by questioning them, and (b) that the investigation is being conducted diligently and expeditiously, the CO may authorise the keeping of that person in service custody.[55]

There are limited circumstances in which a review of continuing custody may be postponed if, having regard to all the circumstances prevailing at the expiry of the last authorisation under section 99(4), it is not practicable to carry out the review at that time.[56] However, detention without charge is subject to the maximum period of 48 hours after arrest, as stipulated by section 99(6), although this may be extended by a judge advocate (s. 101) or in exceptional circumstances where it is either not practicable to carry out the review prior to the end of the period for which custody was authorised, or the person is being questioned, and carrying out the review would prejudice the investigation, or the CO is not available (s. 100). See Review, on page 138.

The *authorising officer* for the purpose of detention without charge is the CO or a person exercising functions delegated[57] under Regulation 3 AF(CWC)R 2009, namely, being an officer under their command not below the rank of naval lieutenant, military or marine captain or flight lieutenant, service police of no lower rank than naval lieutenant, military or marine captain or flight lieutenant, or if no such person is reasonably available, service police, so long as the person delegated to is not the arresting officer.[58]

This delegation might be necessary because some of the CO's functions may need to be performed urgently when they are unavailable. The suspect in custody before charge must be informed of the reasons for their detention and given an opportunity to make representations regarding their custody.

---

54 *Ibid.*, s. 99(3).
55 *Ibid.*, s. 99(4).
56 *Ibid.*, s. 100(3).
57 CO may delegate any or all of their functions under ss. 98–102 of the AFA 2006.
58 AF(CWC)R, r. 3, made under powers conferred by s. 104(1) AFA 2006.

Written records relating to the custody are to be kept and a copy given to the arrested person on release.[59]

The information to be provided to the person in custody is set out in Regulation 6 AF(CWC)R, and includes,

(a) the name, rank or rate of the authorising officer;
(b) if he was arrested
  (i) the reason for that arrest; and
  (ii) the date and location of the arrest;
(c) the relevant time;
(d) that he is to be retained in custody;
(e) the grounds (under section 99(4)(a)) for his retention in custody;
(f) the period of custody authorised.
(g) that he may make representations, orally or in writing, to the authorising officer—
  (i) requesting his release and giving his reasons why he should be released from custody; or
  (ii) making any other representations regarding his confinement in custody; and
(h) that, subject to regulation 7, he may nominate an assisting officer in preparing and making any representation under paragraph (g).

The person in custody may have an assisting officer appointed[60] to make a representation requesting their release and giving their reasons why they should be released from custody or to assist in making any other representations regarding their confinement in custody.

A custody record must be maintained containing the information required in Regulation 10 AF(CWC)R.

Under section 99, once the report has been received by the CO, in order to authorise the retention of the person in custody the CO must determine

a. whether continued custody is necessary to secure or preserve evidence relating to the Service offence for which

---

59 Reg. 6. Summary of the EN to the Regulations, paras 7.1 and 7.2.

60 AF(CWC)R, r. 7 – a volunteer of at least the rank of petty officer, military or marine sergeant, or air force sergeant.

person is under arrest, or to obtain such evidence by questioning him, and

b. that the investigation is being conducted diligently and expeditiously.

Such an authorisation ends not more than 12 hours after it is given and, unless it is authorised by a judge advocate, in any event cannot exceed 48 hours without charge (save in the exceptional circumstances mentioned above and below).

Review

The CO must keep the custody of the individual the under review[61] no later than the end of the period for which it is authorised (i.e. every 12 hours).[62] However, a review may be postponed if,

(a) having regard to all the circumstances prevailing at the expiry of the last authorisation under section 99(4), it is not practicable to carry out the review at that time [s. 100(3)] , or

(b) the person in Service custody is being questioned and the CO is satisfied that an interruption of the questioning for the purpose of carrying out the review would prejudice the investigation, or

(c) the CO is not readily available [s. 100(4)].

While this last provision looks troubling, it caters for a number of situations encountered in service life where the relevant person is unavailable in the unit. All the same, if a review is postponed, it must be carried out as soon as practicable after the expiry of the last authorisation under section 99(4).[63] The continued service custody of the person to whom the review relates is authorised by virtue of section 100(6) until that time.

As will be noted, the provisions principally place the decision for detaining in custody on the CO, who may also make the decision to release without charge at any time. There is no power for the service police themselves to retain a person in custody. Therefore, in an extreme case, there

61 s. 100 AFA 2006.

62 S. 99(5).

63 S. 100(6)(a).

is nothing to prevent a CO from releasing someone whom the service police would rather were detained in custody. In other words, the service police may not, themselves, apply to a judge advocate for an extension of time, in spite of the provision in section 99(2) that, pending preparation of a report to the CO, the person making the arrest is authorised to detain the person in service custody to secure evidence, etc. It is only after receiving the report that the CO determines whether to keep the individual in service custody. If they do keep the person in service custody, then the duties that flow from that decision apply. Only the CO can apply for an extension of time.

Extension by a Judge Advocate[64]

Where a person has been arrested under section 67, custody without charge may be further extended by a judge advocate at a custody hearing, up to a maximum of 96 hours.[65] The onus rests upon the military authorities to justify why retention in custody is necessary. Consequently, it is incumbent upon the CO to make an application to a judge advocate[66] for continued custody without charge at any time before the expiry of the first 48 hours after the person's arrest, or, if it is not practicable to make the application within this time, as soon as practicable and no later than 96 hours after the arrest. This takes into account any operational or other proper military exigencies. However, if an application is made after the initial 48 hour period and it appears to the judge advocate that it would have been reasonable for the CO to make it before the expiry of that period, the judge advocate must refuse the application (s. 102(3)).

The person being kept in custody is entitled to legal representation at the hearing.[67] Rule 24 Armed Forces (Custody Proceedings) Rules 2009 (AF(CP)R 2009) stipulates that where a CO proposes to make an application for an extension of custody without charge under section 101(1) they, or someone acting on their behalf, 'shall notify the court administration officer of that fact and the necessary information' and inform the person arrested in writing. The CO must inform the person being kept in custody of their

64 AFA 2006, s. 101.

65 S. 101(4).

66 *Ibid.*, s. 102 and Explanatory Notes.

67 The legal representative must comply with the formal professional requirements of r. 18 AF(CP)R 2009.

entitlement to legal representation and provide the 'necessary information' listed in rule 24(2).

As a matter of practice, the CO is represented by a legal officer on the command staff of the superior headquarters, whose office will have been involved in providing the CO with advice.

The judge advocate may authorise the keeping of a person in service custody for such period, ending not more than 96 hours after 'the arrest', if they consider it appropriate, having regard to the evidence before them.[68] Where a person, while kept in service custody without being charged is arrested under section 67 for another service offence, 'the arrest' is to be read as a reference to the arrest for the service offence for which they were originally arrested. In the event that the custody hearing is adjourned, the person to whom the application relates may be kept in service custody during that adjournment, in accordance with the power in section 102(5). These provisions also apply where a person is transferred to or taken into service custody following arrest by civilian police under warrant of a judge advocate or is arrested or surrenders to police while a deserter or illegal absentee. See section 103.

Time spent in any type of custody will count as time served towards any subsequent sentence. The CO will specifically set this out in their award, announcing the sentence, the days already spent in custody and the remainder after deducting that figure from the sentence awarded. Furthermore, time in custody counts even if it is for a related offence.[69]

How the Provisions Work in Practice

A good example of how the law operates is to be found in the MSL, Chapter 5, paragraphs 26 and 27:

> 26. … A person is arrested for offence A at 0830 hrs on 12 January. The maximum period for which they may be held in custody without charge, on the authority of their CO, is 48 hours and they must therefore be released by 0830 hrs on 14 January at the latest unless their continued custody, either without charge or after charge, is authorised by a judge advocate. If, following

68 *Ibid.*, s. 101(4) ACA 2006.

69 *Ibid.*, s. 247(2) and see below, on page 450.

the person's arrest for offence A, the CO authorises custody without charge at 0900 hrs on 12 January, the CO must review that custody no later than 2100 hrs on 12 January, assuming they have authorised the maximum of 12 hours' custody. Where the person held in custody without charge for offence A is subsequently arrested for offence B at, say, 1900 hrs on 12 January, their CO may authorise their custody without charge in respect of offence B. If that authorisation takes place at 1930 hrs, there is no requirement for the CO to carry out the review of custody for offence A at 2100 hrs on 12 January. Assuming the CO had authorised the full period of 12 hours in relation to offence B, the next occasion on which a review of custody without charge will fall due is 0730 hrs on 13 January.

27. Subsequent reviews of custody without charge will fall due at intervals of no more than 12 hours in respect of both offences A and B. Unless the CO applies to a judge advocate for an extension of custody without charge ... or the person is charged and held in custody after charge ..., the person must be released by 0830 hrs on 14 January at the latest, which is 48 hours after their arrest for offence A.

Custody After Charge

Main Provisions: sections 105–109 AFA 2006; AF(CP)R 2009

The relevant law is to be found in sections 105–109 of the AFA 2006 and the AF(CP)R 2009, replicated below:

105 Custody after charge

(1) Where a person (referred to in this section and sections 106 to 109 as 'the accused') is kept in service custody after being charged with a service offence, he must be brought before a judge advocate as soon as practicable;

(2) At a hearing under subsection (1), the judge advocate may by order authorise the keeping of the accused in service custody, but only if one or more of conditions A to C in section 106 are met;

(3) The period for which a judge advocate may, by an order under subsection (2) , authorise the keeping of the accused in service custody is such period, ending not later than eight days after the day on which the order is made, as he considers appropriate having regard to the evidence before him;

(4) For the purpose of deciding whether condition A in section 106 is met, the judge advocate must have regard to such of the following considerations as appear to him to be relevant

  (a) the nature and seriousness of the offence with which the accused is charged (and the probable method of dealing with him for it),

  (b) the character, antecedents, associations and social ties of the accused,

  (c) the accused's behaviour on previous occasions while charged with a Service Offence and released from service custody or while on bail in criminal proceedings, and

  (d) the strength of the evidence that the accused committed the offence, as well as to any other considerations which appear to be relevant.[70]

(5) If—

  (a) the accused is charged with an offence under section 42 as respects which the corresponding offence under the law of England and Wales is

    (i) murder,

    (ii) manslaughter,

    (iii) an offence under section 1 of the Sexual Offences Act 2003 (c. 42) (rape), or

    (iv) an attempt to commit an offence within sub-paragraph (i) or (iii) ,

70 *Cf* the civilian case of *R (E) v Wood Green Crown Court* [2012] EWHC 1869 (Admin), where the trial judge applied the test of taking the 'prosecution case at its highest'. The Divisional Court said this was wrong, as conceded by the Crown, and held that s. 9 of the Bail Act 1976 requires the court considering an application for bail to consider 'the strength of the evidence of his having committed the offence'.

(b) representations are made as to any of the matters mentioned in condition A in section 106, and

(c) the judge advocate decides not to authorise the keeping of the accused in service custody, the judge advocate must state the reasons for his decision and must cause those reasons to be included in the record of the proceedings.

(6) An order under subsection (2) does not authorise the keeping of the accused in service custody

(a) if the accused is subsequently released from service custody, at any time after his release; or

(b) at any time after he is sentenced in respect of the offence with which he is charged.

(7) Subsection (1) does not apply where the accused is charged with a Service Offence at a time when he is kept in service custody by reason of a sentence passed in respect of a Service Offence or of an order under subsection (2) , unless that reason ceases to apply.

106 Conditions A to D

(1) Condition A is that the judge advocate is satisfied that there are substantial grounds for believing that the accused, if released from service custody, would

(a) fail to attend any hearing in the proceedings against him;

(b) commit an offence while released; or

(c) interfere with witnesses or otherwise obstruct the course of justice, whether in relation to himself or any other person.

(2) Condition B is that the judge advocate is satisfied that the accused should be kept in service custody for his own protection or, if he is aged under 17, for his own welfare or in his own interests.

(3) Condition C is that the judge advocate is satisfied that, because of lack of time since the accused was charged, it has not been practicable to obtain sufficient information for the purpose of deciding whether condition A or B is met.

(4) Condition D is that the accused's case has been adjourned for inquiries or a report and it appears to the judge advocate that it would be impracticable to complete the inquiries or make the report without keeping the accused in service custody.

107 Release from custody after charge

(1) Subsections (2) and (3) apply where, at a hearing under section 105(1), the judge advocate does not authorise keeping the accused in service custody.

(2) Subject to subsection (3), the accused must be released from service custody without delay.

(3) The accused may be required to comply, before release or later, with such requirements as appear to the judge advocate to be necessary

- (a) to secure his attendance at any hearing in the proceedings against him;
- (b) to secure that he does not commit an offence while released from custody;
- (c) to secure that he does not interfere with witnesses or otherwise obstruct the course of justice, whether in relation to himself or any other person; or
- (d) for his own protection or, if he is aged under 17, for his own welfare or in his own interests.

(4) On an application made

- (a) by or on behalf of the accused, or
- (b) by the commanding officer of the accused, any requirement imposed under subsection (3) (including such a requirement as previously varied under this subsection) may be varied or discharged by a judge advocate.

(5) A person on whom a requirement has been imposed by virtue of subsection (3)(a) commits an offence if, without reasonable excuse, he fails to attend any hearing to which the requirement relates.

(6) A person guilty of an offence under this section is liable to any punishment mentioned in the Table in section 164,

but any sentence of imprisonment imposed in respect of the offence must not exceed two years.

108 Review of custody after charge

(1) Where the keeping of the accused in service custody is authorised by an order under section 105(2), it must be reviewed by a judge advocate not later than the end of the period for which it is authorised.

(2) If at any time it appears to the accused's commanding officer that the grounds on which such an order was made have ceased to exist, he must

(a) release the accused from service custody; or

(b) request a review.

(3) Where a request is made under subsection (2) (b), a review must be carried out as soon as practicable.

(4) Sections 105(2) to (6), 106 and 107 apply in relation to a review as they apply in relation to a hearing under section 105(1 ; but the application of section 105(3) is subject to subsection (7).

(5) At the first review the accused may support an application for release from service custody with any argument as to fact or law (whether or not he has advanced that argument previously).

(6) At subsequent reviews the judge advocate need not hear arguments as to fact or law which have been heard previously.

(7) On a review at a hearing at which the accused is legally represented, the judge advocate may, if the accused consents, authorise the keeping of the accused in service custody for a period of not more than 28 days.

(8) In this section 'review' means a review under subsection (1).

109 Custody during proceedings of Court Martial or Service Civilian Court

(1) In relation to a review under section 108(1) which takes place between—

(a) the arraignment of the accused before the Court Martial or the Service Civilian Court, and
(b) the conclusion of proceedings before the court,
(c) section 105(2) to (6) (as applied by section 108(4)) apply with the following modifications.

(2) The reference in section 105(2) to conditions A to C is to be read as a reference to conditions A to D.

(3) Where the accused is awaiting sentence—
(a) references in section 105(4) (a) and (5) to an offence with which the accused is charged are to be read as references to the offence for which he is awaiting sentence; and
(b) section 105(4) (d) does not apply.

At a hearing before a judge advocate, both sides are entitled to legal representation. In the case of the accused, the AF(CP)R 2009 stipulate that the accused's legal representative must be a qualified lawyer.

An accused kept in service custody after being charged with a service offence must be brought before a judge advocate as soon as practicable[71] who may by order authorise the keeping of the accused in service custody, but only if one or more of conditions A to C in section 106 are met.[72]

Section 105(3) stipulates that the judge advocate may only authorise detention, or further detention, for a maximum period of eight days, whereafter a further hearing must be held. For the purpose of the custody hearing, the judge advocate shall not be bound by any enactment or rule of law relating to the admissibility of evidence in proceedings before courts of law (see rule 35, AF(CP)R 2009).

If the accused is charged with a section 42 criminal offence of

a. murder,
b. manslaughter,
c. an offence under section 1 of the Sexual Offences Act 2003 (c. 42) (rape), or
d. an attempt to commit an offence of murder or rape,

71 Bail Act 1976, s. 105.

72 Condition D is that the accused's case has been adjourned for inquiries or a report and it appears to the judge advocate that it would be impracticable to complete the inquiries or make the report without keeping the accused in service custody.

and the judge advocate determines to release the accused from custody, the judge advocate must state the reasons for the decision and must cause those reasons to be included in the record of the proceedings (s. 105(5)).

Custody hearings may take place via a live link if a judge advocate so directs. They may also be held in the absence of any person to whom they relate, if the judge advocate so directs. [73]

Who May Release the Accused from Service Custody

Where the accused was retained in custody after charge, they may be released by the judge advocate subject to a requirement to comply, before release or later, with such conditions as appear to the judge advocate to be necessary[74]

a. to secure their attendance at any hearing in the proceedings against them;
b. to secure that they do not commit an offence while released from custody;
c. to secure that they do not interfere with witnesses or otherwise obstruct the course of justice, whether in relation to themself or any other person; or
d. for their own protection or, if they are aged under 17, for their own welfare or in their own interests.

An application for variation of any such requirement imposed by the judge advocate may be made by or on behalf of the accused or by their CO.

Under section 107(5), it is an offence if, without reasonable excuse, a person on whom a requirement has been imposed fails to attend any hearing to which the requirement relates. The offence is punishable by imprisonment for up to two years.[75] If a judge advocate has authorised custody, a review must be carried out by a judge advocate (not necessarily the same one) no later than the end of the authorised period. However, if at any time the *accused's CO considers* that the reasons for authorising custody (conditions A to C in section 106) have ceased to exist they are under a duty either to release the accused or request that a judge advocate reviews the

73 AF(CP)R 2009, rr. 16 and 17.
74 *Ibid.*, s. 107(3).
75 *Ibid.*, s. 107(5) and (6).

matter, and such a review has to be done as soon as practicable. The existence of an option here ('or') gives the CO the power to effectively overrule the judge advocate's earlier decision by substituting their judgement for that of the judge advocate.[76] It is a course fraught with danger for a CO and it can only be in the most extreme circumstances that they would exercise the power without referring the matter to a judge advocate for review.

When a judge advocate is conducting such a further review of custody after charge where the accused is legally represented, if the accused gives their consent the judge advocate may authorise up to 28 days in custody. At their first review hearing the accused is permitted to advance any argument of fact or law to support their application for release, but at any further review hearings the judge advocate is not obliged to hear any such arguments that they have already heard.[77]

Custody During Trial by Court Martial or Service Civilian Court[78]

A defendant may be retained in custody until the conclusion of proceedings in accordance with Conditions A–C[79] and the additional Condition D, in section 106, which is concerned with the situation where:

> the accused's case has been adjourned for inquiries or a report and it appears to the judge advocate that it would be impracticable to complete the inquiries or make the report without keeping the accused in Service custody.

Interpreter

Rule 20 of the AF(CP)R 2009, as amended by the Armed Forces (Interpretation, Translation and Alcohol and Drugs Tests) Rules 2013 (AF(ITADT)R 2013), and the Armed Forces (Service Courts Rules) (Amendment) Rules 2015 set out the procedure for the appointment of an interpreter. Where a person to whom proceedings relate is due to attend a hearing, the court administration

76 See s. 108(2)(a).
77 S. 108(6) and Explanatory Notes thereto.
78 S. 109.
79 See above, p. 143.

officer, unless satisfied that the person does not need interpretation, shall appoint an interpreter to act at the hearing.

Amendments are also made in respect of the SAC Rules 29; where an appellant is due to attend a hearing, the court administration officer, unless satisfied that the appellant does not need interpretation, shall appoint an interpreter to act at the hearing. On application or on their own initiative the judge advocate may require a written translation of any document or part of a document to be provided for an appellant who needs interpretation. The 2015 Rules add a new section 29A into the Armed Forces (Summary Appeal Court) Rules 2009 (AF(SAC)R 2009), concerning interpretation and translation for persons other than the appellant.

7 COMMONLY USED SERVICE OFFENCES

The Manual of service law (MSL) deals with non-criminal conduct (disciplinary) offences in Chapter 7, some of which are very serious and which are unsuitable for summary hearing, for example mutiny (s. 6), failure to suppress mutiny (s. 7), desertion (s. 8), prize offences by an officer in command of a ship or aircraft (s. 37). Other offences, which appear to be very serious, for example disclosure of information useful to an enemy (s. 17), can be heard summarily. This chapter looks at only the most commonly used offences, many dealt with by the Commanding Officer (CO).

Criminal conduct offences are dealt with through section 42 Armed Forces Act 2006 (AFA 2006). Criminal conduct of a service person or service civilian can be charged as an offence contrary to section 42. By way of example, a charge of theft, contrary to section 1 of the Theft Act 1968 would be charged as

Committing a criminal conduct offence contrary to section 42 of the Armed Forces Act 2006 namely theft contrary to section 1(1) of the Theft Act 1968.

As mentioned above, on page 59, 'service offence' is defined in section 50 AFA 2006, as

a. any offence under Part 1 (that is, s. 1–s. 49);
b. an offence under section 107 (breach of requirement imposed on release from custody);
c. an offence under section 229 (breach of service restraining order);
d. an offence under section 266 (failure to comply with financial statement order);
e. any offence under Chapter 1 of Part 13 (testing for alcohol and drugs);

f. any offence under regulations under section 328 (false answer during enlistment in a regular force) or section 343 (service inquiries) that the regulations provide is a service offence;
g. an offence under section 18 or 20 of the Armed Forces Act 1991 (orders for the protection of children);
h. an offence under any of sections 95 to 97 of the Reserve Forces Act 1996 (reserve forces offences); or
i. an offence under paragraph 5(1) of Schedule 1 to that Act (false answer during enlistment in a reserve force) committed by a person within paragraph 5(3) of that Schedule.

The disciplinary offences are all those offences listed in Part 1, less the provisions of section 42, which is concerned with substantive criminal offences under the law of England and Wales.

To many civilian practitioners, some of the service disciplinary offences appear to be similar to offences they are familiar with in the criminal courts, for example fighting (assault) and damage to or loss of service property (criminal damage). So why are there special offences for the services? The answer is that service life imposes special requirements on service personnel and these offences are regarded as essential for the maintenance of discipline, recognising in their provisions the exigencies and nuances of service life.

Recognising the needs of our expeditionary forces, the Act ensures parity in dealing with service persons, under a familiar system of law, wherever they may be serving. The services' disciplinary system operates effectively anywhere in the world and applies to conduct both within and outside the United Kingdom.[1]

Mens Rea

Service offences require one of the three mental elements: intent, recklessness or negligence. A specific intent is the intention to bring about

1 MSL, Version 2.0 at p. 1-1-4.

the consequences of an act, for example to wound X with the intention of causing them grievous bodily harm.

Some examples of service offences of specific intent are:[2]

a. Assisting an enemy (s. 1)
b. Obstructing operations (s. 3)
c. Mutiny (s. 6 within 6(2)(a))
d. Desertion (s. 8)
e. Absence without leave (s. 9 within 9(2))
f. Disobedience to lawful commands (s. 12)
g. Making false records (s. 18 (3) and (4))
h. Ill treatment of subordinates (s. 22)
i. Damage to or loss of public or service property (s. 24)
j. Obstructing or failing to assist a service policeman (s. 27)
k. Allowing escape, or unlawful release, of prisoners, etc. (s. 30(1))
l. Hazarding a ship (s. 31(1)(a))
m. Giving false air signals (s. 32)
n. Dangerous flying (s. 33(1)(b)(i))
o. Low flying (s. 34) – see below, *R v Jackson* (Robert Valentine); Annoyance by flying (s. 35).

Other offences involve allegations of negligence or recklessness, and are illustrated in the following service disciplinary provisions:

a. Obstructing operations (s. 3)
b. Absence without leave (s. 9)
c. Disobedience to lawful commands (s. 12)
d. Failure to attend for, or perform, duty (s. 15)
e. Damage to or loss of public or service property (ss. 24(1)(b)(ii) and 24(2))
f. Allowing escape, or unlawful release, of prisoners, etc. (s. 30)
g. Hazarding of ship (s. 31(1)(b))
h. Dangerous flying (s. 33(1)(b)(ii) and 33(2))
i. Low flying (s. 34)
j. Annoyance by flying (s. 35).

2 Some offences can be committed intentionally or recklessly, e.g. obstructing operations.

Negligence and recklessness are well-known legal concepts throughout the criminal justice system, although they are not without difficulty. They impose subjective standards, as, in the case of recklessness where D, knowing that there is a risk that an event may result from their conduct or that a circumstance may exist, goes on to take that risk, and it is unreasonable for them to take it having regard to the degree and nature of the risk which they know to be present.[3] Negligence, on the other hand, is truly an objective test measured against the yardstick of the 'reasonable person' – has the accused failed to exercise such care, skill or foresight as a reasonable person in the situation of the accused would have exercised?[4] The MSL points out:

> Few criminal conduct offences can be committed negligently … although a significant number of non-criminal conduct offences can be … (non-criminal conduct (disciplinary) offences). This is because non-criminal conduct offences relate wholly to a Service person's professional responsibilities, where certain basic (or reasonable) standards of performance can be expected. Where a Service person fails to meet these standards, his failure to do so may be negligent and the charge against him may therefore be proved.
>
> For non-criminal offences, in the case of an offence where negligence suffices to find the charge proved, liability will be avoided if the accused behaves as a reasonable person, with the same skills, professional training, knowledge and experience would have done in the circumstances. A person is negligent if he fails to exercise such care, skill or foresight as a reasonable person would exercise in the same situation. This is an objective test for the officer hearing the charge to apply.[5]

Some offences may entail strict liability. Certainly, the previous Service Discipline Acts (SDAs) contained such an offence. In the case of *R v Jackson*

---

3 See *R v G and another* [2003] UKHL 50, at para, 13, per Lord Bingham, citing with approval the Law Commission Working Paper No 31 (*General Principles: The Mental Element in Crime*) proposed definition of recklessness (on page 48).

4 Law Commission Working Paper No 31 (*General Principles: The Mental Element in Crime*), at p. 7.

5 JSP 830 MSL Version 2.0, §37 & 38 on p. 1-12-10.

(Robert Valentine),[6] decided under the former Air Force Act 1955, D pleaded guilty to unlawful low flying contrary to section 51[7] (charge 1) and negligently causing damage to aircraft belonging to Her Majesty contrary to section 44A(1)(d) of the Air force Act 1955 (charge 2). He had flown a Jaguar plane at a height lower than 100 feet, hit a tower and caused substantial damage to the plane. Fortunately, he survived. Flying at under 100 feet was forbidden by regulations. It was submitted to the court that the offence of low flying is an offence of full *mens rea* and that the prosecution had to prove that the appellant either knew that he was flying under 100 feet or that he was reckless.

The court took the view that offences of strict liability are usually creatures of statute – the common law finding the notion of guilt without proof of a guilty mind repugnant.

> One is always reluctant to conclude that Parliament has created an absolute criminal offence or one of strict liability, but there are occasions when it has done so. In this case I have come to the conclusion that the Prosecution was correct in their assertion that it has done so in the case of section 51 AFA 1955. [*per* Hooper, LJ]

In the event, the AFA 2006 ensured there is a requirement in section 34(1)(b) for intention or recklessness in the new version of the offence of low flying contained in section 34.

Some service offences are offences of basic intent, where the

> mental element is that which is generally described as the 'basic' *mens rea*. Thus the accused must be shown either to have intended to act (or omit to act) as he did or to have been reckless whether he was so acting (or omitting to act). The *mens rea* therefore requires intentional or reckless acts (or omissions).[8]

Conduct prejudicial to good order and discipline is a classic example of such an offence of basic intent.

6 [2006] EWCA Crim 2380.

7 'Any person subject to Military Law who, being the pilot of one of Her Majesty's aircraft, flies it at a height less than such height as may be provided by any regulations issued under the authority of the Defence Council except- (a) while taking off or alighting, or (b) in such other circumstances as may be so provided shall, on conviction by court-martial be liable to imprisonment for a term not exceeding two years or any less punishment provided by this Act.'

8 See below, the case of *Dodman*, page 157.

Air Navigation Order Offences

Main Provisions: section 49 AFA 2006; ANO2005; AF(PANOO)O2009; AF(MA)R 2019

These are offences under the Air Navigation Order 2005 relating to civil aircraft and which are criminal offences for the purposes of section 49 of the AFA 2006 and set out in the Armed Forces (Prescribed Air Navigation Order Offences) Order 2009.[9] The effect is to make the same behaviour a service offence if it relates to a military aircraft:

a. recklessly or negligently acting in a manner likely to endanger an aircraft (Article 240)
b. recklessly or negligently causing or permitting an aircraft to endanger any person or property (Article 241)
c. entering an aircraft when drunk, or being drunk in an aircraft (Article 242(1))
d. when acting as a member of the crew of an aircraft or while carried in an aircraft for the purpose of so acting, being under the influence of drink or a drug to such an extent as to impair his capacity so to act (Article 242(2))
e. while in an aircraft, failing to obey a lawful command which the pilot in command of an aircraft may give for the purpose of securing the safety of the aircraft and of persons or property carried in it, or the safety, efficiency or regularity of air navigation under Article 77 of the Order[10] (Article 244 (authority of pilot in command of an aircraft))
f. while in an aircraft, using threatening, abusive or insulting words towards a member of the crew (Article 245(a) (acting in a disruptive manner))
g. while in an aircraft, behaving in a threatening, abusive, insulting or disorderly manner towards a member of the crew (Article 245(b))
h. while in an aircraft, intentionally interfering with the performance by a member of the crew of the aircraft of the crew member's duties (Article 245(c)).

9 AF(PANOO)O2009, as amended by AF(MA)R 2019, ss 9 and 10.
10 https://www.legislation.gov.uk/uksi/2009/1094/article/3/made.

Conduct Prejudicial to Good Order and Discipline

Main Provision: section 19 Armed Forces Act 2006

Section 19 of the Armed Forces Act 2006[11] provides:

Conduct prejudicial to good order and discipline

(1) A person Subject to Service Law commits an offence if he does an act that is prejudicial to good order and service discipline.
(2) In this section 'act' includes an omission and the reference to the doing of an act is to be read accordingly.
(3) A person guilty of an offence under this section is liable to any punishment mentioned in the Table in section 164, but any sentence of imprisonment imposed in respect of the offence must not exceed two years.

The section makes it clear that the conduct alleged must be prejudicial to both good order *and* service discipline, although an act which is prejudicial to service discipline is, *ipso facto*, prejudicial to good conduct. The Manual of Military Law stated:

> The words good order as used in this section are wide enough to include good order in the sense in which the word would be understood in civilian life as applicable to civilians, and in the sense in which they would be understood in military life as applicable to members of a military force. Conduct which is to the prejudice of good order is not necessarily to the prejudice of military discipline, e.g. an officer who creates a disturbance when he is away from his unit and in civilian clothes may be guilty of conduct to the prejudice of good order but not necessarily of military discipline. On the other hand, conduct which is prejudicial to military discipline is also prejudicial to good order in the military sense as applicable to members of a military force.[12]

11 See also, s. 69 of AFA 1955.

12 Footnote 3 to section 69, set out in the Manual of Military Law, 12th edition, Amendment No. 21, and Manual of Air Force Law.

The offence is not limited to an act that actually prejudices good order and discipline. It is sufficient if it has the potential, or likelihood, to do so. 'Conduct which is "likely to be harmful" or "potentially harmful" can thereby be actually prejudicial.'[13] The officer who creates a disturbance when away from their unit and while in civilian clothes may have displayed conduct to the prejudice of good order but not necessarily of service discipline, as there is nothing that obviously identifies the individual as a member of a service. The words 'prejudicial to' mean that the prosecution does not have to show that good order and military discipline were actually affected, but only that the accused's conduct was potentially harmful to good order and military discipline.

The test is an objective one as to whether conduct is or is likely to prejudice good order and discipline. The *Dodman* case (see also page 169 *et seq* for a discussion on dishonesty) laid to rest the former idea under section 69 of the Air Force Act and Army Act that the conduct had to be blameworthy. It is no defence, nor does it negative the basic *mens rea,* that the accused did not realise that their conduct was prejudicial to good order and service discipline. Prosecutors and Judge Advocates were enjoined not to use the expression 'blameworthy' when addressing or summing up to courts martial.

The offence charges the doing of an act and does not specify a requirement for intention or recklessness. It is clearly a statutory offence, unlike the nearest civilian equivalent of misconduct in a public office, which is an offence at common law triable only on indictment. That common law offence carries a maximum sentence of life imprisonment, is confined to those who are public office holders and is committed when the office holder acts (or fails to act) in a way that constitutes a breach of the duties of that office. The Court of Appeal has made it clear that the offence of misconduct in public office should be strictly confined. It can raise complex and sometimes sensitive issues.[14]

13 Practice In The Service Courts Collected Memoranda, Version 6, § 1.4.

14 Attorney General's Reference No 3 of 2003 [2004] EWCA Crim 868: 'the offence should be strictly confined but we do not propose to develop the point or to consider further the question of what, for present purposes, constitutes a public office.' See CPS Guidance, 'Misconduct in Public Office'. In *R v Whitaker* (1914) KB 1283 the court said: 'A public office holder is an officer who discharges any duty in the discharge of which the public are interested, more clearly so if he is paid out of a fund provided by the public.'

In contrast, the use of 'conduct to the prejudice' within the military system is quite widespread. In fact, many defendants prefer to be charged with the offence as a conviction does not carry the obvious stigma of a criminal conviction for a substantive offence. Moreover, the sentencing options are limited to two years imprisonment/detention upon conviction by court martial. Under section 19 it is necessary to particularise the conduct of the accused that is alleged to amount to an offence the section. Since the offence is essentially a disciplinary offence, it does not follow that the conduct will itself be criminal under civilian law. Indeed it ordinarily will not be since such conduct should be charged under section 42.

It can cover a multitude of conduct that, to the discerning eye, appears on the face of it to be criminal. For example, the soldier who breaks into the cookhouse and eats food from the refrigerator technically commits an offence of burglary (enters as a trespasser and steals). The charge could instead be laid under section 19 as a disciplinary infraction for conduct to the prejudice. While breaking into the cookhouse would be regarded as a burglary, the CO might not see it that way. They might view it as not being in the public (including the service) interest to send the accused for trial on the more serious offence but, rather, prefer to deal swiftly with the offending under section 19 as a breach of unit discipline.[15] In contrast, the charge under section 19 would not be justified in the more serious event of the unlawful entry into a married quarter, or civilian house, where there are wider public interest considerations.

The offence has been misused in the past to enable inappropriate summary dealing to avoid the administrative burden of a court martial, especially when there are other pressures on the unit (such as training, deployment or personnel constraints). Of course, this acts to the advantage of the accused service person, although it is a clear failure to deal properly with criminal conduct. However, it is not permissible to use section 19 to charge someone with conduct that amounts to a criminal offence as a means of obtaining a conviction without proving all the ingredients of the

15 See below on p. 345 *et seq*, where the service interest is considered to militate in favour of prosecution rather than against.

substantive criminal offence.[16] Diluting criminal conduct in such a way would undermine confidence in the service justice system.[17]

The particulars of offence should identify the conduct alleged. No specific state of mind is required or need be alleged. Proof of an intentional or reckless state of mind in relation to that conduct will accordingly satisfy the requirement of *mens rea*. However, in a given case the situation may not be as simple as that. It may be that the conduct alleged only has the requisite quality of prejudicing good order and discipline if it is accompanied by a specific state of mind. Thus, an allegation of making an inaccurate report may be accompanied by an allegation that it was known or ought to have been known to be inaccurate. Therefore, if the charge against an accused is that they conducted themself prejudicially to good order and service discipline in that they made a dishonest representation, it then becomes necessary for the prosecution to prove in relation to that charge that at the relevant time the accused was acting dishonestly; hence, the fact that there may be a need for a dishonesty direction.[18]

The leading case on conduct prejudicial to good order is *Dodman*.[19] Although decided under the Air Force Act 1955, it was held in *Armstrong*[20] to be a binding decision. The *actus reus* and *mens rea* requirements are the same under the 2006 AFA.

When discussing the issue of dishonesty, the *Dodman* case ought now to be read in the light of the decisions since *Ghosh*[21] was decided, as mentioned below, on page 231. In *Dodman* it was held:

> 15. The *actus reus* under s. [19] is conduct (whether it be an act or omission) which has the character of being prejudicial to good order and Air Force discipline. The conduct having been proved it is necessary also for the prosecution to satisfy

---

16 *Armstrong* [2012] EWCA Crim 83; see p. 162 below.

17 Several years ago, a charge of conduct to the prejudice was misused to prosecute what should have been a very serious criminal offence. For some unknown reason, the offender was prosecuted for that offence rather than rape under the 1956 Sexual Offences Act. Such charging practice actually undermines public confidence in the Service disciplinary system. Fortunately, it is a thing of the past.

18 *Dodman, op cit.* at §20.

19 Case No 97/3007/S2, 13th March 1998, Courts Martial Appeal Court. See also *R v Miller* (CMAC) The Times, 6 June 1983 (unreported) referred to in the judgment.

20 [2012] EWCA Crim 83. See also below, on p. 161, for the facts.

21 *R v Ghosh* [1982] EWCA Crim 2, cited in *Dodman*, at §20.

> the Court that there was *mens rea* on the part of the accused at the time he acted or omitted to act. No specific intention is required. The mental element is that which is generally described as the 'basic' *mens rea* or as 'basic' intent. Thus the accused must be shown either to have intended to act (or omit to act) as he did or to have been reckless whether he was so acting (or omitting to act). The *mens rea* therefore requires intentional or reckless acts (or omissions). If the act was deliberate then no question of recklessness (or, indeed, any other question of *mens rea*) will arise at that stage. If the act was not deliberate then it must be proved that when he acted or omitted to act his state of mind was reckless in the sense that he either closed his mind to the risk of his doing or omitting to [*sic*] the act in question or, appreciating the risk, he chose to run that risk.
>
> 16. The proof of the offence thus involves the proof of conduct of the accused having the requisite character objectively assessed (the *actus reus*) and the proof that at the material time his state of mind was that he either intended to do, or omit to do, the act in question or, if he did not so intend, he was reckless whether he would do or omit to do the act.

The *Dodman* decision was considered in two subsequent cases: where there were challenges to the lawfulness of the charge; and whether the charge sufficiently identified the offence a defendant was alleged to have committed.

The first case, *Ainsworth v United Kingdom*,[22] concerned a Royal Marines Lieutenant Troop Commander at the Commando Training Centre, who was charged with two counts of conduct to the prejudice of good order and military discipline contrary to section 69 of the Army Act 1955,[23] alleging he had neglected to supervise adequately, as his duty required him to do,[24] the Troop Party by failing to ensure that proper safeguards were

22 (1998) Application no. 35095/97.

23 'Any person subject to military law who is guilty, whether by any act or omission or otherwise, of conduct to the prejudice of good order and military discipline shall, on conviction by court-martial, be liable to imprisonment for a term not exceeding two years or any less punishment provided by this Act.'

24 *Quaere*, he could have been charged with negligently performing a duty, under section 29A of the Army Act 1955. This is now an offence under section 15(2) of AFA 2006.

imposed to prevent consumption of alcohol by under-age recruits in his troop and by failing to ensure that under-age recruits in his troop did not at that function consume alcohol contrary to the law. One of the under-age recruits died from an excessive intake of alcohol at the Troop Party held in June 1992.

In the European Court of Human Rights (ECtHR), two principal issues were argued:

The first concerned *Findlay* arguments about Article 6 compatibility. The second was a challenge to the unforeseeable nature of the conviction under section 69 of the Army Act 1955 invoking Article 6(2) and (3)(a)[25] and Article 7. The Commission held that the first ground was admissible, on the basis of *Findlay* and *Coyne*.[26]

In relation to the second argument, it was held that,

> a. Given the potential sentence under section 69 of the 1955 Act (two years imprisonment) and the nature of the charges against the applicant, the applicant was found guilty of a 'criminal offence' within the meaning of Article 7.
> b. The criminal offence in question must be clearly defined in the law and that this requirement is satisfied where the individual can know from the wording of the relevant provision and, if need be, with the assistance of the courts' interpretation of it, what acts and omissions will make him criminally liable. Section 69 of the 1955 Act, read in the light of the detailed and precise provisions of the Standing Orders, satisfied the requirement of foreseeability under Article 7 of the Convention.

---

25 Art. 6.2: 'Everyone charged with a criminal offence shall be presumed innocent until proved guilty according to law.'
Art. 6.3: 'Everyone charged with a criminal offence has the following minimum rights: (a) to be informed promptly, in a language which he understands and in detail, of the nature and cause of the accusation against him.'
Art. 7: 'No punishment without law 1. No one shall be held guilty of any criminal offence on account of any act or omission which did not constitute a criminal offence under national or international law at the time when it was committed. Nor shall a heavier penalty be imposed than the one that was applicable at the time the criminal offence was committed. 2. This Article shall not prejudice the trial and punishment of any person for any act or omission which, at the time when it was committed, was criminal according to the general principles of law recognized by civilised nations.'
26 Application Number 25942/94.

The ambit of the charge was again challenged in the case of *Armstrong*,[27] which looked at the scope of section 69, re-enacted in materially similar terms in the offence in section 19 of the AFA 2006. Armstrong was a serving army officer who served in Iraq, Afghanistan and Northern Ireland. On 1 May 2009, when serving in Northern Ireland, his service quarters were searched by the Royal Military Police (RMP). They discovered the following:

i) A damaged Glock 9 mm self-loading pistol which had belonged to an Iraqi policeman and which he had kept as a 'trophy';
ii) A considerable quantity of ammunition that would be of considerable utility to anyone who acquired it for an ulterior purpose;
iii) A morphine auto-injector; and
iv) 189 top secret and other classified material for which, if the documents had fallen into the wrong hands, there were potentially grave consequences for national security.

In the Court Martial Appeal Court's (CMAC) view, each of these four matters could have formed the basis of charges of substantive offences under the criminal law of England and Wales. The first two represented offences under the Firearms Act 1968, the third represented offences under the Misuse of Drugs Act 1971 and the last could have been charged as a breach of the Official Secrets Act 1989.

In this case, which was a prosecutor's appeal against a terminating ruling made at a court martial, the court held that the decision in *Dodman*, which is binding on the court martial, set out the *mens rea* of the offence under section [19]. The ingredients of the offence under section [19] are a matter of statutory construction. A distillation of the decision as it relates to section [19] shows that the Crown must establish:

a. that the defendant in fact conducted himself in the way alleged

27 [2012] EWCA Crim 83. See above, on p. 159. This decision is also important with regard to the duties of the prosecutor in circumstances where the judge does not agree with the course a prosecutor wishes to follow. In the present case, he stated his concern was directed not at the level of charge, but the form of the charge. The appeal court held that the procedure set out in the *Farquharson Guidelines* (published in 1986) should be followed. They directed that at the resumed trial of the guilty plea, the judge should recuse himself.

b. that he did so either intentionally or recklessly
c. that the effect or potential effect of the conduct was to adversely affect military discipline
d. that the effect or potential effect of the conduct was to adversely affect good order.

The CMAC confirmed that *Dodman* is a binding decision that stated the *mens rea* or mental element of the offence was one of basic intent; and that conduct that constituted a specific criminal offence under the law of England and Wales or a specific offence under military law should be charged as such and not under section 19. The CMAC reiterated that the requirements of prejudice of good order and service discipline are conjunctive and not disjunctive. Subject to that, the Act does not seek to define conduct. In the CMAC's view, 'where conduct constitutes an offence under the ordinary criminal law, it must be charged as such save in wholly exceptional circumstances'. 'The Crown also accepted that it was not permissible to use section 69 to charge someone with conduct which amounts to a criminal offence as a means of obtaining a conviction without proving all the ingredients of the substantive criminal offence.' The CMAC said that the Crown's approach was consistent with *Dodman*.

It is for the Crown to prove the conduct in question and to prove that it is to the prejudice of good order and military discipline. That is an objective test to be undertaken by the tribunal of fact. In the CMAC's judgment, it is irrelevant that the conduct did not have the potential to become known to others within the military or that it was not in fact known until it was discovered on investigation. If a person takes the utmost precautions to conceal conduct so that it is unlikely to be discovered, the steps taken to conceal the conduct ordinarily would aggravate the prejudice that such conduct is likely to cause.

The CMAC confirmed that contrary to the view of the trial judge, 'there is nothing in the statutory language that imposes any requirement for proof of specific intent. There is a simple question of fact in each case which must be objectively determined on the evidence.'

Moreover, on the basis of the legal authorities, the CMAC held it is clear that the offence under section 69 (s. 19) satisfies the principles of legal certainty within Article 7 of the European Convention on Human Rights and Fundamental Freedom (ECHR), containing as it does the four ingredients set out and stated above.

In allowing the prosecutor's appeal, the CMAC did not enquire whether there were 'wholly exceptional circumstances' that entitled the military prosecuting authorities to have taken the course they did in this case. It was not within the ambit or subject matter of the appeal. The CMAC said, at paragraph 49 of its judgment:

> Moreover, given the fact that the respondent is being tried upon charges under s. 69, and that, in the light of our ruling, he has a case to answer on those charges where the powers of the court are, on the unusual facts, substantial, it is far too late to consider whether charges for substantive offences were in fact the charges that should have been brought on a lawful exercise of the prosecutor's discretion. That step should have been taken at the plea and case management hearing and the request to the Director of Service Prosecutions or the Attorney-General should have been made at that stage.

Clearly, the charges selected in this case, as in all cases of prejudicial conduct, have a sentencing ceiling of two years rather than the statutory maximums for comparable substantive offences, and are therefore more attractive to a defendant.

The cases of *Dodman* and *Armstrong* are extremely important decisions. In *Armstrong*, the trial judge began by setting out his view that there was 'something of a misconception in certain quarters within the military jurisdiction as to the true scope or ambit of s. 69 and that that misconception was evident in this case'.

The CMAC and ECtHR have now hopefully laid to rest these misconceptions.

Damage to or Loss of Public or Service Property

Main Provision: section 24 AFA 2006

> (1) A person Subject to Service Law commits an offence if
>
> (a) he does an act that causes damage to or the loss of any public or service property or any property belonging to another person Subject to Service Law; and

(b) either

(i) he intends to cause damage to or the loss of the property, and there is no lawful excuse for his act; or

(ii) he is reckless as to whether he causes damage to or the loss of the property.

(2) A person Subject to Service Law commits an offence if

(a) negligently, he does an act that causes damage to or the loss of any public or service property; or

(b) he does an act that is likely to cause damage to or the loss of any public or service property and

(i) he is reckless as to whether he causes damage to or the loss of the property; or

(ii) he is negligent.

(3) For the purposes of this section

(a) 'act' includes an omission and references to the doing of an act are to be read accordingly;

(b) references to causing include allowing;

(c) 'loss' includes temporary loss;

(d) 'property' means property of a tangible nature, and references to public or service property are to be read accordingly.

(4) A person guilty of an offence under this section is liable to any punishment mentioned in the Table in section 164, but any sentence of imprisonment imposed in respect of the offence must not exceed

(a) in the case of an offence under subsection (1), ten years;

(b) in the case of an offence under subsection (2), two years.

Subsection (1) creates offences of unlawfully intending to cause damage or loss, and recklessly causing damage or loss. Subsection (2) creates offences of negligently causing damage or loss, recklessly doing an act likely to cause damage or loss, or negligently doing an act likely to cause damage or loss. The offences are focused upon 'public or service property', which are defined as follows:[28]

---

28 AFA 2006, s. 26.

Public Property: means property belonging to or held for the purposes of

(a) a department of the Government of the United Kingdom;
(b) any part of the Scottish Administration;
(c) a Northern Ireland department; or
(d) the National Assembly for Wales.

Service Property: means property

(a) belonging to or used for the purposes of any of Her Majesty's forces;
(b) belonging to a Navy, Army and Air Force Institute; or
(c) belonging to an association established, or having effect as if established, under section 110 of the Reserve Forces Act 1996 (reserve associations).

These are serious offences. While they may be dealt with summarily,[29] on conviction after trial by court martial for an offence under subsection (1), an offender is liable to a sentence of up to 10 years' imprisonment. For offences under subsection (2), the maximum penalty on conviction after trial is two years imprisonment.

The offence under subsection (1)(a) includes intentionally or recklessly causing damage or loss to 'any property belonging to another person subject to Service law'. In other words, damaging, etc. property belonging to another service person would fall within this provision. This offence is similar to the civil offence of criminal damage. As will be seen, it does not include damage, etc. to the property of service civilians. Any such offence would have to be charged under the criminal law.

The offence under section 24(2) is less serious than under the previous subsection. It covers a multitude of situations where negligence causes loss or damage. For example, a soldier driving a service vehicle in a negligent manner and causing damage to it would be susceptible to a charge under this section. He could also be made subject to a service compensation order for the cost of that damage.

29 See s. 164(1)(d).

Misapplying or Wasting Public or Service Property

25 Misapplying or wasting public or service property

(1) A person subject to Service law commits an offence if he misapplies or wastes any public or service property.
(2) A person guilty of an offence under this section is liable to any punishment mentioned in rows 2 to 12 of the Table in section 164.

The lesser offence of misapplying or wasting public or service property is found in section 25 AFA. The offence may be dealt with summarily but, if dealt with by court martial, no sentence of imprisonment may be imposed. According to the MSL, 'misapplies' requires misconduct on the part of the accused. It is not sufficient for the accused to merely use the property in question they must also use it for an improper purpose. It is irrelevant whether the accused or any other person has benefited from the misapplication. The misapplication need not result in any actual loss to the service. 'Wasting' is a term that should be given its normal dictionary meaning. The act requires some form of misconduct. For example, a legitimate disposal would not be wasting; there must be some unnecessary loss to the service.[30]

The MSL includes a specimen charge that depicts one offence of misapplying or wasting public or service property. In other words, 'misapplying' and 'wasting' are not separate heads of conduct. However, the particulars of the offence will specify whether public or service property was either misapplied or wasted.

Absence from Duty and Conscientious Objection

Main Provisions: sections 8, 9, 15, 16, 316(3), 330 and 344 AFA 2006; AF(EIATSC)R 2009; AF(FS)(2)R 2009

Absence from duty in the armed force can take one of four forms:

(1) Desertion

30 JSP 830 MSL Version 2.0, p. 1-7-82.

(2) Absence without leave (s. 9)
(3) Failing to attend for / perform a duty (s. 15)
(4) Malingering (s. 16).

Desertion

Main Provisions: sections 8, 53, 344 and Schedule 2(7) AFA 2006; section 3 AFA 2016; AF(FS)R 2009

An offence of desertion may only be dealt with by court martial. Although the offence under section 8(1) was not originally contained in Schedule 2, it is now an offence that must be referred to the Service Prosecuting Authority (SPA). Section 3 (1)(a) of AFA 2016 provides that any offence that is not one that may be dealt with summarily by the CO ('a CO offence') is a Schedule 2 offence. An offence of desertion with intent to avoid a period of active service has always been a Schedule 2 offence under section 8(2)(b).

Section 8

(1) A person subject to Service law commits an offence if he deserts.
(2) For the purposes of this Act a person deserts if he is absent without leave and
  (a) he intends to remain permanently absent without leave; or
  (b) he intends to avoid a period of active service.
(3) In this section 'active service' means service in
  (a) (an action or operation against an enemy;
  (b) an operation outside the British Islands for the protection of life or property; or
  (c) the military occupation of a foreign country or territory.
(4) A person guilty of an offence under this section is liable to any punishment mentioned in the Table in section 164, and any sentence of imprisonment imposed in respect of the offence
  (a) if the offender intended to avoid a period of active service, may be for life;
  (b) otherwise, must not exceed two years.

Desertion is a serious offence.[31] It is committed when a service person goes absent without leave with the intention of remaining permanently absent without leave or with intent to avoid a period of active service. Being labelled a deserter still carries opprobrium. The maximum penalty is two years imprisonment, save where the offence is committed with intent to avoid a period of active service, when the maximum sentence is life imprisonment.[32] A consequence of being convicted is that the service person will forfeit service for the time spent absent from their unit. See further note 36 below on page 171.

The Court of Appeal in *Mahoney*[33] decided that, where there was evidence that a soldier was absent from 9 April until his arrest on 13 April of the following year, this was *prima facie* evidence of desertion and in the absence of any satisfactory explanation by the accused the Court could properly convict of that offence. In other words, they were able to draw an inference that, due to the period the soldier was absent, he had formed an intention not to return, that is, to desert.

The intention to desert may arise after the offender has left their unit without permission. What might start out as a period of absent without leave (AWOL) turns into an offence of desertion the moment the offender forms the relevant intention. In the event, it is the intention of the offender that is crucial, save that where it is not expressed, it can be inferred from the length of time before arrest or surrender, as *Mahoney* provides.

The *Tointon*[34] case, where the applicant pleaded guilty to desertion, on count 1, and being AWOL, on count 2 (for a period of 100 days), should be considered. On 16 April 2009, Tointon was deployed to Afghanistan. On 10 September 2009, that is after approximately five months' service on that tour, he qualified for rest and recuperation. He received verbal and written instructions to return to Afghanistan on a flight due to leave Brize Norton on 23 September 2009. He failed to turn up for the flight. He was arrested on 10 October 2009 by the civilian police for an unconnected matter. His period of absence therefore was 18 days. Because his intention was to avoid operational duty overseas, his behaviour constituted desertion, as

31 See below, Chapter 5, p. 63 for provisions relating to arrest and detention by civil authorities.
32 See s. 8(4)(a).
33 *R v Mahoney*, 40 Cr App R.172; (1957) 1 WLR. The case was decided under Army Act 1881 s. 12(1).
34 [2010] EWCA Crim 1781.

charged in count 1. He was sentenced to dismissal and a total of 15 months' imprisonment; the individual sentences being 15 months' imprisonment and dismissal on count 1 and two months' imprisonment concurrent on count 2. Under the 'Sentencing in Courts Martial: A Short Guide' guidelines, the entry point for desertion is dismissal plus 12 months' imprisonment. Leave to appeal his convictions was refused.

Termination

Illegal absence, whether desertion or AWOL, is terminated in one of two circumstances:

a. by arrest of the absentee, whether by the civilian or service police, or
b. surrender by the absentee to the police or the military authorities.

Rule 85 of the Armed Forces (Court Martial) Rules 2009 (AF(CM)R 2009) provide that a record purporting to be made in any service record in pursuance of any Act or of Queen's Regulations (QRs), or otherwise in pursuance of naval, military or air force duty, and signed by the CO of the person to whom the record relates or by a person whose duty it was to make or keep the record, shall be evidence of the matters stated in the record. This means that a Part 1 Order issued by the CO, stipulating the commencement and termination dates of absence from the unit, is admissible in itself to prove the absence.

A document purporting to be a copy of such a record (including the signature) as is mentioned in paragraph (1) and purporting to be certified to be a true copy by a person stated in the certificate to have custody of the record shall also be evidence of the matters stated in the document.

The Armed Forces (Evidence of Illegal Absence and Transfer to Service Custody) Regulations 2009 are concerned with section 316(3) AFA 2006, which requires a civilian court to transfer to service custody, or to release with conditions, a person subject to service law who is believed to be illegally absent, provided that such person either admits to being subject to service law and being illegally absent or the court has a certificate signed by the person's CO detailing specified information about the person. The regulations also set out the information to be included in the certificate issued by the CO of the person subject to service law and who is believed to be illegally absent. The certificate of transfer to service custody from the civilian police is to be found in regulation 4.

Custody

If at any time the CO of a person who is kept in service custody without being charged with a service offence becomes aware that the grounds for keeping that person in service custody have ceased to apply, and is not aware of any other grounds on which continuing to keep that person in service custody could be justified under the Act, the CO must order the person's immediate release from service custody (s. 98(2)). However, this does not apply where a person who appears to the CO to have been unlawfully at large when they were arrested. In such a case, the person may not be released under section 98(2).[35] The same applies under section 301(4) where a person who had been unlawfully at large is then released subject to complying with a condition, and they fail to comply with the condition; or they are at large after the end of the period for which they were released.

Long-Term Absentees

Long-term absentees may be discharged from the service rather than prosecuted, where there is a confession to desertion and the surrounding circumstances make discharge more appropriate.[36] The effect of desertion is that a service person forfeits the period of service[37] for all time spent absent. This means that those days for which they are unlawfully absent do not count towards their overall period of service. This impacts upon pension and terminal benefits as, once service has been forfeited, the service person may not qualify. If a service person is convicted by court martial of desertion, forfeiture of service will follow, although the Defence Council may at its discretion restore the whole or part of the forfeited service[38] where, for example, since the end of the period of desertion, the service person has engaged in 'distinguished, gallant or other conspicuous conduct'.[39] The competent service authorities for this purpose are set out in Annex A to MSL Chapter 18. The Armed Forces (Forfeiture of Service) (No 2) Regulations 2009 (AF(FS)(2)R 2009) provide a discretion to dispense with a trial where a deserter has made a written confession, which they have signed, saying they intended to avoid a period of active service. In

35 S. 98(3).

36 See: s. 330 AFA 2006 and The Armed Forces (Forfeiture of Service) (No 2) Regulations 2009, reg. 3, which replaced The Armed Forces (Forfeiture of Service) Regulations 2009.

37 *Ibid.*, s. 330.

38 AF(FS2)R 2009, reg. 5.

39 MSL, Ch. 18.

the latter case (avoiding active service), the CO must consult the Director, Service Prosecuting Authority (DSP).[40]

Aiding and Abetting Desertion or Absence

This is an offence under section 344, punishable by the civilian courts in the case of any civilian who 'aids, abets, counsels or procures' the commission by another of an offence of desertion or absence without leave. The offence is committed where the defendant

a. knows that another person is subject to Service law;
b. does an act intending to cause that other person to be absent without leave; and
c. it causes that other person to be absent without leave.

If a service person has already absented themselves without leave, or deserted, and the person assisting, etc. them knows or believes this to be the case, it is an offence to do an act which may impeded their apprehension or prosecution (s. 344(3)).

The maximum punishment on summary conviction by a magistrates' court is imprisonment for a term not exceeding 12 months in England and Wales (otherwise, six months) or a fine not exceeding the statutory maximum, or both. On conviction on indictment, by the crown court, they may be sentenced to imprisonment for a term not exceeding two years or a fine, or both (s. 344(7)).

Conscientious Objection

Article 9 of the ECHR gives everyone 'the right to freedom of thought, conscience and religion'.[41] The Advisory Committee on Conscientious Objectors advises the Secretary of State for Defence on all conscientious objection claims. These claims are made by those in the armed forces whose

40 AF(FS)(2)R 2009, reg. 3(3).

41 See the interesting case of *Papavasilakis v Greece*, (Application no. 66899/14), of 15 September 2016, which concerned a case where a country (Greece) had a domestic legal mechanism for assessing conscientious objectors. The ECtHR emphasised the fundamental requirement for independence (para. 60). The Court unanimously found for the applicant that there had been a breach of Article 9.

application for permission to retire, resign or be discharged have been refused by the service authorities.[42]

Conscientious objection has been raised in courts martial in recent years, perhaps as a result of the regular involvement of HM forces in armed conflicts / peace support operations around the world in response to foreign policy decisions rather than the protection of vital national interests. Each service has its policy on how to treat genuine cases of conscientious objection.[43] When such a defence is raised in a court martial, it is often the first time the phrase is uttered out loud by the service person. In addition to the case of *Glenton*, cited above,[44] which was held *not* to be a case of conscientious objection, two cases will serve to illustrate how the defence has been raised.

The first is an RAF case, concerning an airman named Kahn,[45] who enlisted in the RAF in December 1999 as a medical assistant. On 15 January 2001, he applied for premature voluntary release, and he was discharged from the RAF on 24 April 2001 with a reserve liability of six years. On 7 January 2003, he was called up for the purposes of Op Telic (Iraq) and sought to avoid his call up liability by citing his mother's mental health problems and the difficulties for the family restaurant running at a profit if he was not able to work there. He later cited his unwillingness to take arms against fellow Muslims. He based his case on a breach of Article 14 ECHR and argued that if a person held a genuine objection on the grounds of conscience based on religious beliefs, then were that not to form a defence in law in relation to a charge of disobeying a lawful order under the law of

42 More information can be found about this Committee at its website on the government server: https://www.gov.uk/government/organisations/advisory-committee-on-conscientious-objectors.

43 The Army actually sets out the procedure in Queen's Regulations, para. 9.424, under the heading of 'Termination of Service – Free on Compassionate Grounds'. This is augmented by AGAI 55. Similar provision appears in QRs or the RAF, para. 10-41. In Ministry of Defence Freedom of Information Act 2000 response 06-01-2011-170837-002, of 24 January 2011, states, 'Such cases are handled administratively by the Service concerned and are dealt with first by the individual's chain of command. If the conscientious objection is considered genuine, arrangements are made for the applicant to be discharged from the Services on compassionate grounds.' None of the legislation passed from 2006 onwards contains any provision relating to a right to conscientious objection nor makes its procedures accessible and transparent, as was recommended by Forces Watch briefing, February 2011. Another organisation that offers help and guidance is the Peace Pledge Union: https://www.ppu.org.uk/.

44 See p. 176.

45 *Kahn v R* [2004] EWHC 2230 (Admin).

England and Wales, it would be incompatible with the rights guaranteed by Article 9[46] and Article 14. The main plank of Kahn's argument was that there was nothing in jurisprudence to support an absolute and unfettered right of refusal to compulsory service on the grounds of conscience or religion. The Summary Appeal Court's (SAC) view was that rights under Article 9 are closely confined. A conscientious objector cannot pick and choose in relation to particular conflicts. Under QRs for the Air Force, current at the time, air personnel (aircrew) who apply for discharge on grounds of conscience are dealt with in accordance with a publication known as AP 3392, Vol. 5, Leaflet 113. The decision of the SAC was that:

(1) The appellant did not know until after he had been arrested on 5 March, in his interview with an officer, that there was a possibility of claiming that his conscience did not allow him to obey the call up notice he had received by means of a formal procedure for making such a claim.

(2) The reasons why the appellant refused to report for duty on 24 February were a combination of factors which included concerns for his family, especially his mother, dislike of his trade, and a genuine and deep belief that the impending invasion of Iraq was wrong and that to be in any way a participant in it would be contrary to his religion. Without that belief he probably would not have refused to return to duty.

He appealed from the SAC decision, by way of case stated, to the Divisional Court. The case stated set two questions for the court:

(1) Whether, in proceedings under the service discipline Acts (the Army Act 1955, the Air Force Act 1955 or the Naval

46 'Article 9 – Freedom of thought, conscience and religion:
1. Everyone has the right to freedom of thought, conscience and religion; this right includes freedom to change his religion or belief and freedom, either alone or in community with others and in public or private, to manifest his religion or belief, in worship, teaching, practice and observance.
2. Freedom to manifest one's religion or beliefs shall be subject only to such limitations as are prescribed by law and are necessary in a democratic society in the interests of public safety, for the protection of public order, health or morals, or for the protection of the rights and freedoms of others.'

Discipline Act 1957) for the offence of desertion or absence without leave, a genuine conscientious objection, general or particular, to the continued performance of military service can amount to a defence in law under domestic legal principles or the European Convention on Human Rights.

(2) Whether, in the particular circumstances of Leading Aircraftman Khan's case, the Royal Air Force authorities acted in breach of either domestic legal principles or the European Convention on Human Rights by omitting to bring to his attention at the time he was recalled for service the fact that if he had a genuine conscientious objection to renewed service he had a right to claim exemption to recall as a conscientious objector.

The Divisional Court considered the ECHR authorities and without hesitation described the appellant's right to invoke a claim to conscientious objection, whether at the time of his call-out, or at any time thereafter, as prescribed by law. The QRs plainly had the status of law, being a form of delegated legislation, and the Leaflet formed part of them, in amplification of a procedure for claiming conscientious objection as a 'compassionate' ground for discharge. The QRs showed conclusively that, for the RAF, compassionate reasons included a claim to conscientious objection, and the Divisional Court saw no reason why, as a matter of language, on an objective view of the 1997 Regulations, the 'compassionate reasons' could not properly include conscientious objection.

Even on the assumption that a person claiming conscientious objection is entitled to protection under Article 9, there still remained the issues under that Article's first paragraph of whether there had been any manifestation of belief entitled to protection and whether there had been any interference with the manifestation of the belief in question.[47]

The next question the court considered was when the appellant manifested his belief. The line of Strasbourg jurisprudence, discussed in *Williamson*, differentiated between the motivation and the manifestation of a belief. Kahn did not manifest his belief until he told the RAF about it as the

47 *Regina (Williamson and others) v Secretary of State for Education and Employment* [2002] EWCA 1926, [2003] 2B 1300.

reason for his absence, although it should not be said that a conscientious objector has manifested their belief until they have expressed it in some way connected to their service. Remarkably, the Divisional Court held that, in some circumstances, it could conceive that the mere act of absence or desertion could be such an expression.

The Divisional Court answered the two questions as follows:

> 1. [w]e do not consider that Article 9 can ... provide a defence to the appellant's charge of being absent without leave. We prefer to limit our answer to the case under consideration.
> 2. We again confine our answer to the Convention, because no separate reliance has been placed on domestic legal principles ... [W]e conclude that, although the appellant's recall papers did not expressly refer to conscientious objection as a ground for claiming exemption, it did sufficiently identify a relevant ground, namely 'any other grounds ... for compassionate reasons'. We also consider that even if there had been any contingent possibility of engaging Article 9(1) of the Convention by reason of any deficiency in the information given to the appellant in his recall papers, nevertheless for the reasons set out in this judgment, there was no relevant manifestation of conscientious objection and no interference with any such manifestation by reason of the appellant's recall, arrest, prosecution or conviction.

It is important to also consider the widely reported decision in *Glenton*.[48] D was sent on pre-deployment leave on 1 June 2007. He was due to return to duty on 11 June at 0800 hours. Instead of returning to duty on that date he flew out of UK to Bangkok. He was found to be absent. He remained abroad for two years and six days before returning to UK from Australia. He took part in a number of public meetings denouncing the war in Afghanistan and claiming conscientious objection. He was quite properly charged with desertion on the basis of the length of absence and the fact that he was to be deployed to Afghanistan on the very day he was due to return from leave,

48 [2010] EWCA Crim 930.

where he had indicated he did not wish to be deployed. He considered this to be unfair. The judge advocate ordered psychiatric reports and the defence psychiatrist determined Glenton suffered from post-traumatic stress disorder (PTSD). In contrast, the prosecution psychiatrist found that there was no evidence of any disturbance of the appellant's mental state such as would make him not responsible for his actions at the time of the offence, although he could not entirely discount PTSD. On that basis, the prosecution decided to reduce the charge to one of absence without leave, to which Glenton pleaded guilty, as he had always intended to do to that charge. Although the press represented the case as one relating to AWOL, the Lord Chief Justice clearly stated, at paragraph 23 of the judgment: 'This is not a case of a conscientious objection. The explanation for the appellant's misconduct is not said to have been a desire to alleviate a concern of conscience, but a consequence of illness caused by the appellant's undoubtedly nasty experiences of war.'

The defence was also raised in *Lyons*.[49] The appellant, a leading medical assistant in the Navy, was found guilty by General Court Martial of intentionally disobeying a lawful command contrary to section 12(1)(a) AFA 2006. The command was to 'accompany Warrant Officer Robert Bainbridge to the armoury, draw a weapon and start weapon training when ordered to do so'. He was sentenced to seven months' military detention, reduction from the rank of Leading Medical Assistant to Able Seaman and dismissal from the service. He appealed against his conviction and sentence on the following grounds:

a. the order was not a lawful order because it contravened his rights under Article 9 of the European Convention on Human Rights as a person who had made a claim to be treated as a conscientious objector;
b. the order was unlawful because as a non-combatant within in the meaning of the Geneva Convention, being a medical assistant, he was not required to carry arms and could therefore not be lawfully required to undergo weapons training; and/or
c. He honestly believed the order to be unlawful.

49 *Lyons v R* [2011] EWCA Crim 2808.

The CMAC found that conscientious objection to military service was first formally recognised in Britain by the Militia Act 1757, but it did not become a major issue until the introduction of full conscription by the Military Service Act 1916. The UK became the first state to legislate for recognition of conscientious objection, simultaneously with full-time conscription. Section 2(1)(d) of the 1916 Act made provision for exemption from conscription on the ground of a 'conscientious objection to the undertaking of combatant service'. The Act established a procedure for claims of conscientious objection to be considered by local tribunals from whom appeal lay to an Appellate Tribunal. The National Service (Armed Forces) Act 1939 contained broadly similar provisions. In 1970 the Advisory Committee on Conscientious Objectors (ACCO) was formed as a non-departmental public body with responsibility to advise the Secretary of State for Defence regarding all claims to conscientious objection to further service in the armed forces.

At the time relevant to this appeal, the procedure for a member of the Navy seeking to be discharged on grounds of conscientious objection was set out in Personnel, Legal, Administrative and General Orders 0801 (PLAGO 0801). In the CMAC's judgment, the procedure for dealing with claims of conscientious objection satisfied the requirements of being prescribed by law and being necessary in a democratic society in the interest of public safety, the protection of public order and the protection of others. The appeal was dismissed by the CMAC in relation to both conviction and sentence.

These decisions should be contrasted with the decision of the ECtHR in the case of *Mammadov and Others v Azerbaidjan*,[50] which concerned applicants who claimed to be Jehovah's Witnesses and provided documents attesting to their membership in that faith. The ECtHR held that there had been a violation of Article 9 of the Convention. The system of alternative service based on domestic regulations could not be said to strike a fair balance between the interests of society as a whole and that of objectors of conscience. Accordingly, the contentious measures, up to and including imprisonment, which were taken against the applicants as a result of their refusal to perform military service and resulting from the lack of alternative service providing the persons concerned with the status of conscientious objector, amounted to an interference which was not necessary in a democratic society within the meaning of Article 9 of the Convention.

50 (2019), Application 14604/08.

Absence Without Leave
Main Provisions: sections 9 and 53(1)(b) AFA 2006

Section 9

(1) A person Subject to Service Law commits an offence if subsection (2) or (3) applies to him.
(2) This subsection applies to a person if he is intentionally or negligently absent without leave.
(3) This subsection applies to a person if
  (a) he does an act, being reckless as to whether it will cause him to be absent without leave; and
  (b) it causes him to be absent without leave.
(4) In subsection (3) 'act' includes an omission and the reference to the doing of an act is to be read accordingly.
(5) A person guilty of an offence under this section is liable to any punishment mentioned in the Table in section 164, but any sentence of imprisonment imposed in respect of the offence must not exceed two years.

This is a more serious offence than failing to attend a duty, as the absence tends to be for a longer period of time.[51] As can be seen from the particulars of the section, AWOL is not an offence of strict liability. The possible mental elements of the offence are stated in section 9(2), namely, that the service person is 'intentionally or negligently absent without leave' or is 'reckless' and encompass acts or omissions[52] that have the effect of causing a service person to be absent without leave. However, it is trite to remark that 'negligence' and 'recklessness' are not straightforward legal concepts.

It is important to remember that even though this offence may be committed negligently or recklessly, it is an offence that potentially carries a criminal penalty, as per the criteria in *Engel*, of loss of liberty for up to 90 days and is therefore more properly classed as a criminal offence.

'Negligence' imports the concept of the 'reasonable man or woman', which is an objective criterion and does not depend on the subjective appraisal of the defendant. This means that a person can be AWOL without intending to be so if, objectively, their conduct or omission is viewed as

51 See p. 63 for provisions relating to arrest and detention by civil authorities.
52 AFA 2006, s. 9(4).

negligent. Negligence within the criminal law does not usually constitute a criminal offence, save in cases of gross negligence, where the act or omission is so severe that it is rightly seen as criminal. An example is the standard set out in *Adomako* [1994] 3 WLR 288,[53] relating to the offence of gross negligence manslaughter.

Negligently Absent

To establish that the accused was negligently AWOL under subsection (2) it must be proved that the accused either did something that a member of the armed forces of their experience, age, training and seniority should not have done, or failed to do something that a member of the armed forces of their experience, age, training and seniority should have done; and that, as a result of their negligent act or omission, they were absent and that their absence was not authorised. Examples would include where an accused oversleeps because they failed to make arrangements to be woken; where the accused loses a rail warrant and has insufficient money so that they cannot arrive at the unit before leave has expired; or where they fail to read unit orders setting out leave dates and so overstay their leave.[54]

'Recklessness' imposes a higher standard than carelessness (which entails a momentary lack of concentration).[55] Since the decision in *Caldwell*[56] was overruled by the House of Lords in *R v G*,[57] the test for recklessness is a subjective one, as set out by Lord Bingham, that a person acts 'recklessly' with respect to:

> (a) a circumstance when he is aware of a risk that it exists or will exist;
> (b) a result when he is aware of a risk that it will occur;
> and it is, in the circumstances known to him, unreasonable to take the risk.

53 Available on BAILII website, https://www.bailii.org/uk/cases/UKHL/1994/6.html.
54 MSL, JSP 830 MSL Version 2.0 at p. 1-7-36.
55 In the context of driving offences, the Road Traffic Act 1988, s. 3ZA(2), defines careless, or inconsiderate, driving where the manner of driving 'falls below what would be expected of a competent and careful driver' taking into account 'not only to the circumstances of which he could be expected to be aware but also to any circumstances shown to have been within the knowledge of the accused' s. 3ZA(3).
56 [1982] AC 341.
57 [2003] UKHL 50, *per* Lord Bingham, at para. 41.

In spite of the subjective nature of the test, 'it is not to be supposed that the tribunal of fact will accept a defendant's assertion that he never thought of a certain risk when all the circumstances and probabilities and evidence of what he did and said at the time show that he did or must have done'.[58] This is an objective standard. However, Lord Steyn commented, at paragraphs 54 and 55 of the judgment '[A] subjective state of mind is to be inferred "by reference to all the evidence, drawing such inferences from the evidence as appear proper in the circumstances" ... The shift is towards adopting a subjective approach. It is generally necessary to look at the matter in the light of how it would have appeared to the defendant.'

The decision shows that the test retains an element of objectivity in its application in that the credibility of the accused's denial of knowledge and understanding of the risk will always be judged against an objective standard of what a person of the same general age and abilities as the accused would be expected to have known.[59]

How one might be absent through negligence by a reckless act or omission will be interesting to argue, bearing in mind the different tests applied in respect of each. Negligence, after all, is gauged against the conduct of a reasonably prudent person. That is, it is an objective test, whereas, as noted, recklessness is subjective *and* objective.

Suggested guidance is provided by the MSL, which contemplates an element of foresight of risk:

> [t]o establish that the accused was recklessly absent without leave under subsection (3) it must be proved that the accused was aware that by acting or failing to act in a given manner, there was a risk that the accused would be absent without leave, yet they unreasonably went on to take that risk and in so doing was absent without leave. Examples would include where an accused renders himself incapable of returning to their unit through drink, or where the accused returns to their unit by inappropriate means (e.g. hitch-hiking) and so returns late. If there were evidence that the accused intended to remain

58 *Ibid.*, Lord Bingham, at para. 39.

59 See *ibid.*, para. 54, for the concerns expressed by Lord Steyn concerning the 'the special characteristics of children and mentally disabled people' in the context of recklessness.

> permanently absent, the proper charge would be desertion under section 8 (desertion).[60]

To the untrained eye, it is difficult to distinguish recklessness from negligence, whereas the distinction between carelessness and negligence poses fewer problems.[61]

In *Martin (Alan James)*,[62] the defendant gave, as a secondary reason for his absence, his anxieties about his domestic situation arising from family proceedings in either the county court or family proceedings court concerning residence and the applicant's contact with the children, and felt that he needed to be at home with the children. The Court considered that excuse to be a mitigating factor, although it found his prime reason (regret of his decision to re-enlist) an aggravating one. It held: 'The Services are used to considering the domestic circumstances of servicemen, if they are properly put before them ...' Importantly in such cases, the SPA is entitled to take into account, as part of the service interest test in determining whether it is in the public interest to proceed, any domestic and compassionate circumstances considered to be properly applicable.

In the case of *Foley*,[63] the CMAC held that AWOL cases impinge on military discipline and the efficient running of Her Majesty's forces. They are not criminal offences in the sense normally understood in civilian life (although they are when one applies the *Engel* criteria). An intention to be absent may be formed even where there are compassionate reasons.

In relation to how AWOL is treated in the three services, the CMAC stated, in *McKendry*,[64] that it was not for the CMAC to impose a standard

---

60 JSP 830 MSL Version 2.0, at p. 1-7-33.

61 For a lay assessment, applying common sense to the equation, see Nick Morgan, 'Tax Hell', 13 November 2017: 'When somebody says to me, "Oh, careless mistake!" I'm 7 and I'm back in a maths class and I've forgotten to carry the one over. When my radar picks up "negligence" my head is full of hospital x-rays where the forceps have been left in. The words "careless" and "negligence" will trigger different associations in the minds of people reading this. But nobody – in their right minds – will say that carelessness and negligence are the same things.'

62 [2007] EWCA Crim 3377.

63 [2012] EWCA Crim 71. See also *Harris (KA)* [2011] EWCA Crim 1242 where the court bore in mind that AWOL along with its more serious counterpart of desertion are military offences with no civilian equivalent. 'Accordingly, the CMAC, which lacks the same feel for the needs of the Services that the Court Martial will have, will exercise caution before concluding that a Court Martial has been wrong in principle or failed to give proper weight to the relevant factors in sentencing in a particular case.' *Ibid.*, para. 7.

64 *McKendry* [2001] EWCA CRIM 578.

policy across the different armed forces with respect to sentencing for offences of being AWOL. The CMAC said that the problem of service personnel being AWOL arose in different ways in each service and might require different solutions. Although a service person in the Army might serve a longer period of detention, one in the Royal Navy was more likely to be dismissed from the service for being AWOL for a shorter period. That was a difference rather than a disparity.

That decision should now be treated with some caution as, since the passing of the AFA 2006, and the creation of the tri-service disciplinary system, it is unlikely that the CMAC will now apply different sentencing policies, because there are no longer any single-service courts martial, just the court martial for all three services, which may comprise members of a service different to that of the defendant. However, it was an enlightened understanding by the CMAC of the separate service interest for each service and serves as a reminder for those that argue the contrary.

Arrest A judge advocate may issue a warrant for the arrest of a person AWOL. See on page 74 above the Armed Forces (Warrants of Arrest for Service Offences) Rules 2009 (AF(WASO)R 2009).

Minor Administrative Action

While very short-term absences may be dealt with by Minor Administrative Action (MAA), army personnel are subject to Army General and Administrative Instruction 67 (AGAI 67) (67/2B-1) which provides that '[c]ases of unauthorised absence for a period over 48 hours may NOT be dealt with by Minor AA. For periods up to 48 hours, where appropriate, an award may be made but only by a CO or an officer who has been specifically designated by the CO to carry out this task.' It is therefore advisable to check the single service policy of any service person being advised in connection with absence.

Failing to Attend/Perform a Duty/Negligent Performance

Main Provisions: sections 15 and 53(1)(b) AFA 2006

Offences under section 15 may be dealt with summarily by the CO.

(1) A person Subject to Service Law commits an offence if, without reasonable excuse, he
   (a) fails to attend for any duty;
   (b) leaves any duty before he is permitted to do so; or
   (c) fails to perform any duty.

(2) A person Subject to Service Law commits an offence if he performs any duty negligently.

(3) A person guilty of an offence under this section is liable to any punishment mentioned in the Table in section 164, but any sentence of imprisonment imposed in respect of the offence must not exceed two years.

When an individual fails to turn up for work in civilian life, it is unusual for a charge to follow, unless they have committed a criminal offence against their employer. The worst consequence from the employer is that they will be dismissed. In military life it is quite another matter and, depending on the circumstances, can lead to court martial proceedings. For example, falling asleep on guard duty when on operations will be viewed very seriously as it imperils the safety of the unit.

The charge must clearly specify the conduct alleged against the defendant, especially as the charge of negligently performing a duty under section 15(2) raises specific legal issues about what constitutes negligence as a matter of law, as discussed below. It is also necessary to particularise the duty, for example to report to a specific place or person.

The service person who oversleeps and is late for work could face a charge of failing to attend for duty, under section 15. In fact, they are more likely to be dealt with by way of MAA rather than formal discipline, especially if it is a first-time offence. MAA is to be contrasted with major administrative action under AGAI 67.[65]

However, where a unit is preparing for an exercise, or a military deployment, oversleeping can have more serious consequences, as a

65 Army General and Administrative Instructions, Vol. 67, and similar policy documents in the other two services, which regulate, e.g. social misconduct, and provide for an administrative course of action against an individual which could, ultimately, lead to their discharge from the service. However, this is outwith the remit of this work, which is focused on the formal statutory disciplinary system. It is nevertheless the case that AGAI documents outline policy. Therefore, as with all policy documents, they do not overrule the statutory powers or duties that rest upon, e.g., COs. If they do, they are *ultra vires*.

service person's absence from duty may delay the necessary preparations and/or departure. This is a 'service interest' factor that militates in favour of disciplinary action. In such a case, a formal charge for summary dealing may be seen as being in the overriding public interest.

Where the individual does not perform their duty properly, or leaves before they should, their conduct also falls within section 15. A more serious form of this failure is one of 'negligently performing a duty' under section 15(2). This offence requires the prosecution to prove the defendant's acts or omissions to an objective standard. The MSL gives the example of the soldier who hands in his rifle without first checking that it had been unloaded properly and made safe, whereby a round is unintentionally discharged.[66] The decision in *Price and Bell*[67] confirmed that the standard of care required to avoid the service offence of negligent performance of a duty, contrary to section 15(2) AFA 2006 was to be measured against the standard to be expected of the reasonable service person having similar training, knowledge and experience as the accused. The CMAC noted the remark by the trial judge advocate that:

> There is no definition of negligence in the Armed Forces Act 2006 but that does not mean to say that is an end of the matter. The general law, that is as applied by the relevant parts of the Armed Forces Act applying English law to service law, the general law is that a person is negligent if he fails to exercise such care, skill or foresight as a reasonable man in his situation would exercise and the test is objective, it is set out there, the care, skill or foresight as a reasonable man in the accused situation would exercise … To put it in a service context if you are sure that the accused behaved as no reasonable man with the same skills, professional training, knowledge and experience would have done in the circumstances, then he is guilty, nothing less will do. If you think that he either was or may have been acting as any such reasonable man would have behaved in all the circumstances of this case, then he is not guilty and must be acquitted.

66 See the specimen charge in JSP 830 MSL Version 2.0, at p. 1-7-50.
67 [2014] EWCA Crim 229; [2014] WLR (D) 86.

Pitchford, LJ stated:[68]

> In the somewhat special circumstances of the service context we agree with the Judge Advocate that the standard of care required to avoid the service offence of negligent performance of a duty is to be measured against the standard to be expected of the reasonable serviceman having similar training, knowledge and experience as the accused .... In our judgment, a subjective consideration of the defendant's 'skills' or 'weaknesses' has no place in the objective judgement whether the defendant reached the appropriate standard of care.

In *R v Townshend*,[69] D was piloting an RAF aircraft carrying nearly 200 passengers and crew from England to Afghanistan. Another experienced officer, Flt Lt J, was the co-pilot. Townshend had his camera on the flight deck, which he used to take pictures. The practice of taking photographs from the cockpit by flight crew was 'frowned upon' as a camera, or anything else that should not really be on the flight deck, was a potential problem because it could fall into a position where it obstructed the controls. This is what happened, as the camera fell and disengaged the autopilot, which caused the aircraft to go into a steep dive, falling some 4,400 feet in about half a minute. Flt Lt J suffered serious spinal injuries. The RAF conducted a detailed examination, costing many millions of pounds. Townshend was charged with and convicted of negligent performance of a duty. He was sentenced to four months' imprisonment suspended for 12 months (as varied on appeal).

To ensure maintenance of good order and discipline, the charge can also be laid where there is a wider responsibility by those who hold rank. In *Melia*,[70] the defendant failed to intervene in the assault of four junior marines following a celebratory company function. He pleaded guilty to four charges of failing to perform a duty (to stop the assault on junior ranks).

---

68 *Ibid.*, para. 20.
69 [2018] EWCA 430 (Crim).
70 [2018] EWCA Crim 602.

While most cases under section 15 would be dealt with otherwise than by court martial, serious offences so triable are subject to the maximum penalty of two years imprisonment.[71]

Malingering

Main Provisions: sections 16, 53(1)(c) and 345 AFA 2006

Only an offence under section 16(1)(a), or an offence under section 16(1)(c) committed by omission may be dealt with summarily but, if tried by court martial, is punishable with up to two years' imprisonment or detention.[72]

Section 16 states:

(1) A person Subject to Service Law commits an offence if, to avoid service
  (a) he pretends to have an injury;
  (b) by any act he causes himself an injury;
  (c) by any act or omission he aggravates or prolongs any injury of his; or
  (d) he causes another person to injure him.

(2) A person Subject to Service Law commits an offence if, at the request of another person Subject to Service Law ('B') and with the intention of enabling B to avoid service
  (a) by any act he causes B an injury; or
  (b) by any act or omission he aggravates or prolongs any injury of B.

(3) In this section
injury' includes any disease and any impairment of a person's physical or mental condition, and the reference to injuring is to be read accordingly;
'service' includes any particular duty or kind of duty.

(4) A person guilty of an offence under this section is liable to any punishment mentioned in the Table in section 164, but any sentence of imprisonment imposed in respect of the offence must not exceed two years.

71 AFA 2006, s. 15(3).
72 See AFA 2006, s. 53(1)(c).

This is another offence without a direct civilian equivalent. Although one might regard this, in its milder form, as what is commonly termed 'skiving', it can have a serious dimension when connected with operational demands on units. The offence set out in section 16 AFA 2006 is concerned with the service person who, in order to avoid service,

(a) pretends to have an injury;
(b) by any act causes himself an injury;
(c) by any act or omission aggravates or prolongs any injury of his; or
(d) causes another person to injure him.

This offence in subsection (2) covers those cases where the offender injures a colleague or aggravates an existing injury, at the colleague's request, in order that the colleague may avoid service.

In the case of *Cross*,[73] D was a young soldier who served in Afghanistan where his unit sustained a number of casualties. He was sent home on 25 May 2010 for a two-week period of rest and recuperation, just after his 19th birthday. He was due to return on 10 June 2010. While he was away on leave, two of his colleagues were killed in Afghanistan. A few days before he was due to go back, he persuaded a friend to run over his leg as he did not want to return to Afghanistan. This did not result in any fractures but there was a considerable tenderness to the leg and it was clear that he would not be returned to active service. He admitted his actions to the RMP and was subsequently charged with malingering. He was sentenced to detention and dismissal from the service. It was held by the CMAC that, on the basis of unchallenged psychiatric evidence obtained by the appellant, the sentence of detention was quashed and a community order with a supervision requirement substituted. In what it described as 'very much a one-off case', the CMAC found the defendant had a 'genuine psychiatric problem'. It nevertheless accepted that:

> [a] Court Martial has to take the view that for the maintenance of discipline that a deterrent sentence is necessary for those who in such circumstances absent themselves from duty.

73 *R v Cross* [2010] EWCA Crim 3273.

Aiding or Abetting Etc. and Malingering

The offence under section 345 may be committed by another service person or a civilian, in which latter case, depending upon the severity of the offence, it will be prosecuted either summarily in the magistrates' court, or on indictment at the crown court. The punishment in the case of summary conviction by magistrates' court is 12 months' imprisonment (in England and Wales, otherwise 6 months) or a fine not exceeding the statutory maximum, or both. On indictment, the maximum sentence is imprisonment for a term not exceeding two years or a fine, or both. There is a jurisdictional limit to the offence, in that it must occur in the UK, Isle of Man or in a British Overseas Territory. The offence can be committed in one of two ways:

a. Section 345(2) – Deliberately injuring, etc.: The offender must be aware that the person they are helping is someone subject to service law. They will commit an offence if they do any act to cause the service person an injury – or they aggravate an existing injury, with the intention of causing the service person to avoid service.
b. Section 345(3) – Causing others to believe: The offence may also be committed where the offender does any act intending to cause a person subject to service law to avoid service, or that either causes 'B' to believe that 'B' has an injury, or causes another person to believe that 'B' has an injury. This would be committed, for example, where a civilian medical practitioner in England, who was opposed on political or philosophical grounds to military operations overseas, falsely certified that 'B' was too unwell to return to his parent unit which was about to deploy on operations. In this situation, if 'B' was unaware of the falsity of the statement 'B' will have committed no offence of malingering.

Contravention of Standing Orders

Main Provision: section 13 AFA 2006

(1) A person Subject to Service law, or a civilian subject to service discipline, commits an offence if

(a) he contravenes a lawful order to which this section applies; and
(b) he knows or could reasonably be expected to know of the order.

(2) This section applies to standing orders, and other routine orders of a continuing nature, of any of Her Majesty's forces, made for any
(a) part of Her Majesty's forces;
(b) area or place; or
(c) ship, train or aircraft;
but paragraph (a) of this subsection does not apply in relation to a civilian subject to service discipline.

(4) A person guilty of an offence under this section is liable to any punishment mentioned in the Table in section 164, but any sentence of imprisonment imposed in respect of the offence must not exceed two years.

Standing orders (SOs) are issued under the authority of section 13 AFA 2006 and are described as 'standing orders, and other routine orders of a continuing nature, made for any (a) part of Her Majesty's forces; (b) area or place; or (c) ship, train or aircraft'. They must be contrasted with orders relating to a specific duty given to a service person, for example to drive a military vehicle from A to B. A breach of a specific or direct order would fall within section 12, as disobedience to a lawful command.

Importantly, section 13 makes it clear that SOs apply to 'a person subject to Service law, or a civilian subject to Service discipline'. In this manner it is possible to introduce orders to regulate conduct within barracks. Moreover, in Germany, for example, SOs are issued to create disciplinary offences for driving offences within barracks and certain other offences relating to the status of members of the force, such as the requirement to hold a valid British Forces Germany driving licence,[74] or to possess insurance. In *Rheines (Robert)*,[75] the defendant, a Sergeant in the Royal Electrical and Mechanical Engineers, was involved in a minor road traffic accident on 29 September 2009 at a British

74 The right to issue such licences derives its authority from Art. IV(1) of the Status of Forces Agreement and in Germany, as modified by Art. 9 of the Supplementary Agreement, as amended. It is therefore a right that may be withdrawn by the military authorities.

75 *Rheines (Robert) v The Queen* [2011] EWCA Crim 2397.

Army barracks in Germany and also contravened a stop sign. His vehicle had been uninsured since January 2004 and, further, since 2004 the car had not been registered with the British Forces Germany (BFG) Vehicle Licensing Office. All of these matters were requirements under BFG SOs.

It is a requirement that the SOs are published and the prosecution must not only prove this fact, by production of a certified copy of them, but also show that the defendant knew, or ought reasonably to have known, of their existence. This is especially important when alleging that a dependant or a member of the civilian component has breached SOs. On conviction by court martial, any sentence of imprisonment imposed in respect of the offence must not exceed two years.

Rule 87 of the AF(CM)R 2009 provides that:

> A certificate purporting to be signed by a person's commanding officer or an officer authorised by the commanding officer to give the certificate, and stating the contents of, or of any part of, standing orders, or other routine orders of a continuing nature, of any of Her Majesty's forces, made for (a) any part of Her Majesty's forces, (b) any area or place, or (c) any ship, train or aircraft, shall be evidence of the matters stated in the certificate.

This certification is therefore admissible in evidence to establish the provenance of the order.[76]

SOs are published within the unit or command and placed on unit notice boards, which every member of a unit is obligated to read / consult. It is no defence to say that they were unaware of SOs.[77] They may place areas out of bounds, give instructions relating to matters to be complied with within certain time limits or give routine instructions relating to sub-unit responsibilities (e.g. guard duties). Overseas, where the content is

---

76 See also, AF(CM)R 2009, r. 86 makes a similar provision in relation to 'Defence Council instructions, regulations and certificates' and covers a certificate purporting to be issued by or on behalf of the Defence Council or by a person authorised by them stating that a decoration is or is not a naval, military or air force decoration, or that a badge or emblem of a description specified in, or as annexed to, the certificate is or is not one supplied or authorised by the Defence Council.

77 Although it may be a defence to show there was an insufficient opportunity to be aware, e.g. where the SO is published while an individual is away from the unit and the breach occurs before they have had an opportunity to read them. *Quaere: ignorantia lex non excusat* (ignorance of the law is no excuse).

more widely applicable to the whole service community, standing orders are published in Navy, Army, and Air Force Institutes (NAAFIs) or sports facilities used by members of the force and the civilian component.

The unreported case of *Wittekind*[78] concerned an interesting and novel challenge to SOs. The defendant was charged with disobedience to standing orders that specified the offence as having a 'member of the opposite sex' in single-sex accommodation after 2359hrs. After the RAF Prosecuting Authority was notified that a pre-trial hearing was required on the discriminatory nature of the charge following *Lustig-Prean v UK*,[79] the charge was dropped. The SO was of a standard format that had not been changed to reflect the fact that gays now serve; the way the order was worded effectively expressly approved homosexual and lesbian contact after 2359 hours, and expressly forbade heterosexual contact. It therefore appeared to offend Articles 8 and 14.

Misconduct Towards a Superior Officer

Main Provision: section 11 AFA 2006

> (1) A person Subject to Service Law commits an offence if
> (a) he uses violence against a superior officer ('B'); and
> (b) he knows or has reasonable cause to believe that B is a superior officer.
> (2) A person Subject to Service Law commits an offence if
> (a) his behaviour towards a superior officer ('B') is threatening or disrespectful; and
> (b) he knows or has reasonable cause to believe that B is a superior officer.
> (3) For the purposes of this section
> (a) the behaviour of a person ('A') towards another person ('B') includes any communication made by A to B (whether or not in B's presence);

78 *R v Wittekind* (2001) – RAF DCM at RAF Aldergrove; 14 Aug. 2001 (unreported).
79 *Lustig-Prean and Beckett v The United Kingdom*, Applications nos. 31417 /96 and 32377/96, 27 September 1999.

(b) 'threatening' behaviour is not limited to behaviour that threatens violence.

(4) A person guilty of an offence under this section is liable to any punishment mentioned in the Table in section 164, but any sentence of imprisonment imposed in respect of the offence must not exceed

(a) in the case of an offence under subsection (1), or an offence under subsection (2) of behaviour that is threatening, ten years;

(b) in any other case, two years.

Respect for those superior in the chain of command is essential for the maintenance of good order and military discipline. Consequently, behaviour that impinges adversely on that relationship, through violence, disobedience or disrespectful language or behaviour, constitutes a disciplinary offence and is regarded seriously within the services.

Section 11 creates two offences. The offence under subsection (1) is committed where a person subject to service law uses violence against a superior officer, and knows or has reasonable cause to believe that the victim is their superior officer. The offence under subsection (2) is that, knowing or having reasonable cause to believe that a person is their superior officer, their behaviour towards their superior officer is threatening or disrespectful. The behaviour includes any communication made by the accused to the superior officer (whether or not in the superior's presence). The section makes clear that 'threatening' behaviour is not limited to behaviour that threatens violence. In practice, problems arise when a convivial occasion is marred by excessive drinking and barriers break down. JSP 835, Chapter 1 paragraph 3 points out that 'Social drinking may play a part in group bonding', but it can lead to blurring of lines of authority.

The offence can be dealt with summarily, save that where tried by court martial for disrespectful behaviour, a maximum sentence of two years imprisonment may be imposed. However, for a more serious offence of using violence, or using behaviour that is threatening, the maximum sentence is 10 years' imprisonment.

Although there is no definition of the term 'superior officer' in section 11, the term is defined in section 374, which is to the effect that '"superior officer" means in relation to a person ("A"), an officer, warrant officer or non-commissioned officer who is subject to Service law and is of superior rank or rate to A or is of equal rank or rate to A and is exercising authority as A's superior'.

In *R v Rabouhi*,[80] the appellant, a corporal in the RAF, was training at Offut Air Base in Omaha, Nebraska, and lodging at a local hotel. On the evening in question, he had gone out with three colleagues to another colleague's house for a meal and they drank alcohol. He went back to his hotel in the early hours and stayed in the foyer, talking to Sergeant 'D' and to SAC 'H'. Flight Lieutenants 'M', 'B' and 'A' came in and a conversation between the men developed into a verbal confrontation between the appellant and 'M', who accepted he himself behaved like 'an obnoxious twat'. The appellant allegedly pushed or punched 'M'. At his trial the appellant's defence was that he had drunk far less than others who were at fault. Furthermore, he said had not been disorderly and he had acted in lawful self-defence. However, even after the offence against 'M', the appellant was plainly still annoyed, had to be led away, then elected to return so as to continue the matter, and when Sgt 'D' tried to stop him had not accepted that. He struck 'D' so that 'D' went to the ground. Lines of authority, it was accepted, had been blurred by his friendship with 'D', but 'D' was his superior. He was lucky 'D' suffered no injury of note. Thankfully, there was no involvement with the host nation and no one involved wanted to complain or to take the matter further.

He was acquitted of the assault on Flt Lt 'M', but found guilty of what was effectively an assault on Sgt 'D'. He was reprimanded and fined £1,000. He appealed his sentence. The CMAC held that the court martial is a specialist tribunal entitled to deference. The CMAC concluded that on these facts, when so much was conceded by the prosecutor, the appropriate disposition was a reprimand. The financial penalty was neither appropriate nor necessary. The CMAC even agreed that these facts 'were unusual as coming before a Court Martial'.

The seriousness of the offence is underlined in *Bell v UK*,[81] decided under the former Army Act 1955. D was charged with an offence of using

80 [2014] EWCA Crim 1517.

81 *Bell v United Kingdom (Appeal): (Application no. 41534/98)* 16 January 2007. The ECtHR held at para. 53 of its decision that the summary procedure before the CO violated the independence, impartiality and, consequently, fairness aspects of Article 6(1) and that the exclusion of legal representation from the applicant's summary trial also constituted a violation of Article 6(3)(c)

insubordinate language to a superior officer on the evening of 19 December 1997. In challenging the summary process before his CO, the case proceeded to the ECtHR, where it was held that, in the Court's view, the offence was disciplinary in domestic law and in nature. The presumption was that the charge against him was 'criminal'. That presumption could be rebutted entirely exceptionally and only if his deprivation of liberty could not be considered 'appreciably detrimental' given its nature, duration or manner of execution.

In the case of *Singh*,[82] the defendant, a corporal (*sic*) in the Royal Navy, had been on a cross-country ski trip with his fleet. They were returning by ferry, and some members, in plain clothes, visited the bar. During the course of an altercation, during which the victim, an officer, told D he was making a fool of himself, the D poured beer over the victim. There had been friction between the parties arising from the victim making repeated derogatory remarks about the defendant's wife's superior rank. He pleaded guilty to two charges of using violence to a superior officer. He stated that the services had not been brought into disrepute in the light of fact that the parties were not in uniform at the material time, that he was of good character and had an exemplary record. Moreover, he had decided to end his service a result of the incident. He was fined £500 and reduced in rank to private (*sic*). In the CMAC, the defendant submitted that the sentence was manifestly excessive in that his net loss at the end of his service, by virtue of the reduced earnings at a reduced rank, was around £1,760. Accordingly, he contended, the reduction in rank had been a financial punishment and it had been wrong to impose an additional fine in the circumstances of the instant case. The CMAC agreed that the sentence was manifestly excessive in the light of the financial consequences of the defendant's reduced rank. The fine of £500 was quashed.

Problems may arise when personnel are off duty. The section requires that D 'knows or has reasonable cause to believe that B is a superior officer'. When personnel are wearing civilian clothes, a defence often raised is that the defendant did not recognise the individual as a person of superior rank. This defence is used to excuse all sorts of egregious behaviour, as

---

of the Convention. The creation of the Summary Appeal Court provides Article 6 guarantees to the service person.

82 *R v Singh* [2005] AllER (D) 243 (Apr).

evidenced by the case of *Griffin*,[83] where the defendant sexually assaulted a female officer in civilian clothes, later saying (as if it excused his odious behaviour) he did not know she was a 'ma'am'.

Disobedience to a Lawful Command

Main Provision: section 12 AFA 2006

> (1) A person Subject to Service Law commits an offence if
> (a) he disobeys a lawful command; and
> (b) he intends to disobey, or is reckless as to whether he disobeys, the command.
>
> (2) A person guilty of an offence under this section is liable to any punishment mentioned in the Table in section 164, but any sentence of imprisonment imposed in respect of the offence must not exceed ten years.

The need for discipline and for service persons to obey lawful commands is a cornerstone of the three services. Consequently, disobedience to commands strikes at the very core of the chain of command. The maximum sentence on conviction by court martial of an offence under section 12 is 10 years' imprisonment and reflects the fact that obedience to some commands may be vital to the success of an operation.[84]

The offence contains three essential elements that the prosecution must prove: there must be a command; that command must be lawful; and it must be intentionally or recklessly disobeyed. A genuine misunderstanding of the order may provide a defence.

Part of the lawfulness test requires the command to have a military purpose. Examples of a lawful command would be to order a service person to perform a military task, for example attend for training or report to a particular place for duty. The order must be a lawful order that is not be contrary to domestic or international law. It must have a military purpose. For example, an order to a soldier to go and fetch personal shopping for an officer has no military purpose and would not be lawful. Similarly, an order to drive the spouse of a superior to a particular place would not be

83 *R v Griffin* [2018] EWCA Crim 267.
84 EN §69.

lawful unless it were connected with an official duty, for example attending a Mess function or dinner, etc. But an order to move military equipment or to attend a guard duty has a clear military purpose.

The order does not have to be given directly, but could be passed on through another individual, for example 'Tell Pte Smith to bring the staff car to my office at 14.00', which in itself constitutes an order to the intermediary. If Pte Smith fails to attend at 14.00 hours, he might have a defence if, for example, the individual tasked with passing the order did not do so. If such a failure on the intermediary's part were culpable, they too might find themselves facing a charge of disobedience to a lawful command. There might also be a defence if they were genuinely unable to carry out the order, for example because of illness, and had informed the intermediary of the reason, but the intermediary had failed to pass on that message to the authority who issued the order. In that case the intermediary might well find themselves charged with an offence under section 15(2) of negligently performing a duty, or of negligence by omission, under section 19.

In *R v James Brown*,[85] there was discontent in a unit about the CO and Colour Sergeant, who were thought to be mismanaging them. Matters came to a head when, on 12 February 2013, the CO called a battle group parade. There was a discussion among the platoon and it was decided that they should make some sort of protest to the senior officers in order to bring the platoon's concerns to their attention. The form of protest ultimately chosen was for the platoon to sit down at the forthcoming parade. It would appear that the idea for this form of protest had originated with the applicant. He was the most senior NCO present at the meeting. The CSM ordered the members of the platoon to stand up, whereupon the applicant was heard to tell them to stay steady and remain sitting down, which they did. Brown was found guilty by court martial of what the judge advocate called, 'a serious disciplinary offence. Its gravity is reflected by the fact that the maximum sentence is one of ten years' imprisonment.'

So long as a person remains subject to service law, they must obey all lawful orders. In the earlier Royal Navy case of *Brown (Kirsty)*,[86] decided under the SDAs, the accused failed to return to HMS Drake from authorised

85 [2014] EWCA Crim 1160. Sentenced by court martial to 60 days' service detention, reduction in rank and dismissal from Her Majesty's Service.
86 *Brown (Kirsty Elizabeth), R v*: [2007] EWCA Crim 2632 19 July 2007.

leave on 8 January 2007. She was charged with two offences: absence without leave contrary to section 17(1)(a) of the Naval Discipline Act 1957 and disobedience to a lawful command contrary to section 12 of the same Act. Before the trial started, the judge advocate, of his own motion, objected to the joinder of the two charges, stating that when AWOL a person could not be guilty of disobeying a lawful command, that it was not right for two charges to be based upon the same facts and that the facts surrounding the wilful disobedience were an aggravating feature of the absence without leave rather than a separate offence. Consequently, he quashed the wilful disobedience on the ground that it was oppressive. The Naval Prosecuting Authority appealed this ruling under the Courts Martial (Prosecution Appeals) Order 2006.[87] The CMAC held that the judge advocate was wrong to hold that when a service person was AWOL they were no longer under an obligation to obey a lawful order. Nor was it correct to say that two charges cannot be based upon the same facts. On the contrary, the fact that two offences arise out of the same facts can be a good reason for joining both charges. The offence of disobedience to lawful commands contrary to section 12 of the 1957 Act is a different offence from that of absence without leave under section 17. Furthermore, the former offence is punishable by imprisonment without limit, whereas the maximum term is two years for the latter offence. For this reason alone it was legitimate to add the second charge. Furthermore, the particulars of the first charge (absence without leave) are quite different from those of the second (disobedience to an order). Accordingly, joinder was proper and was not oppressive.[88]

See also, *Lyons*[89] and *Wisely*.[90] In the latter case, the SAC upheld a submission by the appellant's advocate that an order given by a Staff Sergeant to the accused, at about 18.00 hours in the following terms, namely 'Parade outside the Guardroom at 20.00 hours in combat 95 and sober',

87 See now, Courts Martial (Prosecution Appeals) Order 2009.

88 By way of interest, this was the first appeal to be brought under the Courts Martial (Prosecution Appeals) Order 2006. At the outset, the CMAC dealt with a conundrum as to its jurisdiction. How was Article 1, which provides that the Order only applies to trials that commence after 5 July 2006, to be reconciled with Articles 4(1) and 4(13) that purport to make Article 4 applicable to a ruling that is made before the trial commences? The CMAC stated that Article 1(2) must be read as if it provides 'this Order shall apply only to trials by courts martial which commence or are to commence on or after 5 Jul 06'.

89 Referred to in detail at FN 49 on page 177, above.

90 *Wisely; R v*: Summary Appeal Court at Bulford, 18/19 October 2001, Deputy Judge Advocate Hill (unreported; facts provided by the appellant's advocate).

was not an order pursuant to section 34 of the Army Act 1955 which the appellant had wilfully or through neglect disobeyed. The finding by the CO and the sentence of seven days' detention was quashed. The order by the Staff Sergeant had been given at a time when the appellant was under the influence of intoxicating liquor, and in civilian clothes, and there was no evidence that could satisfy the court that the appellant would have been sober by 20.00 hours, and therefore he could not be lawfully required to parade in uniform and sober.

Fighting or Threatening Behaviour

Main Provision: section 21 AFA 2006

Section 21 states:

> (1) A person Subject to Service Law commits an offence if, without reasonable excuse, he fights another person.
> (2) A person Subject to Service Law commits an offence if
>   (a) without reasonable excuse, his behaviour is
>   (i) threatening, abusive, insulting or provocative; and
>   (ii) to cause a disturbance; and
>   (b) he intends to be, or is aware that his behaviour may be, threatening, abusive, insulting or provocative.
> (3) For the purposes of this section a person's 'behaviour' includes anything said by him.
> (4) A person guilty of an offence under this section is liable to any punishment mentioned in the Table in section 164, but any sentence of imprisonment imposed in respect of the offence must not exceed two years.

The offence of fighting is committed where, without reasonable excuse, a service person fights another person. The evidential burden shifts to the defendant in such a case to establish the reasonable excuse. The defence of self-defence is also available. The maximum penalty on conviction by court martial is two years' imprisonment.

It is, in reality, a difficult offence to prove where the allegation is contested and relates to a fight between two individuals, as both parties generally rely on the violence of the other to justify their own conduct.

Fighting tends to be charged to enable a CO to deal summarily with assaults, without the stigma of a substantive charge. However, it can be used against all participants – or more than just one. Fighting connotes the participation of others, whereas an assault is more usually one against one other. If a defendant elects trial by court martial, on a charge of fighting, the chance of a successful defence to the charge is greatly increased wherever there is the suggestion that the other person fought too and the defendant alleges they were merely defending themselves. The MSL, at 1-7-67 states:

> Fight is given its normal dictionary meaning, (i.e. a person taking part in a struggle or contention), and as to whether a fight has occurred is a question of fact for the officer hearing the charge. It must be proved that an accused intended to fight. If both parties were merely engaging in 'horseplay' this may not amount to an offence under subsection (1). If the accused struck another who did not retaliate there would be no fight (but a charge of assault may be brought under section 42). A fight between the accused and any other person is covered by the section (it is not necessary that the other person should be another member of Her Majesty's forces therefore one person can be charged with fighting).

The charge could be used as a military alternative to affray, under section 3 of the Public Order Act 1986, where several individuals are involved in a fight outside a unit bar or mess. In the case of fighting, there is no requirement that a person of reasonable firmness, if present at the scene, would fear for their personal safety.

The other offence, created by subsection (2), is that of threatening, abusive, insulting or provocative words or behaviour. This is committed when a person subject to service law, without reasonable excuse, intentionally behaves in a threatening, abusive, insulting or provocative manner that is likely to cause a disturbance. As there is no requirement for the behaviour to take place or the words to be uttered in public, this distinguishes the offence under section 21(2) AFA from the offences in Part I of the Public Order Act 1986 (sections 4, 4A and 5).[91] In other words, the offence could be committed in a dwelling, for example a married quarter.

91 No offence under these Public Order Act sections is committed where the conduct takes place inside a dwelling and the person to whom it is directed is inside that or another dwelling.

Ill Treatment of Subordinates

Main Provision: section 22 AFA 2006

> (1) A person Subject to Service Law who is an officer, warrant officer or non-commissioned officer commits an offence if
> (a) he ill-treats a subordinate ('B');
> (b) he intends to ill-treat B or is reckless as to whether he is ill-treating B; and
> (c) he knows or has reasonable cause to believe that B is a subordinate.
>
> (2) For the purposes of this section a person ('B') is a subordinate of another person ('A') if
> (a) B is Subject to Service Law; and
> (b) A is a superior officer of B.
>
> (3) A person guilty of an offence under this section is liable to any punishment mentioned in the Table in section 164, but any sentence of imprisonment imposed in respect of the offence must not exceed two years.

The three services take an extremely dim view of those who abuse their rank to mistreat service personnel of junior rank. The relationship within the hierarchy should be one of trust and respect, not fear and disrespect.

In *R v Skeet*,[92] the applicant, a Bombardier (Corporal), and the co-accused were instructors at the Army Foundation College in Harrogate. They were so disappointed with the state of Hut 1, where 11 Platoon was accommodated, that privileges such as television were withdrawn. The following day, the platoon excelled and the privileges were installed and they were given permission to get pizzas. Everyone had a free evening and the instructors themselves went into town for a drink. When they returned, the offences occurred as the accused became annoyed that the state of the hut had deteriorated. He pulled one man out of bed, injuring his ankle, and ordered the occupants of the room to do exercises as punishment. At his trial D did not accept that he had done wrong. He was reduced to the rank of Gunner (Private), sentenced to six months' detention and to ordered pay £1,000 compensation to the soldier he injured. He appealed conviction

92 *R v Skeet* [2005] EWCA Crim.

and sentence. In rejecting leave to appeal, the CMAC took the view that a strong element of deterrence was justified to reflect the service's abhorrence of bullying and its determination to stamp it out. The reasons given for the sentence demonstrate that the trial court, the court martial, had regard to all relevant matters.

There is no definition in the Act of what amounts to ill treatment for the purposes of this section or section 38,[93] where it is mentioned in the context of prize[94] offences. It is suggested that the words should be given their ordinary meaning, viewed by the objective standards expected of those having responsibility for junior personnel. Nevertheless, there is a requirement for specific intention or recklessness to ill-treat the individual, coupled with a reasonable cause to believe that they are a subordinate.

An officer who struck a soldier, who was asleep on guard duty, on the helmet with a pickaxe helve to wake him up, was charged and convicted by General Court Martial of ill treatment.[95] Many might have agreed with the officer's conduct, bearing in mind the severity of sleeping while on guard duty. However, the gravamen of the ill treatment was clearly the intentional physical assault of one of his soldiers.

Section 22 is sometimes used to discipline trainers of recruits who mistreat those under their care for any conduct that is considered to be ill treatment. Substantive assault charges remain available for more specific violence.[96] It is regrettable that the Act does not define ill treatment, leaving it open to cover a wide range of conduct from verbal abuse and humiliation through to overt and systematic violence. Perhaps that is the intention,

---

93 S. 38 (1) 'A person subject to Service law commits an offence if— (a) he ill-treats a person who is on board a ship or aircraft when it is taken as prize; or (b) he unlawfully takes anything in the possession of such a person.'

94 EN to AFA 2006, para. 110: 'During an armed conflict, COs are entitled to capture (as 'prize') most enemy ships and aircraft, and any goods in them. Under international law they must bring the captured enemy ship, aircraft or goods to an appropriate place for a proper adjudication on whether they were lawfully seized (and can therefore properly be deemed as 'prize').'

95 Unreported case from 1981, Chelsea Barracks. The officer, a Lieutenant, was dismissed the service. The author was an observer at the proceedings.

96 In *R v Birch* [2011] EWCA Crim, an instructor was convicted of offences of violence against recruits and sentenced to six months' detention and reduced to the ranks. Although it is not clear from the judgment what specific offences the appellant was convicted of, whether assault or ill treatment, the principle remains the same.

thereby enabling the conduct to be dealt with summarily.[97] The specimen charge in the MSL demonstrates how widely a charge can be worded.[98] The MSL goes on to suggest that:

> Ill treatment is to be given its normal dictionary meaning of treating badly or cruelly. A series of minor actions against a victim, which would in themselves not amount to offences of assault, might amount to ill treatment. Ill treatment includes but is not limited to physical force. Bullying or frightening may suffice, as may behaviour or treatment that is unnecessarily harsh or which degrades or humiliates another person. Whether conduct amounts to ill treatment is a question of fact for the person hearing the charge to decide on the circumstances of each case.

Subjecting recruits to mock executions, making them stand in the rain with arms outstretched and making them carry a bucket of sand in each hand may be regarded as treatment that is inhuman and degrading under human rights case law, which must amount to ill treatment.[99] While this conduct can be quite serious, it seems to fall between two stools. Ill treatment is not a Schedule 2 offence, therefore there is no requirement under section 116(2) for the service police to refer the matter directly to the SPA. Indeed, section 116(3) requires them to refer the case to the person's CO. Moreover, insofar as the CO is concerned, there is no statutory duty upon them to call in the service police to investigate if there is no allegation or circumstances that would indicate to a reasonable person that a Schedule 2 offence has or may have been committed by a relevant person (see section 113(2)). It may be accepted that the matter would fall within the 'prescribed circumstances' mentioned in section 114, where it is shown there is a course of conduct by a person subject to service law, involving on at least two occasions an assault in which that individual participated as a principal offender or as a secondary party; or an allegation has been made that would indicate to a reasonable person, or there are other circumstances which would indicate

97 See s. 53(1)(d).

98 '[AB] on ..., ill-treated [CD], whom he knew or had reasonable cause to believe was a subordinate.' See JSP 830 MSL Version 2.0, p. 1-7-69'

99 See for example, *Ireland v United Kingdom* [1978] ECHR 1; *Al-Skeini and others v Secretary of State for Defence* [2007], UKHL 26.

to a reasonable person, that a person subject to service law has or may have been the victim of an assault causing serious injury, inflicted by a person of superior rank or rate while the assailant was otherwise carrying out their duties.[100] In such cases, the CO must inform the service police.

Resistance to Arrest

Section 28 provides:

> (1) A person Subject to Service Law ('A') commits an offence if another person ('B'), in the exercise of a power conferred by or under this Act, orders A into arrest and
> (a) A disobeys the order;
> (b) A uses violence against B; or
> (c) A's behaviour towards B is threatening.
>
> (2) A person Subject to Service Law, or a civilian subject to service discipline, commits an offence if
> (a) he uses violence against a person who has a duty to apprehend him, or his behaviour towards such a person is threatening; and
> (b) he knows or has reasonable cause to believe that the person has a duty to apprehend him.
>
> (3) For the purposes of this section
> (a) a person's 'behaviour' includes anything said by him;
> (b) 'threatening' behaviour is not limited to behaviour that threatens violence;
> (c) a 'duty' to apprehend a person means such a duty arising under Service law.
>
> (4) A person guilty of an offence under this section is liable to any punishment mentioned in the Table in section 164, but any sentence of imprisonment imposed in respect of the offence must not exceed two years.

Section 67 sets out the main power of arrest provided for in the Act and describes the manner in which an arrest can be carried out. Broadly

100 See AF(P5AFA06)R 09, reg. 3.

speaking, under that section a person may simply be arrested, or (if they are a member of the Armed Forces) be ordered to regard themselves as under arrest (and perhaps to report to a certain person or place).

Section 28 applies to service personnel who disobey an order that requires them to succumb to arrest or use or threaten violence towards a person who has ordered them into arrest in the exercise of a power granted under the Act.

Dangerous Flying

Main Provision: section 33(1) and Schedule 2 AFA 2006

Section 33(1) states:

> (1) A person Subject to Service Law commits an offence if
> (a) he does an act
> (i) when flying or using an aircraft, or
> (ii) in relation to an aircraft or aircraft material,
> that causes or is likely to cause loss of life or injury to any person; and
> (b) either
> (i) he intends to cause loss of life or injury to any person, and there is no lawful excuse for his act; or
> (ii) he is reckless as to whether he causes loss of life or injury to any person.
> (2) A person Subject to Service Law commits an offence if, negligently, he does an act—
> (a) when flying or using an aircraft, or
> (b) in relation to an aircraft or aircraft material,
> that causes or is likely to cause loss of life or injury to any person.
> (3) In this section
> 'act' includes an omission and the reference to the doing of an act is to be read accordingly;
> 'aircraft material' includes
> (a) parts of and accessories for aircraft (whether or not for the time being in aircraft);
> (b) armaments in or for use in aircraft;

(c) any other equipment or instrument in or for use in aircraft;
(d) any equipment for use in connection with the taking-off or landing of aircraft or for detecting the movement of aircraft;
(e) any fuel for the propulsion of aircraft; and
(f) any lubricant for aircraft or for anything within any of paragraphs (a) to (d).

(4) A person guilty of an offence under this section is liable to any punishment mentioned in the Table in section 164, and any sentence of imprisonment imposed in respect of the offence
(a) in the case of an offence under subsection (1), may be for life;
(b) in the case of an offence under subsection (2), must not exceed two years.

This offence under ss (1) is committed by a person subject to Service law when he either intentionally or recklessly does an act

(a) when flying or using an aircraft, or
(b) in relation to an aircraft or aircraft material, that causes or is likely to cause loss of life or injury to any person.

This offence cannot be dealt with summarily. It is specified in Schedule 2. The maximum sentence in the case of conviction under subsection (1) is life imprisonment. The term 'aircraft' is wide enough to include 'aircraft which are not only Her Majesty's aircraft. For example, it would include British military personnel who are controlling foreign military or civilian aircraft or for example, where foreign aircraft use military airfields in the UK.'[101]

Subsection (2) creates the lesser offence where the defendant negligently does an act that causes or is likely to cause loss of life or injury to anyone when the defendant is flying or using an aircraft, or in relation to an aircraft or aircraft material. The maximum punishment for this offence

101 MSL, JSP 830 Version 2.0, at p. 1-7-103.

under subsection (2) is two years imprisonment. It cannot be dealt with summarily (see s. 53(1) AFA 2006 and below, page 285).

The MSL suggest that defences to these charges may exist where the incident was either beyond the accused's control or where they made a mere error of judgement, not amounting to negligence, following a mechanical defect such as an instrument failure.[102] In relation to 'lawful excuse', the section makes a negative averment of a matter peculiarly within the defendant's knowledge, which means the defence bears an evidential burden to show sufficient evidence that raises an issue as to whether they had such an excuse. Once the issue has been raised, the court hearing the charge must be satisfied beyond reasonable doubt that the accused did not have a lawful excuse before they can convict.[103]

Annoyance by Flying

Main Provision: section 35 AFA 2006

Section 35 states:

> (1) A person Subject to Service Law commits an offence if
> (a) he flies an aircraft so as to annoy or be likely to annoy any person;
> (b) he can reasonably avoid flying the aircraft as mentioned in paragraph (a); and
> (c) he intends to fly, or is reckless as to whether he flies, the aircraft so as to annoy any person, or he is negligent.
>
> (2) If a person flies an aircraft in contravention of subsection (1) on the orders of another person who is in command of the aircraft, that other person is for the purposes of this section to be treated as flying the aircraft.
>
> (3) A person guilty of an offence under this section is liable to any punishment mentioned in rows 3 to 12 of the Table in section 164.

This is a lesser offence than dangerous flying.

102 MSL JSP 830 Version 2.0, at p. 1-7-10

103 See the discussion on the burden of proof on p. 298 and at FN 77 thereon.

Where a person flies an aircraft in such a manner on the orders of the aircraft commander, that other person is treated as flying the aircraft. The commander might be a senior pilot (the aircraft 'captain'), a flying instructor or, in the case of a Navy aircraft, the observer (navigation and weapons systems officer).

This offence may be dealt with summarily, but, if dealt with by court martial, is subject to the penalties set out in section 164, save that imprisonment, dismissal with disgrace and dismissal from HM Service may not be imposed.

The MSL helpfully points out that, in connection with the phrase 'reasonably avoid',

> A pilot who flies an aircraft in a manner which causes annoyance or is likely to cause annoyance does not commit an offence under this section unless the manner of their flying could be reasonably avoided. Even if a pilot is authorised to fly at a low level in a particular area, if they were to fly several times over their girlfriend's house just to impress her and in doing so they caused annoyance to others in that area then they would commit an offence under this section. However, if the manner of the flying is due to an exigency such as bad weather within the reasonable scope of the authorisation of the flight no offence would be committed.[104]

Unfitness Through Drugs or Alcohol

Main Provisions: sections 20, 20A, and 93A–93I, and Chapter 3A of AFA 2006; sections 9–11 AFA 2011; AF(ITADT)R 2013; AFA2011(C4)O 2013;[105] AF(ALPSCD)R 2013; section 2 AFA 2016; AGAI 64; AC 64243 Booklet; JSP 835; Appendix N Alcohol and Drugs Testing

104 JSP 830 MSL Version 2.0, at p. 1-7-109.

105 In force, 1 November 2013. AF(ITADT)R 2013 also fully in force by 1 November 2013. AF(ALPSCD)R 2013 in force from 1 November 2013.

It is a trite, though a sad fact of service life that alcohol continues to play a significant part in the leisure activities of service personnel, in spite of efforts by the services to discourage it.

> The nature of the Services' role demands the highest standards from its personnel, who are required to perform exacting duties, which often directly affect the lives of their colleagues. Social drinking may play a part in group bonding; however, it is recognised that those who misuse alcohol or suffer from alcohol dependency are a potential hazard to themselves, their families and their colleagues. Personnel are liable to be called for duty at any time. Therefore, the excessive consumption of alcohol and, in some situations, any alcoholic consumption, may adversely affect their capability to perform their duties safely and accurately. Misbehaviour, unfitness for duty due to alcohol, and drinking and driving offences may be dealt with as offences under the Armed Forces Act 2006 (the Act).[106]

Drinking games and general excessive consumption have in the past led to tragic loss of life.[107] JSP 835 points out at paragraph 4 on page 3-1, that '[d]irect harm from alcohol poisoning can occur when very large quantities are consumed, and death may result from respiratory depression or inhalation of vomit.' Its combination with other circumstances is often at the root of ill discipline and criminal conduct. It is, therefore, not surprising that the three services disapprove of excessive alcohol consumption.

A useful pamphlet used to be available for Army personnel (although its principles were germane to all three services), entitled 'Alcohol and Drugs – the facts' (AC 64243).[108] This pointed out that the 'Army's message is not "anti-alcohol" and does recognise that moderate use of alcohol has

---

106 See JSP 830 MSL Version 2.0 at p. 1-21-2.

107 An example is the tragic death of 21-year-old Pte Murgatroyd, who died in February 2013 during an in-barracks drinking challenge. He consumed 16 shots of assorted spirits within 30 minutes. He died after choking on his own vomit. Another case involved a female soldier in the 160 Military Provost Company, Royal Military Police, Mhairi Burnet, aged 28, who died after binge-drinking gin when she returned from a posting in Germany. She was found dead in her home in Hampshire in December 2018 with a blood alcohol level of 360 milligrammes per 100ml. See also the case of *Ainsworth*, discussed below, on p. 160.

108 Formerly included in AGAI, Volume 2, Chapter 64 (Alcohol Misuse), but omitted from the latest version as at April 2019.

a social function. [However], it must not be at the expense of operational capability. Responsible drinking is what the Army expects and what the Army demands.' Current policy is in AGAI 64 (Substance Misuse) and JSP 835.

Illegal recreational drugs are a more recent phenomenon, which present less obvious signs to the observer, but in some cases hallucinogenic substances have the potential for more profound and dangerous effects on the individual and those with whom they have contact.[109] In each case they may render an individual unfit for duty or any duty they are reasonably expected to perform, by impairing their faculties. While turning up for work in a state of intoxication is generally dealt with as a minor disciplinary matter by the CO, at a more serious extreme the intoxicated service person poses a potential danger to themselves and others when handling a weapon or equipment, or driving a vehicle or flying an aircraft, etc.

There are, essentially, four circumstances where the service person lawfully may be tested for alcohol or drugs:

a. Under the Road Traffic Act, where they drive a vehicle on a public road. This type of testing will be conducted by UK civilian police.
b. In barracks or overseas, where the offence is essentially a breach of SOs and the requirement for testing is made under those orders (e.g. driving while under the influence of alcohol or drugs within barracks).
c. Where the service person is performing a safety-critical duty,[110] as defined below on page 219.
d. Where the requirement is made in connection with a random compulsory drug-testing (CDT) (JSP 835).

Missing from this list is the situation where the service person is charged with unfitness under section 20(1)(a) where the duty that they might reasonably expect to perform is a non-safety-critical one, or under section 20(1)(b), alleging behaviour that is disorderly or likely to bring discredit to Her Majesty's forces. There is no lawful right to test for the presence of alcohol or drugs in those circumstances, that is, to provide incriminating

109 See the example cited in p. 349, FN 35.
110 See also, JSP 835 Alcohol and Substance Misuse and Testing: https://www.gov.uk/government/uploads/system/uploads/attachment_data/file/256943/20131111_msl_ch21.pdf.

evidence against oneself. The test of unfitness is an objective one. However, submitting to a voluntary test would seem unobjectionable.

The relevant statutory provision is as follows:[111]

> **20 Unfitness or misconduct through alcohol or drugs**
>
> (1) A person Subject to Service Law commits an offence if, due to the influence of alcohol or any drug
>
> (a) he is unfit to be entrusted with his duty or any duty which he might reasonably expect to be called upon to perform; or
>
> (b) his behaviour is disorderly or likely to bring discredit to Her Majesty's forces.
>
> (2) Subsection (1) does not apply to the influence of a drug on a person ('A') if
>
> (a) the drug was taken or administered on medical advice and A complied with any directions given as part of that advice;
>
> (b) the drug was taken or administered for a medicinal purpose, and A had no reason to believe that the drug might impair his ability to carry out the duties mentioned in subsection (1)(a) or (as the case may be) result in his behaving in a way mentioned in subsection (1)(b);
>
> (c) the drug was taken on the orders of a superior officer of A; or
>
> (d) the drug was administered to A on the orders of a superior officer of the person administering it.
>
> (3) In this section
>
> (a) 'drug' includes any intoxicant other than alcohol;
>
> (b) a person's 'behaviour' includes anything said by him.
>
> (4) In proceedings for an offence under this section, any paragraph of subsection (2) is to be treated as not having applied in relation to the defendant unless sufficient evidence is adduced to raise an issue as to whether it did.

111 AFA 2006, s. 20.

(5) A person guilty of an offence under this section is liable to any punishment mentioned in the Table in section 164, but any sentence of imprisonment imposed in respect of the offence must not exceed two years.

The proof of alcohol consumption must be combined with one other element. The wording of the section does not provide a specific consumption limit necessary to commit the offence. So, the consumption of any alcohol, in combination with the disorderly, etc. behaviour, might be argued to suffice. But the key wording to the section is 'if, *due to* the influence of alcohol or any drug', which links the consumption of alcohol or drugs to their effect on the behaviour or fitness of the accused, rather than the mere presence of alcohol. This is further clarified in subsection (1A), which specifically refers to impairment in the ability to carry out the duty in question, not mere consumption.[112] So it is the impact (*effect*) on the service person's fitness for duty or behaviour that is critical.

It would be difficult to argue, for example, that a fight or loud altercation in a restaurant between two service persons, A and B, would be sufficient for a charge of drunkenness against A, merely because A had consumed half a pint of beer. It would be better for the prosecutor to choose a different charge in such circumstances. If the two people were in civilian clothes, it may also be difficult for the prosecution to establish that their behaviour was likely to bring discredit to HM forces.

What is likely to bring discredit to Her Majesty's forces is determined as behaviour that falls below that standard that is expected of a service person. It has the potential to be reputationally damaging, especially when the circumstances of the misconduct occurs where the public are likely to witness it and the public is aware that the accused is a member of Her Majesty's forces.[113] The *actus reus* therefore requires more than just the presence of alcohol or drugs. It must combine with one or more of the four circumstances specified in section 20(1).

112 Inserted as a new s. 20(1A) by s. 9 of the AFA 2011, in force from 1 November 2013. See Art. 3, AFA2011(C4)O 2013.

113 MSL Pt 1, at p. 1-7-65.

The MSL explains that 'duty' mean the normal professional duties of the person subject to service law and any other duties incidental to service life. This includes, but is not limited to:

a. Any duty to attend at a particular place or muster/parade.
b. Any duty imposed on an accused because they were acting in some specific capacity (e.g., mess treasurer or officer of the watch).
c. Any duty arising from some order given to the accused or applicable to them.
d. Any duty arising from the accused's rank or rating.[114]

This section places a general duty on the service person to avoid being drunk when it is reasonably anticipated they will be required to report for duty.

What amounts to unfitness for these purposes is gauged through the prism of unfitness for duty or misconduct occasioned through alcohol or drugs. It does not mean that an individual has to have more than a prescribed amount of alcohol in their blood. It is an objective assessment by a superior or the police and would be guided by the common symptoms of glazed eyes, unsteady gait, slurred speech, presence of alcohol on the breath and so on. Evidence of other persons would be admissible on the state of the accused and what they were doing at the time, for example unable to stand, shouting out, being aggressive.

Exceeding Alcohol Limit for Prescribed Safety-Critical Duties

The very nature of service life means that service personnel are responsible for operating dangerous equipment and weapons systems. Those who do so are likely to fall within the provisions relating to safety-critical duties. Here, the consumption of even small amounts of alcohol may risk some impairment in the performance of their safety-critical function.

The test of unfitness through alcohol under section 20 is ostensibly a higher standard than that for offences under section 20A, where the proportion of alcohol in breath, blood or urine must exceed the relevant prescribed limit in respect of safety-critical duties. Merely drinking a glass of wine may be sufficient to create an offence under section 20A, depending

114 JSP 830 MSL Version 2.0, p. 1-7-48.

upon the duty to be undertaken, but it would be highly unlikely to do so in the case of an offence under section 20, which requires unfitness.

Section 20A:

Exceeding alcohol limit for prescribed safety-critical duties

(1) A person subject to service law ('P') commits an offence if the proportion of alcohol in P's breath, blood or urine exceeds the relevant limit at a time when P
   (a) is performing, or purporting to perform, a prescribed duty; or
   (b) might reasonably expect to be called on to perform such a duty.

(2) In subsection (1) 'prescribed duty' means a duty specified, or of a description specified, by regulations;[115] but a duty or description may be specified only if performing that duty (or a duty of that description) with ability impaired by alcohol would result in a risk of
   (a) death;
   (b) serious injury to any person;
   (c) serious damage to property; or
   (d) serious environmental harm.[116]

(3) In this section 'the relevant limit', in relation to a duty specified or of a description specified by regulations, means the limit prescribed by regulations in relation to that duty or duties of that description.

(4) In this section 'regulations' means regulations made by the Defence Council for the purposes of this section.

(5) A person guilty of an offence under this section is liable to any punishment mentioned in the Table in section 164, but any sentence of imprisonment imposed in respect of the offence must not exceed two years.

---

115 AF(ALPSCD)R 2013.

116 AFA2006, s. 20A, inserted by s. 10 AFA 2011.

Safety-Critical Duty
The 'prescribed duty' explained in section 20(2) above corresponds with the definition of 'safety-critical duty' found in section 93I(1), being a duty that the CO of the person mentioned in section 93A(1), 93AA(6)(a) or 93E(1) reasonably believes is such that performing the duty with ability impaired by alcohol or drugs would result in a risk of those harms arising.

The AFA 2011[117] inserted section 20A creating the offence of exceeding the alcohol limit for prescribed safety-critical duties. The offence can only be committed by a person subject to service law performing such a safety-critical duty, provided the proportion of alcohol in their breath, blood or urine exceeds the relevant limit (published in Armed Forces (Alcohol Limits for Prescribed Safety-Critical Duties) Regulations 2013). Different limits apply to each category of function.

Section 20A also applies when a person might reasonably *be expected* to carry out such a duty. A duty may only be prescribed if its performance (while the ability to do so is impaired through alcohol) would carry a risk of death, serious injury, serious damage to property or serious environmental harm. It is important to understand that this section does not cover the use of drugs. The prescribed duty alluded to in subsection (2) is a prescribed safety-critical duty, as contained in the Armed Forces (Alcohol Limits for Prescribed Safety-Critical Duties) Regulations 2013 (AF(ALPSCD)R 2013). Safety-critical duties are listed in Regulations 4 and 5 respectively as any duty:

(a) of an officer of one of Her Majesty's ships afloat to secure the safe conduct of the ship;
(b) of the quartermaster, bosun's mate, helmsman or planesman of one of Her Majesty's ships afloat to execute, or secure the execution of, orders as to the navigation of the ship;
(c) of a member of the crew of any of the following which belongs to, or is being used for the purpose of, any of Her Majesty's forces
    (i) hovercraft;
    (ii) landing craft;
    (iii) rigid-hulled inflatable boats;

117 *Ibid.*, section 10 which inserted a new s. 20A into the 2006 Act, 'Exceeding alcohol limit for prescribed safety-critical duties', in force from 1 November 2013. See Art. 3(b) of AFA2011(C4)O 2013 and the Armed Forces (Interpretation, Translation and Alcohol and Drug Tests) Rules 2013.

(iv) raiding craft;
(v) any other vessel, if it belongs to one of Her Majesty's ships afloat;

(d) as the ship control officer of the watch of one of Her Majesty's submarines afloat;
(e) as a member of the standing sea emergency party of one of Her Majesty's ships afloat;
(f) as a member of the harbour duty watch of one of Her Majesty's ships';
(g) as a member of the strategic weapons system harbour duty watch of one of Her Majesty's submarines;
(h) to supervise a hazardous seamanship evolution;
(i) to supervise a maritime lifting operation;
(j) as a member of a boarding party;
(k) as a member of the flight deck crew of one of Her Majesty's ships;
(l) as a diver;
(m) to supervise a diver;
(n) as a driver or commander of a mechanically propelled vehicle;
(o) to supervise, certify or carry out maintenance on an aircraft;
(p) to authorise a military flight.

and

(a) of a pilot of an aircraft during flight or when conducting a ground run;
(b) of any other member of the crew of an aircraft during flight;
(c) of a person on board an aircraft during flight to give or supervise training or to administer a test;
(d) as an air traffic controller;
(e) of an operator of a remotely-piloted aerial system in flight;
(f) of any person in relation to the handling and use of a firearm when he has in his possession the firearm and ammunition capable of being discharged from the firearm;
(g) of a person doing any of the following in relation to the operation of a depth-charge launcher, firearm, mortar, rocket, or torpedo—

(i) handling missiles, fuses, charges or propellants;
(ii) firing or giving orders to fire;
(iii) calculating or setting a target area;
(iv) giving instructions about the location and bearing of the target;
(v) deciding or setting the explosive effect;
(vi) ensuring that the missile or explosive functions correctly in the target area;

(h) to supervise a person carrying out a duty within paragraph (f) or (g) ;

(i) of a person handling, or supervising the handling, of explosives.

The relevant duties and limits for breath, blood and urine are prescribed in the Armed Forces (Alcohol Limits for Prescribed Safety-Critical Duties) Regulations 2013.[118] Regulation 3 specifies the following:

(1) In relation to a duty prescribed in regulation 4, the relevant limit is
  (a) in the case of breath, 35 microgrammes of alcohol in 100 millilitres;
  (b) in the case of blood, 80 milligrammes of alcohol in 100 millilitres; and
  (c) in the case of urine, 107 milligrammes of alcohol in 100 millilitres.

(2) In respect of a duty prescribed in regulation 5, the relevant limit is—
  (a) in the case of breath, 9 microgrammes of alcohol in 100 millilitres;
  (b) in the case of blood, 20 milligrammes of alcohol in 100 millilitres; and
  (c) in the case of urine, 27 milligrammes of alcohol in 100 millilitres.

118 These regulations came into force on 1 November 2013.

A person guilty of an offence under section 20A will be liable to any punishment mentioned in the Table in section 164, but any sentence of imprisonment imposed in respect of the offence must not exceed two years.[119]

Joint Service Publication (JSP) 835 makes reference to non-prescribed safety-critical duties, explaining that the list of prescribed duties is not intended to take account of each and every safety-critical duty that a person may be required to perform and should not be taken as an exhaustive list:

> There may be occasions when a person is called upon to complete a duty that is not prescribed, yet still fulfils the definition of a safety-critical duty. In such situations the CO has flexibility to require preliminary alcohol and drugs testing if a duty is safety-critical in nature and he has reasonable cause to believe that a person subject to Service law is committing an offence under section 20(1)(a) of the Armed Forces Act 2006 (or has committed such an offence and still has alcohol or a drug in the body or is still under the influence of a drug). Before determining that a duty is safety-critical, a CO should be satisfied that the duty meets the definition of safety-critical. A CO must avoid placing unnecessary liability for testing on any other person who may be required to perform any related or ancillary duty merely because it is convenient to do so (e.g. all members of a work party or team) but which is not genuinely safety-critical in nature. When personnel are required to complete an activity that is not within their normal daily routine, or where a CO considers that members of his command may need to carry out tasks which in the circumstances he thinks are safety-critical, the CO should provide advance warning that their particular duty on a given day or for a given period is safety critical. This will help guide personnel as to the amount of alcohol, if any, they may consume ahead of the duty, and ensure all relevant personnel understand their potential liability for testing.[120]

---

119 It is only the power to make regulations under section 20A of the 2006 Act that is in force as from 8 March 2012. See, The Armed Forces Act 2011 (Commencement No. 1, Transitional and Transitory Provisions) Order 2012.

120 JSP 835, p. 6-3, §8.

Consequently, being unfit under section 20(1)(a) through the consumption of drugs may also engage the safety-critical testing procedures in these circumstances where the duty is one that the offender *might reasonably expect* to be called upon to perform.

The Testing Regime

The highest standards are expected of those who operate the complex and dangerous equipment used by the services. Just as in civilian life, those who, for example, fly aircraft or operate dangerous machinery are subject to drug testing as part of the safety regime for an airline or engineering company, the military has a system of stringent checks for those who operate equipment of a safety-critical nature. The CO has power to order a preliminary test in the form of a breath test procedure (section 93B), preliminary impairment test (section 93C) and impairment test, and a preliminary drug test (section 93D). Appendix M of this book should be consulted for the applicable legislation and procedure. It is not proposed to duplicate the provisions here, merely to provide an overview.

The CO's power to require preliminary tests arises in five situations in relation to safety-critical duties under section 20A. Section 93A(1) provides for the CO to require preliminary tests[121] where they have reasonable cause to believe that a person either is committing a safety-critical offence or has committed a such an offence and still has alcohol or a drug in their body or is still under the influence of a drug. 'Relevant offence' means an offence under section 20A or an offence of unfitness under section 20(1)(a) in respect of a safety-critical duty. A safety-critical duty is defined by section 93I, namely, a duty that the CO reasonably believes is such that performing the duty with ability impaired by alcohol or drugs would result in a risk of:

a. death
b. serious injury to any person
c. serious damage to property, or
d. serious environmental harm.

The second situation, set out in section 93A(3), applies to a civilian subject to service discipline, where the CO has reasonable cause to believe that

121 AFA 2011, s. 11, inserts a new s. 93A into the AFA 2006. See, also, Appendix M Alcohol and Drugs Testing.

person is committing, or has committed, an offence under section 42 for which the corresponding offence under the law of England and Wales is an offence under sections 78, 79, 92 or 93 of the Railways and Transport Safety Act 2003 (maritime and aviation offences) and still has alcohol or a drug in their body or is still under the influence of a drug.[122]

The third situation, set out in section 93A(3A), covers the circumstances where the CO of a person subject to service law or a person who is a civilian subject to service discipline has reasonable cause to believe there has been an accident involving an aircraft or a ship and at the time of the accident the person was carrying out an aviation function or a marine function.

Section 93A(3B) sets out the fourth situation that applies where the CO has reasonable cause to believe that:

(a) there has been an accident involving an aircraft or a ship;
(b) before the accident, the person carried out an aviation function in relation to the aircraft or a marine function in relation to the ship; and
(c) it is possible that the carrying out of the function by the person may have caused or contributed to
  (i) the occurrence of the accident;
  (ii) any death, injury to a person, damage to property or environmental harm resulting from the accident; or
  (iii) any risk of death or of such injury, damage or harm created by the accident.

Finally, the fifth situation, set out in section 93A(3C), deals with the circumstances where the CO of a person subject to service law or of a person who is a civilian subject to service discipline has reasonable cause to believe that there has been an accident which resulted in or created a risk of death, or serious injury to any person, or serious damage to property, or serious environmental harm, and the person was carrying out a safety-critical function at the time or carried out a safety-critical function before the accident where the safety-critical function may have caused, or contributed to, the accident and resulting consequences.

---

122 AFA 2006, s. 93A(3) and (4) inserted by s. 11 AFA 2011, in force 1 November 2013 (AFA2011(C4)O 2013).

Specimens for analysis may be required during the course of an investigation by service police under section 93E when investigating an offence under section 20A, or an offence under section 20(1)(a) in respect of a safety-critical duty (as defined by section 93I), or an offence under section 42 for which the corresponding offence under the law of England and Wales is an offence under section 78, 79, 92 or 93 of the Railways and Transport Safety Act 2003.

In a situation described in section 93A(1) or (3), a preliminary breath test is a procedure administered by service police through which the affected person provides a breath specimen and the specimen is used for the purpose of obtaining, by means of an approved device, an indication whether the proportion of alcohol in the person's breath or blood is likely to be such as is necessary for the commission of the suspected offence. There are certain circumstances when a preliminary breath test is a procedure administered by a service police officer and in the locations specified in section 93B.

Other tests may also be undertaken by service police, such as a preliminary impairment test under section 93C and a preliminary drug test under section 93D.

Prescribed Proportions of Alcohol for Replacement of Specimens of Breath

If the specimen with the lower proportion of alcohol contains no more than a prescribed proportion of alcohol, the person who provided it may claim that it should be replaced by such a specimen of blood or urine as may be required under section 93E. The prescribed proportion of alcohol is determined by regulation 6 AF(ALPSCD)R 2013 as follows:

- (a) in the case of an offence under section 20(1)(a) of the Act in respect of a safety-critical duty, 50 microgrammes of alcohol in 100 millilitres of breath;
- (b) in the case of an offence under section 20A(1) of the Act (performing, or purporting to perform, or might reasonably expect to be called on to perform a prescribed duty),
  - (i) in respect of a duty prescribed in regulation 4, 50 microgrammes of alcohol in 100 millilitres of breath;

(ii) in respect of a duty prescribed in regulation 5, 15 microgrammes of alcohol in 100 millilitres of breath;

(c) in the case of an offence under section 42 of the Act as respects which the corresponding offence under the law of England and Wales is an offence under section 78 or 79 of the Railways and Transport Safety Act 2003, 50 microgrammes of alcohol in 100 millilitres of breath;

(d) subject to regulation 6(e), in the case of an offence under section 42 of the Act as respects which the corresponding offence under the law of England and Wales is an offence under section 92 or 93 of the Railways and Transport Safety Act 2003, 15 microgrammes of alcohol in 100 millilitres of breath; and

(e) where the offence mentioned in regulation 6(d) is in relation to the aviation function specified in section 94(1)(h) of the Railways and Transport Safety Act 2003, 50 microgrammes of alcohol in 100 millilitres of breath.

Under section 93A(6),[123] a person commits an offence if without reasonable excuse they fail to comply with a requirement to provide a sample for analysis where the CO suspects they are committing or have committed one of the offences specified in section 93A. If found guilty of an offence under this section, they are liable to any punishment mentioned in the Table in section 164, but any sentence of imprisonment or service detention imposed in respect of the offence must not exceed two years (s. 93A(7)).

Misuse of Drugs – Random Drug-Testing

Other cases of deliberate misuse of drugs, especially the illicit sale of drugs, are likely to proceed to trial by court martial. Under Schedule 1 AFA 2006, COs may only deal summarily with simple possession of controlled drugs under section 5(2) of the Misuse of Drugs Act 1971.

In addition to those occasions where the disciplinary authorities have reasonable cause to believe that a person is committing a service offence, the service authorities occasionally carry out random tests upon service personnel. The authorisation is in section 305 AFA, which states:

123 See below, Appendix M.

> A drug testing officer may, in order for it to be ascertained whether or to what extent a person subject to Service law has or has had drugs in his body, require the person to provide a sample of his urine for analysis.

These tests are compulsory and without notice. The powers of a drug-testing officer are in the Drug Testing Regulations 2009; hence, the word 'require' in the Act. The compulsory drug-testing (CDT) team will turn up at a unit and everyone present on camp must attend for testing. No exceptions are made for rank. The idea is to catch unauthorised drug use within a unit. As section 308(3) of AFA 2006 makes clear, the results of compulsory tests are not admissible in evidence against any person in proceedings in respect of a service offence.

Refusal

Failure to provide a sample for random drug-testing is an offence contrary to section 305(3) AFA 2006, and is subject to a sentence upon conviction by court martial of any punishment mentioned in the Table in section 164, but any sentence of imprisonment or service detention imposed in respect of the offence must not exceed 51 weeks. After conviction, it is quite likely that administrative discharge of the individual will be sought. The MSL points out that '[a]n offence of failing to provide a sample is also capable of being dealt with summarily, but the powers of punishment of the CO do not include dismissal. Administrative discharge will need to be considered where an offence is dealt with summarily.'[124]

Use of Specimens

The Armed Forces (Interpretation, Translation and Alcohol and Drug Tests) Rules 2013 (AF(ITADT)R 2013) inserted a new Chapter 7 into the AF(CM) R 2009, which appears after rule 100. These new rules provide for how specimens in proceedings for offences relating to alcohol and drugs are used. The rules apply only to offences under section 20(1)(a) of the 2006 Act (unfitness for duty through alcohol or drugs), section 20A (exceeding alcohol limit for prescribed safety-critical duties), or an offence under

---

124 Under QRRN 3626, Army QR Chapter 9 section 3, and RAF QR 1028. See Chapter 5 of JSP 835, at p. 6-7.

section 42 (criminal conduct relating to an offence under section 78, 79, 92 or 93 of the Railways and Transport Safety Act 2003(1)). Similar provisions apply to the SAC,[125] although the trigger offences are different for trial by Service Civilian Court (SCC).[126]

Fraud/False Records

Main Provisions: section 1, Forgery & Counterfeiting Act 1981; sections 1 and 11 Fraud Act 2006; section 18, Schedule 1 and Schedule 2 AFA 2006

There are specific provisions in the Act to deal with entries in official documents. The offence under section 18 is a service offence of fraud in relation to 'official documents'. The maximum punishment that can be imposed must not exceed two years' imprisonment. The section also embraces elements of the offences of false accounting under the Theft Act 1968 and forgery under the Forgery & Counterfeiting Act 1981 Act. In essence, the offence requires knowledge of the official nature of the document and an intent to make or alter it so it tells a lie, or to adopt it in that knowledge. Failure, with intent to deceive, to make a record that the accused is under a duty to make falls under subsection (4):

Making false records, etc., section 18:[127]

(1) A person Subject to Service Law commits an offence if
  (a) he **makes** an official record, knowing that it is false in a material respect; and
  (b) he knows or has reasonable cause to believe that the record is official.

125 The AF(SAC)R 2009 have a new 'Chapter 6' inserted by reg. 8 AF(ATADT)R 2013 in respect of the use of specimens in relation to offences under ss 20(1)(a) (unfitness for duty through alcohol or drugs) and 20A (exceeding alcohol limit for prescribed safety-critical duties) of the AFA 2006.

126 AFA(ITADT)R 2013, rule 22 inserts 'Chapter 7' after rule 84 of the AF(SCC)R 2009. These are aimed at offences under the Railways and Transport Safety Act 2003 (shipping and aviation staff: offences relating to alcohol and drugs) and general criminal offences under s. 42.

127 Formerly s. 62 of the 1955 Acts and s. 35 Naval Discipline Act 1957. Emphasis added.

(2) A person who **adopts** as his own a record made by another person is for the purposes of subsection (1) to be treated, as well as that other person, as making the record.

(3) A person **Subject to Service Law** commits an offence if
  (a) with *intent to deceive*, he **tampers** with or **suppresses** an official document; and
  (b) he **knows or has reasonable cause to believe** that the document is official.

(4) A person **Subject to Service Law** commits an offence if
  (a) with **intent to deceive**, he **fails to make** a record which he is under a duty to make; and
  (b) he knows or has reasonable cause to believe that the record would, if made, be official.

(5) For the purposes of this section
  (a) '**record**' means a document or an entry in a document;
  (b) '**document**' means anything in which information is recorded;
  (c) a record or document is **official** if it is or is likely to be made use of, in connection with the performance of his functions as such, by a person who holds office under the Crown or is in the service of the Crown.

(6) A person guilty of an offence under this section is liable to any punishment mentioned in the Table in section 164, but any sentence of imprisonment imposed in respect of the offence must not exceed two years.

This offence may only be charged against a person subject to service law. Service civilians would have to be charged under other provisions. Whereas the substantive criminal law offences of fraud or forgery would be available for the more serious cases, where a sentence in excess of two years is thought appropriate,[128] the services have the alternative of charging what would otherwise be regarded as such conduct under section 18. Persons subject to service law are often required to make financial and other returns

---

128 As in the case of *Knock* [2014] EWCA Crim 1986, where the appellant, a member of the Territorial Army who held the rank of Lieutenant Colonel, pleaded guilty at court martial to five charges of forgery and five charges of using a false instrument, under ss 1 and 3 of the Forgery and Counterfeiting Act 1981 respectively.

and create or make entries in official documents on a wide range of matters, but especially when claiming allowances or recovery of expenses.

The section importantly refers to the situation where the accused tampers with an official document already in existence, with an intent to deceive, so that it tells a lie about itself. Suppressing the document, in the sense of preventing other people from learning of it, would be a deliberate failure to produce the document for the official purpose for which it was intended, giving the impression that it did not exist. This might be a record of a military operation, which, because of its incriminating nature, was not disclosed to the military authorities.[129]

Mens Rea

The requirement for *mens rea* covers doing an act intentionally ('knows', 'knowing', 'intent') with a second, objective element – a 'reasonable cause to believe'. Both elements must be present for the charge to be proved. In approaching the matter of what constitutes a reasonable cause to believe, the MSL states, at 1-7-56, that

Belief should be given its normal dictionary meaning, and what is a reasonable cause to believe is for the officer hearing the charge to decide. The officer hearing the charge can imply from the surrounding circumstances of the case, what would have been within the knowledge of the accused. In doing so they would wish to consider the Service experience of the accused, relevant training and time in post.

No authority is cited for the proposition, but it is helpful insofar as it explains how the decider of fact should approach the objective element of 'reasonable belief', but it does not explain if it is necessary to consider the defence from 'the facts as the defendant honestly believed them to be, whether or not he was mistaken, but, if he made a mistake of fact, [whether] he can rely on that fact only if the mistake was a reasonable one for him to have made'.[130] That would seem to be the proper approach.

129 If such conduct were committed during the currency of a police investigation it would be a more serious matter of doing an act tending and intending to pervert the course of public justice.

130 See *Ashley (FC) and another v Chief Constable of Sussex Police* [2008] UKHL 25: https://publications.parliament.uk/pa/ld200708/ldjudgmt/jd080423/ashley.pdf and s. 76 CJIA (Self Defence).

Definitions

Official Document and Official Record

There is a distinction between an official record and an official document. Both are defined in section 18(5). A 'record' means a document or an entry in a document; 'document' means anything in which information is recorded. A record or document is official if it is or is likely to be made use of in connection with the performance of the functions of such, by a person who holds office under the Crown or is in the service of the Crown. Making false entries on Joint Personnel Administration (JPA) would fall foul of section 18 as such a record would be defined as a document, being 'anything in which information is recorded'.

Claims forms for allowances or repayment of expenses are official documents. Thus, offences have been charged under this heading where allowances have been wrongly claimed. In certain circumstances, a service spouse is entitled to claim Boarding School Allowance (BSA) for children sent away to school. Claims for BSA are particularly fraught with potential difficulty as sometimes families split their residence between a married quarter on a base and a home where, for example, the non-service spouse works during the week. Similarly, there can be difficulties where the parents are divorced or in the process of divorcing or are separated and children spend time with one or other parent during holidays and exeats. Anyone conversant with military regulations for allowances and charges will attest that they are not always the easiest or clearest of documents to read.[131] The issue of dishonesty, therefore, tends to feature prominently in these cases.

Official Purposes

The section concerns improper conduct in relation to documents to be used for official purposes. Use for official purposes includes use by the Armed Forces themselves or by Crown servants such as Ministry of Defence officials. 'Documents' are not just paper documents, but any record of information in any form, including computer records [EN §75]. In today's modern armed forces, personnel are able to access what is known as the Joint Personnel Administration System (JPA or JPAS), which replaced the need for paper transactions in many areas. Among the information that

131 JSP 752 is the policy document for expenses and allowances for regular and reserve UK armed forces personnel. The guidance runs to over 700 pages.

is accessible is that relating to pay records, allowances and leave records. There is no requirement that the falsification or tampering should be undertaken purely for financial gain. Changing dates on a leave pass, an official document within section (5) authorised for specified dates to show that the end date was longer than that originally authorised, would be an offence under section 18(3). Moreover, the offence under ss (1) is complete even if the document is never actually used.

An example of the utility of section 18 is the case where a service medic at the medical centre makes an entry in the medical notes of a patient to falsely indicate that they had seen him or given specific instructions to him, which they had not. This is not fraud, as there is no financial gain. Section 5 of the Fraud Act defines 'gain' and 'loss' as:

> (a) extend[ing] only to gain or loss in money or other property; (b) includ[ing] any such gain or loss whether temporary or permanent; (3) 'Gain' includes a gain by keeping what one has, as well as a gain by getting what one does not have. (4) 'Loss' includes a loss by not getting what one might get, as well as a loss by parting with what one has.

The medical record is an official document as it is or is likely to be made use of in connection with the performance of a person's functions as such, by a person who holds office under the Crown or is in the service of the Crown and, being a person subject to service law. The medic commits an offence if they (a) make an official record, knowing that it is false in a material respect; and (b) know or have reasonable cause to believe that the record is official.

Overseas, where service dependants are often employed within units, and those who are medically qualified may work within the medical facilities provided, any such offence committed by a dependant would have to be charged under the Forgery & Counterfeiting Act and be tried by either the SCC or court martial. The requirements of forgery (makes a false instrument, with the intention that they or another shall use it to induce somebody to accept it as genuine, and by reason of so accepting it to do or not to do some act to his own or any other person's prejudice) would appear to be satisfied if a medical record were changed. An 'instrument' includes any document, whether of a formal or informal character; and any disk, tape, soundtrack or other device on or in which information is recorded or stored by mechanical, electronic or other means. The medical documents would fall within this definition.

Case Law

In the case of *R v I*,[132] a case charged under provisions similar to section 18, namely section 62 of the Army Act 1955 (making in an official document an entry which to his knowledge was false in a material particular), the D was an officer. On promotion the D was required to complete a mobility certificate to justify payment of an educational allowance towards the cost of private education for one of his children, confirming that the child resided with the claimant when not at school. Although his son had never lived with his father following the D's separation from his wife, D completed the form stating that his son had accompanied him to his new posting, identifying his 'family address' adding the words 'I have legal responsibility following separation for [this son] only'. This falsity in fact was unnecessary because had the appellant made enquiries, he would have discovered that as his son was in the last two years of his education, he was eligible for a BSA, whether or not he resided with the defendant. This was the essence of the offence under section 62. He also faced fraud charges relating to submission of false pay statements by accessing his pay slips electronically and altering the amount by a total of £6,500 over a six-month period, printing off the documents and submitting them as originals to mislead his wife who was making a financial claim against him. He subsequently swore an affidavit to confirm the position. For the fraud offence he was sentenced to a term of six months' imprisonment, the operation of which was suspended for two years. For the military offence, he was dismissed from Her Majesty's service. He appealed against sentence, limited to the order of dismissal. The CMAC held, referring to the Guidance on Sentencing in the Court Martial, that the financial loss flowed from the loss of his career. In order to succeed in an appeal against sentence it is not sufficient to cherry-pick the single offence to which the sentence has been attached even though the Act requires a

132 *R v I* [2010] EWCA Crim 1645, concerned with an offence under s. 62(1)(b) of the Army Act 1955.

sentence for each offence.[133] Rather it is necessary to look at the overall picture and determine whether the sentence for the offending was, in the round, wrong in principle or manifestly excessive. The CMAC concluded that he was 'fortunate to escape an immediate custodial sentence'.

By contrast, in the case of *Gray*,[134] the SPA took a more robust approach and charged a naval officer with nine counts of the substantive offence of fraud for misrepresenting that he was entitled to a Service Education Allowance for the fees of his children at boarding school.

> 'The children were at a local fee-paying school. Because they were at that time undoubtedly part of a family unit which would be affected by the postings by the service of their father, there is no doubt that at that point the defendant was entitled to and had been claiming what is called "continuity of education allowance" (CEA). As its name suggests, the rationale of that allowance is to provide continuity of education for children whose education would otherwise be disrupted because one or other of their parents, as a service person, is likely to be posted successively here and there. When they came to separate, the defendant and his wife arranged for the children to become boarders. That was no doubt in a sensible effort from a parental point of view to maintain some consistency in their education and perhaps to protect them, so far as one could, from the inevitable disruption of the separation of their parents. However, the question which inevitably arose, and was on any view seen to arise by the defendant, was whether the Service allowance which hitherto had been payable remained payable or not.'

The sole issue at the trial was whether the defendant had been dishonest. What was abundantly clear was that the principal thing that affected the Court Martial Board was the Child Support Agency order for maintenance. The CMAC observed that whether or not financial negotiations had finally broken down by the time of the Child Support Agency order, the fact that it was made could only indicate that the parents were at loggerheads. He was convicted and sentenced on each count to be dismissed from the service

133 S. 255.
134 *R v Gray* [2012] EWCA 252 Crim.

and sentenced to six months' immediate imprisonment. He was made to pay a compensation order in the sum of nearly £53,000. The conviction and sentence were confirmed by the CMAC. The loss of a military career is an anticipated consequence of a conviction for fraudulent conduct.[135]

Theft

Main Provisions: section 1 Theft Act 1968; section 3 Theft Act 1978; Schedule 1 and Schedule 2 AFA 2006

There is no military equivalent offence in the AFA 2006 for theft, although it has been known in the past for offenders to be charged with being in 'improper possession' of items (see below). While theft offences involving appropriation from civilian shops and stores is corrosive to service reputation, intra-unit theft is corrosive to trust. It is a disappointing fact of life that cases of theft from fellow service persons are not unknown in the services. Trust is essential to any cohesive force and nothing undermines unit cohesion so fundamentally than a breach of that trust. Theft from fellow unit members is, consequently, particularly serious. Rightly, the services take a very dim view of this offence. There is a discretion to deal summarily with minor instances of the offence, with the more serious cases proceeding to trial by court martial, as in the case of theft from colleagues.

The maximum punishment under section 7 of the Theft Act on trial by court martial is imprisonment not exceeding seven years.

Dishonesty

The decision in *Ghosh*[136] has been superseded by the test for dishonesty set out by the Supreme Court in *Ivey v Genting Casinos*[137] as the correct test to be applied in determining the issue of dishonesty:

---

135 See: *R v Wetherall* [2011] EWCA Crim 990; *R v Downing* [2010] EWCA Crim 739; *R v Blyth (Stephen)* GCM, Gütersloh, 12 Oct. 2001 (unreported); *R v Coull (Dawn Marie)* [2003] EWCA Crim 967.

136 *R v Ghosh* [1982] EWCA Crim 2, cited in *Dodman*, at para. 20.

137 *Ivey v Genting Casinos* [2018] AC 391, at para. 74. The CPS report that the Divisional Court adopted this as the correct approach in *DPP v Patterson* [2017] EWHC 2820 (Admin), Sir Brian Leveson observing that it was 'difficult to imagine the Court of Appeal preferring *Ghosh to Ivey* in the future.' In *Ivey*, the UKSC adopted with approval the test of dishonesty set out by Lord Nicholls in *Royal Brunei Airlines Sdn Bhd v Tan* [1995] 2 AC 378, and by Lord Hoffmann in

> These several considerations provide convincing grounds for holding that the second leg of the test propounded in *Ghosh* does not correctly represent the law and that directions based upon it ought no longer to be given. The test of dishonesty is [Although a dishonest state of mind is a subjective mental state, the standard by which the law determines whether it is dishonest is objective. If by ordinary standards a defendant's mental state would be characterised as dishonest, it is irrelevant that the defendant judges by different standards]. When dishonesty is in question the fact-finding tribunal must first ascertain (subjectively) the actual state of the individual's knowledge or belief as to the facts. The reasonableness or otherwise of his belief is a matter of evidence (often in practice determinative) going to whether he held the belief, but it is not an additional requirement that his belief must be reasonable; the question is whether it is genuinely held. When once his actual state of mind as to knowledge or belief as to facts is established, the question whether his conduct was honest or dishonest is to be determined by the fact-finder by applying the (objective) standards of ordinary decent people. There is no requirement that the defendant must appreciate that what he has done is, by those standards, dishonest.

Since the decision in *Ivey v Genting Casinos*, the Court of Appeal, in *Raychaudhuri v GMC*,[138] took a more nuanced approach. The Court held that the Medical Tribunal in Dr Raychaudhuri's case had made a defensible and legitimate assessment of the facts, which provided a legitimate basis on which the Tribunal was entitled to find that the appellant had not been dishonest. 'That is so whether the relevant approach to dishonesty is that in Ghosh or, *as has now been clarified*, that in Ivey.'[139] In agreeing with Sales LJ, Bean LJ said that the tribunal in *Raychaudhuri* 'were well placed to make an evaluative judgment of the nuances of how the various individuals had interacted and that judgment should have been accorded great weight...'. Nevertheless, it was expressly accepted by Sales LJ that

International Ltd v Eurotrust International Ltd [2005] UKPC 37. The test now clearly established was explained in *Barlow Clowes* by Lord Hoffmann, cited at para. 62 of *Ivey*.

138 [2017] EWHC 2116, Court of Appeal.

139 *Ibid.*, para. 64, per Sales LJ. Emphasis added.

Ivey had 'clarified' *Ghosh*. However, this may be a decision on its own special facts as the case was one relating to medical professional conduct – there appeared to be a great deal of sympathy from the Court to the plight of Dr Raychaudhuri.[140] Or, it could open the door to a more contextualised evaluation.

The Court of Appeal, in *Barton and Booth v The Queen*,[141] followed the UK Supreme Court (UKSC) *obiter* remarks about dishonesty in *Ivey*, highlighting that Lord Hughes explained there are serious problems with the second leg of the test in *Ghosh* (whether the defendant himself must have realised that what he was doing was by those standards dishonest) and that it involved a departure from pre-1968 law.

> The undoubted reality is that in *Ivey* the Supreme Court altered the established common law approach to precedent in the criminal courts by stating that the test for dishonesty they identified, albeit strictly contained in obiter dicta, should be followed in preference to an otherwise binding authority of the Court of Appeal ... In the result, the test for dishonesty in all criminal cases is that established in *Ivey*.

Although it is normal prosecution policy to charge the substantive offence where the ingredients of the offence are present, as stated above there are occasions when an offence of 'improper possession' is substituted for an offence of theft and charged under section 19 AFA. Where it is, for example, difficult to prove an element such as the intention to permanently deprive, a charge of improper possession would be appropriate. Clearly, in such a case a substantive charge would fail, but it might still be a disciplinary matter to be in possession of service property – or an item belonging to a colleague – without authority, even absent an intention to permanently deprive.

In *Armstrong*, instead of being charged with four substantive offences under the ordinary criminal law, in respect of weapons, ammunition, drugs and classified material found in his possession, Armstrong was charged with four separate charges in respect of the four separate matters under section 69 of the Army Act 1955 (now s. 19 of the AFA 2006), which carried a maximum sentence of two years imprisonment. This caused some concern

140 See, in particular, *ibid.* at paras 65–68.
141 [2020] EWCA Crim 575.

to the presiding judge advocate, who was of the view that substantive criminal offences should have been laid and who voiced discontent with the charges that the Crown sought to prefer. The CMAC indicated, at paragraphs 39–41:

> Outside ... well-known and clearly circumscribed procedures [of the *Farquharson* Guidelines (published in 1986)], the decision of a prosecutor is final. Given the prosecutor's independent position within the executive branch of the State and the position of the judiciary as a separate branch of the State, as the authorities explain, there are good constitutional reasons why a judge must abide the decision of the prosecutor once the methods described have been exhausted.
>
> In the system of military justice the position is in fact similar although this may not be as widely known ... Her Majesty's Attorney General superintends the service prosecuting authorities. This is the manifestation of the general constitutional principle that the independence of prosecutors as part of the executive branch of the state is subject in the case of all prosecutors to accountability to Parliament through the Attorney General.
>
> It is therefore clear that if in the military courts a judge does not approve of a course that a military prosecutor intends to take, the proper course for the judge is to ask for the matter to be referred either to the Director of Service Prosecutions, the equivalent to the Director of Public Prosecutions, or to the Attorney General as may be appropriate.

Although this may not accord with the understood charging policy of using the substantive offence wherever it is made out, as it benefits a defendant, they are unlikely to complain of a section 19 charge instead. It is where such a charge is used in a victim crime dealt with by the CO that a problem may arise, as compensation orders are only available as an award at summary hearing for an offence which has caused any personal injury, loss or damage, provided they do not exceed £1,000 in aggregate.[142]

142 AFA 2006, s. 137 and, in respect of trial by court martial, s. 175(1).

Penalty
Offences of theft, whether from a fellow service person or from the service (e.g., through misuse of fuel cards, theft of monies, theft of equipment) may very well result in a sentence by court martial of dismissal from the service, with the consequent financial loss that flows from it. An offence under section 3 of the Theft Act 1978 of making off without payment, where the payment required or expected did not exceed £100, may be dealt with summarily by the CO. It is listed in Schedule 1. Where the offence is dealt with by court martial, on conviction it is subject to a maximum term of imprisonment of seven years under section 7 of the Theft Act.

In the case of *R v Liddell*,[143] a training NCO was found guilty of the theft of £100 from a recruit. He was sentenced to eight months' detention and reduced to the ranks. On appeal, he argued that the sentence was wrong in principle, as non-custodial alternatives were available and, in any event, the sentence was too long. It was held that such offences attract imprisonment in civilian life, which most likely means the loss of the civilian's job. Had imprisonment been awarded here, the Appellant would have been dismissed the Army. A merciful course had been taken. One of the benefits of detention was the training that the individual received. In this case it had clearly borne fruit as the appellant had been promoted immediately upon release to LCpl and was back up to the rank of Cpl within nine months. The sentence of detention was not wrong in principle, neither was it manifestly excessive.

Case Illustrations
Conviction for an offence of theft, fraud/making a false entry can have serious consequences for the defendant's career and future pension. The following selected illustrations demonstrate how the courts have approached sentencing when confronted with the prospect of a loss of significant sums by way of pension and gratuity.

Cusack:[144] A 38-year-old Warrant Officer Class 1 with a hitherto impeccable military career financed his gambling addiction by acquiring a credit card using the identity details of a military senior non-commissioned officer (SNCO) colleague, having stolen mail from his home. He wrote

143 CMAC, 3 Oct. 2003.
144 [2010] EWCA Crim 2663.

cheques for several thousand pounds to pay gambling debts. He had completed 19 of the 22 years. If he had stayed on until he had completed 22 years, when he would have been aged 41, he would have received some £13,500 a year from age 41 to 60. Taking into account his loss of gratuity, the financial loss consequent upon dismissal was close to £300,000. The sentencing court considered his behaviour as disgraceful and it seriously jeopardised the financial integrity of his colleagues and friends and, in particular, of his military SNCO victim. It was a seriously aggravating factor that he had stolen from his colleague. In the civil world, theft of this kind would almost always result, said the court, in instant dismissal. Dishonesty of the kind in which this appellant engaged had the potential to break the bond of trust that was an essential part of service life and also had the potential to disrupt operational effectiveness. His appeal was dismissed.

Earle (Stephen):[145] The appellant was a Sergeant in the Fleet Forward Support. In the early part of 2008 he was sent on a course to St Athan. 'He was allowed to use an MoD hire car for part of the journey. To refill the car he was given an MoD credit card.' On 16 occasions he used it to refuel his own car and a private car he had hired, and twice also claimed motor mileage allowance. The total amount he obtained was £693. He was charged with fraud. The Judge Advocate General's Sentencing Guide provides that theft from an employer and breach of trust are serious aggravating factors and that even a minor theft is often treated more severely within the service context. Rank is another relevant factor. It was held that the appellant had served 18 years out his 22-year engagement, so the sentence imposed severe financial consequences in respect of his pension reduction and loss of earnings. A sentence of custody was wrong in principle. He had served his sentence by the date of appeal. The court substituted a reduction in rank to the rank of Corporal. That in itself carried quite serious financial consequences from the reduced salary.

Owusu:[146] This was an appeal against a sentence of 90 days' detention and dismissal for using a friend's bank card to obtain €400 from bank. The appellant gave a different explanation on appeal to that given at trial. The CO said he would have given a different recommendation had he known of the later mitigation. The court took note of the considerations

145 [2009] No 200904059 D5.
146 [2010] EWCA Crim 3098, [2010] EWCA Crim 2816.

for dismissal from the service and the views of the CO. At his first appeal it was alleged the court martial did not have before it the full story. The case was adjourned to the second hearing for the defendant's current CO to express his view on the question of his retention. At the resumed hearing the CMAC had the views of both COs, and their shared conclusion was that retention in the Army was not indicated in the public interest. The court held that this case did not differ significantly from the kind of barrack room theft that is explained in the Sentence Guidance for Court Martial. His appeal was denied. The court was unable to see anything wrong in principle with the imposition of what is, as often as not, the normal sentence for a case of this type, at least in the absence of positive assertion from the defendant's service line of command that his retention is indicated in the public interest.

Peters:[147] An RAF policeman who was serving with NATO pleaded guilty to five offences of obtaining property by deception, three offences of obtaining services by deception and three offences of theft (he stole a number of items and took them back to the service stores, saying he had bought them but had no receipts. He received both cash refunds and credit gift cards, which gave a credit facility to purchase items at one of the stores). He asked for 60 similar offences to be taken into consideration. He was sentenced to be reduced to the rank of Senior Aircraftman, to be dismissed from Her Majesty's service, to undergo six months' detention and to pay compensation in the sum of £3,000. *R v Cooney, Allam and Wood* [1999] 3 AER 173 stressed the importance, when considering the sentence for a person towards the end of his 22-year period, of the financial impact dismissal would have upon him. The Court held that the loss of pension in these circumstances is not peculiar to the Armed Forces; it also applies to police forces. Furthermore, the loss of money on dismissal is not peculiar to either of those two organisations. An employee earning £30,000 a year, who is found to have stolen something from their employer, is very likely to be dismissed, after which they will not then earn what they had been earning and their future employment chances are obviously reduced by the fact that they have been dismissed for theft from a previous employer.

147 [2005] EWCA Crim 3096.

Disgraceful Conduct of a Cruel or Indecent Kind

Main Provision: sections 23 and 53(1)(b) AFA 2006

Section 23 states:

> (1) A person Subject to Service Law commits an offence if
> (a) he does an act which is cruel or indecent; and
> (b) his doing so is disgraceful.
> (2) In this section 'act' includes an omission and the reference to the doing of an act is to be read accordingly.
> (3) A person guilty of an offence under this section is liable to any punishment mentioned in the Table in section 164, but any sentence of imprisonment imposed in respect of the offence must not exceed two years.

This offence is committed where 'a person subject to service law ... does an act which is cruel or indecent and his doing so is disgraceful'. In effect, there are two offences created: one alleging cruel conduct and the other alleging indecent conduct. This offence may also be heard summarily, under section 53(1)(b). The maximum sentence which may be imposed after conviction by court martial is two years' imprisonment.

What amounts to cruel or indecent conduct will be a matter of fact, to be assessed by an objective test, but the terms are to be given their ordinary dictionary meanings.

> No amount of mental gymnastics can enable one to interpret the word 'disgraceful' or the words 'disgraceful conduct' as being subject to a subjective test in law. It must be objective.[148]

Subsection (2) includes a failure (omission) to act when aware of the consequences. Locking a dog in a car on a hot day might constitute an offence of commission or omission, depending upon the state of mind. However, whether conduct is indecent or cruel conduct, for it to constitute an offence it must be regarded as disgraceful. These terms are given an ordinary dictionary meaning. There is the possibility of an overlap with civilian offences under the 2003 Act and conduct that should be dealt with

148 *R v Walsh* (1980) CMAC, CM/12/80 (unreported), per Comin J.

by court martial might be dealt with summarily by the CO under these provisions. This raises victim issues.

The charge is available to deal with quite serious conduct, for example, initiation ceremonies. It is, nevertheless, appropriate where conduct occurs in a military environment and though not sufficient to amount to a civilian offence, is contrary to discipline. For example, where two warrant officers at a mess function engage in consensual kissing and/or intimate touching in full view of others present. Even though consensual, that could be regarded as disgraceful conduct of indecent kind, whereas the same consensual conduct in private would not.

Case Illustrations

In spite of the decision having been made some time ago, the case of *Walsh* remains an important one as it declared the test to be an objective one. It was described by Comin J as a 'pathetic case' concerning an RAF officer who indecently exposed himself to passengers on passing trains. He was convicted on 10 charges of disgraceful conduct. The appellant had argued that the test should have been a subjective one. However, the CMAC held that '[w]e are afraid that no amount of mental gymnastics can enable one to interpret the word "disgraceful" or the words "disgraceful conduct" as being subject to a subjective test in law.'[149]

In *Lee*,[150] the accused was an RMP NCO, who was serving in Iraq at the time of the offence. He kicked a dog and stamped it to death. He was charged with disgraceful conduct of a cruel kind (contrary to section 66 of the Army Act 1955). He was convicted and sentenced to dismissal with disgrace, six months detention and reduction to the ranks. The reviewing authority varied the sentence to one of dismissal and reduced the detention to three months, upheld by CMAC.

In *Summers*,[151] the appellant pleaded guilty at a court martial to two offences of disgraceful conduct of an indecent kind, in relation to a 17-year-old female recruit. One charge related to an allegation that he had tapped her on the head with his penis while she was crouched down, and the second indecent incident a few days later occurred when the complainant

149 *R v Walsh* (1980) CMAC, CM/12/80 (unreported), per Comin J.

150 *Lee, R v* [2005] AllER, 17 Jan. (D) 96.

151 *Summers, R v* [2012] EWCA Crim 2073 CMAC.

placed a chocolate sweet into her mouth. The appellant made a sexually inappropriate remark to her, which she took as nothing more than banter, but the appellant immediately got up, ran after her and grabbed her jacket. She tried to get away but slipped to the floor. While on her knees she put her hand over the top of her head, fearing that he would repeat what he had done a few days earlier. She then felt his penis being placed on top of her hands for about five seconds before he removed it. The allegation of battery related to the appellant punching the complainant in the back. He was sentenced to reduction in rank and four months' military detention on each charge of disgraceful conduct of an indecent kind, to run consecutively; and for one offence of battery, the sentence for which was reduction in rank and 60 days' military detention, to run concurrently. Thus the total sentence was one of eight months' military detention and reduction in rank from Lance Corporal to Private. The sentences were quashed on appeal and the CMAC substituted sentences of three months' detention on each of the disgraceful conduct charges, to run consecutively, and a sentence of two months' detention, to run concurrently, on the charge of battery. The total sentence was therefore one of six months' military detention. The net effect of the CMAC's intervention was to reduce somewhat the unanticipated financial loss which flowed from the use of the mistaken figures in the calculation of financial loss. These figures had been compiled on a false premise that the loss was £1,187 per annum, whereas, in fact, as it had subsequently emerged, it was £5,233 per annum.

Attempts

Main Provisions: sections 39, 43, 44, 53(2), 54 and Schedules 1 and 2 AFA 2006; section 1 Criminal Attempts Act 1981

An attempt is committed where a person, with the intention of committing an offence, does an act that is more than merely preparatory to the commission of that offence. Even an attempt to commit an offence where the facts are such that the commission of the offence is impossible will still be an offence of attempt. These will be questions of fact for the court to decide. Merely discussing a crime, as a theoretical exercise, would not be sufficient to classify as an attempt. Discussing a burglary, drawing up plans, buying tools and making reconnaissance trips to the target premises, on the other hand, may cross the rubicon. This is the stage where the accused has properly embarked upon a criminal enterprise or equipped themselves to do so. It should be said that this conduct may also amount to a conspiracy, as discussed on page 252, as it is a joint criminal enterprise with one or more other persons.

Under section 1(4) of the Criminal Attempts Act 1981, there are a number of criminal offences that cannot be the subject of an attempt. These are: conspiracy at common law or under section 1 of the Criminal Law Act 1977 or any other enactment; aiding, abetting, counselling, procuring or suborning the commission of an offence or offences under section 4(1) (assisting offenders) or 5(1) of the Criminal Law Act 1967 (accepting or agreeing to accept consideration for not disclosing information about an arrestable offence).[152] An attempt to commit an offence is another inchoate offence.

Service Offences

Under section 39 AFA 2006 it is an offence for a person subject to service law to attempt to commit a service offence, or for a civilian subject to service discipline to attempt to commit one of the offences mentioned below on page 244 and set out in section 39(4). Offences under section 41 – aiding and abetting, and criminal conduct offences under section 42 (subject to exceptions explained below) – are excluded from summary dealing under this provision. Section 39(5) replicates section 1(4) of the 1981 Criminal Attempts Act by stating that the accused must intend to commit the offence and do an act which is more than merely preparatory to the commission of that offence.

Section 53(2)[153] makes it clear that an attempt to commit a section 42 criminal conduct offence is not capable of being heard summarily under section 39, even if the civil offence falls within Schedule 1. The reason for this particular exception is that an offence of attempting to commit a section 42 civilian criminal offence is itself an offence under section 43 and may be dealt with in the same way as the corresponding section 42 criminal conduct offence.[154] In other words, if the section 42 offence falls within

152 Crown Prosecution Service Guidance (CPSG).
153 Section 53 lists the offences that may be dealt with at a summary hearing.
154 EN, para. 115.

Part 1 of Schedule 1,[155] it is capable of summary dealing[156] and so too is the attempt to commit it. The authorising section for criminal conduct offences is therefore section 43 and not section 39.

A person guilty of an offence under this section is liable to the same punishment as would applicable if found guilty of the offence attempted.

Criminal Offences

As explained above, section 43 deals with attempting criminal conduct (as opposed to a disciplinary offence). Under the Criminal Attempts Act 1981, an attempt is a criminal offence under the law of England and Wales only if it is an attempt to do an act which, if done, would be an indictable offence under that law (other than certain excluded offences). Section 43(2) modified the 1981 Act[157] so that it is sufficient for an offence under section 42 if the intended act would be an indictable offence if it were done in England or Wales. It is immaterial that the act would, in fact, be done outside England and Wales.[158] This essentially now makes those subject to service law and service discipline liable to be tried for attempt wherever the offence is committed.

If there is sufficient evidence to justify a finding that the act was an attempt, it is a question of fact – and therefore, in a trial by the court martial, for the members of the court other than the judge advocate – whether the act was an attempt. The section is in similar terms to section 4(3) of the Criminal Attempts Act 1981. Section 39(8) makes corresponding provision for the trial of a charge of attempt under that section (EN para. 124).

An attempt to commit an offence under section 42 (criminal conduct) may be dealt with at a summary hearing if the corresponding offence under the law of England and Wales is

(a) an offence listed in either Part of Schedule 1 (criminal conduct offences that may be dealt with at a summary hearing); or

155 See Appendix D.

156 See s. 39(2) and s. 53(3).

157 Previously, this meant that a person subject to service law committed no offence under section 42 unless their intention was to do an act which, if done, would in fact have been an indictable offence under English law. This requirement would not normally be satisfied if the intended act had been done outside England and Wales.

158 EN, para. 123.

(b) an offence under section 1 of the Criminal Attempts Act 1981 (c. 47) of attempting to commit an (indictable) offence so listed [AFA 2006, s. 53(3)].

As there is only jurisdiction for a CO to deal summarily with those criminal offences that fall within Schedule 1, the attempt must relate to such an offence. An attempt to commit an offence listed in Part 2 of Schedule 1 or an offence under section 1 of the Criminal Attempts Act 1981 of attempting to commit an (indictable) offence so listed may only be heard summarily by an officer after permission has been granted by higher authority[159] or the officer is of or above the rank of rear admiral, major-general or air vice-marshal.[160]

Schedule 2 paragraph 10 specifies that an attempt to commit certain listed service offences is actually itself a Schedule 2 offence.[161] These would, therefore, not be amenable to summary dealing. The service offences concerned are:

a. assisting an enemy
b. misconduct on operations
c. obstructing operations, which relates to an action or operation against an enemy
d. looting (under section 4(1) or (2) mutiny)
e. failure to suppress mutiny
f. desertion, where the accused intended to avoid a period of active service (within the meaning of that section)
g. hazarding of ship
h. dangerous flying, etc.

The MSL sets out a specimen charge, illustrating how the conduct is alleged:[162]

> COMMITTING A CRIMINAL CONDUCT OFFENCE CONTRARY TO SECTION 42 OF THE ARMED FORCES ACT 2006 NAMELY ATTEMPTING TO COMMIT THEFT

159 In relation to a CO, this means any officer in the CO's disciplinary chain of command who is superior in that chain of command to the commanding officer. See s. 54 , Sch. 1 and s. 361 meaning of 'higher authority'.
160 S. 54(1).
161 See Appendix E.
162 See JSP 830 MSL Version 2.0, at p. 1-8-65.

CONTRARY TO SECTION 1(1) OF THE THEFT ACT 1968 CONTRARY TO SECTION 1(1) OF THE CRIMINAL ATTEMPTS ACT 1981]

[AB] on …, did attempt to steal a watch, the property of …'

Civilians

In relation to civilians, section 39(4) provides that a civilian subject to service discipline commits an offence triable by a service court if the attempt is to commit an offence under

a. section 4 (looting)
b. section 13 (contravention of standing orders);
c. section 27 (obstructing or failing to assist a service police officer)
d. section 28(2) (resistance to arrest)
e. section 29 (offences in relation to service custody)
f. section 107 (compliance with conditions on release from custody after charge)
g. section 93A (testing for alcohol and drugs after serious incident)[163]
h. section 18 of the AFA 1991 (obstructing exercise of an assessment order)
i. section 20 of the AFA 1991(obstructing exercise of a protection order)
j. section 40 AFA (an offence of inciting another person to commit an offence mentioned in paragraphs a–i above).

Attempting to commit an offence is itself an offence, even where its completion may be impossible. The House of Lords in *Shivpuri* made it clear that the only kind of impossibility that is relevant to liability is true legal impossibility. Even if the facts were such as the accused believed them to be, then the defendant would still not be committing any offence, having made a mistake about what the law was.[164] This is to be distinguished from a mistake of pure fact, which does not assist a defendant.

What amounts to an attempt is defined in section 39(5) and section 1(1) of the Criminal Attempts Act 1981 as the doing of an act that is more than merely preparatory to the commission of the offence, with intent to commit

163 Original provisions in ss 306 and 307 were repealed by AFA 2011 and AFA2011(C4)O 2013.
164 [1986] 2 AllER 334, cited in CPS Legal Guidance.

the offence. The difficulties inherent in proving inchoate conduct, such as an attempt, being an offence requiring a specific intent are well known to criminal practitioners, particularly where the matter of when the acts alleged have 'crossed the rubicon', that is, are more than merely preparatory to the commission of the offence, and were done with the intention of completing the act. In *R v Qadir and Khan*[165] it was held that an '[a]ttempt begins at the moment when the defendant embarks upon the crime proper, as opposed to taking steps rightly regarded as merely preparatory'.

While this is clearly a question of fact, as specified in subsection (8), it is not always clear-cut. At a court martial or SCC, where counsel represents a defendant the matter can be thoroughly explored before the court. Legal argument can be presented to the judge advocate and, as relevant, an appropriate direction given to the court. There is no such right to legal representation before a CO.

Encouraging and Assisting

Main Provisions: sections 40, 46 and Schedule 1 AFA 2006; section 94(1), Schedule 5 paragraph 9, Schedule 13 paragraph 5 Serious Crime Act 2007; Article 2(a) Serious Crime Act 2007 (Commencement No. 3) Order 2008

Section 59 of the Serious Crime Act 2007 abolished the common law offence of incitement, with effect from 1 October 2008. The AFA provides for two offences. The first involves encouraging, etc. the commission of a service offence under section 40 and the second relates to encouraging, etc. the commission of a criminal offence under section 46. These provisions allow people who assist another to commit an offence to be prosecuted regardless of whether the underlying substantive offence is actually committed or attempted; they create inchoate offences.

Service Offence

Section 40 states:

165 [1997] 9 Archbold News 1, CA, cited in CPS Legal Guidance.

Encouraging and assisting

(1) A person subject to Service law commits an offence if he encourages or assists the commission of a Service Offence (other than an offence under section 42).

(2) A civilian subject to service discipline commits an offence if he encourages or assists the commission of an offence mentioned in section 39(4).

(3) Reference in this section to encouraging or assisting the commission of an offence is to the doing of an act that would have constituted an offence under Part 2 of the Serious Crime Act 2007 if the offence encouraged or assisted had been an offence under the law of England and Wales.

(4) In determining whether an act would have constituted an offence under that Part, section 49(4) of that Act has effect as if for 'offences under this Part and listed offences' it read 'offences under sections 39 and 40 of the Armed Forces Act 2006'.

(5) Any requirement in that Part to specify matters in an indictment applies for the purposes of this section as it applies for the purposes of that Part, but with references to the indictment being read as references to the charge sheet.

(6) A person guilty of an offence under this section is liable to the same punishment as he would be liable to if guilty of
  (a) the Service Offence encouraged or assisted; or
  (b) if convicted of the offence under this section by reference to more than one such service offence, any one of those service offences.

The offence under section 40 applies where a person subject to service law or a service civilian encourages or assists the commission of a service offence (other than an offence under section 42, that is, non-criminal conduct). The offence under this section is concerned *only with disciplinary offences* (i.e. service offences) under the Act.[166] A person convicted of an offence under

166 Defined on p. 151.

this section is liable to the same punishment as if they were convicted of the offence they incite the other person to commit.

In the case of a service civilian subject to service discipline, an offence will be committed if they encourage or assist the commission of an offence mentioned in section 39(4), namely, an offence of looting; contravention of SOs; obstructing or failing to assist a service police officer; using violence against or threatening behaviour to a person who has a duty to apprehend them; offences in relation to service custody; failing to comply with conditions for release from custody; failing to comply with a CO's requirement regarding preliminary tests for alcohol or a drugs; failing to comply with a protection order; or attempting to encourage or assist the commission of any such an offence mentioned in section 39(4).

The term 'encouraging or assisting the commission of an offence' is defined in subsection (3) as 'the doing of an act that would have constituted an offence under Part 2 of the Serious Crime Act 2007 if the offence encouraged or assisted had been an offence under the law of England and Wales'. However, for the purpose of jurisdiction under the AFA, section 40(4) of the AFA substitutes the offences within sections 39 (attempts) and 40 (encouragement or assistance) of the AFA 2006.

A person guilty of an offence under section 40 is liable to the same punishment as would apply if they were found guilty of the service offence encouraged or assisted; or if convicted of the offence under this section by reference to more than one such service offence, any one of those service offences (s. 40(6)).

Where the service offence encouraged or assisted is one that is specified in Schedule 2(11) of the Act, namely an offence in paragraphs 1 to 9 of that Schedule,[167] then any offence of encouraging or assisting the commission of such an offence will itself be a Schedule 2 offence and not amenable to summary dealing.

Criminal Conduct

Section 46 states:

167 These are the offences of: assisting an enemy; misconduct on operations; obstructing operations which relates to an action or operation against an enemy; looting; mutiny; failing to suppress mutiny; desertion with intent to avoid a period of active service; hazarding of a ship; dangerous flying; or attempting to commit any of these offences.

(1) Subsection (2) applies if a person subject to Service law, or a civilian subject to service discipline, encourages or assists the doing of an act (or one or more of a number of acts) that, if done in England or Wales, would be punishable by the law of England and Wales.

(2) Regardless of where that act (or those acts) might be done and of his state of mind with respect to that question, his encouragement or assistance shall be treated for the purposes of section 42(1) as an act that is punishable by the law of England and Wales (so far as it is not such an act in any event).

(3) Reference in this section to encouraging or assisting is to an act that would constitute an offence under Part 2 of the Serious Crime Act 2007 disregarding any provision in that Part about the place where the act (or acts) being encouraged or assisted might be done or the accused's state of mind with respect to that question.

In relation to both service personnel subject to service law and civilians subject to service discipline, encouraging and assisting a criminal conduct offence (under s. 42 AFA) is itself a criminal offence and would be charged as such, under section 46 AFA. Practitioners will be aware of the specific provisions relating to the abolition of the offence of incitement made by the changes introduced by Part 2 of the Serious Crime Act 2007, which replaced the common law offence with three inchoate offences.

Section 46(3) makes clear that reference to 'encouraging or assisting' is to an act that would constitute an offence under Part 2 of the Serious Crime Act 2007, disregarding any provision in that Part about the place where the act (or acts) being encouraged or assisted might be done or the accused's state of mind with respect to that question:

a. section 44 – intentionally encouraging or assisting an offence
b. section 45 – encouraging or assisting an offence believing it will be committed
c. section 46 – encouraging or assisting offences believing one or more will be committed.

The belief required in sections 45 and 46 is a state of mind that is more than suspicious. The word 'belief' is a word of ordinary usage and does not require any elaboration: *Treacy v DPP* (1971) 55 Cr App R 113. If elaboration is required, a direction approved in *R v Moys* (1984) 79 Cr App R 72 should be given, confirming that suspicion, in addition to the fact that the defendant shut their eyes to the circumstances, is not enough, although such matters were relevant to the jury's determination of the defendant's knowledge or belief.[168]

An offence under section 46 is not capable of being dealt with summarily, as it is omitted from section 53.

Aiding, Abetting, Counselling or Procuring

Main Provisions: sections 41, 42, 47 and Schedule 1 AFA 2006.

A person may commit an offence under either section 41 or section 47. The offence under section 41 relates to the commission of a service offence, while the offence under section 47 relates to criminal conduct. Section 41(3) makes clear that a 'person who by virtue of subsection (1) commits an offence is liable to be charged, tried (including dealt with at a summary hearing) and punished as a principal offender.'

Service Offence

Section 41 states:

> **Aiding, abetting, counselling or procuring**
>
> (1) Where a person subject to service law aids, abets, counsels or procures the commission by another person of an offence to which this subsection applies, he commits that offence.
> (2) Subsection (1) applies to any Service Offence except an offence under section 42.
> (3) A person who by virtue of subsection (1) commits an offence is liable to be charged, tried (including dealt with at a summary hearing) and punished as a principal offender.
> (4) Where a civilian subject to service discipline aids, abets, counsels or procures the commission by another person of an offence mentioned in section 39(4), he commits that

168 CPS Guidance.

offence and is liable to be charged, tried and punished as a principal offender

It will be noted that the offence applies to a person *subject to service discipline* charged with any Service offence except an offence under section 42. This excludes service civilians. It can be dealt with at a summary hearing.

If the substantive offence can be dealt with summarily, being listed in either Part of Schedule 1, then so will be the offence of aiding, etc. its commission.[169] The necessity for permission from higher authority remains in those cases falling within Part 2 of Schedule 1.[170]

Where a civilian subject to service discipline aids, abets, counsels or procures the commission by another person of an offence mentioned in section 39(4), they commit that offence and are liable to be charged, tried and punished as a principal offender. The offences in section 39(4) are:

a. an offence under section 4 (looting), 13 (breach of Standing Orders), 27 (Obstructing or failing to assist a Service policeman), 28(2) (using violence/threats against a person who has a duty to apprehend him), 29 (Offences in relation to Service custody), 107 (failure to release from custody after charge), or 93A of this Act (Testing for alcohol and drugs after serious incident)[171] or under section 18 (offences in relation to assessment orders) or 20 (protection orders) of the Armed Forces Act 1991; and an offence under section 40 of encouraging or assisting the commission of an offence mentioned in paragraph (a).

Criminal Conduct

The offence of aiding, abetting, counselling or procuring criminal conduct is set out in section 47 AFA 2006, which applies to persons subject to service law and those subject to service discipline.

169 See AFA 2006, s. 53(3).

170 See AFA 2006, s. 54.

171 Section 39(4) originally referred to an offence under s. 306. However, Sch. 5 of the 2011 Act repealed ss 306 and 307 of the 2006 Act. These provisions were inserted by s. 11 of the 2011 Act, which presented a new Chapter 3A after s. 93, dealing with testing for alcohol and drugs on suspicion of an offence.

Aiding, abetting, counselling or procuring criminal conduct

(1) Subsection (2) applies if
  (a) any person ('A') does an act that is punishable by the law of England and Wales or would be so punishable if done in England or Wales; and
  (b) a person subject to Service law, or a civilian subject to service discipline, aids, abets, counsels or procures A's doing of that act.

(2) Regardless of where the act aided, abetted, counselled or procured was done, the aiding, abetting, counselling or procuring shall be treated for the purposes of section 42(1) as an act that is punishable by the law of England and Wales.

(3) For the purpose of determining whether an attempt is an act that falls within subsection (1)(a) above, section 1(4) of the Criminal Attempts Act 1981 (c. 47) has effect with the modification made by section 43.

An accused may be charged, tried and punished as if they had personally committed the offence the commission of which they aid, abet, counsel or procure. This means that the offender will be charged under the substantive criminal provision covering the conduct alleged to have been aided or abetted, etc. For example, a person aiding and abetting theft will be charged as a principal under section 42. The offence is not subject to a territorial limit, so long as the principal offence is one contrary to the law of England and Wales (s. 47(2)).

Conspiring to Commit Criminal Conduct

Section 45 Conspiring to commit criminal conduct

(1) For the purpose of determining whether an agreement that a course of conduct be pursued is an offence under section 42
  (a) sections 1(1) and 2 of the Criminal Law Act 1977 (c. 45) (conspiracy) have effect as if any reference to an offence included a reference to an act that, if done in England

or Wales, would be punishable by the law of England and Wales; and

(b) section 1(2) of that Act has effect as if it read

'(2) Where liability for any offence may be incurred without knowledge on the part of the person committing it of any particular fact or circumstance necessary for the commission of it, a person is nevertheless not guilty by virtue of subsection (1) above of conspiracy to commit

(a) that offence, or

(b) an act that would amount to that offence if done in England or Wales,

unless he and at least one other party to the agreement intend or know that that fact or circumstance shall or will exist at the time when the conduct constituting the offence, or the act, is to take place.'

(2) Section 42(6) applies[172] for the purposes of section 1(2) of the Criminal Law Act 1977 as substituted by this section.

In the reference to 'doing an act', section 42(6) AFA clarifies that 'act' includes an omission and that references to the doing of an act are to be read accordingly. This meaning is applicable to offences under sections 45–49.[173]

Under the Criminal Law Act 1977 (CLA 1977), a person is guilty of conspiracy if that person agrees with any other person or persons that a course of conduct shall be pursued which, if the agreement is carried out in accordance with their intentions, either

(a) will necessarily amount to or involve the commission of any offence or offences by one or more of the parties to the agreement, or

(b) would do so but for the existence of facts which render the commission of the offence or any of the offences impossible.

172 'In this section and sections 45 to 49 "act" includes an omission and references to the doing of an act are to be read accordingly.'

173 Conspiring to commit criminal conduct; encouraging or assisting criminal conduct; aiding, abetting, counselling or procuring criminal conduct; Air Navigation Order offences; and the supplementary provisions of s. 48.

The modified text of section 1(2) is to be read as set out above in section 45(1)(b). Thus, section 45 modifies the criminal offence of conspiracy to commit an offence under the CLA 1977 so that an agreement to a course of conduct being pursued outside England and Wales is an offence under section 42 not only if that course of conduct would in fact involve the commission of an offence under English law, but also if the same conduct in England or Wales would involve the commission of such an offence.[174] Therefore, a conspiracy may be charged wherever the agreement or act takes place, provided it would amount to a conspiracy offence under English law.

The essence of a conspiracy is the agreement between the defendant

> with another or others (whether identified or not) that a course of conduct would be pursued which if carried out in accordance with their intentions would necessarily involve the commission of any offence by one or more of the parties to the agreement, or would do so but for the fact that it was an impossible attempt. The *mens area* for conspiracy requires proof of that intention to be a party to an agreement to do an unlawful act and that D and one other party knew or intended that the circumstance element(s) of the intended offence would exist at the time of the offence (even if the substantive offence can be committed without proof of knowledge).[175]

There does not have to be anything done to further the agreement. The offence of conspiracy is by no means of recent origin. Its history is more than 700 years old. In 1304, the Third Ordinance of Conspirators (33 Edw I) set out the first precise definition of conspiracy:

> Conspirators be they that do confeder or bind themselves by Oath, covenant, or other Alliance, that every of them shall aid and support the Enterprise of each other falsely and maliciously to indite or cause to be indited or falsely to move or maintain Pleas.
>
> In the *Poulterers' Case*, 9 Coke 55 b decided in 1611, the Court of Star Chamber held that the conspiracy offence was made out even when no false appeal had in fact been brought against the

174 EN, para. 125.

175 The Crown Court Compendium, Part I: Legal Summaries, Directions and Examples, December 2020, at 7-7.2.

> innocent party provided the defendants had combined for that purpose. It followed that the offence of conspiracy lay once the conspirators had settled on their criminal enterprise and not just when that enterprise was fulfilled in accordance with their intentions.[176]

Mr Guido Fawkes and his colleagues, Thomas Wintour and Robert Catesby, could probably tell us much about the offence, for which they suffered brutal consequences. Torture broke Fawkes' ability to resist his suffering and, after several days of physical abuse, he revealed the names of his co-conspirators in the plot to blow up Parliament. He signed a confession.[177] It is an interesting discussion point in the debate over torture and whether information extracted by such means is ever reliable.

The utility of charging an inchoate offence like conspiracy is that there is no requirement for the offence to be completed, merely that there is a genuine agreement to participate to that effect.[178] Once a conspiracy is in existence others can join it. The requirement in such a case is that 'a defendant can only join an agreement when he shares the designs of those who are already parties to it'.[179] The use of a conspiracy charge is, perhaps, more appropriate where there is no substantive offence, as it was never completed, but there is clear evidence of an agreement between two or more persons to commit it. The evidence of each defendant as to what was said about the nature of the agreement and the stage to which it had advanced would be admissible against the others involved.

The agreement cannot be a mere mental operation; it must involve spoken or written words or other overt acts. If the defendant repents and

---

176 'The Use and Abuse of Conspiracy', Paul Jarvis and Michael Bisgrove, [2014] CrimLR at 260.

177 'Catesby and Percy were shot at Holbeche House, where a few of them were hiding, until being discovered by the Sherriff of Worcester. Fawkes and most of the conspirators survived their torture, to stand trial before the Star Chamber, presided over by the Lord Chief Justice, Sir John Popham. The prosecution was led by the Attorney General of the day, the redoubtable Sir Edward Coke. Coke was a member of the Honourable Society of the Inner Temple. It is said that Fawkes fell and broke his neck as he climbed the scaffold, thereby cheating the hangman's noose. However, his body was still cut into quarters and sent to the "four corners of the kingdom". Antonia Fraser, *The Gunpowder Plot: Terror and Faith in 1605*, ISBN 10-0753814013. The interrogations lasted for 23 days: Humphry W Woolrych, *The Life of the Right Honourable Sir Edward Coke, Knt., Lord Chief Justice of the King's Bench, &c*, p. 69.

178 *Mulcahy v The Queen* (1868) L.R. 3 H.L. 306.

179 *Op cit.*, FN 176.

withdraws immediately after the agreement has been concluded they are still guilty of the offence. Withdrawal from it goes to mitigation only.[180]

The complexity of the offence of conspiracy was expounded in the case of *R v Shillam*:[181]

> Conspiracy requires a single joint design between the conspirators within the terms on the indictment. It is possible, as in *Mehta*, that the evidence may prove a conspiracy of narrower scope and involving fewer people that the prosecution originally alleged, in which case it is not intrinsically wrong for the jury to return guilty verdicts accordingly, but it is always necessary that for two or more persons to be convicted of a single conspiracy each of them must be proved to have shared a common purpose or design.

The matter is complicated even further where conspiracies exist within a conspiracy, as occurred in *Mehta*. [182]

Some conspiracies are charged where the participants do not know all the others involved. For example, a 'wheel conspiracy' would involve a number of individuals who are called upon to deliver a specific function to central organisers at the hub, for example to forge bank note templates, develop the printing plates, print the notes, distribute, etc., organised by a central hub of main conspirators. Yet, the agreement in such a case is not between each 'spoke of the wheel', that is each participant with all other participants (spokes); it is between each individual (spoke) and those at the centre (the hub). In other words, these are separate conspiracies or, more aptly, substantive joint offences, for example fraud or theft in respect of the conduct which the individual (spoke) conspirator participates in (with the hub).[183] A true conspiracy would be for each conspirator to agree to the main objective of the criminal enterprise and each play their part in furthering it: a common unlawful purpose or design being a 'core requirement' of criminal conspiracy. However, in the example given, the forger is paid for their work, and the print plate maker for theirs, and so

180 CPS Legal Guidance, citing also: *R v Gortat and Pirog* [1973] CrimLR 648.
181 [2013] EWCA Crim 160; *op cit.* n 177, FN 176, Jarvis and Bisgrove, p. 267.
182 [2012] EWCA Crim2824.
183 *Griffiths* [1966] 1 QB 589., *op cit* FN 176.

on. They may not know each other or of each other's existence. There is no common intent in such cases of each member sharing in the end of seeing each other succeed in the fraud.[184] But there is an agreement with the central organiser, which could be charged as a separate conspiracy. While it is not necessary for each member of the conspiracy to know the other members, '[i]f it is alleged that the parties to the conspiracy is a "wheel" or "chain" conspiracy, each alleged conspirator must each be shown to be party to a common design, and he/she must be aware that there is a larger scheme to which he/she is attaching him/herself.'[185]

In the case of *R v Ryan Johnson, Paul Casey, Andrew Belford, Robert Anderson and Darren McBride,*[186] Simon LJ explained:

> What are referred to as 'chain' conspiracies and 'wheel' conspiracies are different in structure. In a chain conspiracy, A agrees with B, B with C and C with D. In a wheel conspiracy: A at the hub recruits B, C & D. In each it is necessary that the defendants must be shown to be a party to the common design and aware that they are part of a common design to which they are attaching themselves: see for example Blackstone Criminal Practice 2020 A5.49.

Another difficult arises in the event that the defence allege a qualified or conditional agreement, whereby the agreement to enter or join the conspiracy is based upon some condition or other being met. 'If the matters left outstanding or reserved are of a substantial nature, the arrangement may amount only to negotiations and thus fall short of being a conspiracy.'[187]

As an exception to the hearsay rule, evidentially, the statements of co-conspirators may be adduced as evidence against each other, under the provisions of section 118(1)7 of the Criminal Justice Act 2003, which preserve 'any rule of law under which in criminal proceedings a statement made by a party to a common enterprise is admissible against another party to the enterprise as evidence of any matter stated'.

---

184 *Op cit.*, FN 176, Jarvis and Bisgrove, at p. 266.
185 Crown Court Compendium, at p. 7-7, at para. 7, and *Shillam* [2013] EWCA Crim 160.
186 [2020] EWCA Crim 482, para. 27.
187 CPS Legal Guidance, citing *R v Mills* [1963] 1 QB 522, 47 CrAppR 49, CCA; *R v O'Brien* (P.J.), 59 CrAppR 222, CA.

The rule is subject to what has been described as a 'three-pronged test':[188]

a. the words relied on must have been said by one of the other parties to the common enterprise; and
b. be reasonably open to the interpretation of having been made in furtherance of the alleged agreement; and
c. there must be some further evidence apart from those words to prove that the other defendant (against whom they are to be used in evidence) was a party to the common enterprise: (*R v A Ltd, X, Y* [2017] 1 Cr App R 1).

What is surprising, bearing in mind the difficulty of conspiracy offences, and the absence of legal representation before the CO, is that such an inchoate charge may be dealt with summarily if the offence to which it relates can be dealt with summarily under section 53. In addition, section 53(3) provides that an offence under section 42 (criminal conduct) is within this subsection if the corresponding offence under the law of England and Wales is

(a) an offence listed in either Part of Schedule 1 (criminal conduct offences that may be dealt with at a summary hearing); or
(b) an offence under section 1 of the Criminal Attempts Act 1981 (c. 47) of attempting to commit an (indictable) offence so listed.

There is no specific reference to offences of conspiracy within section 53, which details the sort of offences that a CO may deal with, nor within either part of Schedule 1. This is in stark relief to the case of an offence of attempt, where specific mention is made in section 53(3)(b) of the Act that attempting to commit an offence listed in section (1) may also be dealt with summarily. Section 54(2) sets out the criminal charges falling under section 53(1)(g) that may be heard summarily by the CO, only with permission, or by a senior officer, and expressly stipulates that in addition to an offence listed in Part 2 of Schedule 1, an offence under section 1 of the Criminal

188 CPS Legal Guidance.

Attempts Act 1981 of attempting to commit an (indictable) offence so listed also requires permission from higher authority. Yet no such permission is expressly required for the more difficult offence of conspiracy.

An offence of theft is within Part 1 to Schedule 1 and may be therefore be dealt with summarily by the CO. Consequently, a conspiracy to steal may also be dealt with summarily and service personnel can acquire summary convictions for conspiracy offences without the right to free legal advice or, more importantly, legal representation before a lay adjudicator. Section 54 permits an offence of assault occasioning actual bodily harm, under section 47 of the Offences Against the Person Act 1861, which appears in Part 2 of Schedule 1, to be dealt with summarily if permission from higher authority is provided. A conspiracy to commit ABH could, therefore, also be dealt with summarily.

It is difficult to comprehend circumstances where a conspiracy charge would be considered suitable for summary dealing.[189] Indeed, it is submitted that a conspiracy offence is utterly unsuitable for summary dealing, bearing in mind the complexities inherent in proving the offence and defining the point at which a conspiracy is in place. Even the MSL itself recognises this fact and states:

> Framing of conspiracy charges is especially complex and COs should not bring a conspiracy charge without legal advice. For the Service Police this will be the DSP and for the CO the appropriate staff legal adviser.[190]

It is probably why the MSL contains no specimen charge for the offence of conspiracy.

Court martial statistics are published by the Military Court Service, but the Ministry of Defence do not publish statistics of summary hearings, which, as a non-compliant process, is a little troubling. A freedom of information request for statistics on summary dealing of conspiracy received a response[191] that '[t]here are fewer than five cases of summary dealings involving conspiracy recorded on the Joint Personnel Administration

189 In the latest available statistics for the court martial, for the period January 2010 to December 2018, there is no record of any prosecution by that court for an offence of conspiracy.
190 JSP 830 MSL Version 2.0, p. 1-8-68, para. 1.
191 1 April 2020, Ref: FOI2020/02471.

system (JPA), all recorded under Section 19 of the Armed Forces Act 2006 (conduct prejudicial to good order and service discipline).' This must be a mistake; as conspiracy is a specific substantive offence under section 45 AFA, it would not be charged as a section 19 offence, because do so would breach the *Dodman* case.[192]

---

192 *Op cit.*, p. 159 n 19.

8 ADMINISTRATIVE ACTION BY THE COMMANDING OFFICER

Introduction

As mentioned in Chapter 1, there exists a body of administrative instructions that govern administrative, as opposed to disciplinary, punishment of service people, known in the army as the Army General and Administrative Instructions (AGAI); in the Royal Navy, as publication BR3 (Section 5 – Minor Administrative Action) and Fleet General and Administrative Orders (FLAGOs). In the Royal Air Force these are to be found in Queen's Regulations (QRs) and Air Publications (APs) and Pamphlet Air (Pam(Air)). The single service instructions are consonant with Joint Service Publication (JSP) 833 (Minor Administrative Action) which sets out a tri-service policy for dealing with minor administrative action (MAA). They are loosely similar to employment sanctions.

As a prosecutor, the commanding officer (CO) has a preliminary discretion whether it is in the interests of justice to institute disciplinary proceedings against a service person. If a CO determines not to take formal disciplinary action against a service person, they may nevertheless decide on an administrative punishment, namely, an MAA, which is action intended to set straight professional and personal shortcomings. An example of a situation where such a course might be considered is if a service person is late for duty or has underperformed in their duty role. The CO will usually seek or receive legal advice from the staff legal adviser in such cases.

In determining the appropriate method of dealing with a matter, the CO is to take into account the seriousness of the misconduct and its impact, or potential impact, on the efficiency, operational effectiveness or reputation of their unit and the service, in accordance with their service Values and Standards. Disciplinary or administrative action may be taken against a service person who fails to meet or uphold the service's standards of conduct. In the case of administrative action, the Service Test –

'Have the actions or behaviour of a Service person adversely impacted or are they likely to impact on the efficiency or operational effectiveness of the Service?' – is always applied to determine whether operational effectiveness has been adversely affected. A breach of the Service Test militates in favour of formal action, whether it be disciplinary or administrative (dependent upon the interests of justice).

JSP 833, Minor Administrative Action

MMA tri-service policy is dealt with under JSP 833, the provisions of which are replicated in the single service instructions (e.g. AGAIs). The JSP underlines, at section 3, paragraph 1, that it is to be used as the basis for dealing with minor professional and personal failings of behaviour or performance.[1]

> MAA provides commanders at all levels with a swiftly delivered and reviewed, legal, fair and formally regulated way of dealing with minor failings in standards and performance. MAA is not appropriate for matters that would amount to criminal conduct, to disciplinary offences that have 'criminal' elements or for cases of repeated minor transgressions. Disciplinary action should be considered in the first instance in such cases.[2]

Disciplinary action is action taken by the chain of command, using their statutory powers, to uphold good order and service discipline. It is taken where criminal or other service offences have been committed and where it is considered necessary to punish the individual or to deter others. MAA may only be taken once it has been decided not to charge an individual with a service offence.

It is not just commissioned officers who may issue an MAA sanction; non-commissioned officers may also do so. The authority of personnel to

1 This contrasts with what AGAI Vol. 2 Chap. 67 states, on its cover: 'This Chapter is sponsored by D Pers Cap. In the Pers Cap numbering system it is Edition 7 of AGAI 67. Joint units should use JSP 833 instead of Part 2 of this document for Minor Administrative Action.' Part 2 reports that it is the Army's guide to commanders for a 'a swiftly delivered and reviewed, legal, fair and regulated way of dealing with minor failings in standards and performance'.

2 JSP, Ch. 3, para. 8.

take MAA will be dictated by a CO of a unit and their view of the needs of that unit. Powers appropriate to rank that may be authorised in accordance with the JSP guidance are at Annex A to JSP 833. Moreover, all members of the regular forces up to and including the rank of OF4 (commander, lieutenant colonel, wing commander) may be subject to MAA sanctions at all times, for conduct both on and off duty. Separate processes exist for sanctions against officers above this rank, including oral and written counselling and major administrative action. Members of the reserve forces may be subject to MAA sanctions for conduct while subject to service law. The sanction available will depend on rank or rate and must be in accordance with Annex A.

It is prudent, when advising a service person on MAA, to consult the single service guidance for the service to which they belong and confirm that it replicates this sanction structure. For example, there are some differences in the sanctions available in the case of MAA of ranks OR-9 and OF-1/OF-2 and above.

Any sanction made must be appropriate, proportionate and remedial in relation to the failing identified. It must not be unreasonable or seek to humiliate or subject the individual to sleep deprivation or deliberate infliction of pain. The authorised sanctions include reporting at a specified time, extra duties, informal interview and a formal interview. This last sanction is intended to make clear to a service person their shortcomings and to advise as to what must be done to rectify the behaviour. The formal interview must contain constructive and specific advice on conduct and performance. Another sanction that might be appropriate is to eject the individual from a training course by ordering them to Return to Unit.

Report-back musters/parades, extra tasks and extra work cannot be awarded in combination with each other. Otherwise, any sanction may be combined with others. More detail about these sanctions is to be found in JSP 833.

By contrast, major administrative action (Major AA) is usually a single service matter and is engaged when, for example, a service person has brought the reputation of the service into disrepute; for example because of a civilian conviction for a serious offence or because of an inappropriate relationship with the spouse of someone under the person's immediate command. Because Major AA is essentially a single service matter, an officer from another service *may not* be the Deciding Officer (DO) for administrative action other than MAA or the issue of a Formal Warning (FW) as the other services apply different regimes. Where the CO is Royal Navy (RN) or Royal Air Force (RAF) then an Army CO will be appointed as the DO for the purposes of taking administrative action and the matter will follow their disciplinary chain of command.[3]

Review

MAA includes review at a higher level than that at which it is initiated.[4] The reviewing officer will ask in all cases whether the service person wishes to contest the sanction (requests a formal review) and if so, this will be noted on the form T-WS-MAA01 in the presence of the reviewing officer. The reviewing officer's options are to cancel the sanction or uphold or reduce the original award. The review is to be conducted as soon as practicable and in any event within 24 hours. In addition to this review and a formal review at the individual's request, it is important that regular inspections of the sub-unit minor sanctions folder are carried out by COs. A weekly inspection should be carried out within every unit by the head of department or officer commanding as appropriate; any significant matters should be brought to the attention of the CO and appropriate steps taken. The CO should carry out an inspection at least once a month. The records of MAA should be retained for at least two years and be available for inspection by higher authority at any time.

In addition, a service person who considers themselves to have been wronged in a matter relating to their service has the right to submit a service complaint, seeking redress of individual grievance.

An examination of the various administrative sanctions for each of the three services is outside the scope of this work, which focuses on the disciplinary processes. However, it is important that practitioners are aware of their existence. Moreover, it is imperative that practitioners are alert to a possible abuse of the administrative system, which has few

3 See AGAI Vol 2. Ch. 67.007.

4 See JSP 833, Ch. 3.

of the safeguards mandatory in the disciplinary process. An example, from one such case, graphically highlights the pitfalls.[5]

Examples

A senior NCO (SNCO) service policeman got drunk at a function and allegedly touched the bottom of a female NCO from his unit, junior in rank to him, who subsequently complained. The Professional Standards Department (PSD) took a section 9 statement[6] from the complainant in which she stated that she did not know whether the touching was sexual, but she was content for the matter to be dealt with administratively. Consequently, the SNCO was interviewed by two warrant-card-carrying members of the PSD, but the interview was not under caution and no solicitor was offered. It was decided to deliver an administrative sanction by way of a disciplinary interview, which was duly carried out.

The complainant subsequently stated that she was not content with the sanction and made a criminal complaint, which was handed over to another service police PSD. The service police refused to disclose the complainant's original section 9 statement to the SNCO's legal representative.

Although the case did not proceed further for reasons of the ill health of the accused SNCO, it raised a number of legal issues, not the least being the Police and Criminal Evidence Act (PACE) section 76 and section 78 admissibility of the statement by the SNCO, when interviewed by the PSD officers, which was made without legal representation and without caution. Another potential challenge that might have arisen was that of an abuse of process, bearing in mind the administrative process was not just at the incipient stage but had actually completed, based upon the consent of the complainant, which led to an award of sanction.

Evidence gathered during a non-disciplinary process, lacking PACE safeguards, was being relied upon for a subsequent disciplinary process, after a change of heart by the complainant. The case illustrates the importance of having a legal representative present to establish the proper parameters.

---

5 Brought to the author's attention by the legal representative of the SNCO, on 17 November 2019.

6 This relates to the Criminal Justice Act 1967, proof by written statement in criminal proceedings. To be admissible as evidence to the like extent as oral evidence by the maker, the statement must comply with certain formal criteria.

Although the uncautioned statement made by the accused during the administrative process is, on the face of it, inadmissible in subsequent disciplinary/criminal proceedings, a method that the service police might employ to introduce an uncautioned MAA statement into a PACE-compliant criminal interview under caution is to simply put the document to the accused, ask them if they understands what it is, ask them to identify their signature on that document, and then ask them whether they accept that its contents are true. If they do, then the document arguably becomes admissible.[7]

In another example, a male Warrant Officer (WO) was accused of impropriety against a female Sergeant. The matter was investigated by Royal Military Police (RMP) as a sexual assault and no evidence was found to support a prosecution. However, the unit subsequently took the RMP report, used the administrative process with its lower burden of proof (on the balance of probabilities) and dismissed the WO administratively, thereby ending his career.

Redress of Grievance

Main Provisions: JSP 831, Parts 1 & 2; Armed Forces (Service Complaints and Financial Assistance) Act 2015 (AF(SCMP)R 2015); Part 14A – sections 340A–40O AFA 2016

If a person subject to service law, or who has ceased to be subject to service law, thinks themselves wronged in any matter relating to their service, the person may make a complaint about the matter (s. 340A), seeking redress in accordance with JSP 831 (policy and guidance for all MoD Service personnel when making, responding to, advising on, investigating and deciding service complaints).[8] The legal basis for redress was originally contained in sections 334–339 of the 2006 Act. However, these provisions were repealed by the Armed Forces (Service Complaints and Financial Assistance) Act 2015 which inserted a new Part 14A – sections 340A–40O – setting out in detail who could make a complaint and the procedure for

7 *Abdurahman v R* [2019] EWCA Crim 2239 Court of Appeal.

8 See https://www.gov.uk/government/publications/jsp-831-redress-of-individual-grievances-service-complaints.

doing so. An appeal against a decision of the Defence Council lies to the Service Complaints Ombudsman (who replaced the Service Complaints Commissioner), and it is the Ombudsman who will determine an appeal to a complaint that has been initially determined by the chain of command (s. 340H). The Service Complaints Ombudsman is appointed by Her Majesty on the recommendation of the Secretary of State.

A service person may make a complaint[9] where they allege being the subject of:

a. discrimination;
b. harassment;
c. bullying;
d. dishonest or biased behaviour;
e. a failure of the Ministry of Defence to provide medical, dental or nursing care for which the Ministry of Defence was responsible; or
f. the improper exercise by a service policeman of statutory powers as a service policeman.

Complaints that are excluded from investigation are listed in regulation 3 of the AF(SCMP)R 2015 and the schedule. Listed therein is a prohibition on complaining about a decision by the Ombudsman for the purposes of any provision of Part 14A of the Act and the handling by the Ombudsman of a service complaint.

In the first instance, the complaint is to be investigated by the Defence Council who may authorise a person or panel of persons to investigate a particular service complaint on the Council's behalf, or on behalf of a person appointed to deal with a service complaint, or to determine an appeal relating to a service complaint.[10] JSP 831 (Pt. 1, para, 5) points out that the intent is that,

9 R. 5(2) of the Armed Forces (Service Complaints Miscellaneous Provisions) Regulations 2015.

10 See r. 4 of the Armed Forces (Service Complaints Miscellaneous Provisions) Regulations 2015 (AF(SCMP)R 2015), concerning the appointment of a person or panel of persons by the Defence Council.

> complaints are dealt with quickly and at the most appropriate level within the chain of command. Every effort should be made, where appropriate, to resolve a complaint informally.

This complaint would be a 'Special-To-Type'[11] complaint, first dealt with as a single service matter. There is a right of appeal from a determination to an appeal body appointed by the nominated Defence Council delegate or, following a review, where the Ombudsman decides that an appeal can proceed. Once the determination of the appeal body has been received, if the complainant is dissatisfied with the outcome of the complaint or how it was handled, or both, the complainant has the right to apply to the Ombudsman for an investigation to be carried out.[12]

The Ombudsman's investigations are conducted in accordance with the AF(SCMP)R 2015. Additional guidance is to be found in JSP 831.

11 JSP 831 Redress of Individual Grievances: Service Complaints Part 1, s. 5 and Part 2: Guidance, ss. 1 and 4.
12 JSP 831 Part 1, §20.

9 SUMMARY HEARING

Main Provisions: Part 6, Chapter 1 AFA 2006: sections 129–139 and 376 AFA 2006; AF(SHASSSD)R 2009; AFA 2016(CN3)R 2019; AF(MPLPRR)R 2009

Background

The summary hearing is the process by which charges are dealt with by the accused's commanding officer (CO).[1] The procedure at summary hearing is dealt with on page 288. First, there is an examination of the more general background, the charges suitable for summary dealing and the considerations for an application for extended powers (EP). Service civilians cannot be dealt with by the CO. See more generally below, Chapter 15.

Aside from the service disciplinary offences, a CO may also deal with some substantive criminal offences, too. There are two sets of criminal offences that the CO may deal with at a summary hearing, both of which are detailed in Schedule 1 to the Armed Forces Act 2006 (see Appendix D): those they can deal with without permission from higher authority[2] and those for which permission is needed. Generally speaking, it is the more serious Schedule 1 offences that requires permission from higher authority for summary dealing.

Offences that cannot be dealt with summarily are listed in Schedule 2 (see Appendix E). There were, previously, some troubling omissions from Schedule 2 that left the way open for summary dealing of unsuitable matters. These offences were under the Sexual Offences Act 2003, namely,

1 See on p. 50 above, concerning who is a person's CO.

2 Higher authority is defined as any officer in the CO's disciplinary chain of command who is superior in that chain of command to the CO. See s. 361 AFA 2006.

sexual assault (s. 3), exposure (s. 66) and voyeurism (s. 67).[3] This gap was closed by the Armed Forces Act 2006 (Amendment of Schedule 2) Order 2018 (AFA2006(AOS2)O 2018), which included them in Schedule 2 and made them triable by court martial as Schedule 2 offences. Unfortunately, the offence under section 71, sexual activity in a public lavatory, was not included. The offence is not one that falls within section 53, so the CO lacks any power to deal with it. However, since the provisions of the AFA 2016 were brought into force,[4] amending section 116 of the 2006 Act, the service police are now empowered to directly refer to the Director, Service Prosecuting Authority (DSPA) any non-Schedule 2 case that the CO cannot deal with (a non-CO offence). The section 71 sexual activity offence will now be dealt with as a direct referral by the service police. The 2016 Act amendment in section 166(2) has had the effect of making non-CO offences quasi Schedule 2 offences, requiring them to be referred by the service police to the DSPA. Section 20 wounding is also a non-Schedule 2 and non-CO offence, yet section 20 inflicting grievous bodily harm is a Schedule 2 offence. However, the practical effect of the amendment is to make all non-CO offences Schedule 2 offences that must be referred to the DSPA under section 116(2).

The changes in the 2006 Act ensure the CO remains at the centre of the disciplinary process, but they no longer have the power to prevent serious offences from going to trial. Under section 76(5) of the Army Act 1955, for example, even though they have no power to deal with a charge of murder, they could dismiss such a charge. Fortunately, section 9 of the Offences Against the Person Act 1861 provides concurrent civil jurisdiction which, if necessary, enables the Crown Prosecution Service to assume jurisdiction for murder and manslaughter offences committed overseas by service personnel in any case where military jurisdiction was not available or not exercised. Double jeopardy did not apply, as the CO had no legal power to deal with the charge, other than to dismiss it.[5] But the effect of their decision to dismiss the charge was only to oust military jurisdiction. It had

---

3 See Sch. 2(12).

4 Armed Forces Act 2016 (Commencement No. 3) Regulations 2019, r. 2, 1 July 2019.

5 The decision would have been subject to judicial review, anyway, under the principle in *R v Director of Public Prosecutions Ex parte Kebeline and Others* [1999] HL.

no effect on the concurrent jurisdiction of the UK civil authorities.[6] This has changed. The position under the Armed Forces Act 2006 is that a CO would not now have any legal power to deal with such a serious offence. As murder is a Schedule 2 offence,[7] the CO must, 'as soon as is reasonably practicable ensure that a Service police force is aware of the matter'. They will then investigate it and, if it is considered there is sufficient evidence to charge a person with a Schedule 2 offence, they 'must refer the case to the Director of Service Prosecutions'. Moreover, as the Manual of Service law points out,[8]

> Although the functions of the CO that can be delegated include ensuring that the Service police are aware of a matter where an offence under Schedule 2 may have been committed or where there are prescribed circumstances, as a matter of policy, it is a function that should not be delegated. This does not mean that the CO cannot instruct someone else to notify on his behalf, however, the obligation to ensure the Service police are aware remains with the CO. When the notification is given by someone other than the CO, the authority under which this has taken place is to be recorded, for example as the duty officer.

Staff legal advice to the CO will be provided for all CO offences. This will address the severity of the charge and the suitability for summary dealing or, in the more serious cases, where the CO's powers may be deemed insufficient, whether to refer the case to the SPA for court martial. It may also be provided where the CO is considering applying for EPs.

A Non-Judicial Process

Most modern forces in the world have a non-judicial process before the CO, and concerns about it have been raised before the courts. Concern about

---

6 See the case of *R v Williams* (2005): A charge of murder against this soldier was dismissed by his CO, ousting military jurisdiction and thereby preventing it from being referred to the prosecuting authority. Jurisdiction was exercised by the CPS, under s. 9 OAPA 1861, which charged him with murder. The charge was eventually discontinued. Had the case been referred to the Army Prosecuting Authority, with their understanding of the operational environment, it is quite likely that the case would never have proceeded to trial in the first place.

7 See Appendix C Disciplinary System.

8 See Chapter 6, p. 1-6-23, para. 67.

summary justice first raised its head in the European Court of Human Rights (ECtHR) as long ago as 1976, in the case of *Engel v Netherlands*.[9] The applicants were conscript soldiers serving in different non-commissioned ranks in the Netherlands armed forces. On separate occasions, various penalties had been passed on them by their respective COs for offences against military discipline. The applicants had appealed to the complaints officer and finally to the Supreme Military Court, which, in substance, confirmed the decisions challenged but, in two cases, reduced the punishment imposed. In upholding the general right of states to create and maintain a disciplinary system, the ECtHR held:

- Each State is competent to organise its own system of military discipline and enjoys in the matter a certain margin of appreciation.
- In order to establish whether a disciplinary penalty deviates from the normal conditions of life within the armed forces of the Contracting States, account should be taken of a whole range of factors such as the nature, duration, effects and manner of execution of the penalty or measure in question.
- The Convention without any doubt allows the States, in the performance of their function as guardians of the public interest, to maintain or establish a distinction between criminal law and disciplinary law, and to draw the dividing line, but only subject to certain conditions. The Convention leaves the States free to designate as a criminal offence an act or omission not constituting the normal exercise of one of the rights that it protects. But, if the Contracting States were able at their discretion to classify an offence as disciplinary instead of criminal, or to prosecute the author of a 'mixed' offence on the disciplinary rather than on the criminal plane, the operation of the fundamental clauses of Articles 6 and 7 would be subordinated to their sovereign will.
- The Court must specify, limiting itself to the sphere of military service, how it will determine whether a given 'charge' vested by the State in question – as in the present case – with a disciplinary character nonetheless counts as 'criminal' within the meaning of Article 6.

9 Application no. 5100/71, 1976.

- It is first necessary to know whether the provision(s) defining the offence charged belong, according to the legal system of the respondent State, to criminal law, disciplinary law or both concurrently. This however provides no more than a starting point.
    - The indications so afforded have only a formal and relative value and must be examined in the light of the common denominator of the respective legislation of the various Contracting States. In other words, the categorisation made by domestic law must be judged against the objective standards of other Convention countries in respect of the same conduct.
    - The very nature of the offence is a factor of greater import.
    - The degree of severity of the penalty that the person concerned risks incurring.

The decision is interesting in that, first, it confirms the general right of states to create and maintain a disciplinary system. Thus, in principle, a military system of justice does not offend against the Convention. Secondly, it sets parameters by which states define a criminal charge and, in particular, applies an objective test to that determination by paying regard to the severity of the *potential* penalty ('risks incurring'), rather than leaving it to the state concerned to define what is or is not criminal conduct. Irrespective of the charge used, when one considers the disciplinary charges and punishments that a CO may impose, some have a very clear appearance of being criminal and engaging Article 6, none more so than charges laid under section 42 of the 2006 Act[10] or those service offences where a CO proposes to award detention, especially an extended punishment of up to 90 days.

The table of punishments that a CO may award under section 132 is reproduced on page 301.

10 On the point of the charges laid, see *Ezeh and Connors v UK*, Applications 39665/98 and 40086/98 (Grand Chamber): concerning two prisoners who were denied legal representation before an adjudication hearing with the governor, at which both were awarded additional days' custody. Their application for extension of time to judicially review the decision was denied as the Governor's exercise of his discretion not to allow such representation was not irrational. The ECtHR found that the nature of the charges, together with the nature and severity of the penalties, were such that the charges against the applicants constituted criminal charges within the meaning of Article 6 of the Convention, which Article applied to their adjudication hearings. It also found a violation of Art. 6(3) concerning the right to legal representation.

Legal Advice

Although there is no right to legal representation at the hearing before the CO, nor is there any entitlement to legal aid for the purpose of advice,[11] that does not mean that service personnel may not seek advice at their own expense.[12] The CO or a person authorised by them should inform the accused that they may consider seeking legal advice and that this should be a matter they discuss with the Accused's Assisting Officer (AAO), if they have nominated one. The only assistance the accused is entitled to at the hearing itself is that provided for under the summary hearing rules,[13] which permit the accused to nominate an AAO, to represent them. Moreover, a person may only be nominated as an AAO if they

> (a) [are] subject to Service law (which specifically excludes most civilian lawyers, save those serving with the TA [Territorial Army] or reservists currently on duty with the unit to which the accused belongs);
> (b) [are] of at least the rank or rate of petty officer, military or marine sergeant, or air force sergeant; and
> (c) consent to be nominated.

The AAO must be a service person subject to service law, of at least the rank or rate of petty officer, military or marine sergeant or air force sergeant, and consent to be nominated.[14]

Rule 10(3) of Armed Forces (Summary Hearing and Activation of Suspended Sentences of Service Detention) Rules 2009 (AF(SHASSSD)R 2009) states that a person may only continue to act as an AAO while subject to service law. If a person ceases to act as AAO the accused may nominate another person to do so, provided they satisfy the criteria. The AAO may, on behalf of the accused, exercise any right of the accused to question witnesses or make submissions. Knowing how to do this effectively is a challenge for a skilled advocate: the prospect of three months' detention may hinge on

---

11 There have been grave concerns about the role of the CO acting as investigator, prosecutor, judge and jury, all without legal representation for the accused. *Thompson v UK* (Application N° 00036256/97). The SAC was created to address these concerns.

12 Usually such legal advice is at the service person's own expense unless there are service legal facilities available, e.g. in overseas stations.

13 AF(SHASSSD)R 2009, r. 10.

14 Rule 10 AF(SHASSSD) R 2009.

the credibility of witnesses being undermined by skilful cross-examination. But at a summary hearing on a charge of, for example, section 47 ABH, the task is given to a lay AAO who supports the accused before a lay CO who, although they have received legal advice, has no immediate legal support on hand to deal with the myriad issues that might arise as a result of the evidence heard. If an issue of identification or hearsay arises, the CO is ill-equipped to deal with it. Similar problems appear when a charge of attempt, or aiding and abetting, is advised for summary hearing. As the AAO is not usually legally qualified, it is a very onerous role to assume, bearing in mind the consequences of summary dealing for the accused and the potential for a criminal record.[15] The Summary Appeal Court (SAC) exists as an appeal avenue for those found guilty by their CO. However, understanding what the basis of an appeal to the SAC should be – or whether to appeal at all – is itself dependent upon the accused and AAO themselves understanding the possible legal arguments that may have arisen in the summary hearing and the manner in which they were dealt by the CO. Even explaining this, *ex post facto*, to a legal representative may not reveal the true merit that exists in the appeal. This could be so easily remedied by allowing legal representation before COs when they are considering detention and especially when they propose applying for EPs. It does seem odd that prisoners, who have committed crimes against society, are permitted legal representation for the purpose of their disciplinary proceedings, yet service personnel who serve their country are denied that right.[16]

The CO's Initial Consideration

When a CO receives an allegation against a member of their unit under their command, they have to decide whether formal disciplinary action is necessary or whether the matter is better dealt with by way of a minor administrative punishment.[17] For example, being late on parade, which is akin to being late for work, or being drunk on duty may be considered more appropriate for minor administrative action. This was discussed in Chapter 8. However, where the circumstances are exacerbated by rank,

15 A finding of guilt ranks as a conviction for the purposes of the Rehabilitation of Offenders Act 1974, as amended. See also, on p. 277.
16 See p. 272,FN 10, *Ezeh v Connors*.
17 See JSP 833, Minor Admin Action.

such as where junior recruits or senior officers are present, or where the service person is required to operate essential equipment when loading for deployment, the CO may well decide that the conduct moves out of a mere failing or misjudgement and decides that formal disciplinary action is more appropriate.

Moreover, even where the CO decides that although the conduct is not disciplinary, but it is however of significance, then major administrative action may be warranted. This is particularly important in social misconduct cases under Army General and Administrative Instructions 67 (AGAI 67).[18] It is worth knowing that major administrative action may follow a disciplinary or criminal proceeding, for example, in the case of a service person dealt with by the civilian criminal courts for a serious offence that impinges upon the values and standards of their service. A service person sentenced to a term of imprisonment by a civilian court usually would be administratively discharged from the service, under QR.9.404, even where the sentence is suspended for a period, unless there are exceptional circumstances. See further, Chapter 8.

Where the CO is considering an award of detention and/or reduction in rank, the use of EP or activation of a suspended sentence, special considerations apply. These are separately discussed under their respective headings.

Where a service offence is be investigated, whether by a CO or by the service police, a complainant is entitled to be referred to victim support services, in accordance with JSP 839 and to be notified, without unnecessary delay, of their entitlement to receive information about the following, inter alia:

a. any decision not to proceed with or to end an investigation or not to prosecute a suspect, including a brief summary or reasons for the decision where available;
b. the time and place of the trial or Summary Hearing, and the nature of the charges;
c. against the accused;
d. an explanation, without unreasonable delay, of a decision not to investigate a crime;

18 These procedures are not dealt with in this publication, as they are non-criminal.

e. to be advised, without unreasonable delay, when an investigation into the case has been concluded with no person being charged or referred, and to have the reasons explained to you.

If the CO decides not to take disciplinary proceedings against an accused, that decision is, in principle, subject to judicial review. This is particularly so where the complainant has suffered harm (including physical, mental or emotional harm or economic loss) which was directly caused by a criminal offence committed by a person subject to service law or a civilian subject to service discipline, or has had such an allegation made on their behalf, or are contacted as a victim in the course of investigations.[19] For example, if the offence under report is a victim offence, such as occasioning ABH, contrary to section 46 of the Offences Against the Person Act 1861, the complainant in such a case may consider a challenge to the decision not to prosecute. An expression of dissatisfaction by the victim in relation to a review decision should not automatically be treated as a complaint. The appropriate way for a victim to challenge the reviewing CO's decision is by way of judicial review.[20]

Effect of Summary Hearing

Main Provisions: section 376 AFA 2006; ROA 1974

It is often, mistakenly, thought that summary dealing carries no consequences in terms of a police record. This is of particular importance to those Commonwealth soldiers who serve with our armed forces and who wrongly believe that 'taking it on the chin' before the Officer Commanding (OC) or CO means they can avoid a criminal record and thereby not impede any citizenship or residence application. They would be quite mistaken.[21] Section 2 of the Rehabilitation of Offenders Act 1974

---

19 See, Annex F to JSP 839, Commanding Officers' Victim Right to Review Scheme.
20 *Ibid.*, Annex F, para. 45.
21 A practical example is the case of Lance Corporal Isimeli 'Bale' Baleiwai, who was refused British citizenship after voluntarily leaving the Army and was ordered to be deported on 9 August 2012. He had a disciplinary record for an offence that equated to a criminal offence in that he was involved in an altercation (fighting) with another soldier. His CO dealt with the issue in a summary hearing. There was no jury, no defence and no witnesses. L/Cpl Baleiwai accepted a £1,000 fine. On 5 December 2012, Bale was informed that the Home Office would be granting him citizenship. (MoD Oracle report, July 31, 2012).

(rehabilitation of persons dealt with in service disciplinary proceedings), as amended to take into account the provisions of the AFA 2006, applies to disciplinary proceedings before a CO in respect of a service offence within the meaning of the Armed Forces Act 2006. Any finding that a person is guilty of an offence in respect of any act or omission that was the subject of service disciplinary proceedings shall be treated as a conviction, and any punishment awarded in respect of any such finding shall be treated as a sentence. See, in particular, section 2(5)(a), as amended by Schedule 16 to the 2006 AFA.

Schedule 16(64) of the AFA 2006 added to section 2 of the 1974 Act, by inserting into section 2(5) the additional provisions relating to 'service disciplinary proceedings', defining them as:

> '(za) any proceedings (whether or not before a court) in respect of a Service offence within the meaning of the Armed Forces Act 2006 (except proceedings before a civilian court within the meaning of that Act);' …
>
> '(6) Section 376(1) to (3) of the Armed Forces Act 2006 ('conviction' and 'sentence' in relation to summary hearings and the SAC) apply for the purposes of this Act as they apply for the purposes of that Act.'

The effect of section 376(1) is to treat any finding by the CO of guilt as a conviction. Conversely, that section confirms that any finding that the charge has not been proved (dismissal of the charge) is to be treated as an acquittal.

A recordable offence is *any* offence under section 42 of AFA 2006 that corresponds to an offence under English law, such as theft or common assault/battery, even when dealt with summarily by the CO. At present, the following service offences are also recorded:[22]

22 See MSL, Annex Q to Vol. 1 Ch 9. JSP 830 MSL and s. 2 Police and Criminal Evidence Act 1984 (Armed Forces) Order 2009 which defines 'recordable service offence' as —

> (a) an offence under section 42 of the Act for which the corresponding offence under the law of England and Wales is an offence specified in regulation 3 of the National Police Records (Recordable Offences) Regulations 2000; or (b) an offence listed in the second column of Schedule 1 subject to any restriction stated in relation to the offence in the third column of that Schedule (as listed above, in the main text).

a. section 11(1) – misconduct towards a superior officer
b. section 14 – using force against a sentry, etc.
c. section 24(1) – damage to public or service property
d. section 27 – obstructing a service policeman
e. section 28 – resistance to arrest, etc. (only in relation to a conviction under s. 28(1)(b) or (c) – using violence or threatening behaviour)
f. section 29 – offences in relation to service custody
g. section 30 – allowing escape, or unlawful release of prisoners, etc. (but only where the conviction is under s. 30(4)(a))
h. section 39 – attempts to commit any offences specified above
i. section 40 – encouraging or assisting the commission of any offence specified above in this Schedule (apart from an attempt)
j. section 96 or section 97 of the Reserve Forces Act 1996 – absence without leave
k. section 1 of the Theft Act 1968 – theft
l. section 12 of the Theft Act 1968 – taking vehicle, etc. without consent
m. section 5(2) of the Misuse of Drugs Act 1971 – possession of controlled drug
n. section 1(1) of the Criminal Damage Act 1971 – criminal damage
o. section 3 of the Theft Act 1978 – making off without payment where the payment required or expected did not exceed £100
p. section 9 of the Criminal Attempts Act 1981 – interference with vehicles
q. section 39 of the Criminal Justice Act 1988 – assault and battery
r. section 5 of the Road Traffic Act 1988 – driving a vehicle where driver has consumed excessive amount of alcohol, etc.

It is therefore important that those advising service personnel who are to be dealt with summarily before the CO are fully aware of the consequences for the accused. If an AAO is appointed, and bearing in mind the potential harm that may ensue, it would be sensible to advise the accused to seek independent (self-funded) legal advice if they are thought likely to be affected by a summary conviction. Many solicitors' firms that specialise in advising military personnel and families offer a free initial consultation. There is quite likely to be information available about local firms in garrison publications and notices, in addition to which local newspapers may carry advertisements. Otherwise, AAOs and accuseds can check on the internet and also the with the Law Society.

Finding of Guilt

Section 376 of the AFA 2006, refers to 'conviction' and 'sentence', etc. in relation to summary hearings and the SAC. It provides that, where a charge against a person in respect of an offence is heard summarily by an officer, and the officer records a finding that the charge has been proved, or the SAC substitutes a finding that a charge in respect of another offence has been proved, that shall be treated as a conviction. Any punishment awarded by the officer, or by the SAC , shall be treated as a sentence.

From an accused's perspective, it is important to realise that the effect of a finding of guilt by the CO may amount to a recordable offence[23] in certain circumstances,[24] and will be recorded on the Police National Computer system.[25] In any event, an entry will be made in the unit conduct sheets of the individual concerned, even though the summary hearing is a non-judicial process. This may impact upon promotion and, importantly, whether the accused's service is continued once they reach a 'manning control point'.[26] Internal policy of each service will determine the period for which an entry in a conduct sheet remains. A company conduct sheet would usually be destroyed after a specified period of service without any entry in any conduct sheet. For more serious offences attracting a regimental entry (Army), one would consider the rehabilitation periods applicable. See below on page 606 *et seq*.

Dismissal of Charge

If the officer dismisses the charge under section 131, or the SAC quashes a finding that the charge has been proved, section 376(4) states that the charge shall consequently be treated as an acquittal. See further, above, page 278, which lists the offences concerned.

---

23 See the definition on p. 277.
24 *Supra*, para 277.
25 The sort of offences falling into this category are listed in the MSL at Annex Q to Chapter 9.
26 This is the point at which a decision is made by the service whether the individual is suitable for further service/full career to pension. In the Army, for example, a soldier is engaged on a Notice Engagement or the Open Engagement. Manning Control Points after 3, 6, 9 or 12 years. See QRs (Army), para. 9.413.

The case of *Deacon*[27] is an interesting one. In 2007, Mr Deacon, then Senior Aircraftman Deacon, pleaded guilty before his subordinate commander to an offence of criminal damage. The offence was found proved and he was ordered to pay £40. The 14-day time limit for appealing to the SAC expired. In 2015, he applied for leave to appeal against this conviction out of time. One of the issues he raised was that both he and the subordinate commander were unaware that the conviction before the subordinate commander for a criminal offence would result in him having a criminal record. Had he known that, he said he would have exercised his right to appeal on 21 September 2007. He relied on the civilian case of *Stratton*,[28] which concerned a woman given a police caution without being told of the adverse consequences to her.[29] The Vice Judge Advocate General (VJAG) distinguished Mr Deacon's case on the basis that, while Ms Stratton had a choice of whether or not to accept the caution, when a caution is administered by the police it means that there will be no criminal proceedings provided that the person who is about to be cautioned admits the offence and they understand that this is something that will go on the Police National Computer and will become a public record. Ms Stratton had not been warned of those consequences. While that is an administrative process, the CO or subordinate commander, while not a compliant court, is acting judicially. Consequently, the VJAG refused to grant Mr Deacon's application on the basis that there was no injustice in view of the admissions made by him to the service police and the time that had elapsed since the original hearing.

The guidance contained in Annex G to Volume 1, Chapter 6 of the Manual of service law (MSL), JSP 380, 'Your Rights if You Are Accused of an Offence', now clarifies this and expressly explains the consequences.

---

27 5 May 2015 and 11 June 2015, unreported, HH Judge Hunter, VJAG.

28 *R (on the Application of Stratton) v Chief Constable of Thames Valley Police* [2013] EWHC 1561 (Admin).

29 It was common ground that the policy of the Chief Constable in relation to the decision to issue a caution required that the offender had to admit the offence, that the evidential and public interest test for prosecution had to be met and the consequences of acceptance of the caution had to be made clear so that the offender gave informed consent to the acceptance of a caution. However, the form did not make clear that the consequences to her were much more serious, particularly as she worked with children and she might require a Criminal Records Board check. Far from spelling out the consequences, the form implied that the only consequence was use in other court proceedings. The caution was be quashed.

This document is to be provided to the suspect by the person directed by the CO to investigate the service offence when a CO's investigation is undertaken in relation to a suspect who is not arrested. In addition, the suspect is to be provided with the Notice of Rights to Suspect, which is in Annex I to Chapter 4 of the MSL[30] (but not publicly available on the internet). The Investigating Personnel must note in writing the provision of these documents.

Some convictions referred to in Rehabilitation of Offenders Act 1974 (ROA) section 6(6)(bb) are to be disregarded for the purposes of extending a period of rehabilitation following subsequent conviction. These are listed in Schedule 1 to the 1974 Act and comprise the disciplinary offences.

The AFA 2006 provides, in Schedule 16(66), that this shall also include any 'Service offence within the meaning of the Armed Forces Act 2006 except one punishable in the case of an offender aged 18 or over with imprisonment for more than two years'.

These latter provisions are more relevant to sentencing after conviction on trial by court martial. See further, Rehabilitation Periods on page 606.

The powers the CO exercises at a summary hearing are statutory powers. As described at page 277, a finding of guilt ranks as a conviction under section 376(2). If, on the other hand, the CO is dealing with a case that they have jurisdiction over, as set out in sections 53 and 54, and dismisses a charge, that dismissal will be final, in accordance with section 376(4). It will not, necessarily, preclude subsequent administrative action, as the standard of proof required in the latter case is lower than that required for disciplinary action, but it does preclude further disciplinary action on those facts. See further, Double Jeopardy, on page 375.

Extended Powers

Main Provisions: sections 132, 133, 135 and 135A AFA 2006; section 12 and Schedule 3 of AFA 2011; Rules 5, 6 AF(SHASSSD)R 2009

A CO below the rank of rear admiral, major general or air vice-marshal (sometimes referred to as 2-star officers) who considers that the charge against a service person should be dealt with summarily, and that their

30 JSP 830 MSL Version 2.0 , at p. 1-6-15.

powers of punishment might be insufficient to deal with the accused if the charge is proved, shall make an application to higher authority for EP.[31]

'Higher authority' in relation to a CO means any officer in the CO's disciplinary chain of command who is superior in that chain of command to the CO.[32] COs of and above 2* rank have inherent EP[33] and, consequently, do not have need for a higher authority.

The CO has to make the application as soon as reasonably practicable, but *before the start* of the summary hearing,[34] and it must contain

a. the reasons for considering that their powers of punishment might be insufficient to deal with the accused if the charge is proved unless they have EP;
b. a copy of the charge sheet;
c. a copy of the written evidence relevant to the charge;
d. a copy of any unused written material gathered as part of the investigation of the charge;
e. a copy of any disciplinary record of the accused;
f. details of all provisions for the purposes of which the CO considers they needs EP; and
g. any other material that may, in the opinion of the CO be relevant to the application.

The very fact that the application must be made before the start of the summary hearing tends to suggest an element of pre-determination and bias on the part of the CO. However, it could be argued that, on the basis of the offence charged, should the accused be found guilty, the CO would wish to have the flexibility to award a more severe punishment than their ordinary powers would allow. It follows that, just because EP are granted, it does not mean that they must be used to their full extent. The CO may determine on the basis of the evidence and any mitigation that their ordinary powers are sufficient. Of course, the accused may be acquitted of the charge. It remains the case that the accused can still appeal to the SAC in the event of a finding of guilt, and have the case re-heard before

31 See r. 6 AF(SHASSSD)R 2009. It follows that a CO of the senior rank stated does not need to apply for EPs.
32 AFA 2006, s. 361.
33 EN para. 284 and s. 133(4).
34 S. 135A, inserted by Sch. 3(11) AFA 2011.

that European Convention on Human Rights and Fundamental Freedoms (ECHR)-compliant tribunal.

After the application is granted, the CO must provide the accused with a copy of the notification from higher authority that the application has been granted. The CO will now be able to impose, in the case of a person of or below the rank of leading rate, or LCpl/LBdr or LCpl in the RAF Regt or Cpl in any of Her Majesty's air forces, a period of detention of up to 90 days. See section 132(1) and section 133(1)(a).

If the application for extended powers is denied, on a finding of guilt the CO is limited to their ordinary powers within section 132.

The sort of charges falling within these circumstances are any offence under section 42 as respects which the corresponding offence under the law of England and Wales is

(a) an offence listed in Part 2[35] of Schedule 1;[36] or
(b) an offence under section 1 of the Criminal Attempts Act 1981 of attempting to commit an (indictable) offence so listed, or
(c) he wishes to impose a sentence of detention in excess of 28 days.[37]

The CO cannot make any such award unless they have been notified by higher authority that their application has been granted.[38] Without extended powers, section 133(1) of the Act makes it clear they are limited to 28 days detention.

RAF Regiment

This is an area of some complexity, especially when concerned with junior non-commissioned officers (JNCOs) from different services, as the ability to award detention under the 2006 Act to a corporal only applies to a 'Corporal in any of Her Majesty's air forces'.[39] Matters are further complicated by

35 See Appendix D.
36 R. 5, AF(SHASSSD)R 2009.
37 S. 133(1) AFA 2006.
38 S. 133(3) AFA 2006.
39 S. 132, Table point 1.

section 12 AFA 2011[40] which amended section 132 (punishments on summary dealing) and section 135 AFA 2006 (reduction in rank) to reflect the new rank of lance corporal in the RAF Regiment. Section 135(3) provides as follows

> (3) Where the person being punished is a corporal in any of Her Majesty's air forces, the reduction in rank authorised by subsection (2)(a) or (b)[41] (as the case may be) is reduction to the highest rank the person has held in that force as an airman; but this is subject to subsection (3A).
>
> (3A) In relation to the Royal Air Force Regiment, the reference in subsection (3) to a corporal is to be read as a reference to a lance corporal.

Section 135(1)(b) states that if the person being punished is a lance corporal or lance bombardier, the CO may reduce them to the rank of private soldier without the need to apply for EP. However, if the person being punished is a warrant officer or non-commissioned officer, they may only reduce them in rank in accordance with section 135, namely, by one acting rank or rate, and not to an extent prohibited by regulations under section 135(4).[42] It should be also be noted that a CO who is of or above the rank of rear admiral, major general or air vice-marshal also has extended powers for the purposes of subsection (1) or (2).

Notwithstanding the power to award EP, the Armed Forces (Minor Punishments and Limitation on Power to Reduce in Rank) Regulations 2009 (AF(MPLPRR)R 2009) authorise minor punishment that may be imposed on offenders by CO who have heard charges summarily. The minor punishments authorised are stoppage of leave orders, restriction of privileges orders and admonitions. See below, Minor Punishments, on page 488.

40 In force from 2 April 2012: AFA 2011 (Commencement No. 1, Transitional and Transitory Provisions) Order 2012.

41 S. 135 AFA 2006: '(2)The reduction in rank or disrating that a commanding officer may award is reduction or disrating—(a) by one acting rank or rate; or (b) if the person being punished has no acting rank or rate, by one substantive rank or rate.'

42 Which relate to limitations under Defence Council regulations. See the AF (MPLPRR) R 2009.

Offences That May Be Dealt With at a Summary Hearing

For a charge to be dealt with summarily under section 52, three conditions have to be met:

a. the offence is one that may be dealt with at a summary hearing as set out in section 53[43] and Schedule 1; and
b. the accused is
    (1) an officer of or below the rank of commander, lieutenant colonel or wing commander; or
    (2) a person of or below the rank or rate of warrant officer; and
c. the accused is
    (1) subject to Service law,
    (2) a member of a volunteer reserve force, or
    (3) a member of an ex-regular reserve force who is subject to an additional duties commitment, from the time the offence is committed to the end of the summary hearing of the charge.

Where the offence is committed by a member of the Reserve forces contrary to section 96 of the Reserve Forces Act 1996 (failure to attend for service on call out or recall), condition C is that the accused is

a. liable to recall, or
b. a member of the regular forces

from the time the offence is committed to the end of the summary hearing of the charge.

If the offence is any other Reserve Forces Act offence, condition C is that the accused is a member of a reserve force from the time the offence is committed to the end of the summary hearing of the charge.[44]

It will be noticed that there is no reference to a civilian subject to service discipline. In other words, a CO has no power to deal summarily with a charge against a civilian. That is a matter for the Service Civilian Court (SCC).[45] However, a CO may be appointed when a civilian subject to service discipline is to be charged with an offence, to be tried in either the SCC or by court martial. The CO must be specifically appointed under

43 See below, on p. 286 for offences that may be dealt with summarily.
44 See s. 52 AFA 2006 for more information on Reserve Forces Act 1996.
45 Online full text at http://bit.ly/ZqWQXd.

regulation 7 of the Armed Forces (Meaning of 'Commanding Officer) Regulations 2009 (AF(MCO)R 2009). See above, on page 54.

If an accused is of or above the rank of Lieutenant Colonel, Commander (RN) or Wing Commander, a CO for disciplinary purposes has to be specifically appointed for them under rules 3 and 4(3) of the AF(MCO)R 2009.[46]

Where the offence is listed in Part 2 of Schedule 1, or is an offence of attempting such an offence, section 54[47] requires that the CO must first obtain permission from higher authority before they may hear the charge. If that permission is refused, the charge must be referred to the DSP as there is no authority for it to be dealt with summarily.

The following service offences may be dealt with at a summary hearing:[48]

a. an offence under section 4(3) (any vehicle, equipment or stores abandoned by an enemy);
b. an offence under any of sections 9 to 15; [49]
c. an offence of malingering under section 16(1)(a), of pretending to have an injury or an offence under section 16(1)(c) (aggravating or prolonging an injury) committed by omission;
d. an offence under any of sections 17 to 29; [50]
e. an offence under section 30(1) of negligently doing an act that results in a prisoner's escape, or an offence under section 30(2 in releasing a prisoner without authority);

46 See the extract on p. 51.

47 And r. 5 AF(SHASSSD)R 2009.

48 S. 53 AFA 2006.

49 9. Absence without leave. 10. Failure to cause apprehension of deserters or absentees Insubordination etc. 11. Misconduct towards a superior officer. 12. Disobedience to lawful commands. 13. Contravention of standing orders. 14. Using force against a sentry. 15. Failure to attend for or perform duty.

50 17. Disclosure of information useful to an enemy. 18. Making false records etc. 19. Conduct prejudicial to good order and discipline. 20. Unfitness or misconduct through alcohol or drugs. 21. Fighting or threatening behaviour etc. 22. Ill treatment of subordinates. 23. Disgraceful conduct of a cruel or indecent kind. 24. Damage to or loss of public or service property. 25. Misapplying or wasting public or service property. 26. Defining in sections 24 and 25: 'public property' and 'service property'. 27. Obstructing or failing to assist a service policeman. 28. Resistance to arrest, etc. 29. Offences in relation to service custody.

f. an offence under any of sections 34 to 36;[51]
g. an offence under section 42 (criminal conduct) within subsection (3) and Schedule 1;
    ga. an offence under section 93A, 93E or 93G (testing for alcohol or drugs in connection with a suspected offence or accident);[52]
h. an offence under section 107;[53]
i. an offence under section 305 (random drug testing);[54]
j. any Service offence under regulations under section 328 (Enlistment) or 343 (service inquiries);
k. an offence under section 96 or 97 of the Reserve Forces Act 1996 of absence without leave.

Even though many practitioners appreciate the difficulties of dealing with attempts to commit offences, the CO may deal with attempts to commit any of the listed offences except a Schedule 2 offence[55] charged under section 42 (charging criminal conduct),[56] or an offence of aiding and abetting under section 41. See above, on page 241 *et seq*. This is subject to the caveat contained in section 54(2)(b), that the CO may only deal summarily with an attempt to commit a Schedule 1 Part 2 civil offence (the less serious criminal offences specifically listed) provided they have permission from higher authority, unless they are a senior officer (at least 2* rank).

Election for Trial by Court Martial

Main Provisions: sections 129 and 130 AFA 2006; regulation 12 AF(SHASSSD)R 2009; Schedule 2(8) AF(CM)(A)R 2009; Schedule 5 AFA 2011; regulation 6 AF(CM)(A)R 2013

Before starting the hearing, the CO must satisfy themselves that the accused understands the charge or charges, has had a reasonable time to prepare for

51 34. Low flying. 35. Annoyance by flying. 36. Inaccurate certification.
52 Inserted by Sch. 4(4)(2) AFA 2011. See also AFA2011(C4)O 2013, and further amended by s. 2(1) AFA 2016.
53 Complying with conditions imposed by a judge advocate at a custody hearing.
54 Amended by Sch. 4(4)(3) AFA 2011. See also AFA2011(C4)O 2013.
55 See Appendix E.
56 S. 53(2) AFA 2006.

the hearing,[57] and they must give the accused the opportunity to elect court martial trial before the charge is heard.[58]

If the accused elects court martial[59] trial the CO must refer the charge to the DSP. The DSP may not, without the written consent of the accused, refer that charge to a CO, under section 125(2)(e) (whether or not amended by the Director), or any charge substituted under section 125(2)(b) or additionally brought under section 125(2)(c). In other words, the accused will be tried only on the charge on which they elected trial. Where two or more charges against the accused are to be heard summarily together, an election for court martial trial in respect of any of the charges takes effect as an election in respect of all of them.[60] If a summary hearing has started and any charge is amended, substituted or added after the start of the proceedings, the accused must be given the opportunity to elect trial by the court martial in respect of that charge.[61]

The effect of an election for trial, on the subsequent sentencing powers of the court martial in the event of a conviction, are explained below, on page 457.

The Procedure at Summary Hearing

Main Provisions: sections 53, 54, 129, 130, 193, 194 AFA 2006; AF(SHASSSD) R 2009; AF(ITADT)R 2013

As the summary hearing before the CO takes place without a legal representative acting for the accused, this section has been included so that those called upon to advise an accused privately, at the accused's expense prior to summary hearing, or on appeal to the SAC,[62] will have an understanding of the process. The consequences of a finding of guilt are explained on page 279.

As the MSL reminds us:

---

57 Rule 12 the Armed Forces (Summary Hearing and Activation of Suspended Sentences of Service Detention) Rules 2009.

58 S. 129 AFA 2006 and r. 12 AF(SHASHSD)R 2009.

59 These complex provisions are helpfully explained in the Notes to the 2011 Act.

60 S. 129(3).

61 See s. 129 AFA 2006.

62 When Legal Aid may be available.

> [the] summary hearing is not a court. It is an inquisitorial process in which the CO endeavours to discover facts by actively searching for evidence and questioning the witnesses. No rules of evidence apply as such, but the principles ensure both best practice and fairness to the accused. [63]

The procedure is, in fact, recognised as not compliant with the requirements of a fair trial under Article 6 of the ECHR. The SAC is what makes the process compliant, coupled with the right to elect trial by court martial. This right to elect court martial trial of the charge must be given *before* the CO hears the charge summarily (s. 129).

The offences that may be dealt with at a summary hearing are listed in section 53 – see page 286 above. Section 124 permits the CO to hear charges summarily unless they exercise their powers under section 123(2)(b), (d) or (e) in respect of the charge (substitution of charge, discontinues proceedings or refers the charge to the DSP), or the accused elects court martial trial of the charge (see s. 129).

Certain charges may only be heard summarily with permission from higher authority (s. 54). If the CO requires permission, they must apply to higher authority, stating the reasons and sending a copy of the charge sheet, written evidence, unused material, disciplinary record of the accused and any other material that may in the opinion of the CO be relevant to the application.[64] The offences where authority must be sought are those that fall within Schedule 2 or an offence under section 1 of the Criminal Attempts Act 1981 of attempting to commit an (indictable) offence listed therein.

There is an exception to this requirement to apply for permission, which arises where the CO is of or above the rank of rear admiral, major general or air vice-marshal (2* rank).

The CO is at liberty to amend the charge or substitute another charge against the accused or bring an additional charge against the accused. These charges must be capable of summary hearing and must be connected with the case for which the original charge was brought. Alternatively, the CO may discontinue proceedings on the charge or refer the charge

63 Ch. 11, para. 2.
64 AF(SHASSSD)R 2009, r. 5 and s. 54(2).

to the Director of Service Prosecutions (DSP) (s. 123). These powers may be exercised before or after the start of any summary hearing of the charge.

If the CO does not feel that their summary powers of punishment are sufficient, and they are below the rank of rear admiral, major general or air vice-marshal, they may apply as soon as reasonably practicable to higher authority for EP. The application should include similar documentation to that required for a section 54 application, but should also include details of all provisions for the purposes of which the CO considers they need EP.

Where a CO below 2* rank intends to make an activation order in respect of a suspended sentence, they should also consider whether their powers might be insufficient to deal with the accused if the charge is proved unless they have extended powers for the purposes of section 194 of the Act (the maximum term for applicable) as, ordinarily, the term of a suspended sentence as it takes effect by virtue of an activation order under section 193 must not exceed 28 days, unless the officer has EP for activation of the sentence for the maximum term under section 194.

Preliminary Procedure

The Armed Forces (Summary Hearing and Activation of Suspended Sentences of Service Detention) Rules 2009 (AF(SHASSSD)R 2009) govern the summary procedure in rules 8–28.

As soon as reasonably practicable, the CO, or a person authorised by the CO, shall prepare a summary of the evidence relevant to the charge ('the case summary') and inform the accused in writing of

a. their right under section 129 of the Act to elect court martial trial;
b. their right under rule 10 to representation;
c. their right under rule 15 to question witnesses whose evidence is adduced by the commanding officer;
d. their right under rule 16 to give evidence;
e. their right under rule 17 to adduce the evidence of witnesses;
f. their right of appeal under section 141 of the Act to the SAC.

The service accused is usually provided with the booklet entitled 'Your rights if you are accused of an offence under the Service justice system' (Annex G to Vol. 1 Chapter 6, JSP 830], which covers these matters in some detail.

Unless the accused has already been already provided with the documentation, the CO or a person authorised by them must provide the accused with

a. a copy of the charge sheet;
b. a copy of the case summary;
c. a copy of the written evidence relevant to the charge;
d. details of all exhibits which form part of the evidence relevant to the charge together with notice of where and when they may be inspected by the accused or their AAO;
e. a copy of any unused written material gathered as part of the investigation of the charge;
f. details of any other unused material gathered as part of the investigation of the charge, together with notice of where and when it may be inspected by the accused or his AAO;
g. an opportunity to inspect all exhibits which form part of the evidence relevant to the charge, and any unused non-written material;
h. a copy of any disciplinary record of the accused.

If the offence that is the subject of the charge is alleged to have been committed during the operational period of a suspended sentence of service detention passed on the accused by an officer or the SAC, the CO must inform the accused in writing of:

a. the power to make an activation order under section 193 of the Act. (The nature of activation orders is discussed in the section on Suspended Sentences, at page 307 *et seq*.);
b. the right of appeal under section 141 of the Act and the effect of section 195 of the Act;
c. the right to make submissions under rule 23(3), either orally or in writing, about the appropriateness of making an activation order and the appropriate terms of such an order if it were made.

The CO will then fix a time for the hearing, in accordance with rule 8(4) and allow the accused a reasonable time to prepare for the hearing. Once a hearing date has been fixed, the CO may, at any time before the hearing, fix a different time for the hearing.

The time fixed for the hearing shall be not less than 24 hours after:

a. the CO has complied with the steps described above; or
b. where permission to hear the charge summarily has been granted, under rule 5(4);[65] or
c. where an application for EP under rule 6(1) has been granted,[66] rule 6(3); or
d. where an application for EP in relation to an activation order under rule 7(1) has been granted, rule 7(3).[67]

Rule 10 provides that the accused may nominate someone to act as their AAO who may, on behalf of the accused, exercise any right of the accused to question witnesses or make submissions at the hearing. (See page 273 above. Accused's Assisting Officer).

The Hearing

Rule 12 of the AF(SHASSSD)R 2009 requires that before starting the hearing the CO shall satisfy themselves that the accused understands the charge or charges; has had a reasonable time to prepare for the hearing; and, if the charge is capable of summary dealing, the CO shall give the accused the opportunity to elect court martial trial. Where two or more charges against the accused are to be heard summarily together, an election for court martial in respect of any of the charges takes effect as an election in respect of all of them.[68] If the charge is capable of summary dealing, or if the accused does not elect court martial trial, the CO, or a person authorised by them (for example, the Adjutant), shall start the hearing by reading the charge or charges to the accused and asking them to state whether they admit or deny each charge.

---

65 This relates to a charge in respect of an offence in section 54(2) of the Act, being a civil offence under section 42 listed in Part 2 to Sch. 1, or an offence under section 1 of the Criminal Attempts Act 1981 of attempting to commit an (indictable) offence so listed, and the CO is below the rank of rear admiral, major general or air vice-marshal.
66 That is, an application for EPs of punishment. The CO must provide the accused with a copy of the notification from higher authority that the application has been granted.
67 Where an application for EPs in relation to an activation order is made, the CO shall provide the accused with a copy of the notification from higher authority that the application has been granted.
68 S. 129(3).

Interpreter

Rule 11A and rule 33A were inserted into the AF(SHASSSD)R 2009 by rule 10 and 13 of Armed Forces (Interpretation, Translation and Alcohol and Drug Tests) Rules 2013 (AF(ITADT)R 2013). The CO is required to appoint an interpreter to act at the hearing, unless they are satisfied that the accused does not need interpretation. On application or on their own initiative, the CO may require a written translation of any document or part of a document to be provided for the accused, unless translation of that document, or part, is not needed to explain the case against the accused, or the accused agrees to do without, and the CO is satisfied that the agreement is clear and voluntary and that the accused has had legal advice or otherwise understands the consequences.

The same considerations apply where there the offender suffers from a hearing or speech impediment (r.11A(5)).

Co-accused and Multiple Charges

Where there are co-accused and multiple charges, the CO may[69] at a single hearing hear a charge brought against more than one accused. In addition, the CO may hear charges against more than one accused if the charges are founded on the same facts. If more than one charge is brought against an accused the CO may at a single hearing hear all charges that are founded on the same facts and all charges in respect of offences that form, or are part of, a series of offences of the same or a similar character.

Accused Admits the Charge

If the accused admits all of the charges, the CO or a person authorised by them will read the case summary to the accused and the CO shall ask the accused whether they dispute any of the facts contained in the case summary. If the accused does not dispute any of the facts contained in the case summary the CO shall treat them as the facts of the case for the purposes of sentencing and rule 22 shall apply (which relates to the sentencing procedure) (rule 21).

Where there is a dispute on the facts, and the CO considers that any of the disputed facts is relevant to sentencing, they shall determine the

---

69 R. 11 AF(SHASSSD)R 2009.

facts of the case for the purposes of sentencing and rule 22 shall apply (see Sentencing, below on page 297). In order to determine the facts, the CO shall assume that any undisputed facts contained in the case summary are true and hear evidence on the disputed facts from such witnesses as they consider can give relevant evidence and from the accused, if they wish to give evidence. Any witness called to the hearing shall give evidence orally and the accused will be given an opportunity to question them.

Accused Denies the Charge

If the accused denies any of the charges (they may admit some but not others), and does not elect trial by court martial, the CO shall proceed to hear the evidence in the order set out in rule 13 in relation to those charges. This stipulates the following order:

(a) evidence adduced by the commanding officer under rule 15(1);
(b) evidence of the accused if he chooses to give evidence;
(c) evidence adduced by the accused;
(d) evidence adduced by the commanding officer under rule 15(2).

All oral evidence given during the summary hearing is on oath (r. 14).

The procedure in rule 15 operates in the following manner. First, any evidence adduced by the CO under rule 15(1) is heard, provided the witness made a written statement and that a copy of it was provided to the accused. Where the CO adduces evidence of a witness, they can read out to the accused any statement made by the witness and may question the witness first. The CO will then allow the accused to question the witness, whereafter the CO will be permitted to question the witness once more (this is in the context of matters arising out of the accused's cross-examination) (r. 15(3)).

This evidence is followed by evidence from the accused, if they choose to give evidence. They cannot be compelled to do so.[70] If the accused does gives evidence they may do so orally or in writing and the CO may question them. If the accused gives evidence in writing their evidence

70 R. 16(1).

shall be read to the CO by the accused or their AAO. The accused may not advance evidence unless they notified the CO not less than 24 hours before the hearing that they wish to adduce evidence of the witness or they have the permission of the CO (r. 17). There is a restriction on adduction of evidence by the accused where 48 hours or more have passed after the CO has complied with rule 8(1) concerning the preparation of the summary of evidence and notice to the accused.

If, as a result of hearing evidence from the accused, an issue of fact arises that could not have been foreseen by the CO, and in relation to which no witness whose evidence may be given can give evidence, then the CO may adduce the evidence of a witness under rule 15(3) provided

(a) that witness has made a written statement;
(b) a copy of the statement has been provided to the accused in accordance with rule 8(1)(c)(iii); and
(c) where the witness is to produce an exhibit, the commanding officer has complied with rules 8(1)(c)(iv) and (vii).

Where the accused neither admits nor denies a charge, the CO shall proceed as if the accused had denied that charge (r. 12(5)).

Where the CO proceeds with a hearing after a charge has been amended, substituted or added, they may determine the order in which evidence is heard (r. 13(2)).

While the CO must try to ensure that the best evidence rule is observed, they are not bound to use original documents. They are free to take legal advice from a staff legal officer on the point and how much weight should be given to the evidence. At the hearing, the CO will either receive oral testimony from witnesses (including any called by the accused), witness statements and other documentary evidence and/or the 'real evidence', for example in the form of exhibits (such as recovered stolen or damaged items related to the charge). Indeed, the CO may find themselves facing some quite complex evidential matters, whether from oral testimony or in the witness statements, concerning identification, opinion and/or hearsay evidence which they are not best qualified to distinguish and which may be very prejudicial to the accused.[71] They also have to decide relevance. The MSL recognises that

71 Evidence which, if the decision is appealed to the SAC, would be ruled inadmissible.

> Hearsay is a complex area of the law of evidence and it is important that the officer hearing the charge thinks carefully about seeking staff legal advice if he considers that it may be an issue or he should consider carefully referring the charge to the DSP for Court Martial (CM) trial. [72]

Similar advice is given in the MSL in relation to issues of identification[73] with the rider that whenever an issue regarding identification arises,

> the potential evidential issues surrounding identification will be sufficient to justify referral of the charge to the DSP for CM trial.

These are encouraging statements, but they are predicated on the CO actually understanding that one of these issues has arisen during the proceedings, if there is no accused's legal representative to point out the fact.

The limitations of summary hearing are further highlighted by the fact that a CO cannot compel a civilian witness to attend the hearing. In relation to the accused themselves, unlike in a criminal case in the civilian courts, an adverse inference should not be drawn against them if they do not give evidence. After all, they have no legal representative there to guide them, to test the evidence or to make a submission concerning the inherent weaknesses of the case against them. All the same, it must be said that there is no guarantee that the CO does not in fact draw an adverse inference.

Where the CO considers that it would be in the interests of fairness to the accused to do so, they may at any time before determining whether or not the charge has been proved give the accused a further opportunity to question any witness whose evidence has been adduced. Where the accused has been provided with such an opportunity to question a witness the CO may also question the witness.

Finding

The CO is able to dismiss the charge at any stage of the hearing if the evidence fails to meet the legal standard.

At the conclusion of the evidence the accused may address the CO on any matter, following which the CO shall determine whether or not each

72 MSL, Ch. 11, para 30.
73 *Ibid.*, paras 33 and 34.

charge denied by the accused has been proved and record their finding and may take one of the following courses:[74]

a. If the CO determines that the charge has not been proved, they must dismiss the charge.
b. If the CO determines that the charge has been proved, they must
   (1) record a finding that the charge has been proved; and
   (2) award one or more of the punishments authorised by section 132.

Sentence

The CO has to give the accused the opportunity to adduce evidence as to their character and to make a plea in mitigation (r. 22). Character evidence can be given orally or in writing but, in the latter case, it shall be read to the CO by the accused or the AAO and the accused must provide the CO with a copy of the evidence. The accused must also be given an opportunity to question a witness called on their behalf. The CO may also ask the witness questions. After hearing any evidence as to character and any plea in mitigation the CO shall, if the accused admitted the charge, determine that the charge has been proved and record the finding and award one or more punishments, in accordance with section 131(4)(b) of the Act.

Where the CO records findings that two or more charges against a person have been proved, the award is a single award (consisting of one or more of those punishments) in respect of the charges taken together (s. 131(5)). Nothing in section 131 authorises a CO to include in an award two or more punishments within the same row of the Table in section 132, except where those punishments are

a. different minor punishments[75] which regulations made under row 7 of the Table permit to be awarded together; or
b. service compensation orders.

The CO may also make an activation order where appropriate (r. 23(2))[76] and, must also give reasons for deciding on the sentence passed, in accordance

74 S. 131 AFA 2006.
75 See Minor Punishments, below on p. 488.
76 See the discussion on Suspended Sentences, at p. 307 *et seq*.

with section 252(1)(a) of the Act. The accused must be reminded of the right of appeal under section 141 of the Act; the right to seek independent legal advice on whether to exercise the right of appeal; and if service detention has been awarded, the right to make an election under whichever of sections 290(2) and 291(2) of the Act applies (whether to elect to serve the sentence forthwith, or to defer it for up to 14 days beginning with the date on which the punishment was awarded, while they consider whether to appeal to the SAC).

Burden and Standard of Proof

Main Provisions: sections 1 to 41, 107, 229, 266 and 306 AFA 2006; section 325 AFA 2006;

The legal burden of proof rests with those seeking to prove the accused was guilty of the offence charged. That is, whoever asserts must prove. The standard to be achieved is proof beyond reasonable doubt. What amounts to a reasonable doubt is a matter of common sense. It does not mean absolute certainty and is often described as being sure, based on the evidence, of the accused's guilt. Being 'sure' and being satisfied on the evidence 'beyond reasonable doubt' mean the same thing.[77]

Unless the case against the service person makes a negative averment of a matter specifically within their knowledge (for example, they did *not* have motor insurance), when the service person will bear the evidential burden of proving that fact, the legal burden of proof is on the CO to establish that the charges are proved to the criminal standard, that is, beyond reasonable doubt. The CO acts as investigating magistrate. Clearly, if the CO is a witness in the case they should recuse themselves and pass the case to another CO to deal with. Personal association with an accusation against an accused or which might objectively be viewed as prejudicial would suffice as a reason to recuse.[78]

---

77 Crown Court Compendium, Part I, June 2020, at 5-2.

78 Clear guidance on recusal was given in the cases of *In Re Pinochet* [1999] HL. The majority of the House (3 to 2) held that Senator Pinochet was not entitled to immunity as a former head of state. Lord Hoffmann (one of the majority) did not give a separate judgment. Lord Hoffmann was a director and Chairperson of Amnesty International Charity Limited. The Pinochet team were unaware of that connection between Lord Hoffmann and Amnesty International

However, there are circumstances when the defendant bears an evidential burden to place sufficient information before the court by way of, for example, an excuse or explanation. Where an offence is charged under any of sections 1 to 41, 107, 229, 266 and 306 AFA 2006, a defendant who has a lawful excuse or a reasonable excuse 'is to be treated as not having had a lawful excuse or reasonable excuse (as the case may be) unless sufficient evidence is adduced to raise an issue as to whether he had such an excuse' (s. 325). In other words, once the defence raises sufficient evidence of the existence of an excuse, on the balance of probabilities, the 'prosecution' bears the persuasive burden and must prove the charge before the CO to the legal standard of beyond reasonable doubt. In meeting the evidential burden, the defendant has to show their explanation is more likely than not.[79] Section 107, for example, concerns the release of a service person from custody after charge. If any condition has been imposed on them by virtue of subsection (3)(a), that person commits an offence if, without *reasonable excuse*, they fail to attend any hearing to which the requirement relates. This allows the defence to put forward an excuse. It is for the prosecution then to prove beyond reasonable doubt it was not a reasonable excuse.

There is ample case law on the matter of when the evidential burden arises, of which the House of Lords decision in *R v DPP ex parte Kebeline*[80] remains instructive. It also remains slightly controversial as it is claimed to be contrary to Article 6 of the ECHR, a point considered in the *Kebeline* case.

However, the ECtHR has never made a ruling to the effect that a reverse onus provision will inevitably give rise to a finding of incompatibility with the Convention. The approach of the European Court to reverse onus provisions is clearly set out in *Salabiaku v France* (1988) 13 EHRR 379 (at para. 28):

---

until after the judgment was given on 25 November. Senator Pinochet made no allegation of *actual* bias against Lord Hoffmann; his claim was based on the requirement that justice should be seen to be done as well as actually being done. The order made on 25 November 1998 concerning the warrants for Senator Pinochet's extradition to Spain was set aside. See also the decision in *Findlay v UK* [1997], at paras 75 and 76.

79 *Carr Briant* [1943] KB 607.

80 [1999] UKHL 43 (28 October, 1999): https://publications.parliament.uk/pa/ld199899/ldjudgmt/jd991028/kabel-1.htm. See para. 25 of Points For Prosecutors, issued by the Legal Secretariat to the Law Officers, 2000. See also, *Shirley Porter v United Kingdom, Application no. 15814/02*, where the Court found that neither the fact that an evidential burden of proof was imposed on the applicant nor the fact that the applicant was unable to cross-examine the Auditor disclosed any appearance of unfairness in breach of Article 6 of the Convention.

> Presumptions of fact or of law operate in every legal system. Clearly, the Convention does not prohibit such presumptions in principle. It does, however, require the Contracting States to remain within certain limits in this respect as regards criminal law. ... Article 6 para. 2 (Art. 6-2) does not therefore regard presumptions of fact or of law provided for in the criminal law with indifference. It requires States to confine them within reasonable limits which take into account the importance of what is at stake and maintain the rights of the defence.[81]

Writing in the *Cambridge Law Journal*, David Hamer makes the point that:

> The presumption of innocence ordinarily places far greater weight on the defendant's rights. The reverse burden favours law enforcement. This shift must be justified. In restriking the balance 'reasonable proportionality' must be maintained.[82]

Legal advice to the CO should point out the particular burdens and standards of proof applicable in any case to be heard summarily.

General Sentencing Principles

The service justice system will apply general sentencing principles whenever an offender is sentenced by the CO or by a service court. These principles are helpfully set out in Part 9, Chapter 1 of the Act (ss 237–254). In the case of a CO sentencing at a summary hearing, the purposes of sentencing that are pertinent and should be considered are discussed at page 443 below.

The very informative JAG Sentencing Guidance adds an additional consideration to those listed in section 237, namely, the making of reparation by offenders to persons affected by their offences.[83]

Summary Punishments

If, after considering the evidence against the accused the CO decides that the accused is guilty, and subject to what is said below concerning EP from

81 Prashant Popat, 'Reverse Burden Of Proof: Developments in the Law', Health, Safety & Environment Seminar, 26 February 2003.

82 'The presumption of innocence and reverse burdens: A balancing act', *Cambridge Law Journal*, March 2007, at p. 147.

83 JAG Sentencing Guidance, para. 2.

higher authority, the available choices of punishment are those set out in section 132 of the Act, being limited to 28 days detention. Some of these punishments are explained in more detail below.

1	Detention for a term not exceeding the maximum allowed by s. 133	only if the person being punished is of or below the rank of— (a) leading rate; (b) lance corporal or lance bombardier; (c) corporal in any of Her Majesty's air forces (but see below)
2	Forfeiture of a specified term of seniority or of all seniority[a]	only if the person being punished is an officer, and only in accordance with s. 134
3	Reduction in rank, or disrating	only if the person being punished is a warrant officer or non-commissioned officer, only in accordance with s. 135, and not to an extent prohibited by regulations under s. 135(4)[b]
4	A fine not exceeding the maximum amount allowed by s. 136[c]	
5	A severe reprimand or a reprimand	only if the person being punished is an officer, warrant officer or non-commissioned officer
6	A service supervision and punishment order (defined by s. 173) – see p. 319	only if the person being punished is an able rate, marine, soldier or airman
7	Such minor punishments as may from time to time be authorised by regulations made by the Defence Council	
8	A service compensation order (defined by section 175) of an amount not exceeding the maximum allowed by s. 137	

Table 9.1 Summary Punishment

a. The officer may lose all seniority in his current rank. The effect is that the Incremental Base Date (ICB), which is calculated from the officer's seniority date as the day upon which they were promoted to the rank held, including for pay purposes an acting rank, is set back by the period of the forfeiture. Any further promotion is then assessed from the adjusted ICB.
b. AFA 2011, s. 13 substitutes for subsections (2) and (3) the provisions: '(2) If he awards detention, the only additional punishments he may award are: (a) reduction in rank or disrating (subject to subsection (8)); (b) a service compensation order.'
c. The maximum amount of a fine that a commanding officer may award is 28 days' pay. See s. 136 AFA 2006, amended by Sch. 3(12) AFA 2011. Where all the offences are under s. 42 (criminal conduct) and the maximum fine allowed by s. 136 exceeds the permitted maximum, nothing in ss 131–138 authorises them to award a fine exceeding the permitted maximum.

The caveat on imposing detention on an JNCO listed is that the CO may only do so if they have permission from higher authority and are granted EP.[84] The award of service detention is discussed at page 305.

Reduction in Rank

Main Provisions: sections 132, 135, 135A AFA 2006; JSP 754; AF(MPLPRR) R 2009; see also page 284

Reduction in rank by the CO under section 132 is limited to one rank. See also JSP 754, which sets out the rules governing *Pay on Reduction in Rank and on Loss of Seniority due to Disciplinary Reasons* and explains the financial consequences. These provisions are important not only for sentences awarded by a CO, but also those sentences handed down by the court martial.

If the person being punished is a warrant officer or NCO, section 135 provides that the CO may order the reduction in rank or disrating by one acting rank[85] or rate or, if the person being punished has no acting rank or rate, by one substantive rank or rate. There are limits on the exercise of this power, as set out in section 135. Essentially, the CO may not award reduction in rank, or disrating, unless first obtaining EP before the start of the summary hearing. However, practitioners should note that, exempted from this requirement for EP is the case where it is proposed to reduce the rank of a lance corporal or lance bombardier, in respect of whom section 135(1)(b) clarifies that EP are not required in order to do so.

Special provisions apply in respect of Corporals in the RAF and the RAF Regiment. See further on page 281 *et seq*.

---

84 S. 133(2) AFA 2006.

85 Acting rank carries with it the responsibilities of the rank worn, together with pay and pension benefits, but it has not been substantiated in the *London Gazette*. This is used, for example, where a unit has insufficient personnel formally qualified for a role, so suitable personnel are appointed to acting rank to fulfil these duties. QR RAF §119 points out, *mutatis mutandis*, 'When an officer has relinquished acting rank and is reappointed to such rank at a later date, he will reckon his subsequent seniority from the date of reappointment only. Nothing in this paragraph affects the counting of service in acting rank for the purpose of pay, retired pay or promotion.' They take command and precedence below all officers holding substantive rank corresponding to their acting rank but above all officers holding substantive rank in the rank below. Airmen holding acting rank are to take precedence next after those holding corresponding substantive or temporary ranks and, among themselves, according to the dates they attained their acting ranks (QR RAF, §125(5)).

Forfeiture of Seniority

Forfeiture of a specified term of seniority or of all seniority is a punishment that the CO may only award to an officer. In order to award this sentence, the CO must have applied for EPs to higher authority, which must have been granted. A CO also has extended powers for the purposes of section 134 if they are of or above the rank of rear admiral, or air vice-marshal.

This award should not be thought of as a trivial one. For the officer concerned it may be tantamount to a reduction in rank and carry with it career and financial consequences for retired pay, pension entitlement or gratuity purposes. Section 3.5 of the Sentencing Guide helpfully explains:

> 3.5.1 This punishment applies only to commissioned officers. An officer's seniority date is the day upon which he was promoted to the rank held, including for pay purposes an acting rank – this is the Incremental Base Date (IBD). From that date he builds up seniority. Annual increments of pay (Incremental Levels (IL)) are applied on each anniversary of the IBD up to a specified maximum level. Promotion to the next rank will normally depend partly on achieving a minimum amount of seniority in the current substantive rank. Forfeiture of seniority for an officer can, in some respects be equivalent to reduction in rank or disrating for other ranks. Although it does not carry the same visible stigma, the impacts on pay, pension and promotion prospects can be significant.
>
> 3.5.2 The court may order that an officer forfeits either all seniority in the present rank or some specified lesser amount of seniority (given in years and months). The effect of the punishment is that the officer's IBD is set back by the period of the forfeiture.

The punishment is also available to the court martial.

Fine

The maximum fine that a CO may award is 28 days' pay. A 'day's pay' is defined in section 136 as:

(a) subject to paragraph (b), the gross pay which is (or would apart from any forfeiture be) issuable to the offender in respect of the day when the punishment is awarded;
(b) if the offender is a special member of a reserve force, the gross pay which would have been issuable to him in respect of that day if he had been an ordinary member of that reserve force of the same rank or rate.

If the offender is a member of a reserve force who is not on duty on the day the punishment is awarded, they are deemed to have been on duty then.

A 'special member' is someone who, in pursuance of arrangements between their employers and the Secretary of State for the provision of support services, enters into an employee agreement to become a special member of a reserve force.[86] They are also known as 'sponsored reserves'.[87] Interestingly, special members of a reserve force are not counted in the numbers of officers and personnel for the time being authorised for the force.[88]

An 'ordinary member' is defined in section 49 Reserve Forces Act 1996 (RFA 1996) as a member of a reserve force who is not a special member of that force, and is not a member of that force for the purpose only of becoming a special member. In other words, an ordinary member is somebody who joined up to do the things that a service person usually does when on duty.

Detention

Main Provisions: sections131–133, 141, 189, 193, 290, 291 AFA 2006; Part 9, Chapter 1 of AFA 2006 (ss 237–254); AF(SHASSSD)R 2009

The power to award a punishment of service detention is found in section 132. There are rank restrictions on the power to award detention, as detailed in section 133(2), which provides that a CO may *only* award detention to a person being punished who *is of or below* the rank of leading rate, lance corporal or lance bombardier, or corporal in any of Her Majesty's air forces (or LCpl in the Royal Air Force Regiment) up to a maximum,

86 S. 38 RFA 1996.
87 JSP 753 Regulations for the Mobilisation of UK Reserve Forces, Part 1: Directive, at para. 9.
88 Reserve Forces Act 1996, s. 3(2).

under section 133(1), of 28 days or, if EPs have been granted, 90 days. If the CO awards detention, the only additional punishments that *may* be awarded are (a) reduction in rank or disrating (provided it is not prohibited by s. 138), and (b) a service compensation order.

To be clear, detention is not a punishment that can be awarded to an officer or to a warrant officer or NCO of or above the rank/rate of (RN) petty officer, (Army) corporal or (RAF) sergeant.

The Act contains powers under sections 189, 190 and 193 that authorise, respectively, consecutive sentences of service detention, suspension of a sentence of service detention, and activation by CO of suspended sentence of service detention where a trigger offence is committed during the operational period and that that the suspended sentence shall either take effect immediately or from the end of another sentence of service detention awarded on an earlier, or the same, occasion. However, section 244 states that no consecutive sentence may be awarded the effect of which would be that a person would (at the relevant time) be subject to sentences of service detention the combined term of which exceeds two years. This applies to consecutive sentences of detention and detention consecutive to the activation of a suspended sentence, discussed below.

If a CO finds a charge proved and intends to award a sentence of detention it must be decided whether to make the sentence of detention immediate or to suspend that sentence.[89] Additionally, if there is an existing sentence of detention, the officer may direct that the award shall take effect consecutively, from the end of any other sentence of service detention that was passed on the offender on a previous occasion (s. 189(3)). These provisions overlap with those applicable to cases where a sentence is made consecutive to an activated suspended sentence. This is discussed in the section below on Suspended Sentences (page 307).

If the CO awards a term of service detention (other than a suspended sentence of service detention) and does not make a direction under section 189(3) in respect of the award that it shall be a served consecutively, the sentence is concurrent and the offender may make an election to serve the sentence forthwith, at the time of the award, in which case the award takes effect from the beginning of the day on which the award is made. If,

89 See the section on Suspended Sentences, on p. 307 *et seq*.

on the other hand, the offender does not make that election, the award takes effect from the end of the appeal period (unless an appeal is brought within that period), or, if an appeal is brought within that period, takes effect from the beginning of the day when the appeal is abandoned or determined.

Consequently, before the sentence of detention takes effect, the CO must give the accused the choice: either to appeal, in which case the sentence does not take effect until the end of the appeal period or from the date of abandonment or determination by the SAC, or to elect to serve the sentence pending the appeal, in which case the award takes effect from the beginning of the day on which the award was made. The offender does not have to decide then and there whether to appeal. The 14-day appeal period can be used to consider options and take advice. If the offender has a change of heart and decides not to appeal or abandons an appeal, the sentence will start from that date.

On the other hand, if the accused does elect to serve the sentence, they may at any time until the end of the appeal period withdraw their election. They will then be released from custody. The remainder of the award ceases to have effect from the beginning of the day of withdrawal and resumes effect, as provided by section 290:

a. from the end of the appeal period (unless an appeal is brought within that period); or
b. if an appeal is brought within that period, from the beginning of the day when the appeal is abandoned or determined.

Thus, the effect of section 290, which concerns the commencement of term of service detention awarded by CO, is to postpone the point at which an award of immediate service detention made by a CO takes effect, whether or not the CO suspends the award under section 189(3). It is worth repeating that, unless the offender elects to start the sentence immediately, it does not take effect until the time allowed for bringing an appeal (the 'appeal period') expires or, if an appeal is brought, the appeal is disposed of (unless the SAC disposes of it by quashing the award or substituting another punishment). This contrasts with a sentence awarded by a service court, which takes effect from the beginning of the day on which it is passed (apart from a suspended sentence).[90]

90 S. 289.

The rationale behind the right to elect to serve the sentence is that if the service person wishes to challenge the deprivation of liberty after a finding by the CO in a non-judicial process, an appeal to the SAC is available, in which case the offender does not have to serve any part of the detention awarded by the CO. This provision is therefore an added protection for the service person in the otherwise non-compliant summary process and ensures that, absent that consent, any award of detention will be a matter for the Article 6-compliant SAC. As stated elsewhere, time spent in custody prior to the hearing is to be credited.

Suspended Sentences

Main Provisions: sections 190, 193, 194, 194A, 244, 290–292 AFA 2006; Schedule 3(16) AFA 2011; AF(SHASSSD)R 2009; section 6 AFA 2016

When a CO awards a sentence of detention, they may wish to suspend the operation of it for a specified period. The power to do this is in section 190(1). The MSL advises (at 1-13-14, para. 47) that, in deciding whether to suspend a sentence of detention the following factors should be taken into account, where appropriate:

a. In all cases, is the offender likely to reform without actually undergoing a period of corrective training? This may be the case where there has been very significant delay between the offence and summary hearing during which period the offender appears to have begun reforming, not re-offended, performed his duties above the standard expected and effectively rehabilitated himself
b. Does the offender show genuine remorse for his offence, including, where appropriate, making reparation for his offending?
c. The age of the offender. Due regard must be given to the welfare of personnel under the age of 18 and therefore detention should only be awarded when there is no other appropriate punishment or when other methods of reform have been tried and proved ineffective. For under 18s, an SSPO [Service Supervision and Punishment Order] should be considered

d. The previous character and convictions / disciplinary record of the offender. If the offence is clearly 'out of character', or it is the first offence, suspension may be suitable
e. Any special domestic circumstances of the offender
f. The gravity and type of offence
g. The operational requirements / movements of the unit or personnel. If the offender is likely to rehabilitate sufficiently because the unit is deployed then suspension may be appropriate. The convenience of the Service should not be a consideration. However, when it is considered that a sentence of detention is the correct punishment, deployed units should not recommend suspension solely on the grounds of cost of air transport, difficulty of arranging it to the UK or problems of providing escort.

It is a matter for the CO then to determine the period during which the sentence is to run, which is known as 'the operational period'. This must be for not less than three and not more than twelve months, beginning with the date of the order is made (s. 190(2) and (2A)(b)). The sentence operates in the same way as a suspended sentence handed down by a civilian court and does not have effect unless, during the operational period, the offender commits another service offence or an offence under the law of any part of the British Islands, and that the sentence shall take effect in accordance with section 193 (activation by CO of suspended sentence of service detention).

There is no credit for time spent in custody[91] when a suspended sentence of detention is passed, unless the sentence is activated. See section 246(2A) and below, page 450, applicable to both summary hearing and service courts.

When a suspended sentence of detention is imposed the MSL provides a helpful example of what the CO should do:[92]

> inform the offender at the time the sentence is awarded of the length of time he would expect to serve should the sentence be activated. For example, the CO has decided to award 60 days detention and has decided to suspend that. With automatic

91 *Ibid.*, s. 247(3).
92 JSP 830 MSL Version 2.0, at p. 1-13-16.

remission, the offender would ordinarily serve 40 days. The offender has spent 4 days in custody after charge. The CO may use the following form of words to announce the sentence:

> 'I sentence you to 60 days Service detention. I further order that the sentence is to be suspended for the period of 9 months. If the sentence is activated, unless the period of detention you are ordered to serve is reduced, the balance of the sentence will be 40 days detention. However, you have already served 4 days in Service custody and this will count as time served. You may therefore serve up to 36 days if you are later committed to detention'.

If the suspended sentence order is breached, the CO is faced with the choice under section 193 of whether to activate the sentence immediately or from the end of another sentence of service detention that has been passed on the offender on a previous occasion or, alternatively, which the officer passes on the offender on the same occasion as they make the order (s. 193(4)). Additionally, the CO will consider whether to leave the original term unaltered or to substitute a lesser term for the original. There is some clear overlap with the provisions relating to concurrent and consecutive sentences of detention discussed above on page 304 *et seq*.

Where the CO awards a suspended sentence of service detention, the only additional punishments they may award are reduction in rank or disrating and/or a service compensation order (s. 138(3)).

Activation

In accordance with the provisions of section 193, the CO may activate a suspended sentence of service detention passed at a summary hearing or by the SAC (but not by the court martial), subject to the procedural requirements of rule 23 AF(SHASSSD)R 2009,[93] in one of two ways: concurrently (s. 192(2)) or, consecutively (s. 192(3)). These are discussed separately, below.

93 These provisions include, inter alia, allowing the accused to make submissions on the appropriateness of making an activation order or the appropriate terms of such an order if it were made. There are other requirements, in r. 22, for the CO to give the accused an opportunity to adduce character evidence and make a plea in mitigation.

It should be noted that a CO may not make an activation order in relation to a naval suspended sentence, or include any provision in such an order by virtue of section 193(4) of the Act, without the written approval of the Royal Navy higher authority specified in Part 2 of Schedule 2 of AF(SHASSSD)R 2009.

If the CO wishes to activate the suspended sentence, and the aggregate effect of the sentence for the second offence is such as to take the award they wish to give beyond the maximum of 28 days' detention, an application must be made to higher authority for EPs.[94] The CO must comply with the requirements of rule 7(2) AF(SHASSSD)R 2009 concerning the information to be provided to higher authority. The *relevant time* for making the application for EPs is *before* the start of the summary hearing of the charge allegedly committed during the operational period of the suspended sentence; or if the accused is convicted of an offence in the British Isles during the operational period, and subsequently appears before their CO, the beginning of the hearing as to whether an order under section 193[95] should be made.[96] The CO may not exercise EPs until being notified that they have been granted.[97]

The breach of a suspended sentence places additional responsibilities on the CO, who must comply with the requirements of rule 8(1)(d) and (e) AF(SHASSSD)R 2009 and inform the accused in writing of, inter alia:[98]

(a) the power to make an activation order under section 193;
(b) the right of appeal under section 141 [see p. 298 of this book] of the Act and the effect of section 195 of the Act; [99]
(c) the offender's right to make submissions under rule 23(3), either orally or in writing, about the appropriateness of making an activation order and the appropriate terms of such an order if it were made.

94 See s. 194 AFA 2006 and r. 7 AF(SHASSSD)R 2009.
95 S. 193 is concerned with activation by CO of suspended sentence of service detention.
96 See ss 194 and 194A (inserted by Sch 3(16) AFA 2011).
97 S. 194(4).
98 See JSP 830 MSL Version 2.0 1-9-13, at para. 30 *et seq*, which sets out the procedure and documents to be provided. See also, s. 193 AFA 2006 and r. 23AF(SHASSSD)R 2009 (concerns sentences imposed by the CO or the SAC).
99 This essentially concerns the powers of the SAC where the CO makes an activation order under s. 193.

The CO or a person authorised by them must ensure that the offender is provided with the documents listed in rule 8(1)(e) AF(SHASSSD)R 2009, namely:

(i) a copy of the written record of the summary hearing, or a copy of any record of the proceedings before the Summary Appeal Court, at which the suspended sentence of service detention was awarded;

(ii) such details as are known to the commanding officer of all proven offences committed by the offender during the operational period of the suspended sentence of service detention;

(iii) copies of

(aa) the written records of any summary hearings,

(bb) the written records of any hearings under Part 3 of these Rules, and

(cc) any records of proceedings before any of the Summary Appeal Court, the Court Martial and the Court Martial Appeal Court at which reasons were given for any decision not to make an order under section 191 or 193 of the Act, in relation to the suspended sentence of service detention.

The CO must also fix a time for the hearing and the accused shall be allowed a reasonable time to prepare for it. This is not less than 24 hours after being provided with the information and documents in rule 8(1). If the CO applies for EPs or, permission under section 54 to hear a charge summarily, the CO may not fix a time for the hearing until 24 hours after receipt of the notification of the result of the application or applications (r. 8(5)). [100]

At the hearing, where the CO records a finding that a charge has been proved, and that it was committed during the operational period of a suspended sentence of service detention passed on the accused by an officer or the SAC,[101] the CO shall give the accused an opportunity to

100 The Armed Forces (Summary Hearing and Activation of Suspended Sentences of Service Detention) Rules 2009 – AF(SHASSSD)R 2009 . See Part 3.

101 Rules 22 and 23 need to be read together to ensure that the procedural steps are complied with correctly.

adduce evidence as to character, either in person orally or in writing, and make a plea in mitigation (r. 22). The accused will have an opportunity to question any character witness they call and the CO will also be allowed to ask questions of the witness as necessary.

Once the evidence is complete, the CO will announce the finding and sentence, providing reasons for any punishment awarded. If the CO decides to make an activation order, the offender must be informed of the reasons for the decision and the terms of the order. The CO must explain whether this is to be concurrent or consecutive to any other order for the trigger offence and indicate whether the suspended sentence shall take effect with the original term unaltered, or with the substitution of a lesser term for the original. The offender may then choose to make an election to commence serving any sentence pending an appeal. Otherwise, if that election is not made, any such sentence will not take effect until the expiry of the appeal period.

If the CO decides to make the order they should also remind the accused of the right of appeal under section 141 of the Act (r. 22) and, further, of the right to make an election under whichever of sections 290(2) and 291(2) of the Act, as modified by section 292, applies. The appeal lies to the SAC (see Appeals, on page 526).[102]

It will be seen that section 290 adapts the rules in section 189 (which deals with a court passing a consecutive sentence) for the case where a CO makes an award of service detention consecutive to an existing award. Unless the offender elects to start the new sentence as soon as the old one expires, the new one does not take effect until the old one has expired and the appeal period has expired or any appeal has been disposed of.

Consecutive sentences are subject to the important restriction set out in section 244, which prevents a consecutive sentence of detention being awarded the effect of which would be that a person would (at the relevant time) be subject to sentences of service detention the combined term of which would exceed two years. In any case, the accused will have a right to appeal the finding or sentence awarded on the subsequent trigger offence.

The CO will make a separate finding for each offence against the accused, and the accused has a right to appeal the finding to the SAC. This

102 Even though not available at a summary hearing, legal aid is available for any appeal to SAC, in accordance with JSP 838 .

right of appeal in section 141 relates to each charge heard summarily. It provides a time limit of 14 days (the appeal period) from the date of the award, unless the period is extended by the SAC.

Even though a service offender may not wish to appeal the finding and sentence for the trigger offence, the offender may seek to challenge the activation of the earlier suspended sentence, for example, on the basis that it is of a totally different and/or lesser character to the trigger offence. A challenge might be made that an initial road traffic offence for which a suspended sentence of detention was awarded should not be activated by a subsequent charge of, for example, theft. The Sentencing Guidelines[103] stipulate that the

> court must activate the custodial sentence unless it would be unjust in all the circumstances to do so. The predominant factor in determining whether activation is unjust relates to the level of compliance with the suspended sentence order and the facts/nature of any new offence.
>
> The facts/nature of the new offence is the primary consideration in assessing the action to be taken on the breach.
>
> Where the breach is in the second or third category below, the prior level of compliance is also relevant. In assessing the level of compliance with the order the court should consider:
>
> i the overall attitude and engagement with the order as well as the proportion of elements completed;
> ii the impact of any completed or partially completed requirements on the offender's behaviour;
> iii the proximity of breach to imposition of order; and
> iv evidence of circumstances or offender characteristics, such as disability, mental health issues or learning difficulties which have impeded offender's compliance with the order.

Section 259 places a duty on a service court to have regard to any guidelines that are relevant to the offender's case. However, there are wider duties

103 For more information on the categories of breach involved, see the Sentencing Guidelines Council guidance on suspended sentences, s. 1. Conviction for further offence committed during operational period of order.

placed upon officers and courts who sentence offenders, which are set out in sections 237–271, which require the sentencer to have regard to purposes of sentencing.

Activation Concurrently

Section 292(2) modifies section 290 for any sentence to be served concurrently. Where the CO's activation order does not provide that the suspended sentence shall take effect consecutively with any other sentence, the default position, under section 292(2), is a concurrent sentence. If the offender elects, at the time when the activation order is made, to start serving the sentence, the award takes effect from the beginning of the day on which the activation order is made. In contrast, if no such election is made, the award of service detention to which the activation order relates

(a) takes effect from the end of the appeal period (unless an appeal is brought within that period); [or]
(b) if an appeal is brought within that period, takes effect from the beginning of the day when the appeal is abandoned or determined.

The 'appeal period' means the period mentioned in section 141(2) for the bringing of an appeal against the activation order.

An election may be withdrawn at any time until the end of the appeal period. If it is withdrawn, the remainder of the award of service detention to which the activation order relates ceases to have effect from the beginning of the day of withdrawal and resumes effect

(a) from the end of the appeal period (unless an appeal is brought within that period);
(b) if an appeal is brought within that period, from the beginning of the day when the appeal is abandoned or determined.

If an award takes or resumes effect in any of these circumstances, and an appeal is subsequently brought, the remainder of the award of service detention to which the activation order relates ceases to have effect from the beginning of the day when the appeal is brought. It resumes effect from the beginning of the day when the appeal is abandoned or determined.

While the 'appeal period' means the period mentioned in section 141(2) for the bringing of an appeal against the activation order, an appeal on the award for the trigger offence, should logically include a challenge to the activation order, as that order would, realistically, fall away if the appeal on the trigger offence were successful. The basis of the activation would have been removed.

Activation Consecutively

In the case of a sentence of detention, which the CO makes consecutive to an activated suspended sentence (and subject to the right to defer pending appeal), the suspended sentence takes effect at the end of the 'initial sentence'.[104] The way this is expressed is slightly confusing, but the 'initial sentence', as defined in section 292(3), means the sentence awarded for the trigger offence, that is the subsequent offence to that for which a suspended sentence was awarded. It is 'initial' in the sense that it is the sentence that will be served first. The activated sentence will then be served consecutively to it.[105]

> If the activation order provides that the suspended sentence shall take effect from the end of another sentence (the initial sentence), section 291(2) to (9) apply.

Section 292(3) modifies section 291 (commencement of consecutive sentence) and explains what happens when an activation order provides that the suspended sentence is to take effect consecutively to the 'initial sentence'.

The offender may, at the time when the activation order is made, make an election to serve the sentence and may withdraw it at any time until the end of the appeal period. If the offender does make an election and does not withdraw it, or bring an appeal, before the end of the initial sentence, the award of service detention to which the activation order relates (with any modification of its term made by the activation order) takes effect from the end of the initial sentence.

104 *Ibid.*, s. 291.
105 *Ibid.*, s. 292.

If the offender does not make an election to serve the sentence, the award of service detention to which the activation order relates (with any modification of its term made by the activation order) takes effect from the end of the initial sentence or (if later) the end of the appeal period (unless an appeal is brought before that time). If an appeal is brought before the end of the initial sentence or (if later) the end of the appeal period, the award of service detention to which the activation order relates takes effect from the end of the initial sentence or (if later) from the beginning of the day when the appeal is abandoned or determined.

If an election is withdrawn after the end of the initial sentence, the remainder of the award of service detention to which the activation order relates (with any modification of its term made by the activation order) ceases to have effect from the beginning of the day of withdrawal and resumes effect from the end of the appeal period (unless an appeal is brought within that period). If an appeal is brought within that period, the award of service detention to which the activation order relates ceases effect from the beginning of the day when the appeal is abandoned or determined.

None of these circumstances applies where the appeal is determined and, on that determination, the activation order is quashed or another order under section 193(3) is substituted for it.

Persuading a court away from activating a suspended sentence is a difficult role even for a skilled advocate to accomplish successfully. As there is no right to legal representation at a summary hearing, this onerous task, which significantly impinges upon the liberty of the subject, is left to the AAO.

Extended Powers

Bearing in mind that a CO may not award more than 28 days' detention unless granted EPs,[106] if it is intended to activate a suspended sentence that would exceed this period, the CO must apply for EPs, under section 194. Section 194A[107] deals with the 'relevant time' for obtaining EPs for activation, defined as follows:

---

106 See on p. 281, *et seq*.
107 Inserted by Sch. 3 (16) AFA 2011.

194A Extended powers of activation: time for obtaining

(1) In section 194(4) 'within the relevant time' means
  a. where section 193(2)(a) applies,[108] before the start of the summary hearing of the charge mentioned there (but this is subject to subsections (2) and (3));
  b. where section 193(2)(b) applies,[109] before the start of the hearing as to whether an order under section 193 should be made.

(2) Subsection (3) applies where the summary hearing of the charge mentioned in section 193(2)(a) is one where, after the start of the hearing
  a. a charge is amended under section 123(2)(a);
  b. a charge is substituted for another charge under section 123(2)(b); or
  c. an additional charge is brought under section 123(2)(c).

(3) Any application for or grant of extended powers which is made in the period between
  a. the making of the amendment, substitution or addition, and
  b. the time when the summary hearing is proceeded with after the amendment, substitution or addition,

is to be treated for the purposes of section 194(4) as made within the relevant time.

An interesting point, however, relates to the maximum sentence a CO can award as a consecutive sentence to an existing award. Paraphrasing section 189(3), where an officer awards a person a term of service detention, the officer may direct that the award shall take effect from the end of any other sentence of service detention that has been passed on them on a previous occasion.

Unlike the statutory power the CO has under section 194 to obtain EPs where they activate a suspended sentence, there is no similar provision

108 CO records a finding that a charge against the offender in respect of an offence committed during the operational period of the suspended sentence is proved.
109 The offender is convicted of an offence against British law that was committed during that operational period, and subsequently appears before their CO.

concerning consecutive sentences. This is confusing, as it would appear that they are limited to a maximum aggregate of 28 days.[110]

However, if one then considers the limit on the combined term of sentences of service detention in section 244, it is provided that

> [an] officer may not make a direction under section 189 (consecutive terms of service detention) ... whose effect would be that a person would (at the relevant time) be subject to sentences of service detention the combined term of which *exceeds two years*. (Emphasis added)

In other words, the CO can add to the sentence the accused is already serving in relation to previous charges, provided it does not exceed two years' detention in aggregate. This seems to conflict with the restriction in section 133(1) that states that the maximum term of detention a CO may award is 90 days if EPs have been granted, otherwise 28 days. It is unlikely that a CO will have such a long-term prisoner in the unit detention facility. It seems more likely that this power is designed to be available to the CO of Military Corrective Training Centre (MCTC), where an offender is already serving a sentence of detention[111] and who commits a further offence. The Explanatory Notes (EN) to the section, at § 485, explain that the provision

> prohibits a court or CO from sentencing an offender, or activating a suspended sentence previously imposed on him, if it would result in the offender being subject to sentences of service detention amounting to more than two years in total. If a court or CO purports to do this, the excess period is remitted.

Service Supervision and Punishment Order

Main Provisions: sections 132, 173 and 174 AFA 2006; AF(SSPO)R 2009

This punishment is available to the CO, under section 132. It may only be awarded where the person being punished is an able rate, marine, soldier or airman. The period specified in the order must be 90, 60 or 30 days beginning with the day the order is made.

110 See s. 133(1)(b).

111 The maximum period of service detention is 2 years, as awarded by sentence of a court martial. See s. 164 AFA 2006.

Rule 2 of Armed Forces (Service Supervision and Punishment Orders) Regulations 2009 (AF(SSPO)R 2009) defines the 'initial period' and 'secondary period'. The initial period must be 21 days for a 90-day order, 18 days for a 60-day order or 14 days for a 30-day order, beginning with the day the order is made. 'The secondary period' means the remainder of the duration of a Service Supervision and Punishment Order (SSPO) after the initial period. Thus, on a 90-day order, this will be 69 days. A further consequence of the SSPO is that one-sixth of the offender's gross pay for that period is forfeit.[112]

An SSPO shall require the offender who is subject to it not to use any entitlement to leave without the permission of their CO, and may impose on the offender one or more of the following further requirements, under Regulation 3:

- (a) to perform extra duties as decided under regulation 4(1) of AF(SSPO)R 2009
  - (i) during the initial period, for up to five and a half hours each day; and
  - (ii) during the secondary period, for up to one hour each day;
- (b) not to leave a relevant place without the permission of his commanding officer;
- (c) subject to such conditions as may be specified in the SSPO, not to enter specified places in a relevant place without the permission of his commanding officer.

Regulation 4 governs the management of the SSPO. It is a matter for the offender's CO to decide, in respect of each day during which the SSPO is in force, what extra duties the offender shall perform pursuant to that requirement, for how long, and when, and they will inform the offender accordingly.

The CO may delegate their function of granting permission under regulation 3 to an officer of or above the rank of naval lieutenant, military or marine captain or flight lieutenant.

112 S. 173(1)(b).

Where, on the other hand, an SSPO imposes a requirement under paragraph (a) of regulation 3, the offender's CO shall decide, in respect of each day during which the SSPO is in force, what extra duties the offender shall perform pursuant to that requirement, for how long, and when, and they will inform the offender accordingly. In this case, the offender's CO may delegate to a person of or above the rank or rate of chief petty officer, marine colour sergeant, military staff sergeant or flight sergeant, the ability to decide, in respect of each day during which the SSPO is in force, what extra duties the offender shall perform pursuant to that requirement, for how long, and when.

Regulation 5 requires that the offender's CO consider whether an SSPO should continue in force

(a) on or before the last day of the initial period; and
(b) at intervals of not more than 14 days beginning on the day after his last consideration under paragraph (a) or under this paragraph.

Prohibited Combinations of Punishments

If the CO awards detention, the only additional punishments they may award are:

a. reduction in rank or disrating, restricted to one rank below that held, as explained above;
b. a service compensation order. The maximum a CO can award is £1,000. But see above, for limitations on subordinate commanders on page 321 *et seq*.

If the CO awards forfeiture of seniority, the only additional punishments they may award are:

a. a severe reprimand or a reprimand;
b. a service compensation order.

If the CO awards reduction in rank or disrating, the only additional punishments they may award are:

a. a suspended sentence of service detention (subject to subsection (8));
b. a service compensation order.

If the CO awards a fine, they may not also award a service supervision and punishment order.

Certain combinations of punishments are prohibited under the AF(MPLPRR) R 2009. Regulation 5 details where a CO can award a minor punishment in respect of an offender.

If the CO awards a stoppage of leave order, the only additional punishments they may award are:

a. a fine;
b. a severe reprimand or a reprimand;
c. a restriction of privileges order;
d. a service compensation order.

If they award a restriction of privileges order, the only additional punishments they may award are:

a. a fine;
b. a stoppage of leave order;
c. a service compensation order.

If they award an admonition (see 'Minor Punishments', below, on page 488), the only additional punishment they may award is a service compensation order.

Subordinate Commander

This means an officer under the command of the CO who is of at least the rank of naval lieutenant, military or marine captain or flight lieutenant.[113] The MSL explains that

> [a] CO should make clear and promulgate any delegation of their powers and functions. A CO should be certain that the person to whom powers are delegated is entitled and qualified to hold and exercise those powers. See Chapter 6 (Investigation, charging and mode of trial).[114]

113 Reg. 2(1) AF(SHASSSD)R 2009.
114 JSP 830 MSL Version 2.0 1-2-3, para. 3.c.

A CO may delegate to a subordinate commander any of their relevant functions in relation to a charge, subject to such conditions as the CO considers appropriate.[115] The delegation may be in respect of one or more of the following, and subject to any conditions they impose:

a. the referral of a case to the DSP;
b. bringing a charge (including bringing a charge as a result of a direction from the DSP);
c. amending or substituting a charge, or bringing an additional charge;
d. referral of a charge to the DSP; and
e. discontinuing proceedings on a charge.

A subordinate commander may not award detention.

However, the CO may not delegate in relation to a charge,

1. Being a charge which may be heard summarily only with permission or by senior officer (being an offence under section 42, within Part 2 of Schedule 1, or an offence under section 1 of the Criminal Attempts Act 1981 of attempting to commit an (indictable) offence);
2. In respect of an offence alleged to have been committed by an accused during
    a. the operational period of a suspended sentence of service detention passed on him;
    b. the operational period of a suspended sentence of imprisonment passed on him by the Court Martial, the Court Martial Appeal Court or the Supreme Court on an appeal brought from the Court Martial Appeal Court;
    c. the term of an order made under section 211 of the Act (which deals with detention and training orders for offenders under 21); or
3. Brought against a person above the rank or rate of chief petty officer, marine colour sergeant, military staff sergeant, or flight sergeant.

115 *Ibid.*, r. 3 AF(SHASSSD)R 2009 and r. 16 of AF(P5AFA2006)R 2009.

The limitations on powers that may be exercised by a subordinate commander to whom delegation has been made are set out in rule 3(5) AF(SHASSSD)R 2009, which states they may *not*:

> be exercised by a subordinate commander:
>
> (a) if he is of the rank of naval lieutenant, military or marine captain or flight lieutenant, award any punishment in column 1 of the Table in Schedule 1;
> (b) if he is of the rank of lieutenant commander, major or squadron leader, award any punishment in column 2 of the Table in Schedule 1; or
> (c) if he is of or above the rank of commander, lieutenant-colonel or wing commander, award any punishment in column 3 of the Table in Schedule 1.

This is contrasted with the power of a CO who may award a service compensation order (SCO) of up to £1,000 without the restriction of personal injury placed upon subordinate commanders (section 137).

See Table 9.2, which is set out in Schedule 1 to AF(ASHASSSD)R 2009, on the following page.

Review of Summary Awards

Main Provision: section 152 AFA 2006

While the review process has been abolished for court martial decisions,[116] it remains for summary awards. Where a charge has been heard summarily and a finding that the charge has been proved has been recorded, the finding or the punishment awarded may be reviewed at any time.[117] The actual review may be carried out by the Defence Council or any officer appointed by the Defence Council. The reason for a review is to ensure there are no procedural errors that operate to the service person's disadvantage and which need to be corrected.

Where such a review has been carried out and the person to whom the review relates has not brought an appeal to the SAC under section 141

116 See s. 16 AFA 1996.
117 S. 152 AFA 2006.

Punishments That May Not Be Awarded by Subordinate Commanders of Particular Ranks

Row Number	Column 1	Column 2	Column 3
1.	detention	detention	detention
2.	forfeiture of seniority	forfeiture of seniority	forfeiture of seniority
3.	reduction in rank and disrating	reduction in rank and disrating	reduction in rank and disrating
4.	a fine exceeding 7 days' pay	a fine exceeding 10 days' pay	a fine exceeding 14 days' pay
5.	a severe reprimand	a severe reprimand	a severe reprimand
6.	a service supervision and punishment order	a service supervision and punishment order	a service supervision and punishment order
7.	a stoppage of leave order specifying a number of days exceeding 7	a stoppage of leave order specifying a number of days exceeding 10	
8.	a service compensation order of an amount exceeding £500	a service compensation order of an amount exceeding £750	a service compensation order of an amount exceeding £1,000
9.	a service compensation order where any part of the compensation relates to personal injury	a service compensation order where any part of the compensation relates to personal injury	a service compensation order where any part of the compensation relates to personal injury

Table 9.2 Punishments That May Not Be Awarded by Subordinate Commanders of Particular Ranks

within the period provided by subsection (2) of that section,[118] the reviewer may, with the leave of the SAC, refer the finding or punishment to that court to be considered by it as on an appeal.[119] This power may be exercised even in those circumstances where the offender has not submitted an appeal.

118 Unless extended by the judge, 14 days beginning with the date on which the punishment was awarded.

119 S. 152(4). The Reviewing Officer might detect an error that they consider the court should correct.

It is loosely akin to the slip rule, which operates in the court martial[120] and in the Crown court. It can only operate to the accused's advantage.

Where the offender appeals to the SAC and it has not completed its hearing of the appeal, the reviewer may notify the court of any matter arising at or from the summary hearing which they consider should be brought to the notice of the court (s. 152(6)).

If the SAC has completed its hearing, and the reviewing authority considers that a matter arising at or from the summary hearing of which the court was not aware should have been brought to the notice of the court, they may, with the leave of the court, refer the finding or punishment, including any finding or punishment substituted or awarded by the court, to the court to be considered by it as on an appeal. It is a safeguard to ensure that the service person is fairly treated.

Any such reference by the reviewing authority is to be treated as an appeal under section 141 brought by the person to whom the finding or punishment relates against the finding or punishment.[121]

---

120 Variation Proceedings under Rules 118–124 of AF(CM)R 2009.
121 S. 152(8). See also the EN at paras 313–316.

10 LEGAL AID

Main Provision: JSP 838; ANNEX F; for inquests, see Chapter 20

Applicability

During the course of the investigation, the suspect can often feel isolated and friendless. This is especially the case when the alleged offence occurs overseas where the accused is serving. Legal aid is not applicable until arrest, unless the service person pays privately for advice. However, once the Commanding Officer (CO) has charged them they may apply for legal aid.[1]

The Armed Forces Criminal Legal Aid Authority (AFCLAA)[2] describes itself as a single non-statutory authority dealing with all aspects of criminal legal aid for those prosecuted through the Service Criminal Justice System. The AFCLAA is a part of the Court Martial Administration Office (CMAO). The scheme is based upon the same principles that apply in the civilian system, with additional provision for the unique circumstances of entitled personnel (i.e. they are entitled to legal aid). So, for example, determination (review) and payment of costs and disbursements is calculated using the formulae applied to lawyers defending in the Crown Court, whether the matter relates to cases for service personnel abroad or not.

The application for legal aid is made on MoD Form 2263, which is available from the AFCLAA website. In cases of hardship, an application for review may be made on Form 2263C.

JSP 838 is the authoritative policy and guidance on the Armed Forces Legal Aid scheme available to those personnel who are subject to service

1 See Appendix F.

2 This information is taken from the MoD website and JSP 838 - Armed Forces Criminal Legal Aid Scheme.

discipline or service law, to ensure their access to publicly funded support is maintained.[3] The document is a very comprehensive guide to the entitlement of and process for applying for legal aid and is essential reading for all practitioners representing accused through the scheme. The Joint Service Publication (JSP) is structured in two parts: Part 1 – Directive, which provides the direction that must be followed in accordance with statute or policy mandated by Defence or on Defence by Central Government; Part 2 – Guidance, which provides the guidance and best practice that will assist the user to comply with the Directive(s) detailed in Part 1.

The JSP provides that the following personnel can apply for criminal legal aid in the following circumstances:

a. members of the Armed Forces, worldwide, for all matters dealt with by the service courts, including Court Martial and Summary Appeal Court;
b. members of the UK-based civilian (UKBC) component, including civil servants, teachers and NAAFI staff, based outside mainland UK, for all matters involving service jurisdiction, dealt with by the service courts;
c. dependants of Armed Forces personnel and the UKBC component, based outside mainland UK, for all matters involving service jurisdiction, dealt with by the service courts;
d. members of the Armed Forces, the UKBC component and dependants, for matters dealt with by the civilian criminal court of the country in which the individual is based on permanent or detached duty. Such applications will only be considered where there is no equivalent legal aid scheme available through the civilian court in which the matter is to be heard.[4]

JSP 838 points out that legal aid covers some, or all, of the costs of legal representation for defendants and appellants who:

a. Appeal against findings and/or the award of a punishment following summary hearings at unit level, including applications for extensions of the appeal period by the Summary Appeal

3 *Ibid.*, p. i. of JSP 838 Pt 2 (V3.0 Dec. 14).

4 AFCLAA website: https://www.gov.uk/guidance/armed-forces-criminal-legal-aid-authority-afclaa.

Court, for leave to appeal out of time (service and ex-service personnel only); or

b. Have a case referred to the Director Service Prosecutions (DSP), for a decision on whether the charges will result in a prosecution; this includes offences under Schedule 2 of the Armed Forces Act 2006 which are referred directly to the DSP by the service police, as well as matters referred to the DSP by the CO; or
c. Are to be tried in the Court Martial (CM) or the Service Civilian Court (SCC); or
d. Wish to appeal in the CM, against the finding and/or sentence after trial in the SCC (relevant civilians only); or
e. Are to be tried in a criminal court outside the UK (see Part 1, Section 5, and Part 2 Chapter 4).
f. The legal aid scheme applies equally to all members of the Armed Forces, including the Reserve Forces (when subject to service law) and to civilians who are subject to service discipline (relevant civilians). See JSP 830, Vol. 1 Ch. 3 – Jurisdiction and Time Limits.

AFCLAA also pay for legal advice and assistance to accused personnel at Royal Military Police (RMP) stations, and legal representation at pre-charge custody hearings, outside England and Wales; legal advice and assistance, and legal representation at pre-charge custody hearings held within England and Wales are paid indirectly by AFCLAA, via the Legal Services Commission.

The scheme does not cover those cases where legal aid would not be available to a person working or residing in the UK. Consequently, any service person charged with an offence in the UK must apply for legal aid in the usual way, through a local solicitor. Neither does the scheme cater for civil proceedings, such as contract disputes or family/divorce proceedings. Finally, it is not available to cover off-duty offences committed abroad when the service person is not there as a result of their duty, for example they are on holiday.

Legal aid is also available for the purpose of Adjudication Proceedings at the Military Corrective Training Centre (MCTC), Colchester, concerning a reference to an adjudicator in respect of a charge against a detainee brought under the Service Custody and Service of Relevant Sentences Rules 2009

(SCSRSR 2009) for a breach of discipline that the Commandant of MCTC considers is so serious that additional days may be awarded for the offence.

Where an appellant wishes to apply to the Court Martial Appeal Court (CMAC) for legal aid for an appeal, see JSP 838 (the Armed Forces Legal Aid Scheme) and below, Appendix F. More information and a link to the relevant JSP (838) is available from the AFCLAA website.[5] The current version is version 3.0, issued in December 2014. It contains an extensive glossary of terms and forms used and is a very useful and informative guide. In Part 2, paragraph 31, it is made clear that,

> Throughout the legal aid process, especially at the initial application stage, AFCLAA Case Officers will contact the DO/SSA/P1 for specific assistance to obtain completed parts of the application form, complete with supporting documentation as necessary. Whilst the responsibility for applying for legal aid remains with the applicant, there is also a Command responsibility to provide such encouragement, guidance and support as is necessary to ensure the applicant fully understands the options open to them. Further details on the assistance required and provided are contained within Chapter 2 of this JSP.

Practitioners should note, there is no retrospective grant of legal aid. Thus, any work carried out by a legal representative in advance of a formal grant of legal aid is a private matter between the legal representative and the applicant, who will remain personally liable for all costs incurred.[6]

Overseas Criminal Courts

Applicants who have been arrested and charged to appear in a civilian criminal court overseas should apply for legal aid without delay, using the same application form as those prosecuted through the military justice system, using the same procedures as other types of cases. It should be noted that it is necessary to act with more urgency as overseas civilian cases tend to proceed to trial within shorter time frames. Therefore, it is

5 See *ibid*.

6 *Ibid*., Part 1, §13.

imperative that appropriate legal representation is obtained without delay. Units are advised to contact AFCLAA immediately upon notification of arrest and / or charge, to initiate the process. Chapter 4 to this JSP provides more information for units and should always be referred to in the first instance.[7]

7 *Ibid.*, §4.

11 FITNESS TO STAND TRIAL

Main Provisions: Criminal Procedure (Insanity) Act 1964; Mental Health Act 1983; Part 7, Chapter 4, sections 166 to 172 and Schedule 4 of AFA 2006

It is open to the court martial to make a finding of insanity or that the defendant is otherwise unfit to stand trial.

Unfitness to Stand Trial

A person is unfit to stand trial[1] if they are under a disability such that apart from the Criminal Procedure (Insanity) Act 1964 it would constitute a bar to their being tried on indictment in England and Wales. The question of fitness to stand trial must be determined as soon as it arises, and by the judge advocate alone, on the written or oral evidence of two or more registered medical practitioners, at least one of whom is duly approved for the purposes of section 12 of the Mental Health Act 1983 by the Secretary of State.

If having regard to the nature of the supposed disability the judge advocate is of the opinion that it is expedient to do so and in the interests of the defendant, they may postpone consideration of the question of fitness to stand trial until any time up to the opening of the case for the defence. If, before the question of fitness to stand trial falls to be determined, the court finds the defendant not guilty on the charge or each of the charges on which they are being tried, that question shall not be determined.

If the judge advocate determines that the defendant is unfit to stand trial, the procedure in section 167 of the Armed Forces Act 2006 (AFA 2006) is to be followed. In this event, the trial shall not proceed or continue

1 Ss. 166 and 167 AFA.

further, but the court must go on to determine whether it is satisfied that the accused did the act charged against them. If it is satisfied, then it must make a finding that the defendant did the act charged against them. If the court is not so satisfied, it must find the defendant not guilty as if *on the charge in question* the trial had proceeded to a conclusion.

The question of whether the court is satisfied that the defendant did the act or omission charged against them is to be determined by the lay members alone, by simple majority. Thus, the lay members perform their traditional function of determining the issue of fact: did the defendant do the act? (s. 187(6)).

Findings of Insanity

Where on the trial of a person by the court martial it is satisfied,[2] as respects the charge or any of the charges on which they are being tried, that the defendant did the act charged against them as the offence, but at the time of that act they were insane, the court must find that the defendant was not guilty of that offence by reason of insanity. No finding of insanity can be made unless on the written or oral evidence of two or more registered medical practitioners, at least one of whom is duly approved for the purposes of section 12 of the Mental Health Act 1983 by the Secretary of State.

Powers of the Court

Subject to what is said below, the court must make the defendant subject to a hospital order, with or without a restriction order (under section 41 of the Mental Health Act 1983 as modified by AFA 2006, Sch. 4), or a service supervision order (defined by section 170), or an order discharging them absolutely. The power to make a hospital order under section 169 is not exercisable unless the conditions in section 37(1)(b) and (2) of the 1983 Act, as modified by Schedule 4 to AFA 2006, are satisfied. The applicable provisions, as amended by the AFA 2006, Schedule 4, are:

2 Ibid., ss. 167 and 168.

37 Powers of courts to order hospital admission or guardianship.

(1) Where

(a) section 169 of the Armed Forces Act 2006 applies,

(b) the offence to which the finding relates is an offence punishable with imprisonment, and

(c) the conditions in subsection (2) below are satisfied,

the court may by order authorise the defendant's admission to and detention in such hospital as may be specified in the order;

(2) The conditions referred to in subsection (1) above are that

(a) the court is satisfied, on the written or oral evidence of two registered medical practitioners, that the offender is suffering from mental disorder and that—

(i) the mental disorder from which the offender is suffering is of a nature or degree which makes it appropriate for him to be detained in a hospital for medical treatment and appropriate medical treatment is available for him; and

(b) the court is of the opinion, having regard to all the circumstances including the nature of the offence and the character and antecedents of the offender, and to the other available methods of dealing with him, that the most suitable method of disposing of the case is by means of an order under this section. ...

(4) Where an order is made under this section requiring a person to be admitted to a hospital ('a hospital order'), it shall be the duty of the managers of the hospital specified in the order to admit him in accordance with it.

The functions of the court under section 19, and any functions under the 1983 Act that are exercisable by the court by reason of this section or Schedule 4, are to be exercised by the judge advocate for the trial or, in a case where that judge advocate has made an interim hospital order under section 38 of the 1983 Act as modified by Schedule 4, by that or any other judge advocate.

Service Supervision Order
A service supervision order means an order whereby the person in respect of whom it is made is subject to the supervision of a specified person (the supervising officer) for a specified period which must not exceed the maximum period of a civilian supervision order set out in Schedule 1A(1) of the 1964 Act (currently two years).

The court may not make a service supervision order unless it is satisfied that,

a. having regard to all the circumstances of the case, the making of a service supervision order is the most suitable way of dealing with the defendant;
b. that the supervising officer intended to be specified in the order is willing to undertake the supervision; and
c. that arrangements have been made for any treatment which (under subsection (4)) is intended to be specified in the order.[3]

Where a person is detained in pursuance of a hospital order made by the court martial and the court also made a restriction order that has not ceased to have effect, then the Secretary of State may, if satisfied after consultation with the responsible clinician that the person can properly be tried, remit the person for trial by the court martial (s. 171).

For the purposes of the provisions of sections 166 and 168, which permit a court to act on the written evidence of a registered medical practitioner, or a registered medical practitioner who is duly approved, a report in writing purporting to be signed by a registered medical practitioner or a registered medical practitioner who is duly approved may be received in evidence without proof of the signature of the practitioner and without proof that they have the requisite qualifications or is duly approved. However, the court may require the signatory of any such report to be called to give oral evidence. Copies of such reports must be handed to the defendant's counsel or solicitor. If the defendant is unrepresented, the substance of the report must be disclosed to them or, if they are under 18, to their parent or guardian if present in court. The defendant may require the signatory of the report to be called to give oral evidence and may call evidence to rebut the evidence contained in the report (s. 172).

3. S. 170(3), AFA 2006.

12 PRE-COURT MARTIAL PROCESS

This chapter covers the process from the receipt of the case by the Service Prosecuting Authority (SPA) on referral, the test applied by them, through to referral for trial, and, where relevant, conviction and sentence.

To civilians unacquainted with the services, the disciplinary system may appear remote and obscure – archaic, even farcical. Popular television depictions, such as the famous episode of the Blackadder court martial,[1] only help to cement those views further. Clemenceau once remarked that 'Military justice is to justice what military music is to music'. He was clearly not a fan of it either. He would be astonished at how his words fail to resonate today, where the modern system is closely aligned procedurally with the civilian system and the remaining differences are essentially those which reflect the environment in which the service accused operates. As Dr Kate Grady points out,[2] 'the paucity of academic literature on the practice of the Court Martial is notable'. In the event, the popular myth tends to perpetuate. In her report for *Liberty*,[3] Emma Norton excoriated the military justice system, claiming:

> the more [Liberty] investigated the way in which service personnel were being treated, the more it discovered that some of the most basic principles of fairness that civilians took for granted, did not necessarily apply to them and in ways that could not be justified. Liberty learned that a Commanding Officer could investigate sexual assaults for him/herself and was not legally obliged to report them to the police; that there

1 Corporal Punishment, *BBC*, Blackadder Goes Forth, Episode 2 (1989).

2 'Disciplinary Offences at the Court Martial' [2016] 10 *Criminal Law Review* 714–742.

3 'Military Justice, Second Rate Justice', January 2019, p. 11, https://libertyinfo.wpengine.com/wp-content/uploads/2020/01/LIB-10-Military-Justice-Report-20_01_19.pdf.

> was no reliable data on rates of sexual offending in the armed forces; or that victims of rape were not being told of their right to seek support from expert civilian sources of help. Liberty learned that conviction rates for rape in the Court Martial paled in comparison to civilian conviction rates and that the Service Police were not subject to the same kind of independent oversight as civilian police. This all painted a picture of second-rate justice for service personnel.

These are serious criticisms, indeed. As explained above (page 269), sexual cases may no longer be dealt with by the Commanding Officer (CO) and principally fall within Schedule 2. However, it is difficult to disagree with her statement that 'the rights of service men and women are just as deserving of protection as those of civilians'.[4]

The genesis of the modern court martial system did not, in fact, begin with *Findlay v UK*,[5] although that decision helped speed up the necessary changes and it has to be conceded that the creation of independent service prosecuting authorities was very much a consequence of the case. The ensuing changes introduced through the 2006 Act, and further supplemented in 2011 and 2016, have created a very fair and robust system of justice for the three services, attested to by the absence of any further substantive challenge to it before the European Court of Human Rights (ECtHR) since the decision in *Morris*,[6] in 2002. Critics complain, for example, that the system of majority decisions is unfair, and refuse to accept the decision in *Twaite*,[7] where the Lord Chief Justice rejected the submissions by the Judge Advocate General's counsel. (See below, on page 430.) Attempts are made to criticise the conviction rate, yet any analysis will show it is roughly the same as in civilian practice. Critics have also yet to point to any major miscarriage of justice in the military justice system, upon which to base a challenge in Europe, which has been brought about by the trial system they say is so unfair.

---

4 *Ibid.*, p. 13.
5 [1997] ECHR 8 (25 February 1997).
6 [2002] ECHR 162.
7 [2010] EWCA Crim 2973.

Service Prosecutions

Before a case is heard by court martial, it must go through a legal process of referral for trial by the SPA. For information about the Prosecutors' Convention and determining jurisdiction of prosecuting authorities, please see above, on page 19.

Background

The tri-service prosecuting authority, a creation of the 2006 Act, was officially set up on 1 January 2009 as the SPA.

The idea of tri-service institutions for the provision of legal advice was one that had been in circulation for many years, with the Allen Review in the 1970s, the Bretherton Review in the 1980s and then the Strategic Defence Review of 1998. It was the latter initiative that was essentially implemented, as it looked at the requirement for a tri-service prosecution approach and the harmonisation of the three service discipline acts. The work for this was undertaken from early 2002 to produce the Armed Forces Act 2006 (AFA 2006), which created the SPA.[8]

All prosecutors are either solicitors or barristers and currently come from the legal branches of the three services. They are subject to their respective professional code of conduct and are required to hold a practising certificate issued by their respective professional body. The first Director to be appointed was Bruce Houlder QC, a civilian barrister.[9] Mr Houlder was succeeded in December 2013 by Andrew Cayley CMG QC, who had previous service as an Army legal officer. Mr Cayley also acted as a prosecutor at the International Criminal Court, culminating as International Co-Prosecutor of the Khmer Rouge Tribunal in Cambodia from 27 November 2009 until

---

8 It was not, as reported in the press, a result of the infamous Iraq War case of the death of Baha Mousa, which occurred in September 2003.

9 A very experienced civilian practitioner. Interestingly, Parliament and the three heads of service wanted a service lawyer to head the SPA. See: Select Committee on Armed Forces First Report, paragraph 70 et seq. In particular, paragraph 73. 'It was clear from our discussions, however, that MoD shares our view on the desirability of the DSP having an appreciation of the military environment, and preferably military experience. Mr Morrison said that: "it is very well acknowledged that the importance of a clear understanding of the Service context will be a central element in the selection process". Service personnel must have confidence that the person taking the decisions on prosecutions has sufficient understanding of the context in which the events occurred. Whilst we appreciate the difficulties involved in defining military experience in statute, we do consider it important for the Director of Service Prosecutions to have had military experience.'

16 September 2013. In November 2020, Jonathan Rees QC succeeded Mr Cayley as Director of Service Prosecutions (DSP). Mr Rees is a highly experienced criminal barrister and former Old Bailey Treasury Counsel with a wealth of experience prosecuting grave and complex cases.

Prosecuting officers comprise legally qualified officers from the three service branches. It has been said, in relation to Army lawyers, perhaps more in jest than in earnest, that they are officers first and lawyers second. This cannot be right and was emphatically contradicted by a former Director Army Legal Services who went on to become a Crown Court judge. Lawyers recruited into the Army are needed for the specific purpose of providing legal services, rather than combat skills, and do not undertake a combat role, per se. Indeed, in the event that they, for any reason, lose their right to practise, they would cease to be employable within the Army as a legal officer. The same applies to the other two services. If anything, they are, therefore, lawyers first and officers second, as their professional duty should always be paramount when fulfilling their role. They are lawyers in uniform: bound by the service disciplinary codes and Values and Standards, but only in so far as these do not conflict with their professional code. It would never be a lawful order, for example, for a superior officer to order a military lawyer who acted as a service person's adviser to divulge that client's instructions.

Prosecutors in the SPA serve usually for a term (tour of duty) of two to three years before moving on to another appointment within their service. While serving with the SPA, they are exempt from routine operational postings. However, the scope is there for the appointment of non-service prosecutors, as section 365 of the AFA 2006 outlines the four criteria for appointment. The appointee:

(a) has a general qualification within the meaning of section 71 of the Courts and Legal Services Act 1990;
(b) is an advocate or solicitor in Scotland;
(c) is a member of the Bar of Northern Ireland or a solicitor of the Court of Judicature of Northern Ireland; or
(d) has in a relevant territory rights and duties similar to those of a barrister or solicitor in England and Wales, and is subject to punishment or disability for breach of professional rules.

This has been further amended, by section 21 of the Armed Forces Act 2011, by substituting references to prosecuting 'officer' with 'person'. This makes it clearer that non-service lawyers may be appointed on either an ad hoc basis, for example, as leading counsel in a complex and serious case, or on a more permanent basis, should there be a dearth of service lawyers.[10]

Prosecutors from the Crown Prosecution Service (CPS) may be seconded to the SPA, as well as former legal officers who are employed to act as prosecuting officers.

The SPA mission statement says:[11]

> The Service Prosecuting Authority provides for the independent, efficient and consistent consideration of criminal cases and offences contrary to Service discipline. It will initiate prosecutions where justified and conduct fair and thorough proceedings in the service courts of first instance and the service appellate courts, whilst liaising effectively with the police and dealing with prosecution witnesses and victims of crime with care and sensitivity. Whilst maintaining independence from the service chain of command, the Service Prosecuting Authority fulfils its functions in support of operational effectiveness of the Armed Forces throughout the world.

On 13 March 2013, the then Attorney General, the Right Honourable Dominic Grieve QC MP, delivered a speech to Queen Mary University of London School of Law entitled 'The case for the prosecution: independence and the public interest'. Part of that speech considered 'The road to a national prosecution service'. The Attorney General, quite properly, referred to himself as 'the guardian of the public interest' and he emphasised the need to 'maintain a system of prosecutorial independence' which avoided interference but welcomed consultation where necessary to inform the prosecutorial decision. This is particularly important, it is suggested, in any hierarchical environment where it is easy to misconstrue well-intentioned opinion with interference or pressure. As *Findlay* reminded us very clearly, justice must be seen to be done.

10 For example, in time of service exigency, service lawyers may be needed to deploy to operational theatres as command legal advisers.

11 Version 2.

SPA Casework Quality Standards set out the benchmarks of quality which the SPA strives to deliver in prosecuting service offences.[12] Prosecutors must make decisions in accordance with:

a. The law;
b. The Code for Crown Prosecutors;
c. Guidance issued by the Attorney General; and
d. Relevant Policies and guidance issued by or on behalf of the DSP.

SPA Prosecution Guidance is not yet publicly available, so it is not possible to ascertain how it relates to the CPS guidance, taking into account the uniqueness of service life, the operational environment service personnel operate in, or the severity with which certain offences are regarded within the services and service community in contrast to how they are viewed in civilian life. It is apparent, from paragraph 7c of the Casework Quality Standards, that the SPA applies the Code for Crown Prosecutors, which, of course, makes no reference to the service interest.

Prosecutor's Test

The CPS Full Code Test has two stages. The first stage is the consideration of the evidence: the evidential stage. If the case does not pass the evidential stage, it must not go ahead no matter how important or serious it may be. If the case does pass the evidential stage, prosecutors must proceed to the second stage and decide if a prosecution is needed in the public interest: the public interest stage.

The prosecutor's test set out by the CPS is:

> Prosecutors must be satisfied that there is enough evidence to provide a 'realistic prospect of conviction' (RPOC) against each accused on each charge selected. A RPOC is an objective test. It means that the members of a CM hearing a case properly directed in accordance with the law are more likely than not to convict the accused of the charge alleged.[13]
>
> Is it in the public interest for the accused to be prosecuted?

12 Issued 28 November 2018: http://spa.independent.gov.uk/linkedfiles/spa/test/about_us/publication_scheme/20161128-caseworkqualitystandards-u.pdf.
13 See SPA Guidance, para. 2.039.

Is it in the public interest for that prosecution to be by CM?[14]

In the services, the element of the public interest test (PIT) is varied to the effect that the service interest (explained in detail below, on page 345 *et seq.*) is subsumed into the public interest. This was something that was made clear to the Army Prosecuting Authority (APA) by the then Attorney General when the APA prosecution test was devised and is self-evidently the correct approach. The questions to be posed, therefore, should be: is it in the public, which *includes* the service, interest for the accused to be prosecuted? Is it in the public, which includes the service, interest for that prosecution to be by court martial? That was the prosecution test devised by the APA.

The SPA Legal/Policy Guidance applies a PIT when determining the forum for prosecution, that is, military or civilian. It is explained that, unlike the CPS in the civilian jurisdiction, the jurisdiction of the SPA is not a complete jurisdiction.[15] However, please see below, under the heading Consent to Prosecution and the discussion on section 326 and the Prosecutors' Convention.[16]

Generally speaking, if a charge fails to meet the evidential test, it will proceed no further. However, if it does meet the test, the prosecutor then must go on to consider the PIT. It is also to be remembered that, even if at the time a decision to prosecute is made, there is sufficient evidence to meet the evidential test, it is the prosecutor's duty to keep the prosecution under constant review so that, if any change in circumstances occurs as the

14 It is suggested that the explanation of this test (the evidential sufficiency test) given in the Vol. 1, p. v of the MSL should be read in the light of the test applied by the SPA in conformity with the CPS, namely, 'Prosecutors must be satisfied that there is sufficient evidence to provide a realistic prospect of conviction' which, in turn, 'is an objective test based solely upon the prosecutor's assessment of the evidence and any information that he or she has about the defence that might be put forward by the suspect. It means that an objective, impartial and reasonable jury or bench of magistrates or judge hearing a case alone, properly directed and acting in accordance with the law, is more likely than not to convict the defendant of the charge alleged' – CPS Code for Crown Prosecutors, para. 4.1 *et seq*.

15 Yet s. 50 specifies that the 'Court Martial has jurisdiction to try *any* service offence'. 'Service offence' is defined in s. 50(2). As all offences within the SJS are 'service offences', the court martial would appear to have jurisdiction, subject to arrangements made under the Prosecutors Convention, where a suspect's conduct could be dealt with by criminal or civil/regulatory sanctions and/or where more than one prosecuting authority and investigating body share the power to take action.

16 SPA Legal/Policy Guidance, para. 2.032.

case develops, including what becomes known of the defence case, any further reasonable lines of inquiry that should be pursued, and receipt of any unused material that may undermine the prosecution case or assist the defence case, the case may be altered or discontinued or the prosecution should not proceed.[17] The discontinuance itself does not necessarily mean no prosecution will ever take place, but may mean that there is insufficient evidence at the time the review was carried out (e.g. a previously available witness has since died and it is not thought that their written statement would be admitted under the statutory regime). However, should other probative and admissible evidence come to light subsequently, then the decision may be made to revive the prosecution. Some public interest factors may militate against prosecution where they outweigh factors tending in favour. For example, in a case where the defendant has a serious illness, it may be considered that factor balanced against the gravity of the case and the likely outcome outweighs all others.[18]

The law is clear that a decision not to prosecute can be challenged. Lord Steyn held, in the House of Lords case of *R v Director of Public Prosecutions Ex Parte Kebelene and Others*,[19] in challenges involving

> a decision not to prosecute, judicial review is available: see *Reg. v. Director of Public Prosecutions, Ex parte* C. [1995] 1 Cr App R 136.

In the later decision of the Divisional Court, headed by Lord Justice Judge LCJ, and Fulford and Sweeney JJs, in *The Queen (On the application of F) v The DPP*,[20] the court held that judicial review was available in respect of the Director of Public Prosecution's (DPP) decision not to prosecute a husband, accused by his wife of rape, who had consented to sex with him but on condition he withdrew before ejaculation. The LCJ said:

> 4. Where a decision not to prosecute has followed a painstaking review of the kind which occurred here, and which will be

17 Code For Crown Prosecutors, para. 3.6.

18 During 2016–2017, the Crown Prosecution Service failed to proceed with 1,522 criminal cases at courts in England and Wales due to youth of defendants, elderly defendants or defendants in significant ill health. Source: 'Release 6 dated 13/02/2019 – For years 2016, 2017 and 2018 for each police force in England and Wales provide the number of prosecutions that were dropped or discontinued and the reason for the course of action'.

19 [1999] House of Lords, 28 October 1999, which examined the issue of reverse burden of proof.

20 [2013] EWHC 945 (Admin).

> expected hereafter following the implementation of the Crown Prosecution Service's offer to the victim of a Right of Review (VRR) which, we understand will become effective in May 2013, it will be a very rare case indeed when the court can properly decide that it should interfere with the decision not to prosecute.

In relation to decisions to prosecute, however, Lord Steyn (with whom the rest of the court concurred in *Kebiliene*[21]) held, in regard to a challenge to decisions to prosecute in judicial review proceedings, that:

> [t]he potential for undermining the proper and fair management of our criminal justice system may be considerable ... the DPP may sometimes not have a concluded view of any kind. But he may nonetheless be persuaded that, despite some uncertainty about the law, a prosecution is justified as being in the public interest. There could then be no question of reviewing his decision for error of law ... [A]bsent dishonesty or *mala fides* or an exceptional circumstance, the decision of the DPP to consent to the prosecution of the Respondents is not amenable to judicial review. And I would further rule that the present case falls on the wrong side of that line. While the passing of the Human Rights Act 1998 marked a great advance for our criminal justice system it is in my view vitally important that, so far as the courts are concerned, its application in our law should take place in an orderly manner which recognises the desirability of all challenges taking place in the criminal trial or on appeal. The effect of the judgment of the Divisional Court was to open the door too widely to delay in the conduct of criminal proceedings. Such satellite litigation should rarely be permitted in our criminal justice system. In my view the Divisional Court should have dismissed the Respondents' application.

Lord Hope considered that:

21 Lord Cooke held: 'in a case where it is obvious that the Director wishes a charge to go to trial, I think that the courts should be very slow to allow review of a consent to be used as a device for resolving points which would otherwise be dealt with in the ordinary course of the criminal proceedings.'

> there are strong grounds for the view that, in the absence of dishonesty, bad faith or some other exceptional circumstance, the Director's decisions to consent or not to consent to a prosecution are not amenable to judicial review in that court. To this must be added the fact that the process of judicial review could do no more than require the Director to reconsider his decision. It could not require him to change his view. It would fall short of providing a remedy which is as effective as that which could be provided by the trial judge during the trial process or on appeal.

Consequently, in the absence of dishonesty or bad faith on the part of the Crown the decision to prosecute is not open to challenge. The matter should be decided at trial, on the evidence.[22]

The principle was referred to in the decision in *R (Barons Pub Company Limited) v Staines Magistrates' Court*[23] where one of the issues was whether the BPC Ltd could challenge the decision to prosecute otherwise than through an abuse of process application. It was the view of the Divisional Court that the magistrates' courts have no power to review a prosecutorial decision other than through an abuse of process application. The President of the Queen's Bench Division, John Thomas, held, having considered the authority in *Kebiliene*:

> As was made clear by the court in *R v A(RJ)*, save in an exceptional case, decisions to prosecute are for the prosecutor. The task of a criminal court is, save in an exceptional case, to determine whether the prosecution has proved its case on the merits.

In *R (FNM) Claimant v DPP*,[24] the Claimant sought and was granted judicial review of the CPS decision not to prosecute the person she accused of

22 See the decision of *Nightingale* (preliminary hearing), where the defence challenged a decision to prosecute a retrial after permission to retry the defendant had been granted by the CMAC: on p. 346 below.

23 [2013] EWHC 898 (Admin).

24 [2020] EWHC 870 (Admin). The case involved a 15-year-old vulnerable female complainant who had been subjected to a serious sexual assault (or assaults) and believed she had been raped. In interview under caution, the D admitted that he had had sexual intercourse with her, believing that she consented and was 16 or 17 years old. The case was referred to the CPS who informed the Claimant of the decision not to prosecute. The decision followed the Claimant's exercise of her right to a review under the Victims' Right to Review Scheme ('the VRR Scheme'). The claimant sought judicial review of the decision not to prosecute.

raping and sexually assaulting her. She was given 21 days from the date on which the judgment was handed down to submit to the CPS her reasons in support of the application for review, after which a fresh decision was to be taken by a member of the CPS Appeals and Review Unit not previously involved with the case. (See also on page 412 for a discussion about the views of the trial judge on the appropriateness of the charge brought by the prosecution.)

As a prosecutor, the decision of a CO not to prosecute would also be amenable to judicial review where there is a victim crime, for example an allegation of actual bodily harm (ABH).

The Service Interest

So, what is the service interest? The Manual of service law[25] examines this within the context of a summary hearing and suggests that it is determined by whether 'disciplinary action is consistent with justice and with the fair and efficient maintenance of discipline and operational effectiveness', and whether there are service factors that make disciplinary action *inappropriate*. No guidance is given about what this means in fact and it almost sounds as if it is relevant to reasons not to prosecute, which is quite the opposite to how the test was first formulated.[26] It is suggested the MSL definition would be inconsistent with the view in *Andrews*[27] and might also be viewed as an invitation to the chain of command to justify non-prosecution for 'service reasons', which would then undermine public confidence in the military justice system, suggesting that circumstances within the military might excuse criminal conduct, or in some way place the interests of justice beneath those of military exigencies.

The original concept of how the service interest test would operate was to permit the chain of command to approach the prosecuting authority with information and service *factors* indicating the severity of the matter in the eyes of the chain of command that would be helpful in the prosecutor's deliberations on the application of the PIT, in much the same way as the defence is free to do so; not for the prosecution to proactively approach

25 Vol. 1, page 1-6-29, para. 94.

26 The author defined the original test, as outlined below.

27 (1998) Case No 98/0600/S2.

the chain of command. The case of *Nightingale*[28] sounded a cautionary note. The defence, at a case conference, had made the valid submission that the mischief was in the Director, Service Prosecuting Authority (DSP/DSPA) seeking this view at all and there was a perception that his independence was thereby undermined. The Judge Advocate General, who presided at the hearing, ruled, at paragraph 17, that the DSP 'has done no more than seek an official view to help inform himself of the service interest test and that ultimately he, independently, made the decision to prosecute'.

The point has not been tested in any trial or appellate proceedings. But, the effect of *requesting* information may be to receive a view hostile to the defendant that, apart from the proactive approach, would not have been volunteered by the chain of command. Conversely, where the view canvassed by the prosecutor from the chain of command influences the prosecutor *not to proceed* with a charge, the decision may be challenged by a victim. It is a fine distinction. The essence of Article 6 ECHR is one of a perception of impartiality.

In understanding what is meant by the service interest, some assistance may be gleaned from Joint Service Publication (JSP) 833, which

---

28 *R v Nightingale*, Preliminary Hearing, 1 May 2013, where an email was sent by the SPA to the Adjutant General to ascertain his 'view on a retrial of Sgt Nightingale if the CMAC decide to direct a retrial', which they did. This correspondence was adverted to in the hearing, which subsequently alleged the decision to prosecute the defendant was not in the public interest. There was no suggestion of bad faith or dishonesty in any communication between the SPA and the Adj Gen, nor was it alleged the Adj Gen gave any direction to the SPA about whether or not to prosecute. The mischief, it was alleged, was in the SPA seeking this view at all and there was a perception that his independence was thereby undermined. The JAG did not uphold the submission. This specific issue has not been heard as an appeal point before the CMAC. The case concerned a SNCO in the SAS who was found in possession of a Glock pistol and a large quantity of ammunition during a search of the house he shared with a fellow SNCO. He pleaded guilty at his trial and was sentenced to 18 months detention. He appealed the sentence, which was reduced to 12 months and suspended by the CMAC, and then appealed his conviction. The CMAC considered that, at the trial, the JAG gave an uninvited sentence indication. The question was whether the uninvited indication given by the judge, and its consequent impact on the defendant after considering the advice given to him by his legal advisers on the basis of their professional understanding of the effect of what the judge had said, had created inappropriate additional pressures on the defendant and narrowed the proper ambit of his freedom of choice. His defence had raised the issue of false memory, which gave rise to his admissions to the police and his housemate. Accordingly, the plea of guilty was in effect a nullity and was set aside. The CMAC quashed the conviction and ordered a retrial, which took place in July 2013, at which he was convicted and sentenced to two years' detention, suspended for 12 months.

deals with minor administrative action (MAA). Some of the service *factors* listed[29] in essence are pertinent to the service/public interest test:

a. The context in which the incident occurred and its impact on operational effectiveness (for example, if the unit or ship was about to deploy, an incident of being late would be much more serious than at other times and MAA is likely to be inappropriate)
b. The extent of any adverse impact or likely impact on the individual, unit or Service
c. The rank or rate of the individual or level of responsibility
d. The age and maturity of the individual
e. The individual's previous conduct and warnings
f. Any relevant personal circumstances or mitigating factors
g. The individual's response to the discovery of the incident.

The service interest relates to a single service and not all three services, as it reflects the operational and disciplinary imperatives of that service. This is emphasised in Queen's Regulations (QRs) where, for example, in the matter of trial by court martial, QRs state that a service defendant will ordinarily be tried by lay members from only their own service.[30] However, where a defendant is tried with a co-defendant from a different service, the lay membership of the court may be a mixture of service personnel from different services. Each defendant will usually have at least one lay member of their own service on the board (see *Gunn v The Service Prosecuting Authority & Or*[31]). However, in that case, AFA 2006 and the Court Martial Rules prevailed over the QRs; accordingly, a defendant may be tried by a board consisting entirely of lay members from a different service, as the Act is permissive and does not prohibit court martial boards from being comprised of suitably qualified officers and warrant officers drawn from any service, in keeping with the tri-service philosophy of the Act. The Court Martial Appeal Court (CMAC), however, acknowledged that, nevertheless, the usual practice, embodied in sentence [1] of paragraph 1140 4) of the QRs

29 JSP 833, para. 9 on p. 3-2.

30 See also *McKendry*, where the CMAC accepted that each service may wish to deal with AWOL in its own way dependant upon the ethos of that service. See *supra*, on p. 182 and FN 64.

31 [2019] EWCA Crim 1470.

(RAF)[32] is that a defendant will be tried ordinarily by lay members of their own service.

Of course, one or both of the other two services may share a particular concern in any given case. The service interest is concerned with the maintenance of discipline in an efficient and professional armed force. It is therefore easier to understand the service interest in the context of prosecutions where, for a comparable situation in civilian life, no proceedings would be brought. For example, a soldier who fails to turn up for work, or a sailor who fails to attend their ship, might face disciplinary action[33] rather than a minor administrative punishment if their unit were about to deploy, or ship about to sail, and their absence delayed or hampered the efficient arrangements for deployment, or departure of the ship. This impacts directly upon discipline. The service interest in such a case would militate *in favour* of prosecution and, in that way, it informs the overriding public interest (as the deployment of our troops is part of government foreign and defence policy) that looks holistically at the factors supporting a prosecution, including compassionate reasons which may justify a different view. However, if the reason the accused did not turn up was because their spouse was ill, or there were domestic problems at home, then those facts alone might suggest it may not be in the public interest to prosecute them, even though their absence is 'intentional' within the meaning of section 9(2).[34]

Moreover, it would be inappropriate for the chain of command to say that, where a soldier is required for service overseas, it is in the service interest not to prosecute them as their presence was required in their unit. That is a service factor overridden by the public interest and prosecution may well be the proper course, especially where the offence is a serious one, for example under Schedule 2 of the Act.

32 '[1] A Service defendant will ordinarily be tried by lay members of wholly his own service. [2] However, where a defendant is tried with a co-defendant from a different Service, the lay membership of the court will be a mixture of Service personnel from different Services. [3] Each defendant will always have at least one lay member of his own Service on the board.'

33 That is, be charged with an offence under the AFA 2006, with the possibility of acquiring a criminal record. See p. 276 above.

34 *Regina v Alan James Martin* [2007] EWCA Crim 3377. See also the discussion below on AWOL on p. 178 *et seq*, where the courts have not been so sympathetic in their views.

In relation to criminal conduct, the services take a dim view of drug abuse, although this is changing in respect of the element of dismissal as part of the sentencing process. The concern is that a service person might be unable to function efficiently if under the influence of drugs and, thereby, jeopardise the safety of their colleagues.[35] The service person engaged in the use of dangerous equipment, firearms or explosives, or the piloting of aircraft, presents an obvious danger to their fellow service personnel, which impacts upon trust, unit cohesion and operational efficiency. The service interest clearly leans to prosecution. The 2006 Act now has clear provisions to deal with these circumstances.

Another example would be the theft of personal items from a roommate or colleague. This offence has a deep impact upon trust among service persons living together, working together and, eventually, deploying together on combat operations, where trust is absolutely key. It is very corrosive to have someone you can't trust in the unit or work group upon whom one might depend for one's safety. Once more, the service interest leans to prosecution.

On the other hand, where a service person warned for deployment on operations has a minor traffic accident while driving a service vehicle, the service interest might, in such circumstances, support a prosecution (the reputational consequence of driving an Army vehicle on a public road and involvement in an accident with a civilian vehicle, the need to punish those whose conduct falls below what is required), the overriding public interest (minor offence, urgently needed for operations – a service *factor* – etc.) could be said to militate against prosecution and in favour of a charge suitable for summary dealing or even minor administrative sanction. Service *factors* alone could never justify the non-prosecution of a service person for serious offences on the grounds that they were key personnel who were required for deployment on operations and whose expertise was hard to replace. The service *interest* in such a case would be to deploy with a full complement of personnel. The *public* interest would dictate that those who commit serious crime face trial. The *service* interest may require

35 The author recalls an incident from service in Northern Ireland in 1989, when a soldier on a four-man foot patrol, who had some time previously taken LSD, suffered a 'flashback' and was threatening to fire his rifle. He had to be disarmed by his colleagues. Fortunately, nobody was harmed.

the person's presence with their unit, but so might the public interest, too. Defence of the nation is clearly in the public interest.

Consequently, rather than being reasons not to prosecute, it is suggested that the service interest is impacted whenever the maintenance of discipline is affected and it therefore militates *in favour* of a prosecution, subject to the overriding public interest.[36] In the example cited above, this may justify deferring prosecution to permit deployment.

The hybrid nature of the court martial, to punish service personnel for the criminality of their conduct and to deal with them also on a disciplinary basis, has been recognised by the CMAC.[37]

> The service community is not the same as a civilian community, and offences which may be at the margin of criminality in the civilian community may be far more serious when committed by servicemen and personnel in a service environment. They may in themselves be destructive of service discipline and impair the operation of the unit to which the soldier belongs. Barrack room theft is an obvious example of this. Even petty pilferage may create unacceptable strains within the unit and impair its ability to operate properly and effectively. Similarly, any breach of discipline creates similar strains .... Those responsible for military discipline are entitled to take this view about drug offences. The corrosive effect on discipline of drug taking in itself in military units is notorious.

It is submitted that this is a good definition of the service interest. In essence, this acknowledges the service *factors* that could support the service *interest* of ensuring good order and discipline.

SPA Guidance confirms[38] that service prosecutors are to apply the Code for Crown Prosecutors, together with the SPA additional guidance.

---

36 The author set out such a definition of the 'service interest' during his time with the Army Prosecuting Authority, as there was a need to have a clear understanding of how it operated and its relationship with the public interest. There was to be no suggestion that the service interest ever trumped the public interest. At the time, there was no consensus on what the 'service interest' meant, although practice – and case law – indicated how it should be defined.
37 *Per* the dicta of Hobhouse LJ in *Andrews*. It is submitted that this is a good definition of the service interest.
38 Para. 2.002.

The examples given above therefore are those that it is suggested convey a broad understanding of what the service interest is.

Consent to Prosecution

A unique feature of service prosecutions is that offences that in civilian life require the consent of the Attorney General may be prosecuted without obtaining such consent. This is clearly set out in section 326 AFA 2006:

> Subject to section 61(2), no enactment requiring the consent of the Attorney General or the Director of Public Prosecutions in connection with any proceedings has effect in relation to proceedings under this Act for a service offence.

The proviso in section 61(2) relates to prosecutions in relation to those persons who are no longer subject to service law or subject to service discipline. In such cases, the Attorney General must be consulted. However, apart from those situations, serious cases, especially those where the civil authorities also have jurisdiction[39] (e.g. murder or manslaughter committed abroad), would be discussed with the Attorney General who exercises a general superintendence of the SPA. This is also agreed in the Prosecutors' Convention (see further on page 19), designed to serve the interests of justice. While this is an influential document that establishes coherence and cooperation in prosecutions of cases that have cross-agency interest, it cannot override the statutory authority of the DSPA.[40]

---

39 '[W]here more than one prosecuting authority and investigating body share the power to take some action. In such cases, depending on the extent to which more than one authority is pursuing or wishes to pursue an investigation, decisions will have to be taken about whether and how investigations should proceed and as to the choice of criminal charges or civil or regulatory sanction. The aim is to ensure that cases are conducted in a way that is just and which best serves the overall public interest, so that the public can have confidence in the outcome.' *ibid.*

40 The Lyons Review recommends removing the prosecutions of murder, manslaughter and rape from the SJS 'when these offences are committed in the UK, except when the consent of the Attorney General is given'. See Part 1 Recommendations, *ibid.*, p. 3. Consideration should also 'be given to including either S2 offences (sexual assault with penetration) or both S2 and S3 (sexual assault without penetration) offences in the category of cases that should be proceeded with under the civil jurisdiction when the offences are committed in the UK'. It is suggested that these are not controversial recommendations. However, it has to be said that the SPA is not without experienced prosecutors who can work either alone or with civilian senior counsel in the prosecution of such offenders. See, e.g. the prosecution of *Payne & Others* (2007) General Court Martial held at Military Court Centre Bulford where the Crown was represented by a QC (former No 1 Treasury Counsel), a senior junior, a further junior and a

Case Papers/Advance Information
Main Provisions: sections 121 and 122 AFA 2006; rule 43 AF(CM)R 2009; rule 2(1) AF(P5AFA2006)R 2009

Following an investigation by the service or civilian police, a copy of the case papers will be provided to the DSPA. The case papers are defined in rule 2(1) Armed Forces (Part 5 of the Armed Forces Act 2006) Regulations 2009 (AF(P5AFA2006)R 2009) as:

(a) all reports relating to the case or charge and written by a Service police force;
(b) the following other papers relating to the case or charge prepared or obtained by a Service police force
  (i) all witness statements;
  (ii) all other records of evidence, including a summary or transcript of all tape-recorded interviews;
  (iii) a list of all exhibits and a statement of where any which are not documentary exhibits are held;
  (iv) all documentary exhibits;
  (v) all formal disciplinary records of the suspect maintained and held by any of Her Majesty's forces;
  (vi) if no formal disciplinary record of the suspect is maintained and held by any of Her Majesty's forces, a list of his convictions (if any) for a Service Offence and of his convictions (if any) by a civilian court;
  (vii) all documents to be provided, in accordance with a code of practice made by virtue of an order under section 78(2)(b) of the Criminal Procedure and Investigations Act 1996(a), to a person involved in the prosecution of service offences; and
(c) all papers equivalent to those within paragraphs (a) and (b), prepared by a UK police force or an overseas police force and provided by that force to a Service police force.

Lt Col prosecutor. The prosecution of seven members of 3 Para, for the murder of an Iraqi, was led by a senior silk (former Treasury Counsel), with a Lt Col and Major as juniors.

The SPA will prepare the prosecution papers comprising the evidence against the accused, for service upon the defence, along with any exhibits and unused material in compliance with prosecution duties of disclosure. This is referred to in rule 43(2) Armed Forces (Court Martial) Rules 2009 (AF(CM)R 2009) as 'advance information',[41] and means:

(a) copies of the statements of those witnesses on whom the Director intends to rely against the defendant;
(b) a list of all exhibits which the Director intends to adduce in evidence against the defendant, and a statement of where any non-documentary exhibits are held; and
(c) a transcript of any sound recording of an interview with the defendant.

These documents are also served upon the court administration officer. However, it is the CO who serves them upon the accused, unless other provisions apply (e.g. the accused or their legal adviser agree to service by other means; see Part 2 of AF(CM)R 2009). By convention, the SPA almost always serve the prosecution evidence at the same time they direct charge, but this does not have to be the case of course. In custody cases, formal service of the prosecution evidence, in accordance with rule 43 may take place some time after charge.

Until the 2016 Act, the charge was usually brought against the accused by the CO on the direction of the DSP, in any case allocated for trial by court martial or Service Civilian Court (SCC).[42] However, section 5 AFA 2016 introduced a new section 121(1A) to the 2015 Act under which the DSP may, as an alternative, bring the charge themselves. Charges brought by the Director may now be allocated to the appropriate service court.

After the charge has been brought, the DSP has the power to amend it or substitute another charge or to bring further charges. They may also discontinue proceedings under section 125(2)(d). Alternatively, they may

41 An unfortunate term, which is used in civilian proceedings in relation to details of the prosecution case provided to a defendant charged with an offence that can be tried in the magistrates' court before: (a) deciding whether to elect trial at Crown Court; (b) deciding whether to consent to summary trial; or (c) being put to plea, if they are under 18, as regulated by Part 21 of the Criminal Procedure Rules 2010. The previous term, contained in the pre-2006 Rules, 'prosecution papers' was a better one and avoided confusion, it is suggested.
42 See ss 121 and 122 AFA 2006.

refer the charge to the CO for summary dealing where it is capable of being so dealt with, or they may make a direction under section 127(1) or (2) that no further proceedings be brought in respect of the offence charged or any offence that could be charged.

The service of the prosecution papers is usually the stage at which the defence solicitor will brief defence counsel, unless the legal representative is a solicitor advocate with higher rights and decides to personally deal with the case. Legal aid may be applied for at the time the individual elects trial, or has a case referred to the DSP for a decision on prosecution. See further Appendix F for information on the process of any application for legal aid. It will be noted that an individual who does not require legal aid, whether they wish to use a service lawyer, or does not require any legal representation, should still complete the application form to that effect. It remains the individual's responsibility to arrange contact with the appropriate service lawyer, if one is required. In this regard, the Defendant's Assisting Officer (DAO) will be of assistance to them.

The case papers/advance information may be added to, under rule 43(3), by the subsequent service of a notice in relation to further evidence it is proposed to adduce. Witness statements must conform to section 9 of the Criminal Justice Act 1967. Usually, typed copies of such statements, rather than photocopies of the originals, are supplied. A copy of this additional evidence is also served upon the court administration officer.

If the charge is such that, on conviction of the defendant on that charge the court would have power to make an activation order, the advance information in relation to the defendant must include a notice to that effect.[43]

Similarly, where a defendant has been conditionally discharged by virtue of Schedule 3 to the 2006 Act,[44] and is charged with an offence committed during the period of conditional discharge, the advance information in relation to them must include a notice that, if they were convicted of the offence, the court would have power to deal with

43 AF(CM)R 2009, rule 44(1).

44 Which sets out modifications of court martial sentencing powers for civilians and ex-service persons.

them under section 186(2) for the original offence for which they were conditionally discharged.[45]

Where an overseas community order is in force in respect of a defendant, the advance information in relation to them must include a notice that, if they were convicted of the offence while the overseas community order is in force, the court would have the powers conferred by paragraph 23 of Schedule 8 to the 2003 Act.[46]

Statutory Powers of the DSP

Main Provisions: sections 125–128, section 130 and 130A AFA 2006; rules 57, 60 AF(CM)R 2009; rules 40, 44 AF(SCC)R 2009; regulations 11 and 14 and Schedule to AF(P5AFA2006)R 2009; section 5 AFA 2016; AF(MA)R 2019

The Act permits the DSP to direct the bringing of charges after referral of a case by either the service police or the CO. Where they decide that it would not be appropriate to make a direction for trial, they may decide to refer it back to the CO, or they may make a direction under section 127(1) or (2) barring further proceedings.[47]

Section 125 sets out the powers of the DSP in respect of a charge allocated for court martial, which includes a charge in respect of which a person has elected trial by court martial.[48] This provides that they may:

(a) amend the charge;
(b) substitute for the charge another charge against the accused;
(c) bring an additional charge against the accused;
(d) discontinue proceedings on the charge;
(e) refer the charge to the accused's commanding officer, but only if the charge is capable of being heard summarily (see

45 F(CM)R 2009, r. 44(2).
46 *Ibid.*, r. 44(3). The Schedule provision referred is that of the CJA 2003, and confers the powers of the Crown court to revoke the order, or both revoke the order, and deal with the offender, for the offence in respect of which the order was made, in any way in which they could have been dealt with for that offence by the court which made the order if the order had not been made. In dealing with an offender, the Crown court must take into account the extent to which the offender has complied with the requirements of the community order.
47 *Ibid.*, s. 121.
48 See s. 129 AFA 2006.

section 52). (S)he may not do this without the accused's written consent in respect of a charge on which the accused elected trial;[49]

(f) allocate the charge for trial by the Service Civilian Court, but only if the charge is one that that court has jurisdiction to try;

(g) make a direction under section 127(1) or (2) (directions barring further proceedings) in respect of the offence charged or any offence that could be charged under paragraph (c) above.[50]

Where a charge is amended or substituted under section 123(2), 125(2) or 126(2), the charge sheet must state the amended or substituted charge. Where an additional charge is to be brought under section 123(2), 125(2) or 126(2), the charge sheet must state the charge to which an addition is to be made and the additional charge. If the accused is subject to service law or a civilian subject to service discipline, the DSP must ensure that a copy of the signed charge sheet is served by hand on the accused. But if the accused has ceased to be subject to service law or a civilian subject to service discipline, the DSP must ensure that a copy of the signed charge sheet is served on the accused either by hand or by leaving it at, or sending it by post to, the accused's last known address.

If the DSP makes any amendment, substitution or addition of charges before the arraignment of an accused, a charge sheet must be prepared in accordance with the Schedule to AF(P5AFA2006)R 2009, the DSP must sign the charge sheet and the DSP must ensure that a copy of the signed charge sheet is served by hand on the accused. As soon as practicable after service of the charge sheet on the accused, the DSP must provide a copy of the charge sheet to the court administration officer. Where the earliest time at which it will be practicable to provide a copy of the charge sheet to the court administration officer is less than 24 hours before the time appointed for a hearing in relation to the accused, the DSP must also provide, as soon as practicable, a copy of the charge sheet to the judge advocate appointed

49 AFA 2006, s. 130(20(a). In other words, they may not refer back to the CO a charge on which the accused has elected, without the written consent of the accused.

50 While at first sight this might appear to be a repetition of the power in s. 121, this is a power that exists *after* direction for trial.

to conduct the hearing (r.14 AF(P5AFA2006)R 2009). This provision applies to both the court martial and the SCC.

When drafting the charge sheet, the Director must ensure it is prepared in accordance with the Schedule to AF(P5AFA2006)R 2009. The requirements are explained above, at page 115.

If at any stage the Director discontinues proceedings on a charge, under section 125(2)(d)[51] or section 126(2)(d),[52] they must as soon as reasonably practicable give written notification of the discontinuation to the accused or their legal representative, the accused's CO and the court administration officer. Aside from the traditional methods of providing notice referred to above, where the accused has agreed to receive it by that method, it may be transmitted it to them by fax or other electronic means, but only if they have agreed (AF(P5AFA2006)R 2009, reg. 15(6)(d)).

Election for Trial

Further restrictions are set on in section 130A on the DSP's powers to substitute or add charges after election where the charge is in respect of an offence which would be a 'relevant offence' for the purposes of Schedule 3A (sentencing powers of court martial where *election for trial* by that court) if the accused were convicted of it. The phrase, 'relevant charge' means:

(a) in relation to a case A offence or a case B offence (within the meaning of Schedule 3A), the charge in respect of which the accused elected Court Martial trial; and
(b) in relation to a case C offence or a case D offence (within the meaning of Schedule 3A), the charge referred as mentioned in paragraph 4(c) of that Schedule.

The phrase 'relevant offence' adverted to in Schedule 3A is an offence that falls within any of cases A to D of that Schedule. The matter becomes even more complicated as the Schedule refers to different definitions of 'relevant charge', defining the phrase according to each case (i.e. A to D). However, the overall effect of the changes does not materially alter the position of the accused.

51 Discontinuance of a charge allocated for Court Martial trial.
52 Discontinuance of a charge allocated for Service Civilian Court trial.

Section 130A Restrictions on DSP's powers to substitute or add charges after election

(1) This section applies where
   (a) a charge is for the time being regarded for the purposes of Part 5 as allocated for Court Martial trial; and
   (b) the charge is in respect of an offence which would be a relevant offence for the purposes of Schedule 3A (sentencing powers of Court Martial where election for trial by that court) if the accused were convicted of it.

(2) The Director of Service Prosecutions (the Director) may not without the written consent of the accused substitute under section 125(2)(b)
   (a) a charge in respect of an offence which is not one that may be dealt with at a summary hearing (see section 53); or
   (b) a charge in respect of an offence within section 54(2) (offences that may be dealt with summarily only with permission or by senior officer), except where the relevant charge was in respect of such an offence.

(3) In subsection (2)(b) 'relevant charge' means
   (a) in relation to a case A offence or a case B offence (within the meaning of Schedule 3A), the charge in respect of which the accused elected Court Martial trial; and
   (b) in relation to a case C offence or a case D offence (within the meaning of Schedule 3A), the charge referred as mentioned in paragraph 4(c) of that Schedule.

(4) The Director may not without the written consent of the accused bring under section 125(2)(c) a charge in addition to the charge.

(5) In construing Part 1 of Schedule 3A (relevant offences) for the purposes of this section, paragraphs 3(b) and 5(c) of that Schedule are to be disregarded.

Rule 57 of the AF(CM)R 2009 provides that where the judge advocate accepts a plea of guilty to any charge, they may (with the DSP's consent) order that any other charge

(a) to which the defendant has not pleaded,
(b) to which the defendant has pleaded not guilty, or
(c) as respects which the judge advocate has recorded a plea of not guilty under rule 56(5), (where the plea of guilty is not accepted by the judge advocate, or the defendant does not plead (or does not plead intelligibly) to the charge)

is to lie on the file, not to be proceeded with without the leave of the court or the Appeal Court.

Rule 60 AF(CM)R provides that after arraignment on a charge, the Director may not exercise any of the powers under section 125(2) without the leave of the court. Where they do so, with leave of the court, they have to serve an amended charge sheet on each defendant. However, if the defendant consents, rule 60(2) provides that leave of the court is not required for the Director to refer a charge to the defendant's CO.

It must also be said that nothing prevents the DSP from offering no evidence in respect of a charge, where a plea of not guilty has been entered. The subsequent ruling by the judge advocate has the same effect as if the accused had been acquitted by the court.[53]

Similar powers exist in relation to cases involving trial before the SCC, under section 126. However, rule 40 of AF(SCC)R 2009 provides that 'where in preliminary proceedings (whether before or after arraignment) or trial proceedings it appears to the court that the charge sheet or any charge is defective, it shall make such order for the amendment of the charge sheet or charge (as the case may be) as appears necessary to meet the circumstances of the case'. The DSP must then serve the amended charge sheet on every defendant

(a) if the amendment had been made otherwise than in accordance with an order made under this rule; and
(b) where the amendment is made after arraignment, if it had been made before arraignment.

Rule 44 provides that where the court accepts a plea of guilty to any charge, it may (with the DSP's consent) order that any other charge

53 See, AF(CM)R 2009, r. 58.

(a) to which the defendant has not pleaded,
(b) to which the defendant has pleaded not guilty, or
(c) as respects which the court has recorded a plea of not guilty under rule 43(5) (where a plea of guilty is not accepted by the court, or the defendant does not plead (or does not plead intelligibly) to the charge),

is to lie on the file, not to be proceeded with without the leave of the court or the court martial.

As with the court martial process, the SPA may offer 'no evidence in respect of any charge to be tried by SCC to which the accused has pleaded not guilty'.[54]

Disclosure Obligation

Main Provisions: CPIA 1996 (AAF)O 2009; CPIA1996(CP)(AF)O 2009

As its title suggests, the Criminal Procedure and Investigations Act 1996 (Application to Armed Forces) Order 2009 (CPIA1996(CP)(AF) O 2009)[55] applied the statutory provisions of the Criminal Procedure and Investigations Act 1996 (CIPA 1996) to the Armed Forces, and made equivalent provision for the service justice system to that which exists in the civilian criminal justice system with regard to investigations, disclosure and certain evidential matters. These provisions apply to all service courts. The specific provisions in Part 1 of the 1996 Act are concerned with the duties of disclosure on the prosecutor and the accused in criminal proceedings. The policy intent is that the service justice system should mirror the civilian justice system unless there are good operational reasons that justify differences.

Article 4, CPIA 1996 (AAF)O 2009 places a specific duty on the Director to comply with initial disclosure to the accused of any prosecution material that has not previously been disclosed to them and which might reasonably be considered capable of undermining the case for the prosecution against the accused or of assisting the case for the accused, or they must give to

54 See AF(CM)R 2009, r. 45.
55 Made under s. 78 of the Criminal Procedure and Investigations Act 1996, as amended by the AFA 2006.

the accused a written statement that there is no such material. 'Prosecution material' means material that is in the DSP's possession and came into their possession in connection with the case for the prosecution against the accused or which, in pursuance of the Code of Practice[56] (see Appendix L Disclosure Code), they have inspected in connection with the case for the prosecution against the accused. Paragraph 2 of the Code of Practice defines 'material' as:

> material of any kind, including information and objects; this includes not only material coming into the possession of the investigator, such as documents seized in the course of a search of premises, but also material generated by him such as records of interview.

The same provision in the Code defines 'prosecution disclosure' as:

> any duty of the Director under Articles 4 and 13 of the Application Order to disclose material which is in his possession or which he has inspected in pursuance of the Code and which might reasonably be considered capable of undermining the case against the accused or of assisting the case for the accused.

If the accused is a person subject to service law or a civilian subject to service discipline, the Director may send any material or written statement or document to the accused's CO in order to comply with their duties under Articles 4 and 5[57] at the same time as they provide the material under Article 4.

Once the Director has complied with their duty under Article 4, the accused must give a defence statement to the Court Administration Officer (CAO) and the Director during the period of 28 days[58] beginning with the day immediately after the day on which the Director complies, or purports to comply, with Article 4,[59]

---

56 See below, on p. 419, The Criminal Procedure and Investigations Act 1996 (Code of Practice) (Armed Forces) Order 2009, (CPIA1996(CP)(AF)O 2009)) made under s. 78(2)(b) of the 1996 Act.

57 Art. 5 (CPIA1996(CP)(AF)O 2009) relates to the code of practice in CPIA1996(CP)(AF)O 2009.

58 Art. 8(1).

59 Art. 7(3).

a. setting out the nature of the accused's defence, including any particular defences on which he intends to rely;
b. indicating the matters of fact on which he takes issue with the prosecution;
c. setting out, in the case of each such matter, why he takes issue with the prosecution; and
d. indicating any point of law (including any point as to the admissibility of evidence or an abuse of process) which he wishes to take, and any authority on which he intends to rely for that purpose.

A defence statement that discloses an alibi must give particulars of it to enable the witness to be traced. Such details would be the name and address of the witness, or names and addresses of those who might assist in tracing the witness.[60] Where it appears to the judge advocate during preliminary proceedings that the accused has failed to comply fully with Article 7, they will warn the accused that there is a risk of comment at trial or inferences being drawn.[61]

The defence statement may be provided on the accused's behalf by their legal representative, in which case it is deemed to be with the accused's authority.[62]

As with the civilian system, even after complying with initial disclosure, the Director has a continuing duty keep disclosure under review[63] and to disclose material which might reasonably be considered capable of undermining the case for the prosecution against the accused or of assisting the case for the accused, and which has not been disclosed to the accused.[64] Indeed, the accused may apply to the judge in accordance with Article 14 if they believe the Director has not complied with their duty of disclosure. However, a failure in the Director's duty of disclosure does not automatically amount to an excuse for staying the proceedings on the basis of an abuse of process, unless the delay has the effect of denying

60 Art. 7(4).
61 Art. 12(2).
62 Art. 12.
63 Art. 13(2).
64 Art. 13.

the accused a fair trial.[65] The accused may also apply to a judge advocate for a review of the question whether, at any given time, it is still in the public interest to disclose material affected by an earlier order made which declared that disclosure of prosecution material would not be in the public interest (Art. 17).

Where there are faults in the disclosure obligation by the accused, as set out in Article 16, the judge advocate in the trial or appeal or any party other than the accused may make such comment as appears appropriate and the court may draw such inferences as appear proper in deciding whether the accused is guilty of the offence concerned.

Disclosed material may only be used for the purpose it was intended, that is, the proceedings. It is an offence under section 309(1) of the 2006 Act for a person (the 'offender') who is subject to service law or a civilian subject to service discipline to knowingly use or disclose an object or information recorded for a purpose not connected with the proceedings in which it was disclosed.[66]

During trial, if the judge advocate is of the opinion that seeing a copy of the defence statement would help the lay members of the court to understand the case or to resolve any issue in the case, they may order that a copy of the defence statement be placed before them, after any necessary editing of it has been done (Art. 12).

Code of Practice

The Criminal Procedure and Investigations Act 1996 (Code of Practice) (Armed Forces) Order 2009 produces a Code of Practice which appears in the Schedule to the Order, and which governs the manner in which service police are to record, retain and reveal to the Director material obtained in service investigations which may be relevant to the investigation and related matters. The Code shall be admissible in evidence in all proceedings before a service court (Art. 5) and a failure by a service police officer to comply with any provision of the Code shall not in itself render them liable to proceedings for a service offence (Art. 4). However, if it appears to the service court that any provision of the Code, or any failure by a service police officer is relevant to any question arising in the proceedings, the

65 Art. 15.
66 Art. 20.

provision or failure shall be taken into account in deciding the question (Art. 6). The Code is set out at Appendix L.

In December 2013, the Attorney General issued updated Guidelines on Disclosure for investigators, prosecutors and defence practitioners.[67] This contains guidance on applications for non-disclosure in the public interest, material not held by the prosecution, and a detailed Annex concerning Supplementary Guidelines on Digitally Stored Material. This Annex points out that digital material falls into two categories: (a) material which is created natively within an electronic environment (e.g. email, office files, system files, digital photographs, audio, etc.); and (b) material which has been digitised from an analogue form (e.g., scanned copy of a document, scanned photograph, a faxed document). Irrespective of the way in which technology changes, the categorisation of digital material will remain the same.

In addition, a revised judicial protocol and revised guidance on the disclosure of unused material in criminal cases was published in December 2013.[68] This protocol covers similar matters to the Attorney General's guidelines. It points out that applications for non-disclosure in the public interest should, whenever possible, be considered by the trial judge. In *R v H and C*,[69] the House of Lords has provided useful guidance as to the proper approach to be applied:

> Fairness ordinarily requires that any material held by the prosecution which weakens its case or strengthens that of the defendant, if not relied on as part of its formal case against the defendant, should be disclosed to the defence. Bitter experience has shown that miscarriages of justice may occur where such material is withheld from disclosure. The golden rule is that full disclosure of such material should be made.

67 Revised 2018. See: https://www.gov.uk/government/publications/attorney-generals-guidelines-on-disclosure-2013. See also, the Crown Prosecution Service Disclosure Manual of 26 February 2018. Of particular interest is the section on 'Triggers for statutory disclosure', p. 5.
68 'Judicial Protocol on the Disclosure of Unused Material in Criminal Cases', https://www.judiciary.uk/wp-content/uploads/JCO/Documents/Protocols/Disclosure+Protocol.pdf.
69 [2004] UKHL 3; [2004] 2 AC 134; [2004]; 2 Cr App R 10.

Defence Preparation

The defendant may nominate a DAO to assist them in preparing for court martial proceedings. The DAO's role is primarily administrative, although they may assist an unrepresented defendant prepare a plea in mitigation. The DAO has no legal standing at the court martial, has no right to speak unless they appear as a witness (for example as a character witness), and is not there to represent the defendant.[70]

The nominee should be subject to service law, at least the rate or rank of petty officer, military, marine or air force sergeant, and consent to be nominated. The defendant does not have to nominate a DAO, but if they dos not wish to do so, they should state in writing that they do not wish an appointment to be made. In the case of relevant civilians, if the defendant requests that they do so, the CO may provide names of people who may be nominated as a DAO and should do so if possible.[71]

Certain people may not act as DAO, for example,[72]

a. Any officer who holds delegated powers to act in custody matters within the defendant's chain of command;
b. Any individual who arrested the defendant for the alleged offence;
c. Any individual otherwise connected with the case under investigation or any other case in which the defendant is under investigation;
d. Any individual who has been involved personally in advising the CO about the case;
e. Legal officers;
f. Discipline staff;
g. The medical officer; and
h. The chaplain / padre.

The DAO's role is that of a facilitator, assisting the defendant to complete forms (e.g. legal aid application), obtain legal advice and representation, attend interviews, receive supporting papers, attend charging procedures and hearings, obtain character references and referees, and generally

70 JSP 830 MSL Version 2.0 2-28-11, §37.
71 JSP 830 MSL Version 2.0 2-29-17, §45.
72 JSP 830 MSL Version 2.0 2-29-B-2, §2.

support the defendant. They should also assist the defendant's legal representative in any way that supports the defendant. The DAO will not be legally qualified and should not become involved in preparing a defence or conducting the defence, although where a defendant chooses to represent themselves the DAO may assist them to prepare a statement in mitigation. The DAO should attend court martial proceedings with the defendant in a support role. The defendant should be advised that the DAO has no right to speak on the defendant's behalf at any proceedings unless they are called as a character witness.[73]

Advocates who are worried about the possibility of regimental loyalty or command influence over the DAO, thereby compromising the confidentiality of discussions with the defendant, should consider their position carefully. The fear of sensitive information relating to the defence finding its way back to the unit may be a major worry. One option would be to make the DAO sign a declaration at the commencement of the legal representative's involvement, in which they undertake not to disclose the confidential discussions that take place or evidence relating to the defendant's case. Another option is to invite the DAO to leave the room when the legal representative is taking instructions. This latter course may have the effect of alienating the DAO, in that they may feel they are not trusted, but it may be necessary to do this if the advocate feels it is unavoidable.

All the same, the service authorities recognise that there is a need for the defendant and their legal representative to be able to trust the DAO and, therefore, an outline of the DAO's duties is contained in the MSL, at Volume 2-29-B-1 (Annex B).

Paragraphs 9 and 10 of Annex B advise the DAO about their duty of confidentiality and about conflict of duty.

> 9. Confidentiality. You should treat anything said to you by the defendant in relation to his defence as confidential. Such confidentiality exists as a matter of policy in order to ensure all defendants are properly advised. You must not, without the defendant's permission, disclose to anyone what the defendant tells you or any other information

73 *Ibid.*, §46.

you discover whilst assisting him to prepare his case. You cannot lawfully be ordered by a senior officer to disclose such information. You should only reveal what has been said to you by the defendant in due process of law (e.g. if required to do so in legal proceedings). Such confidentiality continues after the case has ended and / or your involvement as DAO has ceased for whatever reason. If in doubt, you may seek legal advice from a lawyer, preferably from the defendant's legal representative if he has one, but you should not seek such advice from the chain of command. Additionally you must be particularly careful in what you say to those directly involved in the prosecution of the case, for example, disciplinary staff, Service police, staff legal advisers, prosecutors, etc. and you should not become involved in 'off the record' conversations with them.

10. Conflict of duty. If you find there is a conflict between your duty as a Service person and your duty as DAO then you may withdraw from the case. If a legal representative is involved you should consult him first.

If the DAO breaches their duty of confidentiality, they may well find themselves on the receiving end of disciplinary proceedings and charged with, for example, conduct prejudicial to good order and discipline under section 19 AFA 2006.

Qualifications of Legal Representative

Only the following persons may be appointed as a legal representative[74]—

(a) those with a general qualification within the meaning of section 71 of the Courts and Legal Services Act 1990;
(b) anyone who is an advocate or a solicitor in Scotland;
(c) anyone who is a member of the Bar of Northern Ireland or a solicitor of the Court of Judicature of Northern Ireland; or
(d) anyone who is a person having in any of the Channel Islands, the Isle of Man, a Commonwealth country 'or a

74 AF(CM)R 2009, r. 39 and AF(SCC)R 2009, r. 26.

British overseas territory rights and duties similar to those of a barrister or solicitor in England and Wales, and subject to punishment or disability for breach of professional rules.

Any right conferred on a party to proceedings may be exercised, and any duty imposed on him by the Rules (except pleading to a charge) discharged, by his legal representative on his or her behalf.

A party who appoints a legal representative shall notify the court administration officer of the legal representative's name and address.

Defence Statement

Where the DSP complies with Article 4 of the CPIA 1996 (AAF)O 2009 or purports to comply with it (the initial duty of disclosure),[75] the accused must give a defence statement to

(a) the court administration officer; and
(b) the Director.

The defence statement is a written statement

(a) setting out the nature of the accused's defence, including any particular defences on which he intends to rely;
(b) indicating the matters of fact on which he takes issue with the prosecution;
(c) setting out, in the case of each such matter, why he takes issue with the prosecution; and
(d) indicating any point of law (including any point as to the admissibility of evidence or an abuse of process) which he wishes to take, and any authority on which he intends to rely for that purpose.

Rule 7 CPIA 1996(AAF)O 2009 provides that where a defence statement discloses an alibi, it must give particulars of it, including:

75 See above, on p. 360.

(a) the name, address and date of birth of any witness the accused believes is able to give evidence in support of the alibi, or as many of those details as are known to the accused when the statement is given; and

(b) any information in the accused's possession which might be of material assistance in identifying or finding any such witness in whose case any of the details mentioned in sub-paragraph (a) are not known to the accused when the statement is given.

As there are no official forms to be found in the legislation, a suggested form of defence statement follows, together with a suggested form for requiring further prosecution disclosure. Better Case Management in the Court Martial (BCM(CM)) paragraph 13.32 states that the prosecution response to the defence statement and other defence items will ordinarily be 14 days later, depending on the anticipated date of trial.

Skeleton Arguments

When required to submit a skeleton argument, it is worth bearing in mind the advice the Bar Council gave to litigants in person, which stated that the essence of a skeleton argument is:

> a summary of your key points to bring to the court and to hand to the judge and to the other party. Lawyers call these 'written submissions' or a 'skeleton argument'. Keep this short. It will not usually need to be longer than two or three pages, double spaced, with numbered paragraphs. For a short hearing, a page may be enough. Lawyers will often write longer ones, but that does not mean that you should. Just write enough to get your points across.[76]

In addition to identifying key points and relevant facts, the skeleton should include details of any authorities to be relied on, together with any defences to be raised or points of law to be argued. A suggested form for a skeleton argument in the court martial appears on page 383.

---

76 'A Guide to Representing Yourself in Court', Bar Council publication (April 2013), https://www.thejusticegap.com/wp-content/uploads/2013/07/Bar-Council-guide.pdf.

Plea and Case Management Hearing

Main Provisions: Part 8 of the AF(CM)R 2009; Practice Memoranda; Form DH1

These proceedings are for the orderly management of the trial. The procedure for preliminary proceedings is to be found in Part 8 of the AF(CM)R 2009. Rule 49 provides that, in preliminary proceedings the judge advocate may give such directions as appear to them to be necessary to secure the proper and efficient management of the case. The Judicial Directions of the Judge Advocate, which are to be followed in preparing the case for trial, are set out on JAG DH1 and are binding *only* if the form is signed by the judge advocate and stamped by the court. It may be submitted on paper or (preferably) electronically. Form DH1[77] is drafted by the Judge Advocate General (JAG). No precedent has been provided within the Acts or the rules, so this provides a helpful and sensible guide to ensure smooth pre-trial administration. Defence representative(s) must complete the DH1 and serve it on the prosecution in sufficient time before the plea and case management hearing (PCMH) to allow the prosecution to consider it, make the entries required of them and submit it in accordance with JAG Practice Memoranda section 3 which, at the time of publication, is at Version 6, dated 1 September 2016. Section 3 is entitled, 'Plea and Case Management Hearings (PCMH) and Listing Practice'. The Act does not specifically refer to PCMH, but it is convenient and recognisable term for a preliminary hearing where case management matters are to be resolved. The Memoranda states that a PCMH:

a. enables binding pleas to be taken at the earliest practicable stage;
b. enables substantive resolution of issues where necessary ahead of trial;
c. emphasises the public accountability of charge reductions or withdrawals by prosecutors; and

77 Form DH1 is a substantial document downloadable from the JAG website, https://www.judiciary.gov.uk/wp-content/uploads/2015/05/20141009-dh1_ver7_hqmcs-o.doc.

d. enables orders to be made concerning matters such as reporting restrictions and grant of permission for live link (video) evidence at trial.

Each new case, with the exception of absent without leave (AWOL) cases, reaching the Military Court Service (MCS)[78] is immediately considered for a PCMH.

DSP Direction

No proceedings may be taken against anyone where the DSP has issued a direction that, for the purposes of section 63, the person specified in the direction is to be treated as acquitted of the service offence so specified.[79] This also applies where the person has had the service offence taken into consideration when being sentenced for another offence. The section also applies where the two offences are not identical but are related in one of the ways specified in subsection (3).

A similar provision applies to service proceedings in respect of a charge dealt with by the civilian authorities.[80]

Finally, section 66(5) provides that any offence so barred from trial by court martial cannot be tried by the Service Civilian Court nor dealt with summarily by the CO.

Immunity from Prosecution

Main Provisions: Part 12A, sections 304A–304H AFA 2016; EN to AFA 2016[81]

Note: These provisions are not yet in force, as at end 2020.

The 2016 Act introduced new provisions in respect of immunity from prosecution and the process for dealing with cooperative defendants who assist the prosecution. Sections 304A to 304H AFA 2006 create a

78 Army and RAF court staffs merged to form the MCS in January 2004, and in 2007 the RN was integrated to create a tri-service MCS. MCS is independent of the service chains of command and is staffed by civil servants (some of whom are retired service personnel). Director MCS is also the line manager for the Head of the Armed Forces Criminal Legal Aid Authority, who is responsible for the provision of civilian legal representation for all eligible service personnel and civilians.

79 See s. 65 and s. 127.

80 See s. 66.

81 EN §§43–66.

statutory framework for immunity from prosecution, restrictions on the use of evidence and sentence reductions for offenders who cooperate in investigations and prosecutions. They make provision equivalent to the provision in sections 71 to 75 of the Serious Organised Crime and Police Act 2005 which applies in the civilian criminal justice system.

The section 304A immunity from prosecution provision allows the DSP to give a person immunity from prosecution for any 'relevant service offence' in return for assistance given by that person to the investigation or prosecution. A 'relevant service offence' is one under section 42 where the corresponding offence under the law of England and Wales is an indictable offence; or any other service offence which, if committed by a person aged 18 or over and tried by the court martial, is punishable with imprisonment for more than two years. The effect of subsection (2) is that immunity may not be given for assistance in relation to a minor service offence.

An offence for which immunity is given and any conditions to which it is subject must be specified in a written notice. An immunity notice ceases to have effect in relation to the person to whom it is given if the person fails to comply with any conditions specified in the notice by virtue of section 304A (4).

Undertakings as to Use of Evidence

Section 304B permits the DSP to give a person a written undertaking that information will not be used against them in service proceedings. These proceedings are in respect of a service offence by way of:

- (a) summary hearing, or
- (b) proceedings in respect of a Service Offence before
  - (i) the Court Martial;
  - (ii) the Service Civilian Court;
  - (iii) the Summary Appeal Court;
  - (iv) the Court Martial Appeal Court; or
  - (v) the Supreme Court on an appeal brought from the Court Martial Appeal Court.

Such an undertaking may be given in return for assistance given by that person to the investigation or prosecution of a service offence. The effect of subsection (5) is that such an undertaking may not be given for assistance in relation to a minor service offence. An undertaking may be given in respect of the use of information in service proceedings for any service offence. Section 304B(4) provides that a restricted use undertaking ceases to have effect in relation to the person to whom it is given if the person fails to comply with any conditions specified in the undertaking.

Reduction in Sentence

This is available, under section 304C, where a person who has been convicted of a service offence following a plea of guilty in the court martial has, pursuant to a written agreement with the DSP, assisted or offered to assist an investigator or prosecutor. Assistance can be in relation to any offence, not just service offences. Importantly, the effect of subsections (4) and (5) is that the power to reduce a sentence overrides provisions in other legislation fixing minimum sentences or, in the case of a sentence fixed by law, affecting how a judge advocate determines the minimum period of imprisonment or detention that a person must serve, so the judge advocate could impose a sentence that is less than would otherwise be required.

The power to reduce a sentence is also available to courts that substitute a sentence when dealing with appeals from the court martial, as sections 16A(2) and 41(3) of the Court Martial Appeals Act 1968 give the Court Martial Appeal Court and the Supreme Court the same powers as the court martial had when originally sentencing.

Review of Sentence Following Offer of Assistance

Section 304D provides that a person who has been sentenced by the court martial may have their sentence reviewed to take account of assistance that they have given, or offered to give, to an investigator or prosecutor pursuant to an agreement with the Director. The reviewing court may reduce the sentence in return for the assistance offered or given. Subsections (2)(b) and (3)(b) refer to assistance in relation to any offence (not just service offences).

The effect of subsection (1)(b) is that a person who received a sentence that was fixed by law may only have that sentence reviewed under new section 304D if they pleaded guilty to the offence for which they received that sentence.

A review under 304D may further reduce a sentence that has already been discounted under sections 304C or 304D, if the person sentenced gives or offers to give further assistance. A right of appeal is available against the court martial's decision to reduce, or not to reduce, the sentence. Such an appeal is also available to the DSP (see s. 304D (9)).

Review of a Discounted Sentence Following Failure to Assist

Section 304E makes provision equivalent to the provision in section 74 of the Serious Organised Crime and Police Act 2005 for sentences to be increased on review (section 74(2)(a) and (5) cases). Essentially, it allows a sentence to be reviewed by the court martial which must, if possible, consist of the people who were the members of the court martial when it passed the sentence to which the referral relates, and permits the court martial to take account of a failure by the person sentenced to give the assistance that they offered and in return for which they received a discounted sentence under section 304C or 304D.

If the reviewing court is satisfied that the person knowingly failed to give the assistance, it may increase the sentence to take account of that failure, but must not exceed the sentence which the court would have passed but for the agreement to give assistance. A case may only be sent for review if the person is still serving the sentence and the DSP considers it is in the interests of justice.

A person whose sentence is reviewed under section 304E has a right to appeal to the CMAC against the decision of the reviewing court (s. 304E(7)). The Director may, with the leave of the CMAC, also appeal against the decision (see s. 304E(8)). There is a power for the Secretary of State to make regulations (see 304E(9) and (10)).

Statements in Open Court

Section 304F(1) and (2) provide that a court which reduces a person's sentence under sections 304C or 304D is required to make a statement in open court confirming this and identifying the sentence it would have passed had that reduction not been given. However, this is subject to subsection (3), which provides that the requirement under subsection (2) and the requirement under section 252 of AFA 2006 (for a sentencing court to give reasons for deciding on the sentence passed) do not apply if the court thinks that it would not be in the public interest to disclose that the sentence has been discounted, in which case it must instead make a statement in writing to the prosecutor and the defendant containing the same information as it would otherwise have been required to include in a statement in open court under subsection (2).

A court conducting review proceedings under new section 304D or 304E can make an order excluding members of the public from those proceedings and prohibiting publication of matters relating to the proceedings.

Meaning of 'Sentence'

'Sentence' has been defined in section 304H, for the purposes of the provisions in sections 304A to 304G, to include any order made by a court when dealing with an offender, in other words, all of the types of punishment listed in section 164(1) of AFA 2006, not just imprisonment or service detention. It includes the service-specific punishments, such as forfeiture of seniority.

Double Jeopardy

Main Provisions: Part 10 CJA 2003; Part 2 Chapter 3 AFA 2006 (ss 63–66); AF(RSO)O 2013

These provisions are a statutory reflection of the practice in the civilian jurisdiction to avoid a person being tried twice for the same offence. *Autrefois convict* and *autrefois acquit* protect a defendant against repeated prosecutions for the same offence.[82] In any case where the court rejects the defence, it is still open to the defendant to argue that it would an abuse of the process of the court for the prosecution to continue.[83] The principles of *autrefois convict* and *acquit* are applicable only where there has been a finding by a court of guilt or innocence. They are also subject to the provisions of Part 10 of the CJA 2003, Retrial for serious offences (see below).

Section 63 AFA places a statutory bar on subsequent proceedings where a person has been convicted or acquitted of a service offence or has had a service offence taken into consideration when being sentenced. The service court (CM or SCC) may not try that person for a subsequent offence that is the same in law as the original offence or where the alleged facts on

82 See *Connelly v DPP* [1964] AC 1254 and *DPP v Humphrys (Bruce Edward)* [1977] AC 1.
83 *R v Beedie* [1998] QB 356.

which the charge in respect of the later offence is based are the same, or substantially the same. The same applies to charges laid in the alternative or taken into consideration. In those circumstances where the original offence ('offence A') was an offence taken into consideration which was not charged, the court martial may not try that person for an offence ('offence B') if the alleged facts on which the charge in respect of offence B is based are the same, or substantially the same, as the facts on which a charge in respect of offence A would have been based (s. 63(4)).

Where a person cannot be tried by the court martial for an offence, neither may the service civilian court try the person, nor may the charge be heard summarily by an officer (s. 63(5)). The same principle applies to a civilian court under section 64, barring subsequent civilian proceedings after conviction or acquittal by a service court.

In those cases where the DSP has given a direction under section 127, barring further proceedings against a person, that person is to be treated as if they had been *acquitted* of the offence (s. 65(1)).

A person *acquitted* of an offence in proceedings described above, is treated as also *acquitted* of any qualifying offence of which they could have been convicted in the proceedings because of the first-mentioned offence being charged on the charge sheet, *except* an offence

(a) of which he has been convicted;
(b) of which he has been found not guilty by reason of insanity; or
(c) in respect of which, in proceedings where he has been found to be unfit to stand trial (within the meaning of section 166 of the 2006 Act) a finding has been made that he did the act or made the omission charged against him.[84]

Retrial

There is no provision in the Armed Forces Acts that permits a retrial after acquittal in a court martial, save for the exception below and where ordered by the CMAC under section19 Court Martial (Appeals) Act 1968 (CM(A)A 1968).[85] The reason is that there is no way of inquiring into the

84 Art. 4, Armed Forces (Retrial for Serious Offences) Order 2013.
85 See also, below, p. 552, concerning retrial ordered by the CMAC after quashing a conviction on appeal, where it is in the interests of justice to make such an order.

vote or opinion of the members of the Board. The nature of the decision is unknown, that is, whether it was unanimous or by majority and, if by majority, the size of that majority. The oath, set out in Schedule 1 to the AF(CM)R 2009 makes it clear that members 'will not on any account, at any time whatsoever, disclose the vote or opinion of any member of the Court Martial, unless required to do so in due course of law'. The exception appeared in the 1881 Army Act and is explained as only applying to 'such cases as those where members of the court are charged individually with partiality or bribery, and thus in a court of justice it would, or might, be necessary to make disclosures'.

Serious Cases: New and Compelling Evidence

In the civilian criminal justice system, Part 10 of the Criminal Justice Act 2003 (CJA) reforms the law relating to double jeopardy by creating an exception to the general rule, permitting retrials in serious cases where there has been an acquittal in court, but compelling new evidence subsequently comes to light against the acquitted person. The CJA provision only applies to cases tried on indictment and would, per se, be inapplicable to cases tried by court martial. Fortunately, in the case of the three services, the Armed Forces (Retrial for Serious Offences) Order 2013 (AF(RSO)O 2013) makes equivalent provision to Part 10 of the CJA, but with modifications for the retrial in the services justice system of persons acquitted of serious offences – the qualifying offences – where there is new and compelling evidence.

A qualifying offence is defined in Article 2 of the AF(RSO)O 2013 as:

(a) an offence under section 42 of the 2006 Act (criminal conduct) as respects which the corresponding offence under the law of England and Wales is an offence listed in Schedule 1 (essentially the same as those in Part 10 of the CJA 2003, namely, murder, attempted murder, kidnapping, sexual offences, etc.);
(b) an offence under section 70 of the Army Act 1955, section 70 of the Air Force Act 1955 or section 42 of the Naval Discipline Act 1957 as respects which the corresponding civil offence (within the meaning of the Act in question) is such an offence;

(c) an offence under section 1 of the 2006 Act (assisting an enemy);
(d) an offence under section 2(1) of that Act (surrendering any place or thing to an enemy etc); or
(e) an offence under section 3 of that Act (obstructing operations) where the offence relates to an action or operation against an enemy.

The cases that may be retried under Article 3 of the Order are those where a person has been acquitted of a qualifying offence in proceedings (a) before the court martial; (b) on appeal against a conviction or finding in proceedings before the court martial; or (c) on appeal from a decision on such an appeal.

The Order sets out the procedures to be followed in respect of investigation and retrial. Article 6 prevents the arrest of a previously acquitted person for a qualifying offence, unless a judge advocate has issued a warrant for the arrest. The judge advocate may only do so if satisfied that there is new evidence which would be relevant to an application for a retrial. The CMAC must, under Article 9(1)(a), order a retrial if satisfied that there is new and compelling evidence against the person charged and that it is in the interests of justice for the court to make the order.

Article 10 details the requirement for new and compelling evidence. Evidence is new if it was not adduced in the proceedings in which the person was acquitted (or, if those were appeal proceedings, in earlier proceedings to which the appeal related). Evidence is compelling if it satisfies three conditions:

a. it is reliable;
b. it is substantial; and
c. in the context of the outstanding issues, it appears highly probative of the case against the acquitted person.

Article 11 requires that making the order for retrial must be in the interests of justice. It sets out what is to be determined, including whether existing circumstances make a fair trial unlikely and whether it is likely that the new evidence would have been adduced in the earlier proceedings against the acquitted person but for a failure by a service police officer or by the SPA to act with due diligence or expedition.

The CMAC must consider at a hearing a prosecuting officer's application to quash the acquitted person's acquittal and order for that person to be retried for the qualifying offence. The prosecutor may only make such an application if the Director has provided written consent. If a retrial is ordered, under Article 9(1)(a), it must be based on a charge specified by the CMAC. The person charged may not under such an order be arraigned more than two months after the date of the order, unless the court gives permission (Art. 19).

Article 15 amends the provisions of the CM(A) Act 1968 so as to provide a right of appeal to the Supreme Court against any decision of the CMAC in relation to an application under this Order.

Suggested Defence Skeleton Argument

See the Figure 12.1 Suggested Defence Skeleton Argument Form on page 380.

Suggested Defence Statement

See Figure 12.2 Suggested Defence Statement Form on pages 381–82.

Suggested Application for Prosecution Disclosure

See Figure 12.3 Suggested Application for Prosecution Disclosure Form on pages 383–84.

The next section, The Judge Advocate, begins on page 384.

IN THE COURT MARTIAL at BULFORD

BETWEEN

R

-v-

No 11223344 Rifleman James BLOGGS [Defendant]

SKELETON ARGUMENT FOR THE DEFENCE

---

1. Rifleman Bloggs has been charged with a single offence of malicious wounding, contrary to section 20 of the Offences Against the Person Act 1861. The incident occurred in Maggie's Bar, Detmold, Germany on 5 July 2013. The defendant will plead Not Guilty.

2. The Crown's case rests principally upon the evidence of LCpl Snooks, the alleged victim.

3. While there is no dispute that LCpl Snooks received the injury alleged, the defendant maintains that at all times he acted in self-defence in response to an assault upon him by LCpl Snooks who attacked him with a knife.

4. The defendant challenges the expertise of Dr IM Bright, the doctor who examined the complainant and who offered the opinion that the wound the complainant received was caused by a downward thrust with the right hand. Dr Bright is not accepted as an expert witness on this point.

5. The defendant will call Dr MBCH Bachelor FRCS, of 22b Harley Street, London W1, a specialist of over 30 years' experience in the treatment of wounds, who examined the defendant two days after the alleged incident and subsequently examined the photographic supplement Exh ABC/1, depicting the injury to LCpl Snooks.

6. The defendant proposes to call evidence of a German civilian, Herr Hans Schmidt, the owner of Maggie's Bar, who was present at the time and who witnessed the incident.

Signed: Xavier de Villefort & Co, Solicitors Dated: 12 /ix/2015

Defendant/Legal Representative

Figure 12.1 Suggested Defence Skeleton Argument Form

DEFENCE STATEMENT

Sections 120 or 121 of the 2006 Act (charge allocated for trial); section 141 of the 2006 Act (Appeal to Summary Appeal Court); the Criminal Procedure and Investigations Act 1996 (Application to Armed Forces) Order 2009, Articles 4, 7, 8, 12 and 16

Case details

Number of defendant:

Rank/Rate of defendant:

Name of defendant:

Case reference number (if any):

Charge(s):

The time limit for giving the information

28 days after initial prosecution disclosure (or notice from the prosecutor that there is no material to disclose beginning with the day immediately after the day on which the Director complies, or purports to comply, with Article 4).

How to use this form

1. Complete the case details box above, and Part 1 below.
2. Attach as many sheets as you need to give the information listed in Part 2.
3. Sign and date the completed form.
4. Send a copy of the completed form to:
   (a) the court administration officer, and
   (b) the Director
   before the time limit expires.

If you need more time, you must apply to the court before the time limit expires. You should apply in writing, but no special form is needed.

Part 1: Plea

I confirm that I intend to plead not guilty to [all the charges] [the following charges] against me:

Figure 12.2 Suggested Defence Statement Form

Part 2: Nature of the defence

Attach as many sheets as you need to give the information required.

Under Article 7 of the Criminal Procedure and Investigations Act 1996, you must:

(a) set out the nature of your defence, including any particular defences on which you intend to rely;

(b) indicate the matters of fact on which you take issue with the prosecutor, and in respect of each explain why;

(c) set out, in the case of each such matter, why you takes issue with the prosecution

(d) indicate any point of law (including any point as to the admissibility of evidence or an abuse of process) which you wish to take, and any authority on which you intend to rely for that purpose; and

(e) if your defence statement includes an alibi (i.e. an assertion that you were in a place, at a time, inconsistent with you having committed the offence), give particulars, including –

(i) the name, address and date of birth of any witness who you believe can give evidence in support of that alibi, or as many of those details as are known to you when the statement is given,

(ii) if you do not know all of those details, any information that might help identify or find that witness.

Signed: ______________________ **defendant / defendant's solicitor**

Date: ______________________

WARNING: Under articles 12 and 16(2) of the Criminal Procedure and Investigations Act 1996 (Application to the Armed Forces) Order 2009, if you (a) do not disclose what the Order requires; (b) give a defence statement but do so after the relevant period extended (if at all) in accordance with articles 9 to 11; (c) set out inconsistent defences in the defence statement; (d) put forward a defence which was not mentioned in the defence statement or is different from any defence set out in that statement; (e) rely on a matter which, in breach of the requirements imposed by article 7, was not mentioned in the defence statement; (f) adduce evidence in support of an alibi without having given particulars of the alibi in the defence statement; (g) call a witness to give evidence in support of an alibi without having complied with article 7(4)(a) or (b) as regards the witness in the defence statement, then the court, the prosecutor or another defendant may comment on that, and the court may draw such inferences as it thinks proper in deciding whether you are guilty.

DEFENDANT'S APPLICATION FOR PROSECUTION DISCLOSURE

Sections 120 or 121 of the 2006 Act (charge allocated for trial); section 141 of the 2006 Act (Appeal to Summary Appeal Court); the Criminal Procedure and Investigations Act 1996 (Application to Armed Forces) Order 2009, Articles 7, 13(5) and 14; section 17 of the Regulation of Investigatory Powers Act 2000)

Case details

Number of defendant:

Rank/Rate of defendant:

Name of defendant:

Case reference number (if any):

Charge(s):

Note: You must give a defence statement, and allow the prosecutor time to respond, before you can make an application for prosecution disclosure.

How to use this form

1. Complete the Case details box above and answer the questions set out in the boxes below. If you need more space, you may attach extra sheets.
2. Attach to this form:
   (a) a copy of your defence statement, and
   (b) copies of any correspondence with the Director about disclosure.
3. Sign and date the completed form.
4. Send a copy of the completed form and everything attached to:
   (a) the court administration officer, and
   (b) the Director.

1) What material do you have reasonable cause to believe has not been disclosed by the prosecutor that you wish him to disclose?

2) Why do you think the prosecutor has that material?

Figure 12.3 Suggested Application for Prosecution Disclosure Form
Source: Adapted from the Criminal Procedure Rules (*Crown Copyright*)

3) Why might that material:

(a) undermine the prosecutor's case against you, or
(b) assist your case?

4) Do you want the court to arrange a hearing of this application? YES / NO

If YES, explain why you think a hearing is needed. (If you do not ask for a hearing, court may arrange one anyway.)

Signed: ______________________ **defendant / defendant's solicitor**

Date: ______________________

The Judge Advocate

Main Provisions: section 362 AFA 2006; sections 29–35 of the CM(A)A 1951; section 151 Senior Courts Act 1981; Schedule 2 AFA 2011

Military courts are presided over by civilian judges who are totally independent of the chain of command.[86] Some may have had previous military service, some may not. What they all have is a collective expertise in the service justice system, often gained as service court advocates, coupled with a significant understanding of the civilian criminal justice system. The Office of the JAG falls under the responsibility of the Department for Constitutional Affairs (now part of the Ministry of Justice).[87]

86 Much of the content in this section has been helpfully provided by the Ministry of Justice, in response to a Freedom of Information Act request dated 7 February 2020.

87 'The Office of the Judge Advocate General – Second Report of Session 2005–06', House of Commons Constitutional Affairs Committee, 6 December 2005. The Constitutional Affairs Committee decided to conduct a short inquiry into the work of the Office of the Judge Advocate General, the body responsible for the conduct of proceedings at courts martial for the Army and Royal Air Force, appointing civilian judge advocates and monitoring the military criminal

Section 362 of the Act defines 'judge advocate' as

1) the Judge Advocate General;
2) a person appointed under section 30(1)(a) or (b) or (2) of the Courts-Martial (Appeals) Act 1951 (c. 46) (assistants to the Judge Advocate General); or
3) a puisne judge of the High Court in England and Wales who (following a request by the Judge Advocate General) is nominated by or on behalf of the Lord Chief Justice of England and Wales to sit as a judge advocate.

The judge advocate must be a person recommended to Her Majesty by the Lord Chancellor.[88] In fact, it is the Judicial Appointments Committee that actually finds a suitable appointee. The individual must have been professionally qualified for at least seven years by the time of appointment.[89] Assistant judge advocates must hold a legal qualification within section 31 of the 1951 Act and be of at least five years' standing. They must satisfy a citizenship requirement as citizens of the UK, or a Commonwealth country, or of the Republic of Ireland. The statutory retirement age is 70.[90]

The JAG website explains that the appointment is

> an entirely standalone role meaning that the JAG is not supervised by any other Judge and he does not come under the authority of any Presiding Judge nor the Lord Chief Justice. The Service Justice System as a whole comes under the responsibility of the Ministry of Defence.[91]

This seems to suggest there is no one to whom the JAG was accountable. The position was, however, clarified by the Judicial Office, which explained the process regarding supervision and accountability, as follows:

justice system. This Committee omitted to receive any evidence from the heads of the Service Legal Branches who, at the time, were also the respective Service Prosecuting Authorities.

88 Courts–Martial (Appeals) Act 1951, s. 29.

89 *Ibid.*, section 31.

90 See the Judicial Appointments Commission website for more details.

91 'Judge Advocate General – Who is the Judge Advocate General?', Judiciary UK website, https://www.judiciary.uk/about-the-judiciary/who-are-the-judiciary/judicial-roles/judges/judge-advocate-general/.

> All members of the Service Judiciary (including the Judge Advocate General (JAG)) are subject to the same regime as all other judges for their conduct and can be removed from office for gross misconduct or mental incapacity. The current JAG agreed with the then Lord Chief Justice that complaints about judge advocates (including the JAG) would be handled in the same way as for other judges, via the Senior Presiding Judge. There may be an issue with the JAG himself as he is not supervised by the Lord Chief Justice and is accountable directly to the Crown, but as a matter of policy the JAG is also always a Senior Circuit Judge and in that capacity, he is supervised by the Lord Chief Justice. If the JAG were guilty of gross misconduct when acting in the capacity as JAG (rather than generally as a judge either in or out of court) then he would be subject to the same disciplinary process as a High Court judge …[92] The Lord Chief Justice includes a section on the Office of JAG in his annual report to Parliament.[93]

Schedule 2 of the 2011 Act details the circumstances of judge advocates sitting in civilian courts and amended section 8 of the Senior Courts Act 1981 permits judge advocates to sit in the Crown Court and magistrates' courts. The jurisdiction of the Crown court exercisable by a qualifying judge advocate is the jurisdiction of the court in relation to any criminal cause or matter other than an appeal from a youth court.

Section 151 of the Senior Courts Act was amended by Schedule 2 of the AFA 2011, to include the definition of 'qualifying judge advocate' to mean:

> (a) the Judge Advocate General; or
> (b) a person appointed under section 30(1)(a) or (b) of the Courts-Martial (Appeals) Act 1951 (assistants to the Judge Advocate General).

---

92 Response from the Judicial Office to a request made on 7 February 2020 under the Freedom of Information Act 2000.

93 The Lord Chief Justice's Report 2019, at p.17. In 2017, the Ministry of Defence commissioned a review of the Service Justice System in preparation for the next Armed Forces Bill. The review was conducted by Judge Lyons, a former Naval legal officer. His report available at: https://www.gov.uk/government/publications/service-justice-system-review.

The JAG tickets suitably qualified judge advocates to hear serious cases. All judge advocates undertake exactly the same training as other judges through the Judicial College and must have attended the relevant courses before being ticketed. The JAG is ticketed by the Senior Presiding Judge to sit in murder and manslaughter and serious sex cases in the Crown court. Some of the judge advocates are also ticketed to sit in serous sex cases in the Crown court.

As at 20 January 2020, the service judiciary consisted of the JAG, Deputy Judge Advocate General (DJAG) and five Assistant Judge Advocate Generals (AJAGs). Of these, the JAG is qualified to hear murder, manslaughter and serious sexual offences. The DJAG is qualified to hear manslaughter and serious sexual offences. Two of the five AJAGs are qualified to hear serious sexual offences (and one other is about to attend the serious sexual offences course at the Judicial College and is expected to be qualified shortly).

Since the Armed Forces Act 2006 came into force, a puisne judge has been appointed as judge advocate on one occasion, in the case of *Mendonca and others.*[94] There were a number of offences charged, but the reason for appointing a puisne judge was that this was the first case in England when a person was charged with an offence under the International Criminal Court Act 2001.[95]

94 *R v Payne* (2007), Court Martial at Bulford, unreported. The case involved the unlawful detention and inhuman and degrading treatment by British servicemen of a number of Iraqi civilians, leading to the death of one of the captives, named Baha Mousa.
95 See above, FN 94.

13 TRIAL BY COURT MARTIAL

Jurisdiction

Main Provision: section 50 AFA 2006

Essentially, while it is a rather simple analogy, the court martial is often regarded as the service equivalent of the Crown court. It will hear cases within its jurisdiction against all members of the service community subject to service law or service discipline. Section 50 expressly states that the court martial has jurisdiction to try any service offence, as defined in that section. The fact that a service person has the unfettered right in every case to elect trial also means that the court martial tries cases that are sometimes at the lower end of the scale of seriousness. For example, a charge of failure to attend for or perform a duty because of oversleeping could, theoretically, be tried by court martial.[1]

As mentioned above, in connection with the Judge Advocate, the Lord Chief Justice includes a section on the Office of the Judge Advocate General (JAG) in their annual report to Parliament. While there is no express reference to the judge advocates in his last report, the Lord Chief Justice did relate that,

> The Lyons Review of the Service Justice System reported in February 2019 and Ministry of Defence Ministers are currently considering its recommendations.[2] The recommendations are expected to set the framework for policy changes contributing to the next Armed Forces Act which is currently due to be brought

1 It is, however, far more likely that such a matter would be dealt with through MAA. See more generally Chapter 8.

2 See also, p. 381, FN 93.

> into force in 2021. A number of recommendations will also require procedural rule changes and the Service Justice Rules Review Committee are looking at these. The rationalisation of the Court Martial hearing centres to two centres in Bulford and Catterick has worked well and has enhanced efficiency.[3]

The report by Judge Lyons confirms the fairness of the service justice system. There does, however, appear to be inferential criticism of the service prosecution system, by seeking to remove a number of offences from service jurisdiction, where they are committed in the UK,[4] leading to the inference that the Service Justice System (SJS) delivers an inferior quality of justice, but that such inferior justice suffices where the offence is committed overseas.

As the prosecution of offences is a matter for independent prosecuting authorities, this proposal may encroach upon that very independence. Current arrangements over prosecution of serious offences are explained in the Prosecutors Protocol, between the Director Service Prosecution Authority (SPA) and the Director of Public Prosecutions (DPP).[5] This is issued under the guidance from the Attorney General set out in the Prosecutors' Convention 2009.[6]

The report also makes a proposal to change the manner in which court martial members arrive at their decisions, suggesting that there should be a board of six of which at least five should agree. This is unsupported by Court Martial Appeal Court (CMAC) decisions, and attempts to overrule *Twaite*, which it describes in section 7.9 of the report, Recommendation 4, as 'criticised as being inconsistent with the presumption of innocence and the requirement of proof beyond reasonable doubt'. However, no jurisprudence from UK courts or the European Court of Human Rights (ECtHR) is cited in

---

3 The Lord Chief Justice's Report 2019, at p. 17. Judge Lyons was a former naval legal officer. In 2017, the MoD commissioned a review of the SJS in preparation for the next Armed Forces Bill. His report is available at https://www.gov.uk/government/publications/service-justice-system-review.

4 Murder, manslaughter, rape and serious sexual offences under s. 2 of the Sexual Offences Act 2003 (assault by penetration).

5 http://spa.independent.gov.uk/linkedfiles/spa/test/about_us/publication_scheme/20111007-juris_eng_and_wales.pdf.

6 https://www.gov.uk/guidance/the-prosecutors-convention-2009-updated-2012.

support of this proposition. Indeed, the report refers to the binding nature of the decision (in s. 8.2). Moreover, it cites no examples of any serious miscarriages of justice which have occurred as a result of majority decisions in the SJS. Were there to be one, the report would have then had to include a comparison with the miscarriages of justice in the civilian system which operates on the basis of a unanimity verdict. Consequently, the proposal on majority decision-making does not appear to be evidence-based and is a matter of opinion, in conflict with the acknowledged binding nature of the CMAC decision in *Twaite*.

Under the current system, a case comes within the jurisdiction of the court martial where,

a. The accused, at commencement of a summary hearing before the CO, elects trial; or
b. The case is referred to the SPA by the CO, and the SPA direct trial; or
c. The case is referred by the service police to the SPA and the SPA direct trial.

The court martial may also sit as an appellate tribunal to hear an appeal from the Service Civilian Court (SCC).

The court martial may sit in any place, either in or outside the United Kingdom. There is no geographical limit placed upon the court's jurisdiction, so this would include for offences committed in the UK, where jurisdiction would overlap with the civilian authorities. Consequently, it would include an offence of murder committed by a service person when in the UK. All the same, in the UK the civil authorities have primacy of jurisdiction and can insist upon investigation by the civilian police and trial of a service offender by the civilian courts.

However, as a service spouse is not subject to service discipline when in the UK, any offence committed by then in the UK would be a matter purely for the civilian authorities. Offences committed overseas, when subject to service discipline, would essentially remain a matter for service jurisdiction.

Case Management

Main Provisions: Protocol for Digital Case Files in the Service Justice System; Service Courts Collected Memoranda; AF(CM)R 2009

Better Case Management in the Court Martial (BCM(CM)) is explained in the Judge Advocate General (JAG) Memorandum 13 (Practice in the Service Courts Collected Memoranda (CMPCM)).[7] It is essential for advocates to consult this document, which the JAG describes as 'an invaluable guide to the procedure of the Court Martial and service courts and sets out best practice in many key areas'. There is no statutory basis for practice memoranda. They are ostensibly issued under the authority granted to JAG by letters patent issued by Her Majesty The Queen. The guidance is for judge advocates to follow, but where they decide to depart from it they should explain their reasons for doing so on the record. However, where there is any conflict with the statutes and the Court Martial Rules, the legislation will prevail. This is important as the CMPCM for some reason does not always use the same terminology as the Rules.

The CMPCM sets down a mandatory timeline, following receipt of papers by the Military Court Service (MCS), for court hearings and deadlines in contested cases in the court martial.

An initial hearing (IH) is normally to be listed in the week commencing 28 days after the case papers are received at the MCS. Where the defence indicate, in advance of the IH, that the case is to proceed as a guilty plea, the IH can be replaced by a plea and sentence hearing. At the IH the judge will arraign the defendant, if appropriate and, with the assistance of the parties, identify key issues in the event of a not guilty plea. See the Collected Memoranda for more detail.

The plea and trial preparation hearing (PTPH) is covered in paragraph 13.22 *et seq*. of the CMPCM. The SPA and defence must have fully completed the BCM(CM) form prior to the court hearing. This form will be issued by the service justice authorities. However, the BCM(CM) form is not required in guilty plea cases. PTPH hearings will, unless otherwise directed, take place via a video link (VTC).

---

7 The JAG Collected Memoranda is available online, https://www.judiciary.uk/wp-content/uploads/2015/05/practice-memo-ver-6-1Sep16.pdf. AFCLAA use their own form, Form1, to record relevant information, including defendant, type of proceedings, pleas, and advocate details.

In exceptional cases, there will be a further case management hearing if, for example, it is necessary to give further directions, or there is a change of plea to guilty etc (see CMPCM, para. 13.36). The subject of preliminary hearings is dealt with below, with reference to the relevant legal provisions applicable.

Trial Readiness

The Collected Memoranda require that, no less than 28 days before the date fixed for trial, a trial readiness form must be submitted by the prosecution and defence and delivered to the court. If a completed form is not received, the case will be listed automatically for a pre-trial review. If a form is received, the court will consider if a hearing is required. The Crown Court Certificate of Readiness for Trial will be used in the court martial (CMPCM, para. 13.47).

Preliminary Proceedings

Proceedings in a service court commence when the charge is brought under section 122 of the Armed Forces Act 2006.[8]

The legal rules governing preliminary proceedings are contained in rules 45–49 of the Armed Forces (Miscellaneous Amendments) Regulations 2019 (AF(MA)R 2019). In preliminary proceedings the judge advocate may give such directions as appear to them to be necessary to secure the proper and efficient management of the case and they may make an order or ruling on any question as to the admissibility of evidence, joinder or severance of charges or any other question of law, practice or procedure relating to the case.[9] Any such order made in these preliminary proceedings endures and remains effective for the related proceedings until varied or discharged. It is possible to appeal against any order or ruling to the

8 *RXG v MoJ* [2019] EWHC 2026 (QB). The situation in the services contrasts with the civilian system, where proceedings are commenced in one of three ways: under section 28 CJA 2003, by police arresting and charging the suspect; or the prosecution issuing a written charge under section 29 CJA, together with a requirement requiring the accused to attend court; or the prosecution obtains a summons issued by the court, under section 1 of the MCA 1980 directing the accused to attend court on a specified date and time.

9 AF(CM)R 2009, r. 49.

Court of Appeal, with the leave of that court, and the judge advocate may continue preliminary proceedings notwithstanding that leave to appeal has been granted. But related proceedings (other than further preliminary proceedings) may not commence until the appeal has been determined or abandoned.[10] The proceedings are recorded on form DH1 which sets out the Judicial Directions to be followed in preparing the case for trial. The form states that the directions are binding only if the form is signed by the judge advocate and stamped by the court. A copy of this form appears in the Office of the Judge Advocate General (OJAG) website.[11]

On receipt of the advance information in relation to a charge, the court administration officer must forward it to the JAG and request them to specify a judge advocate for preliminary proceedings in relation to the charge. They will appoint a time and place for the commencement of the preliminary proceedings.[12] The judge advocate for any preliminary proceedings may give a direction under this rule, either on the oral application of the Director, Service Prosecuting Authority (DSPA or DSP) or a defendant; or of their own motion. Similarly, they may give a direction under rule 46, for further directions, on the written application of the DSPA or a defendant, in which case the application:

(a) must be made to the court administration officer;
(b) must specify the reason for which further preliminary proceedings are required;
(c) must include an estimate of the likely length of the further preliminary proceedings; and
(d) subject to rule 47 (preliminary proceedings in chambers without notice), must be served on every other party to the proposed trial proceedings.

Application may be made by the DSP under rule 47 that the judge advocate for any preliminary proceedings directs that the proceedings are, or that any part of the proceedings is, held without notice to a defendant.

One of the powers the judge advocate can exercise under rule 48 is to direct the DSP to serve on each defendant and the court administration

10 AF(CM)R 2009, r. 50.
11 Currently at http://www.justice.gov.uk/courts/judge-advocate-general.
12 AF(CM)R 2009, r. 45.

officer (CAO), before any preliminary proceedings, an outline of the prosecution case.

Under rule 54(1), where it is in the interests of justice, the judge advocate in preliminary proceedings (whether before or after arraignment) or at trial shall make an order for the amendment of the charge sheet or charge (as the case may be) if they consider that it is defective. It is then incumbent upon the DSPA to serve the amended charge sheet on every defendant.

Appeal from Preliminary Hearing

Main Provisions: rule 50 AF(CM)R 2009; rules 27–30 CM(AC)R 2009

Under rule 50 AF(CM)R 2009, the Appeal Court has jurisdiction to hear an appeal against any order or ruling made in preliminary proceedings. It must grant its consent. Any appeal by a defendant or the DSP against an order or ruling made in preliminary proceedings of the court martial must comply with the Court Martial Appeal Court Rules 2009. This requires that the appellant must serve a written appeal notice on the registrar not more than seven days after the date of the order or ruling against which the appellant wants to appeal (r. 28) and comply with the requirements of rule 29, regarding the information and documents to be provided, for example, particulars of the hearing, copy of the transcript and skeleton arguments and the grounds of appeal.

A party on whom an appellant serves an appeal notice may serve a respondent's notice detailing the matters specified in rule 30 and must do so if that party wants to make representations to the court or is directed to do so by the registrar or a judge of the court.[13]

An appeal under rule 50 AF(CM)R 2009 may be brought only with leave of the Appeal Court. Even though leave to appeal has been granted, a judge advocate may continue preliminary proceedings, but related proceedings (other than further preliminary proceedings) may not commence until the appeal has been determined or abandoned.

The case of *R v S* [2013] EWCA Crim 2579 was a successful application by the SPA for leave to appeal against a terminating ruling at a preliminary hearing. The prosecution related to five charges, namely, one charge of inflicting grievous bodily harm, contrary to section 20 of the Offences Against

13 CM(AC)R 2009, r. 30.

the Person Act. The remaining four charges (2 to 5) related to offences of battery. The judge advocate refused the defence submission that charges 2 to 5 were time-barred because the respondent should have been charged within six months of the date of the commission of the offences, following the principles set out in section 127 of the Magistrates' Court Act 1980 (MCA), as that section 127 of the MCA 1980 did not apply in this situation and could not be imported into section 42 AFA 2006. However, he did allow a submission that there would be an abuse of process for the trial to proceed as there was no good reason, when there was no specific military context, for a service person to be in a worse position than their civilian counterpart or for a prosecution to depend on whether it was brought by the Crown Prosecution Service (CPS) or the SPA. The CMAC granted permission to appeal and rejected the defence submission of abuse of process. Referring to the Lord Dyson's *dicta* in *R v Maxwell* [2011] 2 Cr App R 31, at paragraph 13,

> It is well established that the court has the power to stay proceedings in two categories of case, namely ... (ii) where it offends the court's sense of justice and propriety to be asked to try the accused in the particular circumstances of the case. In the first category of case, if the court concludes that an accused cannot receive a fair trial, it will stay the proceedings without more. No question of the balancing of competing interests arises. In the second category of case, the court is concerned to protect the integrity of the criminal justice system. Here a stay will be granted where the court concludes that in all the circumstances a trial will 'offend the court's sense of justice and propriety' ... or will undermine public confidence in the criminal justice system and bring it into disrepute.

The CMAC held that the high threshold indicated by Lord Dyson had simply not been reached in the case of *R v S* and that the judge was wrong to hold that it had been. There was no suggestion that the prosecution in its widest sense had misconducted itself or acted in bad faith. Nor was there any suggestion of improper manipulation of the process of the prosecution. This was a case where there had been some delay, but that was not a matter that went to the second limb of abuse. See also, Prosecution Appeal, on page 553, which highlights the more urgent time constraint operating in court martial proceedings.

Witness Summonsing

Main Provision: Part 11 AF(CM)R 2009 (rr. 62–71)

Securing the attendance of witnesses is not always an easy task even in the civilian system. In the military justice system (MJS), where service witnesses lead very active lives and are subject to deployment around the world, on exercise or operations, the problem is more difficult.

In Germany there is the extra challenge of obtaining the attendance of German witnesses who, of course, are not subject to UK military jurisdiction. In such a case, witness summonsing is carried out through special treaty arrangements, as explained on page 26 above. Identifying potential witnesses early is therefore a good way of ensuring they will be available for trial, without delaying matters to a later date. If the witness is serving on operations, they will have to be flown back to attend the trial, thereby depleting their unit and depriving them of their service. Providing ample warning should ensure the witness is available when needed and provide the unit with sufficient time to make any necessary arrangements to manage their absence.

Where the defence require a witness to attend trial, they may ask the CAO to notify them and, if necessary (following a direction by a judge advocate), serve a summons. The CAO will then serve it on the individual to whom it is addressed.[14] If notification or service is not reasonably practicable, the CAO will notify the judge advocate and the party requiring the witness's attendance. A summons may be obtained by application to a judge advocate.[15] It will be issued in any case where a judge advocate is satisfied that the individual is likely to give valuable evidence or produce material evidence and it is in the interests of justice to issue the summons for their attendance and / or to produce the document or thing required.

The application for a witness summons must be made as soon as is practicable after the applicant becomes aware of the grounds for making it and it may be made orally unless the judge advocate otherwise directs. The application must:

a. identify the proposed witness;

14 AF(CM)R 2009, r. 62.
15 *Ibid.*, r. 63.

b. explain
   (1) what evidence the proposed witness can give or produce;
   (2) why it is likely to be material evidence; and
   (3) why it would be in the interests of justice to issue a witness summons.
c. If the application is in writing it must contain a declaration that the facts stated in it are true to the best of the applicant's knowledge and belief. It must be served then on the court administration officer and as directed by the judge advocate.

Importantly, any witness summons issued must state that failure to comply with the summons may result in the issue of a warrant for the arrest of the person to whom the summons is addressed.

If upon being required to produce a document or thing, the person served with an application for a witness summons objects on the grounds that it is not likely to be material evidence, or even if it is likely to be material evidence, the duties or rights (including rights of confidentiality) of the proposed witness or of any person to whom the document or thing relates outweigh the reasons for issuing a witness summons, the judge advocate may require the proposed witness to make the document or thing available for the purpose of assessing the objection. They may allow witness evidence to be called on the matter to help them assess the objection.[16] It is, of course, open to the judge advocate under rule 65 to issue a witness summons of their own motion.

Any witness summons may be withdrawn under rule 66 upon application by the party who initially applied for it, on the ground that it is no longer needed, or on application by the witness on the basis they were not aware of any application for a witness summons and cannot give or produce evidence likely to be material evidence or, even if they can, their duties or rights (including rights of confidentiality) or those of any person to whom the evidence relates outweigh the reasons for the issue of the witness summons.

In the event that the judge advocate is satisfied by evidence on oath that a witness summons issued under rule 63 or 65 would not procure the

16 *Ibid.*, r. 64.

attendance of a material witness, they may issue a warrant to arrest that individual, under the authority of rule 68.

Constitution of the Court

Main Provisions: sections 154–157 AFA 2006; Part 4 CM(AF)R 2009; rules 29 and 29A AF(CM)R 2009

The Armed Forces (Court Martial) Rules 2009 is the subordinate legislation which deals with court procedure, covering a wide range of matters from administrative procedures to aspects of the law of evidence such as hearsay, bad character and special measures. The Rules make provision with regard to the constitution of the court martial under section 155 of the 2006 Act and the powers of the court on election under section 165 of the Act.[17] As far as practicable, the Rules have been modelled on Crown court practice as regulated by the consolidated Criminal Procedure Rules 2005.[18] However, where there were good operational reasons for departing from this (for example, being able to sit in camera if information that might be useful to an enemy could be disclosed), or where the peculiarities of the service system demand it (such as provision regarding sentencing by judge advocates and lay members), then these Rules depart from civilian practice.[19]

Because the court martial is a standing court, the Courts-Martial Appeal Court has therefore been renamed he Court Martial Appeal Court. Section 155 AFA 2006 contains the relevant and detailed provisions about the constitution of the court martial. The former District Court Martial and General Court Martial have been abolished and replaced by a single court martial, which consists of

(a) a judge advocate and at least three but not more than five other persons ('lay members');[20]
or in the case of proceedings of a prescribed description under the Court Martial Rules, there are to be

17 See the ENs to the Rules.

18 Now the Criminal Procedure Rules 2013, which consolidate the Criminal Procedure Rules 2012.

19 ENs, para. 7.2.

20 Similar to the constitution of a District Court Martial and General Court Martial, under the earlier Acts. See also, on p. 403 for further information about lay members. In relation to variation proceedings, see the exception for the operation of the slip rule on p. 452.

(b) at least five but not more than seven lay members,[21] or no lay members, which permits the judge advocate to sit alone. This latter provision is entirely new and reserved for the circumstances outlined below.

The court will be convened with at least five lay members where[22]

a. the defendant is charged with a serious offence listed in Schedule 2;[23] or
b. being aged 18 or over at the commencement of the proceedings the defendant is charged with an offence for which he could, if convicted, be sentenced to more than seven years' imprisonment; or
c. the defendant is aged under 18 at the commencement of the proceedings and is charged with an offence for which he could, if convicted while under the age of 18, be sentenced to more than seven years' detention under section 209,[24] which provides a power to detain for specified period and the court is of the opinion that none of the other methods by which the offender may legally be dealt with is suitable.
d. The sentencing proceedings of an offender who falls to be sentenced for a Schedule 2 offence. The Armed Forces (Court Martial)(Amendment) Rules 2013 amend the AF(CM)R2009 so that in all cases where a guilty plea is offered before trial, no matter the offence, only 3 members are required for sentencing. Where there is more than one defendant, all must plead guilty for this provision to have effect. A new r. 29A is inserted in the 2009 Rules.[25]

The sentencing proceedings of an offender who falls to be sentenced to more than seven years' detention under section 209. (Emphasis added)

Please see below, on page 409, in respect of lay members in the trial of a civilian defendant who is not subject to service law at the commencement

21 Similar to the constitution of the former General Court Martial.
22 AF(CM)R 2009, r. 29.
23 See Appendix E.
24 See also, on p. 581.
25 This provision also applies to variation proceedings. See below, on p. 492.

of the proceedings, and also below, page 400 in respect of sentencing proceedings where the judge advocate may sit alone.

The judge advocate may sit alone[26] where they are dealing with:

(1) preliminary proceedings to deal with matters similar to those dealt with in the Crown Court, such as the making of any necessary orders and directions in the interests of justice and for the better administration of the trial concerning issues such as admissibility of evidence, joinder or severance of charges, or any other question of law, practice or procedure relating to the case.[27] There is a facility under Rule 47, on application by the Director Service Prosecutions, for the judge advocate to conduct Preliminary proceedings in chambers without notice to the defendant. Where the prosecution or defence is dissatisfied with any order made by the judge advocate in the Preliminary Proceedings, an appeal can be made to the Court of Appeal, with leave of that court. See Practice Memorandum 13: Better Case Management in the Court Martial, issued by JAG, which emphasises the importance of the effective management of proceedings whilst preserving judicial discretion and disposing of guilty plea cases at the earliest opportunity.

(2) sentencing proceedings where every offender who falls to be sentenced is either
    (a) a civilian offender who will be awarded a sentence under Part 1 of Schedule 3 to the 2006 Act;[28] or
    (b) an offender to whom Part 2 of Schedule 3 to the 2006 Act (ex-servicemen etc) applies, and who was convicted of every offence for which he falls to be sentenced either
        (i) in trial proceedings for which no lay member was subject to Service law; or

26 AF(CM)R 2009, r. 27.

27 *Ibid.*, Part 8 and r. 49.

28 Civilian offender: modifications of Court Martial sentencing powers which modifies s. 164 in relation to the sentencing of civilian offenders (Part 1) and ex-service personnel (Part 2), namely, people who committed an offence while subject to service law but are no longer subject to service law when sentenced for the offence.

(ii) on a guilty plea, where the court administration officer had previously notified him in writing that, if trial proceedings were required, none of the lay members would be subject to Service law.

(3) variation proceedings if
  (a) it applied to the proceedings in which the sentence that falls to be varied was imposed; or
  (b) a direction is given under rule 120 that variation proceedings be held without lay members.
(4) activation proceedings under s. 191 AFA 2006, of a suspended sentence of service detention if
  (a) the original sentence was imposed by the court, and this rule applied to the proceedings in which it was imposed; or
  (b) the original sentence was imposed by the Service Civilian Court.
(5) ancillary proceedings[29] to review stoppages or SDA compensation order (Rule 145 applies in relation to an application to the court under section 177).

The award of stoppages under the former Service Discipline Acts (SDAs) was replaced by the service compensation order (SCO). Stoppages were awarded to pay compensation for loss or damage resulting from the offence and were enforced by deductions from the offender's pay. Under section 175 of the 2006 Act the SCO replaced stoppages. It may be made to make the offender pay compensation for personal injury, loss or damage resulting from the offence, and is enforced in the same way as a fine, by deductions from the offender's pay under regulations made by virtue of section 342. The court martial or the Service Civilian Court could also make a compensation order against a civilian offender. The compensation order is available for both service and civilian offenders and closely resembles the compensation order available in civilian courts in England and Wales. The order under section 175 is modelled on section 130 of the Powers of Criminal Courts (Sentencing) Act 2000.

29 'Ancillary proceedings' defined in r. 2(2)(c) CM(AF)R 2009 as '(a) a hearing under rule 127 of an application under section 285(4) for leave to appeal out of time from the Service Civilian Court; and (b) proceedings under any provision of Part 18 (Overseas Community Order or an application for discharge of service compensation order or reduction of amount payable)'.

An example of how stoppages used to operate is provided in *Anson*.[30]

> A senior aircraftman in the RAF was found guilty of four charges of disobedience to standing orders: dangerous driving (charge 1); driving with excess alcohol (charge 2); leaving the scene of an accident (charges 3 and 4). On 9th September 2005 he was sentenced to 60 days' detention and was ordered to be put under stoppages of pay (now, a Service Compensation Order) until he had made good the sum of £1,706.97 by way of compensation to Teach Motors and the sum of £616.28 by way of compensation to SAC 'H'. A challenged the award by way of stoppages, on the basis that the charges omitted any reference to the additional facts upon which reliance was placed if the accused was to be rendered liable to the punishment of stoppages. This was a requirement under the then extant paragraph 8 of the schedule to the Court Martial Royal (Air Forces Rules) 1997.

Section 175 stipulates that the court must consider making a service compensation order in any case where it has power to do so. This effectively renders the decision in *Anson* otiose. See further on page 521 in relation to service civilians. In any case where court has the power to make a SCO on passing sentence, and it does not do so, it must give reasons for that decision.

As explained on page 521, SCOs may not be made in respect of injury, loss or damage due to an accident arising out of the presence of a motor vehicle on a road unless it is in respect of damage resulting from an offence of unlawfully obtaining any property (e.g. theft of a vehicle) or it is in respect of injury, loss or damage as respects which the offender is uninsured in relation to the use of the vehicle; and compensation is not payable under any arrangements to which the Secretary of State is a party. Where a SCO is made in respect of injury, loss or damage due to an accident arising out of the presence of a motor vehicle on a road, the amount to be paid may include an amount representing the whole or part of any loss of or reduction in preferential rates of insurance attributable to the accident.

30 [2006] EWCA Crim 1208, CMAC, 28 April 2006.

No SCO may be made in respect of bereavement, funeral expenses or loss of any other kind suffered by the dependants of a person in consequence of their death. Further guidance and policy on implementing, effecting and enforcing forfeitures and deductions is contained in JSP 754 (tri-service regulations for pay and charges).

Interpreter

Rule 22 of the AF(CM)R 2009 is replaced by a new rule, inserted by rule 16 of the Armed Forces (Interpretation, Translation and Alcohol and Drug Tests) Rules 2013 (AF(ITADT)R 2013), to provide for the appointment of an interpreter to act at the hearing unless the CAO is satisfied that person to whom any proceedings relate does not need interpretation. On application or on their own initiative, the judge advocate may require a written translation of any document or part of a document to be provided for a person to whom any proceedings relate. An interpreter may also be appointed under rule 22A for anyone attending a hearing as a complainant or other witness, other than a person to whom any proceedings relate, who is required to give evidence at a hearing. Such appointment is usually made by the Court Martial Administration Office (CMAO).

Lay Members

Main Provisions: sections 155–157 AFA 2006; rule 34 and Part 4 AF(CM)R 2009

This unfortunate and disrespectful description of members of the board was introduced by the AFA 2006. It has the tendency to diminish their specialism and ignores their role as sentencers. Prior to the AFA 2006, only officers were eligible to sit on a court martial board. This meant that a young lieutenant or pilot officer, with two years experience since commission, could participate in the judicial process. On the other hand, none of the very experienced and knowledgeable senior non-commissioned officers (SNCOs) and warrant officers with several years' service behind them, and with a wealth of understanding of the service personnel they trained and worked with, were able to do so. This changed with the Armed Forces Act 2001 which provided for warrant officers to sit on courts martial. This welcome change was retained in section 156 of the AFA 2006 which provides for officers and warrant officers to be members of the court martial. Sections 156 and 157 of the 2006 Act also specifically identify those who would *not* be qualified for membership, including:[31]

a. an acting warrant officer;
b. an officer or warrant officer who
    (1) is a member of the Military Court Service;
    (2) is a member of or on the staff of the Service Prosecuting Authority;
    (3) is a Service policeman;
    (4) is a member of the Royal Army Chaplains' Department or the Royal Air Force Chaplains' Branch;
    (5) has a general qualification within the meaning of section 71 of the Courts and Legal Services Act 1990;
    (6) is an advocate or solicitor in Scotland;
    (7) is a member of the Bar of Northern Ireland or a solicitor of the Court of Judicature of Northern Ireland; or
    (8) has in a relevant territory rights and duties similar to those of a barrister or solicitor in England and Wales, and is subject to punishment or disability for breach of professional rules.
c. An officer who
    (1) was the commanding officer of the defendant at any time in the period beginning with the date of commission of the offence to which the arraignment relates and ending with the arraignment;
    (2) he has taken part in investigating the subject matter of any charge against the defendant; or
    (3) he has conducted (whether alone or with other persons) an inquiry into the subject matter of any charge against the defendant.
d. A warrant officer who
    (1) has taken part in investigating the subject matter of any charge against the defendant; or

31 This list does not follow the Act. It is a combination of the provisions of the Act and the Rules.

(2) has conducted (whether alone or with other persons) an inquiry into the subject matter of any charge against the defendant.

e. On the trial of the defendant, any person who at any time since the date of the commission of any offence charged, was serving in the same unit as the defendant;[32]

f. For any sentencing proceedings, a person who at any time since the date of the commission of any offence for which an offender falls to be sentenced, has served in the same unit as the offender;

g. For any variation proceedings, any person who was not a member of the court for the proceedings in which the sentence that falls to be varied was imposed.

h. For any appellate proceedings, a person who at any time since the date of the commission of any offence to which the proceedings relate was serving in the same unit as the appellant.

i. For any activation proceedings a person who at any time since the date of the commission of the offence for which the original sentence was passed was serving in the same unit as the offender.

j. If he was a member of the court for any previous proceedings to which any person to whom the proceedings relate was a party, save where

(1) he was a member of the original court of trial or appellate hearing in which any offender was convicted of any offence for which he falls to be sentenced

(2) or, for the purpose of variation proceedings, by virtue of having been a member of the court for

(3) any trial or appellate proceedings in which the offender was convicted of any offence for which the sentence that falls to be varied was imposed; or

(4) the proceedings in which that sentence was imposed.

k. For sentencing proceedings by virtue of having been a member of the court for any trial or appellate proceedings in which any offender was convicted of any offence for which he falls to be sentenced.

32 AF(CM)R 2009, r. 32.

Another situation catered for in the Rules relates to individuals who are not to be regarded as being ineligible where they were members on a non-effective court (e.g. caused through objections to lay members where the successful objections exceeded the number of waiting members).[33] In these circumstances, the other person who is not ineligible is the judge advocate.[34]

Rule 34 AF(CM)R 2009 states that in any court martial proceedings with lay members, at least one officer must be qualified to act as President of the court. The President is a kind of military version of the foreman of the jury.[35] An officer is qualified to be the president of the board if

a. he is of or above the rank of lieutenant commander, major or squadron leader;
b. he is subject to Service law; and
c. where any person to whom the proceedings relate is subject to Service law, he is of superior rank to every such person.

When the AFA 2006 was being drafted, the harmonisation it was to bring about was not intended to be at the expense of the single service interest. The single services wanted what was known as 'stovepipes' whereby each service dealt with its own personnel before court martial boards from their own service. This point arise in the case of *Gunn*,[36] where a member of the Royal Air Force was offered a trial date with an RAF board, but the

33 See rr. 25(2)(b) and 32(9).

34 Although not precisely on this point, but nevertheless of interest to the role of the judge advocate, see the case of *R v Armstrong* [2012] EWCA Crim 83, where the CMAC considered that in the light of observations the judge made at the trial, an impartial and independent observer could not conclude that justice would be seen to be done. The interests of justice therefore required that another judge hear the proceedings in relation to sentence at a different court martial centre.

35 A far cry from the days of the service discipline acts, where not all trials had a judge advocate present and, when they were present, the rulings made by the trial judge advocate were not binding on the President and the court, but purely advisory. The court martial was the President's court. For example, rule 33(2) of the Rules of Procedure (Army) Rules 1972 stipulated, 'the court shall take the charge-sheets in the order indicated by the convening officer or, where he has given no such indication, in such order as the president thinks fit'. Rule 42 provided that a court should not accept a plea of guilty if the president considered that the accused should plead not guilty. Importantly, under r. 64, after the closing addresses, if there was a judge advocate, they summed up the evidence and *advised* the court on the law relating to the case in open court. This was not a binding ruling, as such. The JA '*advised* the court upon all questions of law or procedure'. Under r. 81, it was for the President to direct that the judge advocate should sit alone to hear a submission of no case. It was not was not compulsory for them to so direct.

36 *Gunn v The Service Prosecuting Authority & Or* [2019] EWCA Crim 1470.

date was inconvenient and was declined in the knowledge that the court martial would proceed before an Army board. At trial, being represented by a different advocate, he objected to being tried by a board of lay members comprising only Army personnel and no RAF personnel. It was held that QRs, a species of delegated or subordinate legislation, having the status of law, which stated that 'Ordinarily' a service defendant will be tried by lay members drawn wholly from their own service, did not impose a mandatory requirement and, in any event, were subordinate to the provisions of sections 154-157 AFA 2006, which provide for 'mixed-boards – that is comprising of officers or warrant officers from all three Services'. Section 156(1) provides that lay members are qualified purely on the basis that they are 'subject to service law'. Interestingly, the court added, 'For the avoidance of doubt, this conclusion does not and is not intended to undermine the usual practice of a Service defendant being tried by lay members drawn from his own Service. But that which is desirable and ordinarily the case is not necessarily mandatory–and is not mandatory here.'

Knowing Your Board

A defence advocate who is unfamiliar with the service disciplinary system would be well advised to use the Defence Assisting Officer to identify the experience of the members of the court. It may be regarded as useful to know the nature of medals awarded. Aside from those awarded for gallantry (e.g. the Victoria Cross, Military Cross, Distinguished Service Order, etc.), and those awarded as an honour (OBE, MBE, etc.), the Long Service and Good Conduct Medal is one that denotes a person with a minimum time of blemish-free service within the ranks. They will have had a long career within the ranks prior to being commissioned and will know a lot about service life and the circumstances faced by ordinary soldiers/sailors/airmen, as applicable. In respect of a soldier, QRs stipulate completion of 15 years' reckonable service (from 17½ years of age). Only those soldiers who are in every way worthy of the distinction and whose conduct has been irreproachable throughout their service should be recommended. In relation to officers, the medal is awarded on completion of 15 years' reckonable service, 12 years of which must have been served in the ranks.[37]

37 In other words, the LS&GC medal is not available to officers who have not served in the ranks.

Officers and soldiers are subject to an annual assessment made through a report written by their superiors. It may sometimes be thought advisable by an advocate for the defence to make reference to such report or reports during the course of the trial. Use of the Officer's Joint Appraisal Report (OJAR) and Soldier's Joint Appraisal Report (SJAR) in trials is explained in the Collected Memoranda (No 2). This provides that:

> 2.1 Evidence of a defendant's good character can be used either during the trial proceedings as part of the defence case, or during sentencing proceedings as part of mitigation. Applications to adduce evidence of OJAR/SJAR during the trial will normally be made by the defence and will ordinarily only be excluded if they contain material which would mislead the Court. Data Protection Act (DPA) principles apply to the provision of these documents which are not routinely available to either the SPA or MCS.
>
> 2.2 However, some defence counsel who are new to proceedings in the Court Martial may not be aware that these documents exist or that they may be submitted to the court. At the PCMH [plea and case management hearing] judges should ensure that counsel are aware of the existence of OJAR/SJAR and the use that can be made of them.

Knowing the board helps advocates gauge how to handle witnesses, whether there are regimental or unit sensitivities, and how to broach mitigation. It is normally a good idea to introduce evidence in chief of a defendant's unblemished good character, if they have one, and particularly details of gallantry or competence operationally. This is important to discuss in advance, with the Defendant's Assisting Officer (DAO) and to ensure that this vital information is available before trial. Once the details of the court are known, the AAO should advise defence advocates on the honours and awards of any of the members.

Civilian Defendant

Where any defendant is not subject to service law at the commencement of proceedings (i.e. is no longer serving in one of the Armed Forces) is to be tried by court martial, AF(CM)R 2009, rule 33, provides that each of the lay members must be either:

(a) a person not subject to Service law who is qualified for membership under paragraph (2) and not ineligible by virtue of rule 32 (which sets out the various disqualifications); or
(b) an officer or warrant officer who would be qualified for membership under section 156, and not ineligible by virtue of section 157 or rule 32, if this rule did not apply.

A person not subject to service law is eligible for lay membership, in rule 33(2), *unless*:

(a) (s)he is aged under 18, or has reached the age of 70, at the commencement of the proceedings;
(b) (s)he is not a United Kingdom national;
(c) (s)he is a mentally disordered person;
(d) (s)he is disqualified for jury service;
(e) (s)he is a member of the Military Court Service; or
(f) (s)he is on the staff of the Service Prosecuting Authority.

The effect of these provisions is that for the trial of a person subject to service discipline, the court martial will generally comprise of lay members subject to service discipline (i.e. not service personnel) and, if the defendant is convicted, the judge advocate will sentence alone. (See also, on page 400.) Rule 33 AF(CM)R 2009 clarifies that, in addition to officers and warrant officers qualified as above, the lay members may be civilians ('a person not subject to service law') if they are over 18 and under 70, UK citizens, of sound mind, eligible for jury service, not an employee of the CMAO, and not on the staff of the SPA. Rule 33 also applies to:

(a) trial proceedings, if any defendant is not subject to Service law at the commencement of the proceedings;
(b) sentencing proceedings with lay members, if any offender who falls to be sentenced was not subject to Service law when convicted;

(c) variation proceedings with lay members, if this rule applied to the sentencing proceedings in which the sentence that falls to be varied was imposed;
(d) appellate proceedings; and
(e) activation proceedings with lay members, if the offender is not subject to Service law at the commencement of the proceedings.

At the commencement of proceedings, apart from sentencing proceedings, appellate proceedings or variation proceedings, any party to the proceedings may object to any lay member, on any reasonable ground.

Challenge to a Lay Member

Any party to the proceedings may object to any lay member, on any reasonable ground[38] and the judge advocate shall make their ruling on the objection before the member is sworn. If the objection is upheld, then the member will be discharged and their place taken by a waiting member. There is no right to challenge a member in sentencing proceedings or variation proceedings.[39]

Trial Commencement and Procedure

The commencement of the trial is different to the commencement of proceedings. Whereas the latter is not expressly specifically provided for in legislation,[40] the commencement of the trial is referred to in the AF(CM)R 2009. Rule 2 provides for a number of 'proceedings':

(a) preliminary proceedings,
(b) trial proceedings,
(c) sentencing proceedings,
(d) variation proceedings,
(e) appellate proceedings,

38 AF(CM)R 2009, r. 35.
39 *Ibid.*, Rule 35(5).
40 S. 55 refers to the time limit for charging a *former* member of a regular or reserve force, under the title of 'Time Limits For Commencing Proceedings' and prevents a charge being laid after six months. In other words, proceedings cannot commence with a charge unless it is laid before six months.

(f) activation proceedings, and
(g) ancillary proceedings,

'Trial proceedings' means proceedings for the trial of a charge by the court (including proceedings authorised by an order of the Appeal Court under section 19 of the 1968 Act), and does not include sentencing proceedings; but does not include the exercise of any power of the court otherwise than at a hearing.

AF(CM)R 2009, R 16 provides that proceedings shall commence at such time and place as may be appointed by the CAO save that, where an offender has been convicted in trial or appellate proceedings, the sentencing proceedings in respect of him shall commence immediately after the conclusion of the trial or appellate proceedings, unless the judge advocate for those proceedings appoints some later time. After the commencement of any proceedings, the court shall sit at such times and for such periods each day as the judge advocate may direct.

In certain circumstances, AF(CM)R rule 27 provides that proceedings may be before judge advocate alone, for example in preliminary proceedings, sentencing of a civilian or ex-serviceman or women, variation proceedings (the 'slip rule'), activation proceedings and ancillary proceedings.

It should be noted that the reference in Appendix H states that the trial commences after the judge advocate has taken the defendant's plea, admitted the lay members and sworn them in, gives the court a *Morris* direction (see page 412) and completed some other formalities, However, the Manual of service law (MSL) stipulates that the 'trial commences immediately after the last lay member has been sworn.'[41]

Clearly, a trial cannot actually commence unless it is lawfully constituted in compliance with section 155 AFA 2006 requiring a minimum of a judge advocate and a specified number of lay members, depending on the severity of the charge, and the charge must have been put to the accused (arraigned), so that it is understood how the case is to proceed.

As will be observed, there are various opinions about when a trial commences. However, the AF(CM)R 2009 seem clear. This draws a

41 JSP 830 MSL Version 2.0 2-29-32, para. 99. In the SAC, the Guidance states that the trial commences after the 'Judge Advocate reads out the charge and confirms with Appellant (or his Counsel/legal representative) whether appeal is against both finding and punishment, or punishment only.'

distinction between 'proceedings' and 'trial'. Whereas the trial proceedings may take place on an appointed date and time, and failure to attend may have consequences, the trial will commence when other formalities have been completed.[42]

The court martial trial procedure very closely resembles that of a Crown court, with an added military flavour, and an extract from the Court Martial and the Summary Appeal Court Guidance is to be found at Appendix H (Order of Procedure in the Court Martial). Civilian advocates are robed unless the judge advocate directs otherwise.

In *Morris v United Kingdom*,[43] the court considered the safeguards of independence of a military court. Of particular concern was the plight of junior members who might be seen to be subject to the influence of more senior members of the board, in spite of the oath they take. Consequently, a *Morris* direction is usually given at the beginning of each trial, reminding each member of their duty to act independently and impartially.

A trial will proceed on the basis of a charge sheet issued by the SPA, containing what they consider to be appropriate charges. In the case of *Armstrong*,[44] the CMAC considered the position of the court where it does not agree with the course a prosecutor wishes to follow. In that case the trial judge had considered that the course followed by the Crown was not appropriate, in that they had chosen to proceed on charges of prejudicial conduct, which did not come 'anyway near the true gravamen of conduct', rather than the more serious substantive offences under the Firearms Act 1968, the Misuse of Drugs Act 1971 and Official Secrets Act 1989. The CMAC held where a difference arose between a judge and a prosecutor about the appropriate charge, the difference was normally as to the level of charge; a judge after enquiry should, in their view, accept the position taken by the prosecutor. In the present case, the judge advocate had stated his concern was directed not at the level of charge, but the form of the charge which

42 In *R v Jones* [2002] UKHL 5, it was held that the Crown Court may conduct a trial in the absence, from its commencement, of the defendant. What was key was that the defendant had been arraigned and entered a plea of not guilty. In a unanimous decision, as a general principle Lord Bingham found 'it hard to discern any principled distinction between continuing a trial in the absence, for whatever reason, of a defendant and beginning a trial which has not in law commenced ... one who voluntarily chooses not to exercise a right cannot be heard to complain that he has lost the benefits which he might have expected to enjoy had he exercised it'.
43 [2002] ECHR 162, Application: 38784/97.
44 See p. 159, FN 20.

required him to determine sentence by reference to the minimal prejudice to good order and military discipline that would have been caused by the fact the respondent's possession of the ammunition and the top secret documents would have been unknown to anyone other than the Royal Military Police.[45]

The CMAC held that:

> in prosecutions before the courts of England and Wales, the course that must be followed where the judge considers that the decision of a prosecutor on charge is erroneous is clear. As is laid down in what are known as the *Farquharson Guidelines* (published in 1986), if a prosecutor invites the judge to approve the course he is going to take he must abide the judge's decision. If the prosecutor does not invite the judge's approval of his decision, it is open to the judge to express his dissent to the course proposed and to invite the prosecutor to reconsider the matter. If, after reconsideration and discussion, the judge remains of the opinion that the course proposed might lead to serious injustice, then the judge may decline to proceed until there is consultation either with the Director of Public Prosecutions or the Attorney General as may be appropriate. If a defendant or a victim of a crime seeks to challenge a prosecutor's decision, then the decision is subject to judicial review in carefully circumscribed circumstances: see *Sharma v Brown-Antoine* [2006],[46] *R v DPP ex parte C* [1995].[47] There is also available a right to seek a review of a prosecutor's decision by a more senior prosecutor: see *R v Killick* [2011].[48]
>
> Outside these well-known and clearly circumscribed procedures, the decision of a prosecutor is final. Given the prosecutor's independent position within the executive branch of the State and the position of the judiciary as a separate

45 It is respectfully suggested that the learned judge misdirected himself on the charge. See 'Practice in the Service Courts Collected Memoranda', Version 6, para. 1.4 and the section above on Conduct Prejudicial to Good Order and Discipline at p. 156 *et seq*.

46 [2006] UKPC 57, [2007] 1 WLR 780.

47 [1995] 1 Cr App R 136.

48 [2011] EWCA Crim 1608.

branch of the State, as the authorities explain, there are good constitutional reasons why a judge must abide the decision of the prosecutor once the methods described have been exhausted.

It is therefore clear that if in the military courts a judge does not approve of a course that a military prosecutor intends to take, the proper course for the judge is to ask for the matter to be referred either to the Director of Service Prosecutions, the equivalent to the Director of Public Prosecutions, or to the Attorney General as may be appropriate.

Rules of Evidence

Rule 73(1) AF(CM) R 2009 states that the rules of evidence applicable in a trial on indictment in England and Wales shall apply, to the extent that they

a. are capable of applying; and
b. are not applied, with or without modifications, by any other enactment or subordinate legislation (whenever passed or made).

'Rules of evidence' includes rules conferring or restricting any discretion to exclude admissible evidence.

Oral testimony is to be given on oath. In respect of written statements, these must comply with the provisions of section 9 of the Criminal Justice Act 1967, subject to some minor modifications, set out in rule 75(2).

Proof is permissible by formal admission under section 10 of the 1967 Act, also subject to some minor modifications.[49]

Memory Refreshing
Main Provision: rule 77 AF(CM)R 2009

Rule 77 AF(CM)R 2009 permits a person giving oral evidence about any matter at any stage in the course of doing so, to refresh their memory of it from a document made or verified by them at an earlier time if:

(a) (s)he states in his oral evidence that the document records his or her recollection of that matter at that earlier time; and

49 AF(CM) R 2009, r. 76.

(b) their recollection of the matter is likely to have been significantly better at that time than it is at the time of their oral evidence.

Similarly, a witness may refresh their memory from any sound recording of an oral account.

Defendant's Bad Character

Main Provisions: Chapter 2 and Schedule 6 of CJA 2003; paragraph 234 to Schedule 16 AFA 2006; rules 78–80 AF(CM)R 2009

Schedule 6 to the Criminal Justice Act (CJA) 2003 applied the bad character provisions to service courts,[50] *mutatis mutandis*. That, in turn, was modified by the AFA 2006. Sections 98 to 106, 109, 110 and 112 of CJA 2003, in so far as they are not applied in relation to proceedings before service courts by provision contained in or made under any other Act, have effect in relation to such proceedings (whether in the United Kingdom or elsewhere) as they have effect in relation to criminal proceedings.

The party wishing to adduce bad character evidence, or to cross-examine a witness with a view to eliciting such evidence, must serve a notice on the CAO and all other parties to the proceedings. The form and content which such notice must take is set out in rule 78 but it must, essentially, describe the misconduct to which the evidence relates and state what evidence of the misconduct the party serving the notice intends to adduce or elicit. If served by the Director, it must identify any witness whom they intend to call about the misconduct and identify the paragraph or paragraphs of section 101(1) of the CJA 2003 which the party serving the notice asserts to be applicable to the evidence.

If the notice is served by the Director, it must be served not more than 14 days after the Director serves advance information in respect of the charge to which the evidence relates. If served by a defendant, the notice must be served not more than 14 days after the date on which the Director complies or purports to comply with Article 4 of the Criminal Procedure and Investigations Act (CPIA) Order; or, if later, the date on which the Director

50 In Sch. 6, '"service court" means— (a) the Court Martial; (b) the Summary Appeal Court; (c) the Service Civilian Court; or (d) the Court Martial Appeal Court'.

discloses to the defendant the previous convictions of the co-defendant to whose misconduct the notice relates.

There is some flexibility in that rule 78(5) provides that, if it is not reasonably practicable to serve a notice under this rule within the time prescribed, it must be served as soon as it is reasonably practicable to do so. Furthermore, the court may dispense with the requirement to serve a notice under this rule if it is satisfied that no injustice would result.

An application to exclude evidence of a defendant's bad character is made under rule 79 either in writing, or, where the court dispenses with the requirement, it may be made orally. If made in writing, the application must state whether a notice under rule 78 has been served on the applicant in relation to the evidence, and if so on what date, and must be made and served not more than 14 days after that date (if any), unless the court dispenses with the requirement to serve a notice under rule 78, or such a notice is served but it is not reasonably practicable to make the application within 14 days of the service of the notice. The application must be made as soon as is reasonably practicable.

It is also possible to make an application for leave to adduce evidence of the bad character of a non-defendant. This must comply with the procedure in rule 80, which sets out the information required, which is similar to that described above. It must be made in writing to the CAO and be served on all other parties to the proceedings, unless a judge advocate gives leave for the application to be made orally. If made by the Director, an application under this rule must be made not more than 14 days after the Director serves advance information in respect of the charge to which the evidence relates. If the application under this rule is made by a defendant, it must be made not more than 14 days after the date on which the Director complies or purports to comply with Article 4 of the CPIA Order or, if later, the date on which the Director discloses to the defendant the previous convictions of the person to whose misconduct the application relates. If the time limit for application is not reasonably practicable, the application must be made as soon as it is reasonably practicable to do so.

Where the court makes a relevant ruling, it must state in open court (but, in the case of a ruling by a judge advocate in proceedings before the court martial, in the absence of the other members of the court) its reasons for the ruling. If it is the Summary Appeal Court or the SCC, it must cause the ruling and the reasons for it to be entered in the note of the court's proceedings.[51]

Hearsay Evidence

Main Provisions: Chapter 2 and Schedule 7 CJA 2003; Schedule 16 paragraph 235 AFA 2006; Chapter 3 AF(CM) R2009

Schedule 7 of the CJA 2003 applies sections 114 to 121, 123, 124, 126, 127 to 129 and 133 and 134, of that Act to proceedings before service courts.

A notice of intention to adduce hearsay evidence is made under rule 81, and admissibility is judged according to the criteria of sections 114(1)(d) (interests of justice), 116 (maker of statement unavailable to give oral evidence) and 117 (statement contained in a document) of the CJA 2003. The notice must be served on the CAO and all other parties to the proceedings, giving details of:

a. the statement that the party serving the notice proposes to tender in evidence;
b. the provision of the 2003 Act on which it is based;
c. where the statement is contained in a document which has not already been served on all the other parties, must include a copy of the document;
d. where the notice is served by the Director and oral evidence of the statement is to be given, must identify any witness who is to give it;
e. where he proposes to tender the statement by virtue of section 114(1)(d) of that Act, must specify which of the factors mentioned in section 114(2)[52] of that Act he considers to be relevant, and how they are relevant; and
f. where the statement is evidence that an earlier hearsay statement was made, must specify whether he proposes to

51 Sch. 6 para. 4 amending s. 110 CJA 2003.

52 E.g. probative value, what other evidence can or has been given on the issue, importance, circumstances of the making of the statement, reliability of the maker, reliability of the evidence, whether oral evidence of the matter stated can be given (if not, why not), amount of difficulty involved in challenging the statement, extent to which that difficulty would be likely to prejudice the party facing it.

> tender it by virtue of section 121(1)(a), (b) or (c) of that Act (relating to multiple hearsay).

Of course, where a party serves a notice under rule 81 in relation to a statement, any other party may serve a counter-notice objecting to its admission, under rule 82. Such a counter-notice must state:

(a) the date on which the party serving it was served with the notice under rule 81;
(b) whether (s)he objects to the admission of the whole or only part of the statement, and if only part which part; and
(c) the grounds on which (s)he so objects.

This counter-notice must be served on the CAO within 14 days of the notice of hearsay under rule 81.

The provisions relating to admissibility of business records, contained in section 117, are applicable to the SJS.[53] In the amendment to section to 117 CJA 2003,[54] the amended provision adds a subsection (8) that prescribes the additional requirements that must be satisfied to adduce a statement contained in a document created or received by a person in the course of a trade, business, profession or other occupation, etc., in order for it to become is admissible. It clarifies that the criminal proceedings in which a statement contained in a document is admissible as evidence of any matter stated 'includes proceedings before an officer in respect of a service offence within the meaning of the Armed Forces Act 2006'.

Other Evidential Matters

Chapter 4 of the AF(CM) R 2009 deals with:

a. Evidence of enlistment (r. 83 - enlistment paper is evidence);
b. Evidence as to service, etc. (r .84 – document issued by or on behalf of the Defence Council dealing with service, discharge, posting, rank, pay, appointment, decorations and badges);
c. Service records (r. 85 – made under any Act or QRs, and signed by the person's CO);
d. Defence Council Instructions, regulations and certificates (r. 86);

53 CJA 2003, Sch. 7.

54 *Ibid.*, para. 3, which refers to business and other documents.

e. Standing or routine orders (r. 87 – certificate purporting to be signed by a person's CO or an officer authorised to sign), which in each case shall be admissible evidence of the matters stated in the document.

The provisions of sections 34–37 of the Criminal Justice and Public Order Act 1994 (CJPOA 1994) relating to adverse inferences from silence are also applicable in service courts.[55] Sections 34 and 35 are, per se, not a breach of the European Convention on Human Rights and Fundamental Freedoms (ECHR). In *Condron v UK* (2001) 31 EHRR 1, the Strasbourg Court recognised that the CJPOA aims to strike an appropriate balance between the right to silence and the drawing of adverse inferences. 'Particular caution is required before an adverse inference can be drawn, but in a situation when an explanation is clearly called for from a defendant, then their silence will be relevant in assessing the persuasiveness of the evidence adduced by the prosecution.[56]

In addition, the provisions of the Domestic Violence Crime and Victims Act (DVCVA) 2004, provide that where, in relation to charges of murder or manslaughter or causing or allowing a person's death, a court or jury is permitted, in relation to the section 5 offence (causing or allowing a person's death), to draw such inferences as appear proper from the defendant's failure to give evidence or refusal to answer a question, the court or jury may also draw such inferences in determining whether they are guilty

(a) of murder or manslaughter, or
(b) of any other offence of which he could lawfully be convicted on the charge of murder or manslaughter,

even if there would otherwise be no case for him to answer in relation to that offence.

55 See the CJPOA1994(AAF)O 2009.

56 Judgment of the Grand Chamber, paras 56–61. See also: 'Guide on Article 6 of the Convention – Right to a fair trial (criminal limb)', para. 199, https://www.echr.coe.int/Documents/Guide_Art_6_criminal_ENG.pdf, updated 30 April 2020.

Vulnerable, Etc. Witnesses
Main Provisions: CMASCC(YJCEA1999)R 2009; YJCEA1999(ASC)O 2009; SMCW(SO)R 2013; CMASCC(YJCEA1999)(A)R 2015

The Criminal Evidence Act 1999) Rules 2009 (CMASCC(YJCEA)R 2009) must be read in conjunction with the amendments made by the 2015 Rules, and the Armed Forces (Service Courts Rules) (Amendment) Rules 2015 (AF(SCR)(A)(R) 2015) which can be found at Appendix N, Special Measures Directions. The effect is to make provision for a range of measures designed to help young, disabled, vulnerable or intimidated witnesses give evidence in proceedings before the court martial and the SCC. These measures are contained in Part 2 of the Youth Justice and Criminal Evidence Act 1999 (YJCEA 1999), certain provisions of which are applied with modifications to, or modified in their effect for the purposes of, proceedings before service courts by the Youth Justice and Criminal Evidence Act 1999 (Application to Service Courts) Order 2009.

Part 2 of the CMASCC(YJCEA)R 2009 deals with restrictions on cross-examination of a witness by a defendant acting in person. Rule 4 makes provision for the Director of Service Prosecutions to make an application to the judge advocate to give a direction that the defendant be prevented from cross-examining a witness. Rule 5 sets out the procedure for the judge advocate to explain to the defendant that they are prevented from cross-examining the witness and the procedure required for the defendant to appoint a legal representative to conduct the cross-examination. Rule 6 deals with the appointment of a legal representative by the judge advocate and rule 7 deals with the appointment of a legal representative by the defendant.

Part 3 of the Rules, rule 8 makes provision for an application under section 41(2) of the 1999 Act for leave to adduce evidence or ask questions about a complainant's sexual history.

Part 4 of the Rules[57] contains provisions relating to applications for reporting directions in respect of adult witnesses under section 46 of the

57 Part 4 of the 2009 Rules contains provisions relating to applications for reporting directions in respect of adult witnesses under s. 46 of the YJCEA 1999. Rules 9, 10 and 12 of the 2009 Rules govern applications under s. 46 of the 1999 Act for directions restricting the reporting of proceedings and directions dispensing with such restrictions. Rules 4, 5, 6 and 7 of these Rules replace rules 9, 10, 11 and 12 of the 2009 Rules with new rules which also govern applications

1999 Act. Rule 9, as amended,[58] provides for applications for a reporting direction to be made in writing or orally under rule 11, as amended. Rule 10 (significantly amended by the 2015 rules) makes provision for any party to the proceedings to oppose an application and rule 11 makes provision for urgent applications to be made in exceptional circumstances. In its amended form, it provides that, with the leave of the judge advocate, a person who may make an application in writing under rule 9(2) may instead make an oral application under rule 11. Moreover, an application may be made under rule 11 whether or not an application has already been made under rule 9(2). A person who seeks leave to make an application under rule 11 must explain the circumstances that justify making an oral application.

Rule 12, as amended, contains provision for applications to be made for an excepting direction under section 45(4) or (5), 45A(10) or 46(9) of the 1999 Act to dispense with the restrictions imposed by a reporting direction. Rule 13 contains provisions relating to applications for the variation or revocation of a reporting or excepting direction and rule 14 contains provision for applications for an extension of time. Rule 15 sets out the procedure for determining applications for reporting directions. In Part 5 of the Rules, rule 16 revokes existing evidence rules for court martial proceedings and proceedings in the Standing Civilian Court.[59]

The CMASCC(YJCEA1999)R 2009 were further amended by the Court Martial and Service Civilian Court (Youth Justice and Criminal Evidence Act 1999) (Amendment) Rules 2015.[60] Inter alia, reference is now made, in paragraph 3(1) of the 2009 Rules, to a 'person concerned in the proceedings', which means:

---

under ss 45 and 45A of the 1999 Act for directions restricting the reporting of proceedings and directions dispensing with such restrictions. Rule 8 of these Rules amends rule 13 of the 2009 Rules, which deals with revocation and variation of various directions, so that it also applies in relation to directions under sections 45 and 45A of the 1999 Act.

58 As amended by CMASCC(YJCEA1999)(A)R 2015, r. 4, which substituted a new r. 9 in Part 4 of the 2009 Rules, concerning an application for a direction under s. 45(3) of the 1999 Act or for a reporting direction under s. 45A(2) thereof.

59 Extracted from the EN to the CMASCC(YJCEA1999)R 2009.

60 Giving effect to s. 78 of the Criminal Justice and Courts Act 2015 and in force on 13 April 2015 – see Criminal Justice and Courts Act 2015 (Commencement No. 1, Saving and Transitional Provisions) Order 2015.

(a) in relation to an application in any proceedings for a direction under section 45(3) (Power to restrict reporting of criminal proceedings involving persons under 18), prohibition on identification, or an excepting direction under section 45(4) or (5) of the 1999 Act (direction by the court or an appellate court dispensing with restrictions), a person falling within section 45(7) of that Act, namely, a person against or in respect of whom the proceedings are taken, or (b) someone who is a witness in the proceedings;

(b) in relation to an application in any proceedings for a reporting direction under section 45A(2), or an excepting direction under section 45A(10), of the 1999 Act, (Lifetime reporting restrictions for victims and witnesses under the age of 18) a person falling within section 45A(3) of that Act, namely, a witness other than an accused in the proceedings and/or a victim;

(c) in relation to an application in any proceedings for a reporting direction under section 46(2), or an excepting direction under section 46(9), of the 1999 Act, a witness falling within section 46(1) of that Act;

'reporting direction' means, unless otherwise specified, a reporting direction under section 45A(2) or 46(2) of the 1999 Act.

The applications for a reporting direction must include the 'specified information' detailed in the amended rule 9 of CMASCC(YJCEA1999)R 2009, depending on whether the provision of the 1999 parent Act which the application is made under, that is, section 45(3), or section 45A(2) or section 46(2).

The term 'publication' is defined in section 63 of the YJCEA 1999 and includes:

> any speech, writing, relevant programme or other communication in whatever form, which is addressed to the public at large or any section of the public (and for this purpose every relevant programme shall be taken to be so addressed), but does not

include an indictment or other document prepared for use in particular legal proceedings.[61]

Within the UK, the Special Measures For Child Witnesses (Sexual Offences) Regulations 2013[62] amend section 33 of the YJCEA 1999 by extending the age presumption for determining eligibility for special measures, which applies currently only to victims of human trafficking offences, to relevant offences for the purposes of the Directive. Consequently, a complainant of a relevant offence[63] whose age is uncertain will be presumed to be under the age of 18 if there are reasons to believe that person is under the age of 18. Although there is no statutory instrument (SI) applying these measures to the MJS as yet – even the AF(SCR)(A)R 2015 do not implement this presumption – it is clearly a matter where the interests of justice would dictate these provisions be taken into account and applied in similar circumstances to those of a civilian court. The provision in the MJS to enable this is dealt with in the section entitled Circumstances Not Provided For.

Interpreter

The AF(SCR)(A)R 2015 insert a new provision into paragraph (2) of CM(AF) R 2009, rule 3 (interpretation: general), defining 'complainant' in relation to any offence (or alleged offence) as 'a person against or in relation to whom the offence was (or is alleged to have been) committed'. See also AF(SCR) (A)R 2015, rule 10 which inserts a new rule 22A into the AF(CM)R 2009 to provide interpretation and translation for persons other than a person to whom proceedings relate, that is the complainant and other witnesses. The new provisions are set out at Appendix N Special Measures Directions.

61 See Ministry of Justice Circular No. 2015/02. At Annex A there is a helpful overview of statutory reporting restrictions, as of 13 April 2015.

62 The regulations implement in England and Wales Directive 2011/93/EU of the European Parliament and of the Council of 13 December 2011, on combating the sexual abuse and sexual exploitation of children and child pornography. It replaced Council Framework Decision 2004/68/JHA ('the Directive').

63 (a) a sexual offence; (b) an offence under s. 1 of the Protection of Children Act 1978(4); (c) an offence under s. 160 of the CJA 1988(5); (d) an offence under s. 4 of the Asylum and Immigration (Treatment of Claimants, etc) Act 2004(6) – see Special Measures for Child Witnesses (Sexual Offences) Regulations 2013, r. 2 (SI 2013 No. 2971).

Special Measures

Main Provisions: Chapter 9 AF(CM) R 2009 (rules 89–100); Chapter 6 AF(SCC) R 2009 (rules 4–84); YJCEA 1999 (sections 23–30); AF(SCR)(A)R 2015

This important procedure is available in the MJS and, in the context of the court martial, guidance is contained in Chapter 9 AF(CM) R 2009 – and Chapter 6 AF(SCC) R 2009 for the SCC – concerning special measures directions. In addition, an application for a special measures direction is made under rule 97 AF(CM)R 2009, or rule 83 AF(SCC) R 2009 and must be made in writing to the CAO, unless a judge advocate gives leave for it to be made orally. The content of the application is set out in rule 97 AF(CM) R 2009 (SCC rule 82). Such directions are available for an 'eligible witness', being a witness

a. under the age of 18, who suffers from a mental disorder within the meaning of the Mental Health Act 1983, or otherwise has a significant impairment of intelligence and social functioning; or has a physical disability or is suffering from a physical disorder; [r. 90] [SCC r. 75]
b. afraid or distressed about testifying.

An 'eligible witness' therefore is defined as a witness eligible for assistance by virtue of rule 90 or 91 (SCC, rules 75 or 76).

A special measures direction may also be given under rule 91 where the proceedings relate to a sexual offence and the witness is the complainant to such offence and the witness has not informed the court that they do not wish for a special measures direction (r. 91) (SCC, r. 76).

A sexual offence means an offence under section 42 as respects which the corresponding offence under the law of England and Wales is

a. an offence under Part 1 of the Sexual Offences Act 2003;[64]
b. an offence of attempting or conspiring to commit such an offence; or
c. an offence under Part 2 of the Serious Crime Act 2007(b) (encouraging and assisting crime) where the offence

64 E.g. rape, assault by penetration, sexual assault, sexual offences against children, abusing position of trust, etc.

(or one of the offences) which the offender intended or believed would be committed is an offence under Part 1 of the Sexual Offences Act 2003.

The court may give a special measures direction in relation to an eligible witness where any of the special measures available in relation to the witness (or any combination of them) would, in the court's opinion, be likely to improve the quality of evidence given by the witness. The direction must specify particulars of the provision made by the direction in respect of each special measure which is to apply to the witness's evidence. In determining whether a particular measure is likely to improve the quality of the evidence to be given by the witness, the court must consider all the circumstances of the case, including in particular, any views expressed by the witness; and whether the measure or measures might tend to inhibit such evidence being effectively tested by a party to the proceedings.

Special Measures Available on grounds of age or incapacity:[65]

a. Screening witness from accused (s. 23);
b. Evidence given in private (s. 25);
c. Removal of wigs and gowns (s. 26);
d. Video recorded evidence in chief (s. 27);
e. Examination of witness through intermediary (s. 29, an interpreter or other person approved by the court for the purposes of that section);
f. Aids to communication (s. 30, being such device as the court considers appropriate with a view to enabling questions or answers to be communicated to or by the witness);
g. Where a defendant is eligible for assistance by virtue of rule 90, [SCC r. 75] the special measures available in relation to him are those for which provision is made by sections 29 and 30 of the YJCEA 1999.

Special Measures Available on grounds of fear or distress about testifying

a. Screening witness from accused (s. 23);
b. Evidence given in private (s. 25);

65 The sections in brackets relate to the YJCEA 1999.

c. Removal of wigs and gowns (s. 26);
d. Video recorded evidence in chief (s. 27).

Other Special Measures available are:

a. Giving evidence in private, where the proceedings relate to a sexual offence; or it appears to the court that there are reasonable grounds for believing that any person other than a defendant has sought, or will seek, to intimidate the witness in connection with testifying in the proceedings [r. 94] [SCC r. 79].
b. Video recorded evidence in chief, admitted under section 27 of the 1999 Act if the court is of the opinion, having regard to all the circumstances of the case, that in the interests of justice the recording, or that part of it, should be so admitted. Any prejudice to the defendant will be taken into account. [r. 95] [SCC r. 80]. An application for this special measure is to be made in accordance with r. 98 [SCC r. 83] and must be accompanied by a copy of the video recording which (or part of which) it is proposed to tender in evidence.
c. Examination of the witness through an intermediary [r. 96] [SCC r. 81]. Such examination must take place in the presence of such persons as the direction may provide; and in circumstances in which the court, and legal representatives acting in the proceedings, are able to see and hear the examination of the witness and to communicate with the intermediary (and for this purpose any impairment of eyesight or hearing is to be disregarded).

Live Link

There is the additional power for the judge advocate to order that any person may (and, if in service custody, must) attend any proceedings by live link. This power exists in rule 18 SCC Rules and rule 18 AF(CM)R 2009 and may be exercised by the judge advocate of their own motion.

Special measures directions may be varied or discharged by the judge advocate if it appears to them to be in the interests of justice to do so (r. 96, and SCC r. 84).

Where a special measures direction is given in any proceedings where there are lay members (the military 'jury'), the judge advocate must give them such warning (if any) as they consider necessary to ensure that the fact that the direction which was given in relation to the witness does not prejudice any defendant (r. 100).

Submission of No Case to Answer

Main Provision: rule 104 AF(CM)R 2009; section 6 DVCVA 2004

As with civilian criminal trial procedures, the defence has the possibility of submitting at the close of the Crown's case, in respect of any charge, that the Crown has failed to establish a case for the defence to answer. Rule 104 AF(CM)R 2009 provides that the application is determined in the absence of members of the court and, if the submission is allowed, the judge advocate shall direct the lay members to find the defendant not guilty of the charge. This rule is subject to section 6 of the DVCVA 2004.[66] A similar provision is to be found in rule 88 AF(SCC)R 2009.

Documentary Evidence

Main Provisions: rules 83–87 AF(CM)R 2009

Certain evidence relating to service matters is admissible in the form of a certified document, signed by an authorised officer. These relate to:

a. evidence of enlistment,
b. evidence as to service – see below, on page 167 *et seq.*,

66 This is a special provision relating the evidence and procedure concerning the offence of causing or allowing the death of a child or vulnerable adult under s. 5 of DVCVA. Section 6 is concerned with a defendant who is charged in the same proceedings with an offence of murder or manslaughter and with an offence under s. 5 in respect of the same death. Adverse inferences may be drawn from the defendant's failure to give evidence or refusal to answer a question in determining whether he is guilty – (a) of murder or manslaughter, or (b) of any other offence of which he could lawfully be convicted on the charge of murder or manslaughter, even if there would otherwise be no case for him to answer in relation to that offence. In the event, the charge of murder or manslaughter is not to be dismissed under paragraph 2 of Schedule 3 to the Crime and Disorder Act 1998 (c. 37) (unless the section 5 offence is dismissed); paraphrasing s. 6 DVCVA 2004. A submission of 'no case' on the charge of murder or manslaughter is not to be considered before the close of all the evidence.

c. Service records, Defence Council instructions, regulations and certificates, and
d. standing or routine orders. See r. 87.

Expert Evidence

Main Provisions: Chapter 5 AF(CM) R 2009

Such evidence shall not be adduced without the leave of the judge advocate unless the party proposing to rely on it has served on every other party and the CAO, not less than 14 days before the date appointed for the commencement of the proceedings, a statement in writing (unless every other party consents to it being made orally) of the substance of the expert evidence. Where more than one defendant wishes to introduce expert evidence on an issue, the judge advocate may direct that the evidence on that issue is to be given by one expert only. In the event that the defendants cannot agree who should be the expert, the judge advocate may either select the expert from a list prepared or identified by them; or direct that the expert be selected in such other manner as the judge advocate directs.

Circumstances Not Provided For

Main Provisions: rule 26 AF(CM)R 2009 and rule 25 AF(SCC)R 2009

The AM(CM)R 2009 and the AF(SCC)R 2009 are very comprehensive documents, but they may not cover every situation that arises in a service court. So, a very useful power exists in rule 26 AF(CM)R 2009, in respect of trial by court martial, and rule 25 AF(SCC)R 2009, in respect of trial by SCC, which permits the judge advocate to deal with situations that the rules themselves do not provide for. Subject to any other enactment (including any other provision of these Rules) the judge advocate shall ensure that proceedings are conducted:

(a) in such a way as appears to him most closely to resemble the way in which comparable proceedings of the Crown Court would be conducted in comparable circumstances; and

(b) if he is unable to determine how comparable proceedings of the Crown Court would be conducted in comparable circumstances, in such a way as appears to him to be in the interests of justice.

In the case of *R v H and J*,[67] the defendants applied before arraignment for the charges to be dismissed on the grounds that the Crown's case was insufficient for the matter to proceed. The JAG ruled there was no power within the SJS for a case to be dismissed before arraignment. The applicants argued that this power existed in rule 26(b) AF(CM)R. Their appeal was allowed. Rule 25(3) grants a judge advocate power to terminate any proceedings if they consider it in the interests of justice to do so; but this did not extend to a power to dismiss proceedings before they had begun.[68] Rule 25 must be read *ejusdem generis* (applying only to persons or things of the same general kind or class as those specifically mentioned). The saving represented by rule 26(b) applied as the interests of justice required the judge advocate to apply a comparable process to that in the Crown court.

Consideration of the Evidence by Members

Main Provision: section 160 AGA; rules 109 and 110 AF(CM)R 2009

At the conclusion of the summing up by the judge advocate, the lay members retire to consider the evidence alone. If they require further directions on the law or the evidence, they may request the same from the judge advocate, and such further direction will be given in open court (r. 109). The decision of the lay members is by simple majority. Where two or more of the lay members are subject to service law, the votes of those members on the finding on each charge are to be given in ascending order of rank or rate and seniority. Once the lay members have reached a decision on the evidence their finding on each charge is announced by the President in open court. This has been the subject of some judicial consideration, on the basis that it is out of step with the civilian right to trial by a jury of 12. Objections have been concerned with conviction by simple majority,

67 [2019] WLR(D) 584.

68 See above, on p. 392, as to when proceedings commence.

inability to know what the majority actually is and voting in reverse order of seniority.

The case of *Appleyard*,[69] decided under the similar provisions of the Army Act 1955, concerned a direction by the Judge Advocate that the members of the Board should 'try to reach a verdict upon which each one of you is agreed'. He went on to say that when it became possible for a majority verdict to be accepted, he would give a further direction. The appellant argued this was in breach of section 96 Army Act 1955 (now s. 160 AFA 2006), which provided that verdicts of a court martial 'shall be determined by a majority of the votes of the members of the court'. The matter was raised again in *Twaite*[70] when, subsequent to conviction, the JAG referred the finding of the court martial to the court on, inter alia, the following points of law:

a. Was a finding of guilt by a simple majority of a Board of five or more lay members inherently unsafe?
b. Was a defendant deprived of his right to a fair trial under Article 6 of the Convention?

If the answer to b. was in the affirmative, the CMAC was asked to make a declaration of incompatibility. The judgment was delivered by the Lord Chief Justice who held that it was inappropriate for any inquiry as to voting to follow the return of a guilty verdict. In their Lordships' judgment it was necessary to ascertain first whether the Board had reached a verdict. If so, the simple question, which should then be asked, was 'Do you find the defendant guilty or not guilty?' No further questions should be asked. The answer to that question was conclusive.

The issue came before the courts for a third time in the case of *Blackman*.[71] B had been convicted by court martial of the murder of a badly wounded Taliban fighter in Afghanistan and sentenced to life imprisonment with a minimum tariff of 10 years, reduction to the ranks and dismissal with disgrace. He appealed conviction and sentence, contending it is a fundamental feature of the English criminal justice system that those facing

69 [2005] EWCA Crim 2750 CMAC.
70 [2010] EWCA Crim 2973. Although a majority verdict, it was decided by four in favour of conviction and one against, paras 1 and 12.
71 [2014] EWCA Crim 1029.

serious criminal charges are entitled to be tried before 12 members of the public and can only be convicted by a majority of at least 10 of the 12. He argued that a simple majority conviction is inherently unsafe because it demonstrates sufficient doubt to defeat the criminal standard of proof. Reliance was also placed upon a passage in the third edition of *Rant on the Court Martial and Service Law* at paragraph 5.126. The court was invited to declare section 160(1) of the Armed Forces Act 2006 incompatible with the Convention rights under section 4 of the Human Rights Act 1998. Referring to the judgment of Judge LCJ in *Twaite,* which set out the case law which had subjected the court martial system in England and Wales to rigorous scrutiny, the court concluded that the provision for a majority verdict set out in section 160(1) of the 2006 Act was entirely compatible with the Convention. The court held that, consistent with the system in the courts, no question should be asked as to whether the acquittal had been by majority or not. There were no circumstances in which the way the individual members of the court had voted should be revealed. The appeal against conviction did not succeed.[72]

In the appeal of *Stables,*[73] heard three months earlier, the CMAC rejected a contention that, following the judge advocate's explanation of the basis of sentence, which displayed an inconsistency in the basis of the conviction which rendered it unsafe, the CMAC ought not to be looking at the sentencing remarks at all. The court's view was that the sentencing remarks quite clearly gave the reasons why the Board found the appellant guilty. So, if those reasons revealed an error, it would be right for the court to intervene.

One criticism often made by defence advocates, which has some resonance in civilian criminal trials, is that the finders of fact do not have to give reasons for their finding/verdict and that this is, therefore, not transparent and leads to unfairness, for example in understanding the basis of the conviction.[74] When such a complaint was made in the civilian

---

72 The court did adjust the 20-year minimum tariff set by the court martial and reduced it to 8 years.

73 [2010] EWCA Crim 2405, 22 September 2010. The judgment is in French.

74 'As the decisions in *Cooper* and *Morris* made clear, one of the important safeguards of the independence of the lay members of a Board is "the confidentiality of the deliberations, a point repeated in the member's oath"': *Twaite*.

case of *Lhermitte v Belgium*,[75] it was held that it was necessary to look at the proceedings as a whole, including the subsequent court decisions that clarified the reasons for the jury's guilty verdict. A combined reading of the judgments of the Assize Court and of the Court of Cassation would have enabled the applicant to understand the reasons why she had been found criminally responsible, and therefore guilty, and why she had been sentenced to life imprisonment. So, the ECtHR does not appear averse to trial by jury, or the absence of express reasons on conviction.

A provision exists in rule 110(3) that 'if the judge advocate is satisfied that the findings announced are acceptable in law, the judge advocate and the president of the board shall sign a record of the findings.' It is not clear when a judge advocate may reject the court's findings, or how they would know of the error, bearing in mind they are not present during deliberation – the Explanatory Notes are silent on the matter and the Practice Memoranda merely confirms that the findings may be refused by the judge on the grounds that, in the judgment of the judge, it is contrary to the law[76] and they then shall direct the court to reconsider their findings[77] – but there is no known equivalent in the civilian criminal justice system and, it is suggested, it is a power that should be used in the MJS with the utmost caution. It is a jury's right to reject the prosecution case.

If, on the other hand, at the sentencing stage, when the judge advocate participates in the sentencing process, it is discovered that they convicted on an erroneous basis (as discussed in the above cases),[78] the proper course, it is suggested, is for them to consult both prosecution and defence on the matter of an appeal to the CMAC. Indeed, the JAG has the power to refer to the CMAC, under section 34 of the CM(A)A 1968 and rule 51 of the CMACR 2009,[79] any case where there is a need for clarification of the law. The case will not have proceeded to the sentencing phase and it is open to the judge advocate, under rule 16(2) AF(CM)R 2009, to defer sentencing until some

75 (2015) (application no. 34238/09).
76 See para. 5.2 of Practice Memoranda 2011.
77 *Ibid.*, r. 110(4).
78 However, in *Twaite*, discussed above, the CMAC shared the concerns of the learned trial judge advocate that, once the mistake was known '*she was in an impossible position*' and could not really ignore it.
79 Reference by the JAG or the Secretary of State in respect of a finding under s. 34 of the Courts-Martial (Appeals) Act 1968 in relation to conviction.

later time (e.g. after receipt of reports). He could then grant 'bail' (release from custody) under section 109(3) AFA 2006. This appears to have been the approach taken in *Twaite*.[80]

If the judge advocate proceeds to the sentencing phase[81] with the court where, perhaps, the anomaly was not apparent to them prior to then, at present there does not appear to be any legal provision that permits the judge advocate to grant bail pending an appeal.[82] However, the route the judge advocate may then consider appropriate in such a case is suggested as one of the following:

a. To support the defence in an application for an expedited bail hearing before the CMAC (the Director may choose not to oppose such an application); or
b. As the proceedings before the court martial have not, as a matter of law, been concluded, and rule 5 of the SCSRSR 2009 does not yet apply,[83] the judge advocate might consider making a ruling under rule 26 AF(CM)R 2009 (Circumstances not provided for) (set out above on page 428) to grant release subject to conditions. It is a course that should be supported by the Director and the defence, pending the application to the CMAC. It is also a course that is not without some risk in relation to possible sanctions if the conditions on bail are breached but, if there is a manifest injustice from the court's erroneous decision, then this may well serve the interests of justice and would avoid interfering directly in the finding of the court.

The judge could not be faulted if they opted for the first, safer, course of action.

If the error is not discovered until sentence has been passed, the normal route would be for an appeal to take place. It is difficult to

80 See the proposals for reform, discussed on p. 388.
81 The judge advocate has an equal vote although, in the case of an equality of votes, the court must find 'for the offender': see r. 112(3).
82 See on p. 549 for the jurisdiction of the CMAC to grant bail pending appeal. The AF(CRFC) O 2009 relates to the judge advocate's power to grant bail pending an appeal to the High Court from the SAC, and to the court martial from the SCC. Consequently, after conviction by the court martial, the only statutory power to grant bail lies with the CMAC under the Court Martial Appeal Court (Bail) Order 2009.
83 Grounds upon which a service person may be released from sentence.

invoke section 109, which permits release from custody up to the point of the conclusion of proceedings before the court as, after sentence, the proceedings would be over.

The matter of majority verdicts in the court martial was the subject of a review by Judge Lyons.[84]

Conviction of an Offence Other Than That Charged

Main Provision: sections 161 (Court Martial) and 281 (SCC) AFA 2006

Section 161 refers to the power of the court martial to convict of an offence other than that charged. This may happen where the court acquits a person of an offence specifically charged in the charge sheet, but the allegations in the charge sheet amount to or include (expressly or by implication) an allegation of another service offence. The court may convict the defendant of that other offence. Acquittal of the main offence nevertheless permits the court to convict of an attempt to commit the offence. In relation to civilians tried by SCC, a similar power exists in section 281, in that it applies to the SCC the section 161 power for court martial. An allegation of an offence to which section 39(1)[85] applies (offence X) shall be taken as including an allegation of an offence under that section of attempting to commit offence X; and (b) an allegation of a completed section 42 offence shall be taken as including an allegation of a section 42 offence of attempt. In other words, attempting to commit a service offence or a criminal offence will fall within section 161. Some possible service alternative charges are suggested below, in the case of military offenders.

Possible Service Alternative Charges

The offences listed below are not absolute alternatives and will very much depend upon the individual facts of the case. However, they may assist in steering advocates in the direction of what may be a possible and preferable alternative. An attempt to commit the offence may also be considered.

While a section 19 charge of prejudicial conduct might be thought to be available for substitution for most offences, provided the element

84 See above, p. 386, FN 93 and p. 388, FN 2.

85 Attempting to commit an offence.

Theft Act Theft Burglary	s. 19 conduct prejudicial to good order, etc., e.g. where the essential element of dishonesty is missing, or it is a technical offence, e.g. unlawfully entering the cookhouse for a sandwich
Theft – false accounting	s. 18 making a false record
Fraud	s. 18 making a false record
Forgery	s. 18 making a false record
Public Order Act 1986 s. 2 violent disorder s. 3 affray s. 4 threatening, abusive or insulting words or behavior s. 4A harassment s. 5 harassment, alarm or distress	s. 20 unfitness alcohol s. 21 fighting s. 21(1)(b) threatening, abusive, insulting or provocative words or behavior common assault
Sexual offences	s. 23 disgraceful conduct of an indecent kind
ABH/common assault	s. 22 ill-treatment of subordinates s. 11 misconduct towards a superior officer
GBH/wounding	s. 18 OAPA to s. 20 s.11 misconduct towards a superior officer ABH common assault
Misconduct towards a superior officer	Where the element of knowledge about the victim being an 'officer' or of 'superior rank', but there is disorder and/or alcohol present: s. 20 (1)(b) drunkenness (behaviour is disorderly or likely to bring discredit to Her Majesty's forces) s. 19 conduct prejudicial to good order and military discipline
s. 20 unfitness or misconduct through alcohol or drugs	where the alcohol/drugs element is missing: s. 19 conduct (good order and discipline)

Table 13.1 Possible Alternative Offences

of potential prejudice to good order and military discipline is present, it should generally be avoided, as a matter of fairness, when the option was present at the time when the substantive charge was laid. This does not preclude the addition of the section 19 charge at trial, where there is express agreement between prosecution and defence for example and where the defence indicate a willingness to plead guilty to the section 19 offence. As the Farquharson Guidelines point out:[86] 'While he remains instructed it is for Counsel to take all necessary decisions in the presentation and general conduct of the prosecution ... it is for Prosecution Counsel to decide whether to offer no evidence on a particular count or on the indictment as a whole and whether to accept pleas to a lesser count or counts. 6.1 The prosecution advocate may ask the defence advocate whether a plea will be forthcoming but at this initial stage should not suggest or indicate a plea that might be considered acceptable to the prosecution before a plea is offered' (cited in *Armstrong*, see page 161 above).

Depending on the stage of the proceedings when the alternative plea is offered to the Crown, and accepted, it may be agreed that a new charge sheet be produced with the alternative charge(s) pleaded. This would be the proper course where the charge sheet has not yet been disclosed to the members of the court.

Release from Custody During Proceedings

Main Provisions: sections 105, 108–111 AFA 2006; rule 27 AF(CP)R 2009

These important provisions are set out above, at page 141.

Section 109 explains that the custody provisions in sections 105(2)–(6), which predominantly concern the grounds upon which the judge advocate must be satisfied in order to keep someone in custody, are applied for those who are subject to service court proceedings. The additional Condition D is applicable, where the accused's case has been adjourned for inquiries or a report and it appears to the judge advocate that it would be impracticable to complete the inquiries or make the report without keeping the accused in service custody.

86 Updated guidance now on the CPS Legal Guidance website.

Where the CO requests a review of custody of the accused under section 108(2)(b), after the prosecution has begun to open its case against the accused before the court martial or the SCC, and at a time when that court is sitting, the request shall be made to the judge advocate for the proceedings before that court.[87] This review of custody under section 108(2)(b) should take place where 'at any time' the CO considers that the grounds on which an order to keep the accused in custody was made have ceased to exist.[88] Section 109 extends this power even when the court martial or the SCC proceedings against the accused are under way.

Arrest After Charge or During Proceedings by Order of the CO

Main Provisions: section 110 AFA 2006; rule 28 AF(CP)R 2009

There is a power under section 110 AFA for the CO to order the arrest of a person who has been charged with, or is awaiting sentence for, a service offence and is not in service custody. The CO may exercise this power where they consider that taking that person into service custody is justified. Reasonable force may be used to effect the arrest.[89] The CO must be satisfied that there are reasonable grounds for suspecting that, if not taken into service custody,

> a person would
>
> (a) fail to attend any hearing in the proceedings against him;
> (b) commit an offence; or
> (c) would interfere with witnesses or otherwise obstruct the course of justice, whether in relation to himself or any other person.

The CO may also take a person into service custody where

> (a) he has failed to attend any hearing in the proceedings against him;

87 AF(CP)R 2009, r. 27(3).
88 See also, r. 27 AF(CP)R 2009.
89 See s. 110(5).

(b) there are reasonable grounds for suspecting that he should be taken into Service custody for his own protection or, if he is aged under 17, for his own welfare or in his own interests; or

(c) there are reasonable grounds for suspecting that
  (i) if not taken into Service custody, he would fail to comply with a requirement imposed under section 107(3); or
  (ii) he has failed to comply with such a requirement.

In this event, the CO must, as soon as is practicable, ensure the person is brought before a judge advocate for a review of whether they should continue to be kept in service custody. The CO must also, as soon as practicable, notify the CAO of that fact and the necessary information and inform the accused in writing that they are to be brought before a judge advocate as soon as practicable. If the accused has been arrested under section 110(1), where they have been charged with, or are awaiting sentence for, a service offence and is not in service custody, the CO must give reasons for believing that the keeping of the accused in custody is justified.

Arrest During Proceedings at Direction of Court

Main Provisions: section 111 AFA 2006; rule 28 AF(CP)R 2009

Whereas the judge advocate in a trial before a service court may order a defendant into arrest, after arraignment – and prior to the conclusion of the trial – this power is not without a requirement for proper justification (see s. 110(2) and (3)), nor without a requirement for review. A judge advocate may direct the defendant's arrest, if satisfied that taking them into service custody is justified. After arrest, if the defendant is in service custody, they are to be brought before a judge advocate as soon as is practicable for a review of whether they should continue to be kept in service custody and that review is to be dealt with as on a review under section 108(1) (see ss 108(4)–(8) and 109).

Bail is not available to appellants convicted of certain serious offences after previous conviction of such offences unless the judge advocate considering the grant of bail is satisfied that there are exceptional circumstances that justify it.[90]

Sentencing: General Provisions

Main Provisions: sections 160, 164, 165 (election); Part 9 (sections 237–271) AFA 2006; rule 112 AF(CM)R 2009; JAG Sentencing Guidance

The maximum sentencing power of five-member court martial is whatever sentence is prescribed by law for the offence of which the accused is convicted – there is no maximum expressly set out in section 164. For a three-member court martial, the maximum sentence is two years imprisonment, unless the accused has elected trial, in which case the sentence passed in respect of the offence must be such that the CO could have awarded the punishments if they had heard the charge summarily (see s. 165 and the related provisions in ss 125 and 129).

With the exception of an offence for which the sentence is fixed by law, or provided for in section 237(3), the court martial has a duty to have regard to purposes of sentencing, as stipulated in section 237. The court must not pass a custodial sentence unless it is of the opinion that the offence, or the combination of the offence and one or more offences associated with it, was so serious that no less severe sentence can be justified for the offence (s. 260(2)). The custodial sentence must be for the shortest term (not exceeding the permitted maximum) that in the opinion of the court is commensurate with the seriousness of the offence or the combination of the offence and one or more offences associated with it (s. 261).

It is important to acknowledge, as the CMAC has done often, that the court martial is

> a specialist tribunal and particular respect must be given to its judgments as to the significance of the military context of an

90 AF(CRFC)O 2009, r. 4. This relates to sexual offences (e.g. rape, assault by penetration) and manslaughter and culpable homicide. See also, Sch. 6 of the Sexual Offences Act 2003, which amended ss 25(2) and 32 of the CJPOA 1994.

> offence and as to the implications for the service, as well as for the individual offender, of imposing particular sentences.[91]

The court martial is central to the services' disciplinary systems and its sentencing reflects not only current sentencing practice, but relevant service policy considerations (the service interest) which affect the gravity of offending which, in civilian life, may not be viewed in quite the same way. The court is not bound by policy considerations but is entirely justified in taking them into account if it is considered in the interests of justice to do so. See paragraph 2.18 *et seq.* of the Sentencing Guide.

Unrepresented Defendant

There is a specific provision, contained in section 263, which prevents a service court from imposing a custodial sentence or service detention on an unrepresented offender. Moreover, this is followed by the Summary Appeal Court, in respect of an offender who is not legally represented in that court.

Decisions and Findings

The decisions and findings of the court are determined by a majority of the votes of the members. Section 252 requires the court must give reasons for any sentence it hands down and must explain to the offender in ordinary language:

a. the effect of the sentence;
b. where the offender is required to comply with any order forming part of the sentence, the effects of non-compliance with the order;
c. any power, on the application of the offender or any other person, to vary or review any order forming part of the sentence; and
d. where the sentence consists of or includes a fine, the effects of failure to pay the fine.

91 *Armstrong* [2019] EWCA Crim 1737, per, Holroyde LJ, at para. 23.

Sentencing Guidelines

Section 259 requires the service court to have regard to any sentencing guidelines[92] that are relevant to the offender's case and, in exercising any other function relating to the sentencing of offenders for service offences, have regard to any guidelines which are relevant to the exercise of the function. Nevertheless, it may depart from those guidelines if, in its opinion, the departure is justified by any features of service life or of the service disciplinary system that are relevant to the case.

Guidance is provided by the JAG's Office in a very helpful and publicly available document.[93] It explains that sentencing in the court martial is undertaken by the judge advocate sitting with the board of lay service members of the court; if the lay members are all civilians the judge advocate sentences alone. The judges are experienced in sentencing practice in the civilian courts and have attended Judicial Studies Board training seminars provided for recorders and judges of the Crown court. Most, if not all, judge advocates also sit as recorders. In addition, they are experienced in the general sentencing practices and policies of the Armed Forces, and their needs and requirements. The board of lay service members bring their service background and knowledge of disciplinary issues to bear on the process, but as laypeople with little or no training in sentencing practice (except for those of them who have been involved with summary hearings) they should always follow the directions of the judge advocate and stay within the sentencing parameters they set.[94]

The guidance provided to judge advocates states in its introduction that it is,

> issued by the Office of the Judge Advocate General and contains judicial guidance on the approach to sentencing and appropriate sentencing tariffs while taking account of the Armed Forces discipline policy. It is intended to assist all concerned in the sentencing of offenders in the Court Martial and other Service courts.

92 The JAG has published such guidance, the current version being Version 5, issued in January 2018. This is under regular review.

93 See http://www.justice.gov.uk/downloads/courts/judge-advocate-general/guidance-sentencing-court-martial.pdf.

94 *Ibid.*, para. 2.3.

Its aim is to 'provide an aid to consistency in the Court Martial'. This is augmented by another very useful and freely accessible publication, called 'Practice In The Court Martial: Collected Memoranda'.[95] This document explains that

> [t]here is no statutory basis for these Practice Memoranda. They have been produced following extensive consultation and represent opinions or directions, including interpretations of the law, which provide persuasive guidance. Judge Advocates are encouraged to follow the guidance contained herein, but where they decide to depart from it they should explain their reasons for doing so on the record.

The Guidance on Sentencing makes some important points about service policy considerations in sentencing for offences:

The Court Martial is an independent court but it is also an essential part of the process which maintains discipline within the Armed Forces. It is legitimate for the Court to take account of (but not to be strictly bound by) declared Service policy considerations when deciding appropriate sentences. Service policy considerations, upon which the Services have been consulted, are given in this guide where appropriate.[96]

> The Court Martial does not as a matter of course hear whether an offender's Commanding Officer wishes to retain him (as used to be the case). The offender may introduce evidence from his superior officers in mitigation or as a character reference, and the prosecutor can address the Court Martial on Service policy regarding the relevant offence. The court routinely invites the Commanding Officer of the offender's unit/ship/station to complete an Operational Effectiveness Impact Statement (OEIS) outlining the consequences of the offence, and of the possible loss of the offender's services. In that way, operational effectiveness

95 This is currently available at http://www.justice.gov.uk/downloads/courts/judge-advocate-general/practice-memos-court-martial.pdf. The version available, as at 1 December 2020, is Version 5, issued on 6 March 2015 by the JAG.
96 The Guide on Sentencing, at para. 2.

> can be taken into account as a relevant consideration. The future employability of the offender is a relevant consideration.[97]

There are reminders in the Sentencing Guidance of the services' policy on drugs and that members of the Armed Forces 'who carry lethal weapons, operate and maintain dangerous equipment, or bear responsibility for the safety of others must display higher standards of behaviour than civilians. They must expect to be punished more severely for breaching those high standards.'[98]

The Guidance lists the aggravating and mitigating factors for certain types of offence and the aggravating and mitigating factors.

Aside from the Guidance, there are statutory criteria which the sentencing court must have regard to. These are contained in section 237 and require that, unless the sentence for an offence is fixed by law, a court *or officer* (e.g. CO) dealing with an offender for a service offence must have regard to the following purposes of sentencing:

(a) the punishment of offenders;
(b) the maintenance of discipline;
(c) the reduction of service offences and other crime (including reduction by deterrence);
(d) the reform and rehabilitation of offenders;
(e) the protection of the public;
(f) the making of reparation by offenders to persons affected by their offences;
(g) the welfare of any offender aged under 18.

There are circumstances where the factors in section 237 do not apply, namely, required custodial sentences, certain firearms and other specified offences. See section 237(3).

The reference in section 237 to 'officer' means that COs must also have regard to these principles when sentencing an accused, as the section sets out *general* sentencing principles.

When considering the seriousness of the offence, section 240 enables a service court to increase a sentence for racial or religious aggravation. If

97 *Ibid.*, at para. 3.2.4.
98 *Ibid.*, at para. 5.7.2.

the offence was racially or religiously aggravated, the court or officer must treat that fact as an aggravating factor and must state in open court that the offence was so aggravated. There are similar considerations where the offender demonstrated towards the victim of the offence hostility based on the sexual orientation (or presumed sexual orientation) of the victim, or a disability (or presumed disability) of the victim, or the victim being (or being presumed to be) transgender, which the court or officer must treat the conduct as an aggravating factor (s. 241).

The CMAC has recognised that the court martial is a specialist court, serving the needs of the special military community which is focused on discipline. It has said often that it will not interfere readily in sentences passed by the court martial. See the illustrations on page 619 *et seq*.

The court martial, like any other criminal court, is reminded by the judge that any sentence passed is required to be in accordance with sentencing principles, and be proportionate by reference to its main purposes (s. 237):

> The principles of sentencing are explained as
>
> a. the punishment of offenders;
> b. the maintenance of discipline;
> c. the reduction of Service offences and other crime (including reduction by deterrence);
> d. the reformation and rehabilitation of offenders;
> e. the protection of the public; and
> f. the making of reparation by offenders to persons affected by their offences.

The Sentencing Guide also states at paragraph 2.8 that,

> [t]he AFA 2006, in addition to requiring sentencers in the Court Martial to have regard to the purposes of sentencing sets out considerations in deciding the seriousness of an offence including the treatment of previous convictions [s. 238]. The Act requires the Court Martial to treat racial or religious aggravation of an offence, and hostility related to victims' disability or sexual orientation, as aggravating factors in deciding the seriousness of the offence [ss 240–241]. The Act also provides for mandatory

> sentences in certain circumstances (such as third domestic burglary, firearms offences, etc.) [ss 217–228]. Where an offender has pleaded guilty, the Act requires the Court Martial to take into account the stage in the proceedings when an offender first indicated his intention to plead guilty, and the circumstances in which the indication or the plea was given [s. 239].

As stated by Toulson LJ in *Harris*:[99]

> A Court Martial is a specialist tribunal which is particularly well able to assess the relevant sentencing considerations. Accordingly, this court, which lacks the same feel for the needs of the Services that the Court Martial will have, will exercise caution before concluding that a Court Martial has been wrong in principle or failed to give proper weight to the relevant factors in sentencing in a particular case. That is not to emasculate the court's jurisdiction. The court has a responsibility, if it considers a sentence to be wrong, to correct it; but in determining whether a sentence is wrong, the court must give proper weight to the factors which we have mentioned.

The importance of the Sentencing Guidelines was underlined in *R v L*.[100] The appellant, who was at the time a Royal Military Policeman, was tried by court martial on one charge of battery, contrary to section 39 of the Criminal Justice Act 1988 and was convicted. On 11 March 2016, he was sentenced by the Board to be reduced to the ranks, dismissed from the service and to serve 60 days' detention. He had leave to argue that the sentence passed, in particular his dismissal from the service, was excessive. He argued on appeal that the sentencing court martial failed to consider the Sentencing Guidelines Council, even though this is provided for in the Court Martial Guidance on sentencing. In upholding the appeal, and considering the sentence afresh, Hallett LJ stated, 'Court Martial Guidance is explicit: regard should be had to the Sentencing Council guideline.'

Unless the court is of the opinion that it is unnecessary to obtain a *pre-sentence report* under section 256,[101] it is a requirement that, before passing

99 *Harris (Kyle Alan)* [2011] EWCA Crim 1242.
100 [2017] EWCA Crim 709.
101 As amended by Part 1 to Sch. 22 LASPOA 2012.

certain sentences,[102] the court must obtain and consider a pre-sentence report unless, in the circumstances of the case, the court is of the opinion that it is unnecessary to obtain a pre-sentence report. Where the offender is aged under 18, the court must not form such an opinion unless there exists a previous pre-sentence report obtained in respect of the offender and the court has had regard to the information contained in that report, or, if there is more than one such report, the most recent report. See section 236(5) and (6) for requirements on appeal.

The practice in the Summary Appeal Court has some differences, as explained in section 256(8) and (9). Where, in the circumstances of the case, the court is of the opinion that it is unnecessary to obtain a PSR it can dispense with one, unless the appellant is under 18, where the court must not form such an opinion unless there exists a previous PSR obtained in respect of the offender and the court has had regard to the information contained in that report, or, if there is more than one such report, the most recent report (s. 256(4)).

Sentencing deliberations, over which the judge advocate presides, necessarily take place in closed court, or in a retirement room, and may not be disclosed. The judge and board members often achieve consensus, but if it is necessary for them to vote on sentence, only a simple majority is required with the judge having the casting vote. As is also the case when a finding is announced, there is no public indication whether the decision was unanimous or by majority.[103] The judge explains in open court the reasons for sentence and the president of the board then formally announces it (s. 252). In very exceptional circumstances (e.g. in professional negligence cases) the president of the board may also address the offender, but only having first sought and obtained the judge's permission.[104]

Where there is any *dispute on the facts* after a defendant has pleaded guilty, rule 112 provides that the judge advocate may direct that any issue of fact be tried by the court. The finding of the court on any such issue shall be determined by a majority of the votes of the members of the court. Where there is an equality of votes, the court must find for the offender, in

---

102 Service detention, custodial sentence, dismissal or dismissal with disgrace, community punishment, minimum sentences for dangerous offenders and sexual or violent offences, offender is aged under 18.

103 See the CMAC decision if *R v Twaite* [2010] EWCA Crim, set out on p. 430 above.

104 *Ibid.*, paragraph 2.4.

other words, prefer their version of the facts. The finding is then announced by the judge advocate.

In the case of *Cairn and others v The Queen*,[105] the Court of Appeal was concerned that far too many appeals against sentence are mounted on the basis that the judge has failed to have any, or sufficient, regard to the basis on which a plea of guilty has been entered. The court stated:

> 4. The admission comprised within the guilty plea is to the offence and not necessarily to all the facts or inferences for which the prosecution contend. Once again, however, the responsibility for determining the facts which inform the assessment of the sentence is that of the judge. In the normal course, when the contrary is not suggested, that assessment will be based on the prosecution facts as disclosed by the statements. If, however, the offender seeks to challenge that account, the onus is on him to do so and to identify the areas of dispute in writing, first with the prosecution and then with the court.
> 5. The proper approach of the prosecution to bases of plea was considered in *R v Tolera* [1999] 1 Cr App R 29 and is now set out in the Attorney General's Guidelines on the Acceptance of Pleas and the Prosecutor's Role in the Sentencing Exercise (issued with effect from 1 December 2009). In so far as it deals with the position of the defendant and the court, it can be summarised in this way:
>   (i) A basis of plea must not be agreed on a misleading or untrue set of facts and must take proper account of the victim's interests; in cases involving multiple defendants, the bases of plea for each defendant must be factually consistent with each other (see para C1).
>   (ii) The written basis of plea must be scrutinised by the prosecution with great care. If a defendant seeks to mitigate on the basis of assertions of fact outside the prosecutor's knowledge (for example as to his state of mind), the judge should be invited not to accept

105 [2013] EWCA Civ 467, decided on 16 April 2013.

this version unless given on oath and tested in cross-examination as set out in IV.45.14 of the Consolidated Criminal Practice Directions (CCPD): see para. C3. If evidence is not given in this way, then the judge might draw such inferences as he thought fit from that fact.

(iii) The prosecution advocate must ensure that the defence advocate is aware of the basis on which the plea is accepted and the way in which the case will be opened (para. C5). Where a basis of plea is agreed, having been reduced into writing and signed by advocates for both sides, it should be submitted to the judge prior to the opening. It should not contain matters that are in dispute: see *R v Underwood* [2005] 1 Cr App R 13 replicated in CCPD IV.45.11(c) and (d). If it is not agreed, the basis of plea should be set out in writing identifying what is in issue; if the court decides that the dispute is material to sentence, it may direct further representations or evidence in accordance with the principles set out in *R v Newton* (1982) 77 Cr App R 13.

(iv) Both sides must ensure that the judge is aware of any discrepancy between the basis of plea and the prosecution case that could potentially have a significant effect on sentence so that consideration can be given to holding a Newton hearing. Even where the basis of plea is agreed between the prosecution and the defence, the judge is not bound by such agreement: see paras. C8 and C10, CPR IV.45.12 and *Underwood* (ibid). But if the judge is minded not to accept the basis of plea in a case where that may affect sentence, he should say so.

As the JAG's Sentencing Guidelines point out, in paragraph 2.18, the court martial may also take into account service policy:

> The Court Martial is an independent court but it is also an essential part of the process which maintains discipline within the Armed Forces. It is legitimate for the Court to take account of (but not to be strictly bound by) declared Service policy

> considerations when deciding appropriate sentences. Service policy considerations, upon which the Services have been consulted, are given in this guide where appropriate.

Aside from Newton Hearings, if there is any dispute on facts after a plea of guilty the judge advocate may direct that any issue of fact be tried by the court martial under rule 112 AF(CM(A)R2009.

Discount for Guilty Plea

In section 239, specific provision is made for those who plead guilty to an offence. The section applies to both proceedings before a service court and at a summary hearing. Guidance is given on the approach that sentencers should take, in determining what sentence to pass on the offender. The court or officer dealing with them for their offence must take into account the stage in the proceedings for the offence at which they indicated an intention to plead guilty (or an intention to admit the offence at a summary hearing) and the circumstances in which this indication was given. Where the offender falls to be sentenced for a third domestic burglary or third drug trafficking offence under section 225(2) or 226(2) (required custodial sentences), nothing in sections 110(2)[106] or 111(2)[107] of the Powers of Criminal Courts (Sentencing) Act 2000 (the Sentencing Act) prevents the court from imposing any sentence which is at least 80% of that specified in section 110(2) or 111(2) of that Act. In other words, a 20 per cent discount for a guilty plea may be awarded even for a third specified offence.

JAG's Guidelines follow the guidelines issued by the Sentencing Council, although they also take into account operational factors which are unique to the services and service policy considerations.[108] The court is free to apply any discount for an early guilty plea as recommended in the Sentencing Council Guidelines.[109] The purpose of the guideline is 'to encourage those who are going to plead guilty to do so as early in the court process as possible. Nothing in the guideline should be used to put

106 Minimum of seven years' imprisonment for third-class A-drug-trafficking offence.

107 Minimum of three years' imprisonment for third domestic burglary.

108 *'[M]isuse of drugs is incompatible with the demands of Service life and poses a significant threat to operational effectiveness.'* See para. 5.7.1 'Guidance on Sentencing in the Court Martial', January 2018.

109 See *ibid.*, para. 5.3.6.

pressure on a defendant to plead guilty.'[110] Although a guilty person is entitled not to admit the offence and to put the prosecution to proof of its case, an acceptance of guilt:

a. normally reduces the impact of the crime upon victims;
b. saves victims and witnesses from having to testify; and
c. is in the public interest in that it saves public time and money on investigations and trials.

The Guidance states, at page 5, that the maximum level of reduction in sentence for a guilty plea is one-third where a guilty plea is indicated at the first stage of proceedings. The first stage will normally be the first hearing at which a plea or indication of plea is sought and recorded by the court. When a plea is indicated after the first stage of proceedings, any reduction should be decreased from one-quarter to a maximum of one-tenth on the first day of trial having regard to the time when the guilty plea is first indicated to the court relative to the progress of the case and the trial date. The reduction should normally be decreased further, even to zero, if the guilty plea is entered during the course of the trial.

Credit for Time Served

Main Provisions: sections 246 and 247 AFA 2006, SCSRSR 2009; Schedule 13 LASPOA 2012; ORA 2014; ORA2014(C2)O 2015; Practice Memoranda 2016, paras 6.1 and 6.7; Guidance on Sentencing in the Court Martial

Legal Aid, Sentencing and Punishment of Offenders Act 2012 (LASPOA 2012) made certain amendments to section 246 AFA by substituting new provisions in place of subsections (2)–(5). These provisions have been further 'tweaked' by the Offender Rehabilitation Act 2014 (ORA 2014).

A court or officer sentencing an offender to a term of imprisonment or service detention in respect of a service offence ('the offence in question') who has been kept in service custody,[111] in connection with the offence in

110 'Reduction in Sentence for a Guilty plea Definitive Guideline', Sentencing Council, 2017, Key Principles, *ibid.*, p. 4.
111 AF(CWC)R 2009, r. 2, provides that 'custody' means service custody and is to be construed in accordance with ss 98 to 102.

question or any related offence, for any period since being charged with the offence in question or any related offence, must direct that the number of days for which the offender has been kept in service custody in connection with the offence in question or any related offence since being so charged is to count as time served by him as part of the sentence.[112] This provision also applies to a suspended sentence of imprisonment or a suspended sentence of service detention (s. 247(3)). Paragraph 6.7 of the Memoranda points out that

> [u]nder the current law the Court Martial decides the appropriate sentence for the offence, passes it in full, and explains for the benefit of the offender that he will be required to serve only 2/3 (detention) or 1/2 (imprisonment) of the sentence, less any time, which the judge specifies, already spent in post-charge custody.

However, if, on any day on which the offender was kept in service custody, they were also detained in connection with any other matter, that day is not to count as time served (s. 246(2A)).

Section 247 has the effect that section 246 applies not only where the offender has been kept in service custody when charged with the offence for which they are being sentenced, but also where they have been kept in service custody in connection with a different charge based on the same facts or evidence. Furthermore, it explains that offences are related if the charges for them were founded on the same facts or evidence and that it is immaterial for the purposes of section 246(1) whether the offender has also been kept in service custody in connection with other offences.

Section 247(3) provides that there is no credit for time spent in custody when a suspended sentence is passed, but the provision does apply if the sentence is activated.[113]

Consecutive terms of service detention, and terms of service detention which are wholly or partly concurrent, are to be treated as a single term (s. 247(3)).

---

112 See: *R v GOC 2 Div, The Army & Ano'r, ex p. Buchanan and Falls* [1998] EWHC Admin 955: Time spent in close arrest (i.e. custody) should be taken into account when reviewing sentences imposed by court martial. Judicial Review granted on this basis and also on the refusal of the military authorities to give reasons for their decisions. See also, s. 246 AFA 2006, as amended by Sch. 13(2) LASPOA 2012.

113 See EN para. 493, but subject to the later substantive amendments made by Sch. 13(3) LASPOA 2012.

Where the court does not give a direction, or gives such a direction in relation to a number of days less than that for which the offender was kept in service custody, it must state in open court that it is, in the opinion of the court just in all the circumstances and it must set out its reasons.

Section 247(6) refers to credit being given for 'relevant custodial terms', which are defined as any consecutive custodial sentence within paragraph (a), (b) or (c) of section 188(4). These are references to:

a. a determinate sentence of imprisonment passed in respect of a Service Offence or by a civilian court in England and Wales;
b. a determinate sentence of detention under section 209 of AFA 2006, or section 91 of the PCC(S)A 2000, concerning offenders under 18 convicted of certain serious offences where there is a power to detain for a specified period;
c. a sentence of detention under section 226B of the CJA 2003 (whether or not passed as a result of section 221A of the AFA 2006). This refers to passing an extended sentence for certain violent or sexual offences on a person aged under 18. These provisions reflect amendments by the LASPOA 2012, para 23 of Sch 22.

Consecutive relevant custodial terms, and relevant custodial terms which are wholly or partly concurrent, are to be treated as a single term if the sentences were passed on the same occasion or, where they were passed on different occasions, the person has not been released on licence (under Chapter 6 of Part 12 of the CJA 2003) at any time during the period beginning with the first and ending with the last of those occasions.

The calculation of credit for time spent in custody is carried out administratively at the MCTC or at the civilian prison, where the offender is informed of their expected release date. Where an offender has been held in service custody after charge pending trial, the time thus served is allowed against the time to be served under the sentence passed (para. 3.4.12 Guidance on Sentencing).

Slip Rule

Main Provisions: sections 155 and 163 AFA 2006; rules 2(4)(c), 27(4), 29, 29A, 118–120 AF(CM)R 2009; AF(CM)(A)R 2013

After the conclusion of any proceedings in which a sentence was imposed, if it is considered there may have been a legal error in those proceedings relating to sentencing,[114] there is a possibility for the defendant, or the Director, to apply to the judge advocate for variation proceedings. Rule 118 AF(CM)R provides a period of 56 days in which the court may vary a sentence imposed by it. This power may not be exercised in relation to any sentence if an appeal, or an application for leave to appeal, against the sentence has been determined. The varied sentence takes effect, as so varied, from the beginning of the day on which it was originally imposed, unless the court otherwise directs.

The judge advocate may, of their own motion, also give a direction that such proceedings be held.[115]

Rule 29 AF(CM)R 2009 stipulates that in proceedings against a defendant charged with an offence within Schedule 2 to the 2006 Act, or facing a potential sentence of more than seven years if over the age of 18, or more than seven years detention if convicted while under the age of 18, the court must be comprised of at least five lay members. Rule 29A amended this provision in the event of a guilty plea indicated before trial, so that the sentencing court did not require five lay members in order to sentence. Moreover, in such cases, rule 29A also provides that the variation proceedings do not require five lay members.[116]

In certain circumstances, the judge advocate may direct that the proceedings should be without lay members, where there would be unfair prejudice to the defendant in any delay occasioned by the inability of one or more of the original lay members to attend the proceedings. (r. 120). Where, for example, there has been an error in the form of sentence handed down that can be easily corrected, the Guidance on Sentencing in the Court Martial, at paragraph 2.25, advises a defendant wishing to seek a variation of sentence to 'write to the Judge Advocate General giving reasons for his application. The JAG will forward the application to the judge who sat in the sentencing proceedings for a decision as to whether to list the case for variation proceedings. There is no appeal against a judge's decision not to

114 E.g. by imposing an invalid sentence outside the court's powers.

115 AF(CM)R 2009, r. 119. The power to make such a rule arises in s. 163(h) of the Act, for the variation by the court of a sentence passed by it or the variation or rescission by it of an order made by it.

116 R. 29A, inserted by AF(CM)(A)R 2013.

proceed' with the application. However, any appeal against sentence would thereafter lie to the CMAC. The slip rule is purely a convenient device to correct any legal error in order to save time and expense and to ensure the correct sentence is handed down.

It is interesting that, although this slip rule power is available in the court martial, there is no equivalent in the AF(SCC)R 2009. In section 288, the rule-making power of the Secretary of State makes no reference to variation proceedings for the SCC. See the further discussion on page 523.

14 SENTENCES AVAILABLE TO THE COURT MARTIAL

Main Provisions: section 164 and 289, Part 8 Chapter 4, (sections 196–206); Schedule 3A AFA 2006; section 14(1) AFA 2011; section 12 AFA 2016; section 304H AFA 2006

The sentences available to a court martial are listed below. In general terms, and unless otherwise provided for, a sentence passed by the court martial or the Service Civilian Court takes effect from the beginning of the day on which it is passed.[1] This includes a sentence of reduction in rank.

The sentences available to the court martial are set out in section 164 AFA 2006:

a. Imprisonment (including custody for young offenders)
b. Dismissal with disgrace from Her Majesty's service
c. Dismissal from Her Majesty's service
d. Detention for a term not exceeding two years (normally served in the Military Corrective Training Centre. This sentence is not available if the person being sentenced is an officer or, as detailed below, on page 396, a civilian)
e. Forfeiture of a specified term of seniority or all seniority
f. Reduction in rank or disrating[2]
g. A fine
h. A service community order
i. A severe reprimand or reprimand
j. A service supervision and punishment order

1 AFA 2006, s. 289(1)

2 AFA 2006, s. 293 (automatic reduction of rank or rate of warrant officer or non-commissioned officer given custodial sentence or sentence of service detention) has been repealed to this extent by s. 13(2) AFA 2011. The navy term 'disrating' means reduction in rank.

k. Such minor punishments as authorised by regulations made by the Defence Council
l. A service compensation order.

It should be noted that, in the case of civilians convicted by court martial, certain of the above sentencing options are not available. See, more specifically, below on on page 503, which sets out the modifications to section 164 in these circumstances.

The Armed Forces (Financial Penalty Enforcement Orders) Regulations 2009 permit the making of an order for the registration by a civilian court ('the relevant court') of a financial penalty imposed under the AFA 2006 by the court martial. Regulation 3(5) of the 2009 Regulations provides that the sum certified in the order as being outstanding is to be treated for all purposes as if it had been a fine imposed upon conviction by the relevant court. These Regulations apply to a person who is neither subject to service law nor a civilian subject to service discipline, or is subject to service law because they are a special member, within the meaning of the Reserve Forces Act 1996, of a reserve force. The reason for this is that if a fine, etc. is unpaid by someone who is still serving, or who is a service civilian, the relevant service court[3] would be able to deal with the default. The regulations therefore provide for enforcement of such orders by the civilian courts.

The AFA 2011 inserted sections 269A and 269B in the 2006 Act.[4] Section 269A requires the court martial, when imposing a fine, to make an order fixing a term for which the person liable to pay the fine may be committed to prison if the fine is not paid. Section 269B enables the court martial, when making a service compensation order, to specify a maximum term for which the person liable to pay the compensation may be committed to prison if the compensation is not paid.

3 Defined in Sch. 7, Part 1, §3 AFA 2006 as a 'court' 'for the purposes of— (a) Part 2 of Schedule 12 to the 2003 Act as it applies in relation to a suspended sentence passed by a relevant service court; (b) paragraph 22 of that Schedule (as modified by section 205(1) of this Act) as it applies to an order under Part 2 of that Schedule amending a suspended sentence order with community requirements made by a relevant service court.'

4 In force from 1 November 2013. See Art.3(c) AFA2011(C4)O 2013.

The Regulations were updated in 2013 to reflect these amendments made by the 2011 Act.[5]

Under regulation 3A, an order under section 269A has effect as if it were an order by the Crown Court in England and Wales under section 139(2) of the Powers of Criminal Courts (Sentencing) Act 2000 in respect of a fine imposed by that court (and, if the relevant court is in Scotland or Northern Ireland, as if the enforcement of that had been transferred to Scotland or Northern Ireland under section 90 of the Magistrates Courts Act 1980).

Under new regulation 3B, where a period has been specified under section 269B, that period is treated as if it had been specified by the Crown Court in England and Wales under section 41(8) of the Administration of Justice Act 1970 (c. 31) in respect of a compensation order made by that court.

Regulation 2(2)(b) corrects regulation 3(5) of the 2009 Regulations to refer to the registration of a financial penalty instead of to the registration of the financial penalty enforcement order.[6]

Sentencing When Accused Elected Trial

Main Provisions: sections 54, 129, 130A, 165 and Schedule 3A AFA 2006; rule 6 AF(CM)R 2009; AF(CM)(A)R)2013

The amended Schedule 3A of AFA 2006,[7] which replaced the provisions of section 165 of the 2006 Act, limits the court martial's powers of punishment in the case of an offender who elected under section 129 to be tried by that court rather than being dealt with by their CO.[8] The effect of these provisions means that the court martial may not impose a sentence which is more severe than that which *could have been* imposed by the CO.[9] If the

5 Reg. 2(3) of the Armed Forces (Financial Penalty Enforcement Orders) (Amendment) Regulations 2013 inserts new regs 3A and 3B in the 2009 Regulations so as to provide, respectively, for the effect of an order under s. 269A and of an exercise of the power under se. 269B.

6 See the EN to the 2013 Regulations.

7 AFA 2011, s. 14(1) replaced s. 165 AFA 2006 ('Sentencing powers of Court Martial where election for trial by that court instead of CO') by making a reference to the new Sch. 3A which now has effect.

8 The text is based upon the EN to AFA 2011, paras 124–138

9 See *Scallan* [2005] EWCA Crim 2040 which, concerned a servicewoman who elected trial before her subordinate commander. Although decided pre-AFA 2006, the CMAC makes the point that the full powers of the CO are available to any subsequent court martial on an election. See AFA 2006, Sch. 3A, para 6.

CO was granted the power by higher authority to award an extended punishment,[10] the court martial will be able to sentence in accordance with that authority; otherwise it is limited to the CO's ordinary summary powers, that is, a maximum of 28 days detention.[11] The sentencing court will impose one global sentence, rather than the usual practice under section 255 of a separate sentence for each offence.

The objective is to ensure that summary hearing of a charge by a CO does not infringe the accused's Convention right under the Human Rights Act 1998 to a fair trial by an independent tribunal, by discouraging them from appealing. The accused has a right to be tried by a compliant court, and they are not discouraged from exercising that right by the legislation, as it ensures they do not risk incurring a more severe punishment by electing trial. [12]

Schedule 3A is reproduced below.

Schedule 3A

Court Martial Sentencing Powers Where Election For Trial By That Court Instead Of CO[13]

PART 1

RELEVANT OFFENCES

10 See s. 133 AFA 2006 and also rule 6, the Armed Forces (Summary Hearing and Activation of Suspended Sentences of Service Detention) Rules 2009 which deals with the procedure for applying.

11 In *R v McIntyre* [2020] EWCA Crim 1297, a corporal elected trial where he was sentenced to 30 days detention and reduced to the rank of LCpl. The CMAC held that the court martial had no jurisdiction to sentence and NCO to detention, so that part of the sentence was quashed. S. 132 of the 2006 Act sets out the punishments available to a CO.

12 Notwithstanding, there may be an issue about whether the absence of any right to legal advice prior to making these important decisions by an accused does actually produce an 'informed decision'. This is not so important when the accused elects trial by court martial, but is more relevant to when their decision is to accept the award of the CO. Although a service person has the subsequent right to appeal, when they will receive legal advice and financial assistance, there is some anecdotal evidence that unit pressures dissuade them from this course. Statistically, only a very small number of summary dealings find their way into the SAC. The Lyons Review (above p. 388) cites statistics from 2017 which show, for the Navy, 15 elections for trial and 12 SAC appeals; Army, 13 elections for trial and 62 SAC appeals; and RAF, no elections for trial, 10 appeals to SAC. However, when one examines the number of summary cases each service has, some very interesting questions emerge: from a total of 3,803 summary matters, the Navy had 633, the RAF had 291, while the Army had 2,879, which is more than three times the combined total of the other two services. One might wonder why it is that so few cases progress further from the non-compliant summary process.

13 Para. 13(2) and (3) relating to service sexual offences prevention orders, is repealed: ASBCPA2014, Sch. 11, Part 2, para. 81.

1 For the purposes of this Schedule an offence is 'relevant' if it falls within any of cases A to D (see paragraphs 2 to 5); and references in this Schedule to a particular case of offence are to be read accordingly.

2 An offence of which a person is convicted falls within case A if the charge in respect of the offence is one in respect of which the person elected Court Martial trial under section 129 (whether or not the charge was amended after election).

3 An offence of which a person ('the accused') is convicted falls within case B if—

(a) the charge in respect of the offence was substituted under section 125(2)(b) for a charge in respect of which the accused elected Court Martial trial under section 129; and

(b) the substitution was not one for which the accused's written consent was required by section 130A(2).

4 Where—

(a) a person ('the accused') elects Court Martial trial under section 129 in respect of a charge,

(b) at the time of the election, another charge brought against the accused ('the relevant charge') is regarded for the purposes of Part 5 as allocated for summary hearing,

(c) the relevant charge is referred to the Director of Service Prosecutions under section 123(2)(e) without the accused having been given the opportunity to elect Court Martial trial of the charge, and

(d) the Court Martial convicts the accused of an offence alleged in the relevant charge,

that offence falls within case C.

5 Where—

(a) a person ('the accused') is charged with an offence which, if the accused were convicted of it, would fall within case C,

(b) another charge ('the new charge') is substituted under section 125(2)(b) for the charge,

(c) the substitution is not one for which the accused's written consent is required by section 130A(2), and

(d) the Court Martial convicts the accused of an offence alleged in the new charge,

that offence falls within case D.

PART 2

SENTENCING POWERS ETC

Sentencing powers: single relevant offence

6 (1) This paragraph applies where—

(a) the Court Martial convicts a person ('the offender') of a case A offence or a case B offence; and

(b) paragraph 8 (multiple relevant offences) does not apply.

(2) The sentence passed in respect of the offence must be such that the offender's commanding officer could have awarded the punishments awarded by that sentence if the commanding officer—

(a) had heard summarily the charge in respect of which the offender elected Court Martial trial; and

(b) had recorded a finding that the charge had been proved.

7 (1) This paragraph applies where—

(a) the Court Martial convicts a person ('the offender') of a case C offence or a case D offence; and

(b) paragraph 8 does not apply.

(2) The sentence passed in respect of the offence must be such that the offender's commanding officer could have awarded the punishments awarded by that sentence if the commanding officer—

(a) had heard summarily the charge referred as mentioned in paragraph 4(c); and

(b) had recorded a finding that the charge had been proved.

Multiple relevant offences

8 (1) This paragraph applies where—

(a) the Court Martial convicts a person ('the offender') of two or more relevant offences; and

(b) condition 1 or 2 is met in relation to any two or more of the offences.

(2) Condition 1, in relation to any two or more offences, is that—

(a) each of the offences is a case A offence or a case B offence; and

(b) the relevant charges would have been heard summarily together if the offender had not elected Court Martial trial.

(3) In sub-paragraph (2) 'relevant charge' means—

(a) in relation to a case A offence, the charge in respect of that offence; and

(b) in relation to a case B offence, the charge in respect of which the offender elected Court Martial trial and for which the charge in respect of the case B offence was substituted.

(4) Condition 2, in relation to any two or more offences, is that—

(a) each of the offences is a case C offence or a case D offence; and

(b) the referred charges would have been heard summarily together if they had not been referred as mentioned in paragraph 4(c).

(5) In sub-paragraph (4) 'referred charge' means—

(a) in relation to a case C offence, the charge in respect of that offence; and

(b) in relation to a case D offence, the charge referred as mentioned in paragraph 4(c) for which the charge in respect of the case D offence was substituted.

Sentencing powers: multiple relevant offences

9 (1) This paragraph applies where paragraph 8 applies by virtue of a condition in that paragraph being met in relation to any two or more relevant offences.

(2) The offences in relation to which the condition is met ('the joined offences') are to be treated for the purposes of section 255 (individual sentence for each offence) as a single offence; and references in this paragraph to 'the sentence' are to the sentence passed by the Court Martial in respect of the joined offences.

(3) Where condition 1 in paragraph 8 is met in relation to the joined offences, the sentence must be such that the offender's commanding officer could have awarded the punishments awarded by that sentence if the commanding officer—
(a) had heard the relevant charges (as defined by paragraph 8(3)) summarily together; and
(b) had recorded findings that the charges had been proved.

(4) Where condition 2 in paragraph 8 is met in relation to the joined offences, the sentence must be such that the offender's commanding officer could have awarded the punishments awarded by that sentence if the commanding officer—
(a) had heard the referred charges (as defined by paragraph 8(5)) summarily together; and
(b) had recorded findings that the charges had been proved.

Further provision about sentencing

10 Where paragraph 9 applies, the following provisions apply in relation to the sentence as if it were a sentence being passed by an officer at a summary hearing—

section 242 (service detention: general restriction);
section 243 (length of term of service detention);
section 248 (forfeiture of seniority, reduction in rank or disrating).

11 Where the Court Martial is dealing with an offender for an offence with which a relevant offence is associated, the offences are to be treated for the purposes of Part 9 (sentencing: principles and procedures) as not being associated.

Activation of suspended sentence of detention

12 (1) This paragraph applies where—
(a) the Court Martial, on convicting a person ('the offender') of a relevant offence, makes an order under section 191(3) in relation to a suspended sentence of service detention passed on the offender; and

(b) the suspended sentence was passed by an officer or the Summary Appeal Court.

(2) The term of the suspended sentence as it takes effect by virtue of the order must not exceed 28 days unless the offender's commanding officer would have had extended powers for the purposes of section 194 if—

(a) where the offence is a case A offence or a case B offence, the offender had not elected Court Martial trial; or

(b) where the offence is a case C offence or a case D offence, the charge referred as mentioned in paragraph 4(c) had not been so referred.

(3) If—

(a) the Court Martial awards a term of service detention ('the new sentence') in respect of the offence (or, where paragraph 9 applies, in respect of the offence and one or more other relevant offences), and

(b) the order under section 191(3) provides for the suspended sentence to take effect from the end of the new sentence,

the aggregate of the terms of the two sentences must not exceed 28 days or, where the extended powers condition is met, 90 days.

(4) The extended powers condition is—

(a) where the offence is a case A offence or a case B offence, that the offender's commanding officer would have had extended powers for the purposes of section 194 if the offender had not elected Court Martial trial;

(b) where the offence is a case C offence or a case D offence, that the offender's commanding officer would have had extended powers for the purposes of section 194 if the charge referred as mentioned in paragraph 4(c) had not been so referred.

(5) In determining for the purposes of sub-paragraph (2) or (4) whether the offender's commanding officer would have had extended powers for the purposes of section 194 if, as the case may be—

(a) the offender had not elected Court Martial trial, or

(b) the charge referred as mentioned in paragraph 4(c) had not been so referred,

no account is to be taken of any of the matters mentioned in sub-paragraph (6).

(6) Those matters are—

(a) any change in the commanding officer's rank after the election or referral;

(b) in the case of sub-paragraph (2)(a) or (4)(a), any possibility that, if the offender had declined Court Martial trial, the commanding officer might subsequently have been granted extended powers for the purposes of section 194; and

(c) in the case of sub-paragraph (2)(b) or (4)(b), any possibility that, if the charge referred as mentioned in paragraph 4(c) had not been so referred, the commanding officer might have been granted extended powers for those purpose after the time when the charge was in fact referred.

Court orders other than sentences

13 (1) The Court Martial may not make an order under section 229 (service restraining order) by virtue of—

convicting a person of a relevant offence; or

(a) acquitting a person of an offence which would be a relevant offence if the person were convicted of it.

(2) [Service sexual offences prevention orders – this provision was repealed by para. 81 of Schedule 11 of the Anti-social Behaviour, Crime and Policing Act 2014]

(3) In sub-paragraph (2) 'relevant finding' has the same meaning as in section 232A.

Review of unduly lenient sentences

14 In section 273 (review of unduly lenient sentence by Court Martial Appeal Court), the reference in subsection (1)(a) to an offence under section 42 does not include a relevant offence.

Appeals: application of Court Martial Appeals Act 1968 to multiple relevant offences

15 (1) This paragraph applies where, by virtue of paragraph 9(2), the Court Martial passed a single sentence in respect of two or more relevant offences ('the joined offences'); and references in this paragraph to 'the 1968 Act' are to the Court Martial Appeals Act 1968.

(2) Where—

(a) section 13 of the 1968 Act (power to re-sentence when some but not all convictions successfully appealed) applies in relation to the sentence, but

(b) the appellant remains convicted of two or more of the joined offences,

those offences are to be treated for the purposes of section 13(2) of the 1968 Act as a single offence.

(3) Sub-paragraph (4) applies where section 14 of the 1968 Act (substitution of conviction on different charge after plea of not guilty) applies in relation to a conviction of any of the joined offences.

(4) The reference in section 14(2)(b) to a sentence that the Court Martial would have had power to pass in respect of the offence mentioned in section 14(1)(b) is to be read as a reference to a sentence that the Court Martial would have had power to pass in respect of all the applicable offences.

(5) For the purposes of sub-paragraph (4) an offence is an 'applicable offence' if it is—

(a) a joined offence of which the appellant remains convicted;

(b) an offence a finding of guilty of which has been substituted under section 14 of the 1968 Act for a finding of guilty of a joined offence; or

(c) an offence a plea of guilty of which has been substituted under section 14A of that Act for a plea of guilty of a joined offence.

(6) Sub-paragraph (7) applies where—

(a) section 14A of the 1968 Act (substitution of conviction on different charge after guilty plea) applies in relation to a conviction of any of the joined offences; and

(b) section 14 of that Act does not so apply.

(7) The reference in section 14A(2)(b) to a sentence that the Court Martial would have had power to pass in respect of the offence mentioned in section 14A(1)(b) is to be read as a reference to a sentence that the Court Martial would have had power to pass in respect of—

(a) all the joined offences of which the appellant remains convicted; and

(b) the offence a plea of guilty of which has been substituted under section 14A for a plea of guilty of a joined offence.

(8) Where sub-paragraph (4) or (7) applies and the case also falls within section 13(1) of the 1968 Act, section 13 of that Act shall not apply.

(9) In section 16A of the 1968 Act (appeals against sentence), the reference in subsection (2)(b) to the offence is to be read as a reference to the joined offences.

PART 3

INTERPRETATION

Commanding officer

16 (1) References in this Schedule to a person's commanding officer are to the person's commanding officer at the time the person elected Court Martial trial.

(2) In determining for the purposes of paragraph 6(2), 7(2) or 9(3) or (4) the punishments that a person's commanding officer could have awarded in respect of a charge or charges, no account is to be taken of—

(a) any change in the commanding officer's rank after the person elected Court Martial trial or (as the case may be) after the charge or charges referred as mentioned in paragraph 4(c) were so referred;

(b) in the case of paragraph 6(2) or 9(3), any possibility that, if the person had declined Court Martial trial, the commanding officer might subsequently have been granted extended powers for the purposes of any provision of Chapter 1 of Part 6; or

(c) in the case of paragraph 7(2) or 9(4), any possibility that, if the charge or charges referred as mentioned in paragraph 4(c) had not been so referred, the commanding officer might have been granted extended powers for the purposes of any provision of Chapter 1 of Part 6 after the time when the referral in fact took place.

Substituted charges

17 References in this Schedule to a charge substituted under section 125(2)(b) for another charge ('the original charge') include—

(a) a charge substituted for a charge that was itself substituted for the original charge,

(b) a charge substituted for a charge within paragraph (a),

(c) and so on.

Imprisonment

Main Provisions: sections 164, 189, 190, 200–206, 209, 218, 221 and Schedule 2 AFA 2009; sections 181 and 182 CJA 2003; Part 1 of Schedule 22 LASPOA 2012 (dangerous offenders subject to service law); see also pages 587 *et seq*. (young offenders)

Imprisonment is not an available sentence for a CO to award at a summary hearing. The punishments which the court martial can award for service offences are set out in section 164. Subject to what is said below, unless otherwise stated the maximum sentence that can be imposed is the maximum prescribed by the statute prescribing the offence of which the accused stands convicted. A sentence of imprisonment may be suspended or may be combined with other punishments.

The provisions in sections 197 to 199 relating to custody plus orders, were repealed by Schedule 10, paragraph 41 LASPOA 2012. However,

Part 1 of Schedule 22 LASPOA 2012, which is concerned with required or discretionary sentences for particular offences, inserted a number of further provisions into the 2006 Act. See below, Required Sentences, on page 471.

JAG Sentencing Guidelines must be considered by the court when determining the appropriate sentence.

In respect of Schedule 2 criminal offences, the maximum sentence is that which is set out in the substantive provision charged under section 42 AFA 2006. For example, in the case of a person charged under section 42 AFA with burglary contrary to section 9(1)(a) of the Theft Act 1968 (entering a dwelling as a trespasser with intent to steal), the maximum sentence would be 14 years' imprisonment, as set out in section 9(3)(a).

If the accused has elected trial by court martial, then the restrictions in Schedule 3A apply.

Where an offender has pleaded guilty to a service offence in proceedings before a court (or, at a summary hearing in respect of a service offence, has admitted the offence), section 239 provides for the possibility of a reduction in sentence to reflect the guilty plea(s). In this event, the court (or officer) dealing with the accused must take into account the stage in the proceedings for the offence at which they indicated an intention to plead guilty (or an intention to admit the offence at a summary hearing) and the circumstances in which this indication was given. Where the offender pleaded guilty to an offence to which section 225 or 226 applies (third drug trafficking and third domestic burglary offences), subsections (4) and (5) allow the court to reduce by up to 20 per cent the minimum sentence that would otherwise be required.[14]

A service court passes a separate sentence for each offence charged,[15] save in the case of a person subject to service law who elected trial, in which case the court martial sentences within the maximum powers available to the CO.[16] In other words, the sentence is limited to what the CO could have awarded.[17] This is subject to the caveat that the CO may have been

---

14 *Ibid.*, s. 239(4) and (5) and EN§479.

15 See s. 255 AFA 2006.

16 See s. 165 and rules 158–161 of the Armed Forces (Court Martial) Rules 2009.

17 See s. 165 and Sch. 1 AFA 2011, amending Sch. 3A AFA 2006. NB: The transitional provisions in Part 20 of the Armed Forces (Court Martial) Rules 2009 (Consequences of Election for Court Martial Trial made under the former SDAs) were revoked by AFA 2011 Sch. 5. See also, r. 6 AF(CM(A)R 2013 revoking Sch. 2 to the 2009 Rules.

granted the power by higher authority to award an extended punishment,[18] in which case the court martial will be able to sentence in accordance with that authority.[19]

Under sections 188 and 189 AFA 2006, it is possible for the court martial to pass consecutive and concurrent sentences of imprisonment or detention, depending on the justice of the case and general sentencing principles.

The court has the option to suspend the immediate operation of a sentence of imprisonment, under sections or a sentence of detention under section 190. (See the section on suspended sentence on page 473.)

Under 12 Months Imprisonment

Where the court passes a sentence of under 12 months, section 196 AFA 2006 applies the provisions of sections 181 and 182 CJA 2003 to the SCC and court martial. The sentence must be at least 28 weeks and not exceed 51 weeks in respect of one offence or the maximum term for the offence (whichever is longer).

The provisions relating to the imprisonment with or without 'custody plus' order, contained in sections 197 to 199 were repealed by LASPOA 2012, Schedule 10, paragraph 41. This repealed sections 181 to 188 CJA 2003, in respect of prison sentences of less than 12 months, custody plus orders and intermittent custody orders.

For certain dangerous offenders aged under 18, convicted by the court martial of an offence under section 42 (criminal conduct), where the corresponding offence under the law of England and Wales is a serious offence, and the court is of the required opinion (defined by section 223) that there is a significant risk to members of the public of serious harm occasioned by the commission by the offender of (a) further specified offences; or (b) further acts or omissions that would be specified offences if committed in England or Wales, the court martial may pass a life sentence.

Although the policy may differ from service to service, as prescribed by the respective QRs, an officer sentenced to imprisonment remains subject to service discipline until they cease to hold their commission as

18 See s. 133 AFA 2006 and also r. 6, the Armed Forces (Summary Hearing and Activation of Suspended Sentences of Service Detention) Rules 2009 which deals with the procedure for applying.

19 Section 132 of the 2006 Act sets out the punishments available to a CO.

notified in the *London Gazette*, but subsequently may be treated as if they were still subject to service discipline until the sentence is completed. If the officer is sentenced to imprisonment by a civil court, they remain subject to service discipline until they cease to hold their commission, as notified in the *London Gazette*.

A soldier sentenced by court martial to detention and dismissal or to imprisonment remains subject to service discipline until discharged from the Army under authority of QRs paragraphs 9.396, 9.397 or 9.404. A soldier sentenced by a civil court to imprisonment, whether at home or overseas, remains subject to service discipline until discharged from the Army under authority of QRs paragraph 9.404.

A sentence of imprisonment (including a suspended sentence), detention or youth custody will almost certainly require the Army to discharge a convicted soldier under QR (Army), paragraph 9.404e. Where the individual appears before a civilian court, if the CO recommends that the individual be retained in the service, the sentencing court must be aware that such retention must be approved by the Ministry of Defence (Army), and that this would only occur as an exceptional case. (QRs paragraph 6.174(d)(7).)

Any service person sentenced to a term of imprisonment, whether immediate or suspended, is usually discharged from the service. In the Army, the authority is to be found in QRs 9.404. Similar provisions apply within RAF QRs, paragraph 607.

A statutory power exists, in section 332 for the CO to administratively reduce the rank or rate of a warrant officer or non-commissioned officer. The CO requires the permission of higher authority. If the person affected feels aggrieved by the decision, they have the option of submitting an application for redress (see page 265 *et seq*.).

Required Sentences

Main Provisions: sections 219–228 AFA 2006; Part 1 of Schedule 22 LASPOA 2012; Schedule 2(5) AFA2006(TP)O 2009

LASPOA 2012 makes amendments to the substantive provisions of the 2006 Act concerning dangerous or violent or sexual offenders, and those offenders within these categories who are under 21[20] years of age. Section 220 is omitted (certain violent or sexual offenders aged 21 or over). In addition to amending a number of provisions that change the sections to which the reader is referred, the Act introduces three new sections for dangerous offenders subject to service law:

a. Section 218A: Life sentence for second listed offence by a person aged 18 or over who is convicted by the Court Martial of an offence under section 42 (criminal conduct), and where the corresponding offence under the law of England and Wales is an offence listed in Part 1 of Schedule 15B to the 2003 Act;[21]
b. Section 219A: Extended sentence for certain violent or sexual offenders aged 18 or over convicted of an offence listed in Schedule 15B to the 2003 Act that, if the court were to impose an extended sentence of imprisonment under section 226A of the 2003 Act as a result of this section, the term that it would specify as the appropriate custodial term would be at least 4 years;
c. Section 221A: Extended sentence for certain violent or sexual offenders aged under 18.

Also, within this category of required sentences are those people who commit a third domestic burglary (s. 226) and those who commit firearms offences (s. 227).

Section 219 explains the circumstances where a life sentence for dangerous offenders is to be imposed in the case where a person aged 18 or over is convicted by the court martial of an offence under section 42 and the corresponding offence under the law of England and Wales is a serious offence, and the court is of the required opinion (defined by section 223) that there is a significant risk to members of the public of serious harm occasioned by the commission by the offender of further specified offences;

20 Sch. 2(5) AFA(TP)O 2009 substitutes '21' for '18' in respect of certain provisions relating to the age when prison is available. The range of sentencing options is laid out in Part 8, Chapter 5 AFA, in particular ss 208–214. See further, under Young Offenders, on p. 577 *et seq*.

21 Sch. 15B was inserted by Sch. 18 of LASPOA 2012, brought into force on 3 December 2012 by the Legal Aid, Sentencing and Punishment of Offenders Act 2012 (Commencement No. 4 and Saving Provisions) Order 2012.

or further acts or omissions that would be specified offences if committed in England or Wales.

Service Detention

Main Provisions: sections 164, 239, 244 and 259 AFA 2006

A sentence of service detention is a punishment available to the court martial only in respect of offenders subject to service law.[22] It does not give rise to automatic discharge from HM forces. However, it is not a form of punishment that may be given to a commissioned officer. This is made clear in section 164, which also sets out that the maximum term of detention that may be given by the court martial is two years, whether for one or more offences awarded consecutively or where a suspended sentence is activated. (See section 244.) In *R v Griffin* [2018] EWCA Crim 267, a case involving sexual assault of a female officer, the CMAC held that service detention is not the same as a sentence of imprisonment. It is essentially a kind of basic training. A court martial is not obliged to follow the Sentencing Council's definitive guideline. It has its own guidance, with which the sentence was consistent. The SCC may not sentence civilians subject to service discipline to detention. (See section 282(3).)

Forfeiture of Seniority

This punishment may only be awarded to an officer. (See the discussion on page 303.)

Suspended Sentences

Imprisonment

Main Provisions: section 189 CJA 2003; sections 200–206 and Schedule 7 AFA 2006; section 6 AFA 2016

22 Civilians subject to service discipline are excluded by the Table in Part 1 to Sch. 3.

Section 200 permits the court martial to impose a suspended sentence of imprisonment. The impact of any sentence of imprisonment is usually a career ending measure.

Section 189(1) and Chapter 4 of Part 12 CJA 2003 confer on the court a power either to make a suspended sentence order of imprisonment with or without community requirements. However, there are limitations placed upon the community requirements that may be imposed. Section 202 AFA excludes a condition for mental health treatment.

Service personnel sentenced to imprisonment, whether immediate or suspended, are usually dismissed the service, unless there are exceptional circumstances.[23] Consequently, further offences committed by a service person during the operational period of any suspended sentence order of imprisonment will usually fall to be dealt with by a civilian court which deals with the subsequent matter. No power exists in the legislation for the court martial to activate a suspended sentence of imprisonment. This essentially reflects the services' general view of the unsuitability of anyone to remain serving after conviction of any offence meriting imprisonment.

At the time an offender is sentenced, a sentence of imprisonment will usually be accompanied by an order for dismissal from the service. If that does not happen, through oversight, the service to which the offender belongs will engage an administrative process to effect discharge.

Nevertheless, LASPOA 2012[24] and section 200(5) provide that a suspended sentence order made by a service court can be activated not only if the offender fails to comply with the community requirements (if any) or commits a civilian offence during the operational period of the order, but also if they commit another service offence during that period. While it is possible to impose a suspended sentence with community requirements, any breach of such requirements would fall to be dealt with by the civilian courts.

Section 200(6) AFA 2006 modifies section 189 CJA 2003 and prohibits a court from including a community requirement to be complied with outside the UK. So, if the court passes a suspended sentence of imprisonment

---

23 E.g., AGAI Vol. 2 Ch. 67, at 67/3F-4, states, 'A decision to retain an officer sentenced to imprisonment, found guilty of a drugs offence or who has failed a CDT test should not be taken by an officer below 2* rank [Major General].'

24 AFA, s. 201, provided for a suspended sentence order without community requirements. It was repealed by LASPOA 2012, Sch. 9 Part 1, para. 16.

and expects the offender to reside outside the UK, the order must be one without community requirements.

Section 202 AFA, concerning a suspended sentence order with community requirements, disapplies the following provisions of Chapter 4 of Part 12 of the 2003 Act (which would not make sense in relation to a suspended sentence order made by a service court) from applying to orders such as:

a. section 207(3)(a)(ii) (condition for mental health treatment requirement);
b. section 219(3) (requirement to give copy of order to magistrates' court).

Section 203 allows for a periodic review by a court of a suspended sentence order with community requirements by modifying section 191 of the 2003 Act. Subsection (2) modifies section 210 of the 2003 Act so that, where a service court makes a suspended sentence order which includes a drug rehabilitation requirement and provides for its periodic review, the court required to review it is the Crown court.

Subsection (3) modifies section 211 of the 2003 Act so that the Crown court reviewing a drug rehabilitation requirement imposed under a suspended sentence order can exercise its ordinary sentencing powers rather than those of the service court that made the order. Subsection (4) enables an offender resentenced by the Crown court to appeal to the civilian Court of Appeal.[25]

A breach of the community requirements of a suspended sentence or a conviction of a further offence may lead to the activation of the suspended sentence. This power is in section 206, which gives effect to Schedule 7 AFA 2006 which, in turn, modifies Part 2 of Schedule 12 to the 2003 Act.

By virtue of section 205, which modifies Part 3 of Schedule 12 to the 2003 Act, a suspended sentence order with community requirements may be amended by a Crown court in England and Wales. When there is power to resentence the offender for the original offence, the Crown court has its ordinary sentencing powers (subject to the limits on the powers of the SCC,

25 EN, paras 415–422.

if it was the SCC that made the order). Subsection (4) enables an offender resentenced by the Crown court to appeal to the civilian Court of Appeal.

Under section 246, time spent in service custody in connection with the offence in question or any related offence based on the same facts or evidence since being so charged is to count as time served by the offender as part of the sentence. However, in respect of a suspended sentence, section 246 does not apply when a suspended sentence is passed, unless the sentence is activated.

Detention

Main Provisions: sections 165, 190–195, 302 and Schedule 3A AFA 2006; section 6 AFA 2016

Section 190 AFA 2006, as modified, permits the court martial, the CMAC and the Supreme Court to suspend a sentence of *service detention* for up to 24 months (the operational period) beginning with the date of the order made (see s. 6, AFA 2016, amending s. 190 AFA 2006).

The court martial, the CMAC or the Supreme Court on an appeal from the CMAC, may impose a suspended sentence of detention, under section 190, for an operational period of at least three months and not more than twenty-four months. The exception is where the accused elected trial by court martial under section129 AFA 2006 (right to elect court martial trial) for an offence on which summary dealing was possible, when the maximum 'operational period' remains twelve months. (See s. 165 and Part 2 of Sch. 3A AFA 2006 on page 458.)

Activation

A suspended sentence of service detention can be activated by the court martial if the court martial either passed a suspended sentence of service detention on a person or, during the operational period of the suspended sentence, that person has been convicted of an offence under the law of any part of the British Islands, or the person appears or is brought before the court martial following the issue of a summons or warrant and it appears to the court martial that the offender has not been dealt with in respect of the suspended sentence (s. 191).

The court may activate a suspended sentence of service detention, imposed by a CO or by a court when a trigger offence is committed during the operational period of the suspended sentence. Section 191 of the Act explains how this is done.

Subsection (1) enables the court martial to activate a suspended sentence of service detention, including one passed at a summary hearing, if it convicts the offender of another offence committed during the operational period.

Subsection (2) further enables the court martial to activate a suspended sentence of service detention passed by that court, or the SAC[26] or the CMAC or the Supreme Court,[27] if the offender commits another service offence or a civilian offence during the operational period and is convicted of that offence. In this case the court martial may issue a summons requiring them to appear before it or a warrant for their arrest.

Where the court martial activates a suspended sentence of detention, under subsection (3) it can order that the offender must serve the entire period of the original sentence or some shorter term. Under subsection (4) it can also make the sentence consecutive to another sentence of service detention passed on the same or a previous occasion, provided that this does not take the total period of the sentences above two years (see section 244).

If the court martial activates a suspended sentence on the basis of a conviction by a CO or the SAC, and the punishment awarded by the CO or the SAC was a service supervision and punishment order or a minor punishment, section 191 (5) provides that any outstanding part of that punishment is remitted by the making of the order.[28]

Where the court martial makes an order under section 191 activating a suspended sentence of service detention, the offender may appeal against the order. The CMAC can substitute an order activating the sentence for a shorter period or can quash the court martial's order altogether.

---

26 S. 191(5).

27 S. 191(7).

28 S. 302 provides for the remission of certain sentences on passing of a custodial sentence. Subsection (2) specifies that 'relevant sentence' means (a) a sentence of service detention; (b) a service supervision and punishment order; or (c) a minor punishment.

Dismissal or Discharge from HM Forces

Main Provisions: sections 164 and 265 AFA 2006; AF(DTRF)(2)R 2009; OJAG Sentencing Guidance paragraph 3.4.11

Dismissal

The statutory power of a court martial to order dismissal is to be found in section 164 AFA 2006. Dismissal from HM forces is a serious punishment that rates immediately below imprisonment in the table to section 164. It can be ordered on its own or in conjunction with another sentence. Dismissal is either with disgrace or dismissal *simpliciter*.

Dismissal with disgrace is a sentence reserved for the most egregious conduct unbecoming a member of the Armed Forces. It will be merited where there is a serious departure from the values and standards of the service to which the offender belongs. In the case of *R v H.A*,[29] the defendant was an army Sergeant convicted of two charges of indecency with a child and three charges of indecent assault. He was sentenced to two years' imprisonment and dismissal with disgrace. Another example is found in the case of *R v Oakes*, [2007] EWCA Crim 246, involving a Corporal in the Royal Air Force who pleaded guilty to four offences of selling a firearm and five offences of theft. He asked for a further 153 offences to be taken into consideration. He was sentenced to eight years' imprisonment, to be dismissed from Her Majesty's service with disgrace and to be reduced to the ranks.

A court may not pass a sentence of dismissal or dismissal with disgrace in respect of an offence unless it is of the opinion that the offence, or the combination of the offence and one or more offences associated with it, was serious enough to warrant such a sentence. In forming any such opinion, a court must take into account all such information as is available to it about the circumstances of the offence and any associated offence, including any aggravating or mitigating factors. The court may not pass a sentence of dismissal or dismissal with disgrace unless the defendant is legally represented.[30]

The Sentencing Guidance underlines:

---

29 CMAC, 3 Apr. 2000, Case N°99/4214/S2.

30 AFA, s. 265.

> 3.2.7 Dismissal with disgrace is an exceptional form of punishment for use when the nature and circumstances of the offence make a sentence of dismissal inadequate to reflect the displeasure with which the court regards the defendant's conduct. It marks the fact that the defendant's conduct has disgraced the Service in the sense that it has been dishonoured, shamed, discredited or brought the Service into disrepute. The offence itself need not necessarily be disgraceful. It is used sparingly to avoid diluting its effect, and when the offending conduct is such that the court wishes to draw attention to its gravity. When considering whether dismissal with disgrace is appropriate the court takes into account:
>
> i. The nature of the offence;
> ii. Its surrounding circumstances;
> iii. The rank of the offender and the degree of responsibility that should therefore be expected of him; and
> iv. Whether the sentence is necessary in the interests of the Service.

QRs (Army) paragraph 5.403 and Annex 10B-2 paragraph 3 emphasise the seriousness consequences that must follow dismissal with disgrace (something all sentencing boards are very much aware of, due to their military experience):

> Any officer or member of the military forces or any former officer or former member of the military forces who has been dismissed the Service with disgrace ... shall be liable at the discretion of the Defence Council, to forfeit any war medal and clasp and any medal and clasp awarded for long service, meritorious service (except for gallantry), efficiency or any of the Queen's Medals for Champion Shots (subsequently referred to herein as the Queen's Medals) of which they may be in possession (together with any gratuity appertaining thereto, not already paid) or to which they may be entitled by their service up to the date of their conviction, and the Defence Council will consider such cases accordingly. For the purposes of this provision the expression 'other great infamy' shall include a conviction by a civil court in

respect of which that Court awarded a sentence of more than six months imprisonment.

When an officer is sentenced to dismissal from the service, their commission is forfeit with effect from the beginning of the day on which the sentence is passed. When a warrant office, NCO or soldier, etc. is dismissed the service, they are also reduced to the ranks in accordance with section 295(4). However, there is no requirement to discharge any convicted offender before they have completed their sentence (s. 295(5)).

In the case of *Blackman,*[31] where the accused was charged with the murder of a wounded Taliban fighter, his court martial convicted him of murder and, after sentencing him to life imprisonment, dismissed him with disgrace from the service.[32]

An interesting decision was handed down in the case of *R v N(S),*[33] which involved a sergeant in the Royal Military Police who was found guilty of five charges of making an indecent photograph of a child and was sentenced to 86 days' imprisonment, dismissed with disgrace and reduced to the ranks. The prosecution case was that there was never any indication that the material that the appellant viewed was other than thumbnail images, and those were images that he had simply seen, not downloaded and not deliberately saved. They had all been deleted. The effect was that he was charged with having access to those sites and accessed them deliberately knowing that they must contain indecent material involving children. In interview the appellant denied the offence and alleged others had access to his computer. He continued his denial before the court martial but was disbelieved and convicted. The CMAC allowed the appeal to the extent that it quashed the sentence of dismissal with disgrace and substituted a sentence of dismissal.

It is worthy of note that, in both the *Blackman* case and the decision in *NS*, the court did not seem to give full weight to the military context – a context well understood by the military court that sentenced in each case.

31 [2017] EWCA Crim 190, http://www.bailii.org/cgi-bin/format.cgi?doc=/ew/cases/EWCA/Crim/2017/190.html.

32 The CMAC subsequently quashed the verdict of murder, as it was unsafe, and substituted a verdict of manslaughter by diminished responsibility. An earlier hearing substituted dismissal for dismissal with disgrace.

33 [2008] EWCA Crim 774.

In the case of *Blackman*, aside from the obvious gravity of the offence, this was an act committed on military operations which severely impacted the reputation of the British Army, raising, as it did, the issue of war crimes. In relation to *NS*, of particular import was the senior rank of the individual and the fact that he was a service military policeman operating within a small community who had sought to deny the offence throughout and attribute blame to others. The employability within a small service environment of offenders convicted of cases involving reputational harm does not seem well understood. These decisions seem to be at variance with the view expressed by Holdroyde LJ in *Ashworth*, where he held:[34]

> We remind ourselves that the Court Martial is a specialist tribunal and particular respect must be given to its judgments as to the significance of the military context of an offence and as to the implications for the service, as well as for the individual offender, of imposing particular sentences.

Dismissal from HM forces, without disgrace, will be awarded where, the offence although serious, lacks that additional quality that marks it out as egregious or scandalous and one that effectively shames the service. Dismissal is usually awarded as a punishment where the offender has been sentenced to a term of imprisonment, whether immediate or suspended or to a period of detention where retention of their service is no longer required. It can be awarded on its own.

Section 295 points out other serious consequences of a sentence of dismissal. If the offender is a commissioned officer, they will forfeit their commission with effect from the beginning of the day on which the sentence is passed. If the offender is a warrant officer or non-commissioned officer, they will be reduced to the ranks from the beginning of the day on which the sentence is passed. In relation to an offender who is not a commissioned officer, the competent authority must discharge them from the regular or reserve force of which they are a member. This effectively means that forfeiture of an officer's commission or a non-officer's rank is a consequence of dismissal. However, the offender does not have to be discharged before they are released from any sentence of service detention

34 [2019] EWCA Crim 1737, at para. 23. This was *not* a case of dismissal, but is relevant to the issue of the specialist nature of the tribunal.

they are already serving or is to serve as a result of the sentence imposing dismissal.[35]

Usually, a sentence of imprisonment – even if the operation of the sentence is suspended – will carry with it a dismissal from HM forces. While a warrant officer or non-commissioned officer is in custody following a sentence of imprisonment or detention, they are for all purposes to be treated as being, if they are a member of any of Her Majesty's air forces, of the highest rank they have held in that force as an airman; otherwise, of the rank or rate of able rate, marine or soldier (as the case maybe) (s. 294])

Dismissal with disgrace carries a rehabilitation period of 10 years. For simple dismissal from the service it is seven years, and for detention five years. These periods are halved if the offender was under 18 at the time.[36]

Dismissal does carry with it financial consequences in terms of pension and gratuity. As the Sentencing Memorandum rightly points out, 'it would be wrong to sentence a high-ranking offender to a significantly lower sentence than a junior or low-ranking one for the same offence solely in order to preserve his financial advantages' (para. 3.3.1).

Discharge

Where the court has not ordered the dismissal of a convicted offender, there is nothing to prevent the relevant service from discharging them. However, the Memoranda point out that '[t]he court has no power to forbid or prevent the discharge of any offender, but in fairness the Services ought not to discharge subsequent to a decision of the court not to dismiss unless there exist other relevant factors which were not before the court.' This is on the basis that the court will have taken into account already the factors militating against discharge and reflected them in the sentence.

The Armed Forces (Discharge and Transfer to the Reserve Forces) (No. 2) Regulations 2009[37] make provision as to the discharge of members of the regular forces, and their transfer to a reserve force on completion of their service, including those who, being a warrant officer, are entitled under regulation 7 to discharge following reduction to the ranks or

35 AFA 2006, s. 295.

36 Rehabilitation of Offenders Act 1974, s. 5, Table A.

37 Both made by the SoS under the power in s. 331 AFA 2006. The text relating to discharge has been taken from the EN to the SI.

disrating under sentence of a court martial. The regulation also applies inter alia where their rank or rate has been reduced by a sentence of the CMAC (provided that it did not apply by virtue of the court martial's original sentence), and where a suspended sentence passed by the Appeal Court is activated. Where the regulation applies, the warrant officer may claim discharge within 28 days of the reduction in their rank or rate.

The competent authorities for this purpose will be found in the respective QR's or personnel administration publications.

There is also a power within the regulations to postpone a service person's discharge or transfer to the reserve, subject to the maximum periods stipulated in regulation 8.

Fine

Main Provisions: section 139 PCC(S)A 2000; sections 164, 249(5), 266, 267 AFA 2006; AF(RF)O 2013; CCA 2013, Schedule 16 Part 8; for summary dealing, see above, pages 304 *et seq*.

The court martial may impose a fine on a defendant. In the case of an offence contrary to section 42 (general criminal conduct), subsection (4) provides that any fine imposed must not exceed

(a) if the corresponding offence is a summary offence, the maximum fine that could be imposed by a magistrates' court on summary conviction;
(b) if that corresponding offence is an indictable offence, the maximum fine that could be imposed by the Crown Court on conviction.

Where a court martial imposes a combination of punishments, such as a fine and reduction in rank, it should be aware of the combined effect of the punishment. See the case of *Singh*, on page 195. The amount of any fine fixed by a court or officer in respect of a service offence must be such as, in the opinion of the court or officer, reflects the seriousness of the offence, but must take into account the circumstances of the case including, among other things, the offender's financial circumstances so far as they are known, or appear, to the court or officer (s. 249).

If the court martial imposes a fine on a person aged 18 or over, the court must make an order fixing a term of imprisonment that the person is to undergo if any sum that the person is liable to pay is not duly paid or recovered and an enforcement order is made. The Table in section 139(4) of the Powers of Criminal Courts (Sentencing) Act 2000 (PCC(S)A 2000) (maximum periods of imprisonment for default) applies for the purpose of determining the maximum periods of imprisonment that may be fixed for fines of the amounts set out in that Table.

Fining Persons under 18

In the case of service civilians under 18 convicted of an offence by the court martial or the Service Civilian Court, where the court is of the opinion that the case would best be met by the imposition of a fine or the making of a service compensation order (with or without any other punishment), section 268 of the provides an order can be made for the service parent or service guardian[38] to pay such fine or compensation. It is essential that the offender must be subject to service discipline and that the parent or guardian must be either similarly so subject or subject to service law. The court must give the parent or guardian an opportunity to be heard before making any order, unless they have failed to attend after having been given notice of a hearing.

The court may, and if the offender is under 16 when convicted must, order that the fine or compensation awarded be paid by the service parent or service guardian instead of by the offender themself unless the court is satisfied that no service parent or service guardian can be found, or that it would be unreasonable to make such an order having regard to the circumstances of the case (s. 268).

Modifications were made to subsections (3) and (4) in section 267 of the 2006 Act by the Armed Forces (Remission of Fines) Order 2013 in the case where the court martial imposes a fine on a person aged 18 or over after a term of imprisonment has been fixed for any default. The effect of these amendments is that if the court remits the whole or part of a fine after a term of imprisonment has been fixed under section 269A(b), it must reduce the term by the corresponding proportion. Any fraction of a day is to be ignored.

38 A parent or guardian is a 'service parent' or 'service guardian' for the purposes of s. 268 'if he is a person subject to Service law or a civilian subject to service discipline'.

Financial Statement Order

Section 266 AFA provides that, before it sentences a person, a service court, excluding the Summary Appeal Court, may make a financial statement order (FSO). The FSO requires the person to give to the court, within such period as may be specified in the order, such a statement of their financial circumstances as the court may require. A failure to comply with the requirement is an offence subject to a fine at level 3 on the standard scale. [39]

A service court's power to make a FSO includes power to require information about an offender's assets as well as other financial circumstances. There is a time limit applicable in those cases where an individual fails to comply with a financial statement order. Section 60 explains that no one may be charged with an offence for failure to comply with an FSO after the end of whichever of the following periods ends first

a. two years beginning with the date of commission of the offence;
b. six months beginning with the date the offence becomes known to a member of the Service Prosecuting Authority.

Anyone who, in providing any statement in pursuance of a FSO, makes a statement which they know to be false in a material particular, or recklessly provides a statement which is false in a material particular, or knowingly fails to disclose any material fact, commits an offence and is liable to a fine not exceeding level 4 on the standard scale (s. 266(4)).

Section 267 of the 2006 Act gives the court a power to remit the whole or part of a fine imposed by a service court where, after determining the defendant's financial circumstances under section 249(5) (e.g. the court has insufficient information to make a proper determination of the offender's financial circumstances) it is satisfied that, had it had the results of that inquiry when sentencing the offender, it would have fixed a smaller amount, or not have fined them in the first place.

The SCC has jurisdiction to deal with an offence under section 266 committed in respect of a financial statement order made by a court other than the SCC (s. 51(3)(b)). See also on page 519 under the section on Service

39 PCC(S)A 2000, s. 139, contains a table with the maximum periods of imprisonment or detention for any default of fine or recognizance applicable respectively to the amounts set out opposite them.

Civilian Court, for Fines, Community Orders and Overseas Community Orders.

Reduction in Rank

This is a punishment that can only be awarded to a warrant officer or NCO. Any reduction in rank will carry with it a commensurate impact on pay and pension earning, so it is not a 'soft sentence'. Where a person has acted otherwise than in the manner required of the rank held, or fails to discharge responsibilities incumbent in it, reduction in rank is a punishment which the court or officer will be bound to consider. JSP 754, Tri-Service Regulations for Pay and Charges, defines the changes in pay which apply on reduction to a lower paid rank from a higher paid rank held on a substantive or acting basis, and when seniority in rank is lost as a result of disciplinary action. It is advisable for defence advocates to task the defendant's assisting officer (DAO) with ascertaining what the likely financial impact will be, if it is anticipated that this sentence is likely. Most convictions by the court martial carry this risk. Reduction in rank can have serious career consequences through loss of pay and posting opportunities, slowing – or even effectively blocking – further promotion. With higher rank comes higher pay and higher pension and gratuity.

The court martial retains its existing power to combine custodial sentences with reduction in rank and these provisions place no restriction upon the court other than, in the case of a member of any of Her Majesty's air forces, they may not be reduced under section 164 AFA 2006 to a rank which is lower than the highest rank held in that force as an airman (i.e. ranks below corporal). See regulation 8 AF(MPLPRR)R 2009.

Section 138, as amended by section 13 AFA 2011, enables COs to combine the punishment of service detention with reduction in rank and service compensation order.

See also, Dismissal, on page 477.

Service Community Order

The SCO is dealt with below, at page 519.

Severe Reprimand or Reprimand

These are formal censures, one being more severe than the other. It is a sentencing option only available for an officer, warrant officer or non-commissioned officer. The effect of a 'rep' or 'severe rep', as they are commonly referred to, is an entry on service conduct sheets. This may, in turn, lead to certain administrative consequences, such as preventing promotion for a given period of time. In the case of the army, such a sentence passed by court martial will result in an entry on regimental conduct sheets (RCSs). Each service will determine for what period of time the reprimand remains effective. The RCS is a record of the individual soldier's conduct. Only serious offences are recorded on the RCS. In the RAF, where no regimental system obtains, an airman's conduct sheets record any offences they may have committed.

The Sentencing Guide, at page 3.9.2, qualifies this, as follows:

> At the time of writing (31 December 2017) the Services had introduced a policy that reprimands are not recorded on an individual's service or promotion record because the sentence is spent immediately after having been passed. Work is currently being undertaken to rectify this anomaly by attaching a period during which the reprimand will be active, and therefore taken into account by promotion boards. Further guidance will be issued when this is resolved.

A word of caution is still sounded, that a reprimand of either category is likely to carry career consequences, whether it be to delay promotion or affect a posting which would otherwise be career enhancing. It is a punishment which may be combined with other punishments, such as forfeiture of seniority, fine, stoppage of leave order and or a SCO . Where such a sentence is considered likely, defence advocates should request DAOs to obtain, prior to trial, relevant career forecasts setting out the likely career and financial impact.

Service Supervision and Punishment Order

Main Provisions: sections 164, 173 and 174 AFA 2009; AF(SSPO)R 2009; above page 319

This punishment by the CO (s. 132) or by court martial (s. 164) is only in respect of an able rate, marine, soldier or airman. In other words, NCOs and warrant officers and officers are excluded. Section 173 AFA defines the SSPO as an order that imposes on the offender, for a period specified in the order,[40] such requirements as regulations made by the Defence Council may prescribe. The current regulations, for this purpose, are the Armed Forces (Service Supervision and Punishment Orders) Regulations 2009; and provide that one-sixth of their gross pay for that period is forfeit. The requirements include requirements not to use any entitlement to leave without the permission of their CO and may impose on the offender one or more further requirements. See above, on page 319, for summary hearing.

The regulations make certain adjustments to the periods specified in the parent Act (which specified 90, 60 or 30 days). For example, regulation 2(c) reduces the specified period from 90 days to 21 days beginning with the day the SSPO is made. Similar adjustments are made for the lesser periods. Sixty days is reduced to 18 days, and 30 days is reduced to 14 days beginning with the day the SSPO is made. On or before the last day of the initial period, and at intervals of not more than 14 days beginning on the day after their last consideration since the initial period, regulation 5 requires that the CO must consider whether the SSPO should continue in force.

The 'the initial period' is defined in regulation 2 as follows:

> (a) in relation to an SSPO in which the period specified under section 173(1) of the Act is thirty days, means the period of fourteen days beginning with the day the SSPO is made;
> (b) in relation to an SSPO in which the period specified under section 173(1) of the Act is sixty days, means the period of eighteen days beginning with the day the SSPO is made; and
> (c) in relation to an SSPO in which the period specified under section 173(1) of the Act is ninety days, means the period of twenty-one days beginning with the day the SSPO is made.

40 This was initially set as 90, 60 or 30 days beginning with the day the order was made. However, these periods have been significantly reduced, as mentioned further in this chapter.

See the AF(SSPO)R 2009 for information on the management of the SSPO by the CO, which permits delegation to an officer of or above the rank of naval lieutenant, military or marine captain or flight lieutenant of the decision to decide what extra duties the offender shall perform, when and for how long. They may delegate to a person of or above the rank or rate of chief petty officer, marine colour sergeant, military staff sergeant or flight sergeant the power to decide the nature of the extra duties to be performed, for how long, and when. This permits the functional management of the award to be undertaken by the accused's unit NCOs.

Minor Punishments

These are punishments authorised by Defence Council regulations, namely the AF(MPLPRR)R 2009. See below, on page 615.

A Service Compensation Order

This is dealt with below, on page 521.

Court Orders Other Than Sentences

Main Provisions: section 103A SOA 2003; Chapter 7 AFA 2006; section 82 and Schedule 5 ASBCPA 2014

Service Restraining Orders

Where the court martial or SCC acquits or convicts someone subject to service law/discipline of a service offence, it may nevertheless make a restraining order under the power set out in section 229 prohibiting the defendant from doing anything described in the order. The order has effect for a fixed period specified in the order or until further order. The interpretation provisions of section 7 of the Protection from Harassment Act 1997 in relation to, for example, the meaning of 'harassing', 'conduct' and 'course of conduct' apply to service restraining orders.

The SRO may be revoked or varied on application by the Director, Service Prosecuting Authority (DSP/DSPA), the defendant or any other person mentioned in the order.

Sexual Harm Prevention Orders and Sexual Risk Orders

Main Provisions: section 103A SOA 2003; section 232A AFA 2006; section 17 AFA 2011; section 113, Schedule 5, Schedule 11 paragraph 82 ASBCPA 2014; ASBCPA2014(C8STP)O 2015; CM(A)A 1968

Repeal of Section 232A – SOPOs

The former provision relating to sexual offences prevention orders, introduced by section 17 AFA 2011, which added a new section 232A to the 2006, was repealed by Schedule 11 paragraph 82 of the Anti-social Behaviour, Crime and Policing Act 2014. The relevant provisions in sections 113 to 115 came into force on 8 March 2015. Section 113 and Schedule 5[41] of the 2014 Act introduce sexual harm prevention orders (SHPO) and sexual risk orders. The SHPO provisions in section 107A of the SOA 2003 apply to service courts, as provided for in section 137 Anti-social Behaviour, Crime and Policing Act 2014 (ASBCPA 2014) which states that 'court' includes a service court.

Effect of Repeal

The effect is to import into service courts the power found in section 103A of the amended SOA 2003 to make a SHPO in respect of an offence listed in schedules 3[42] and 5 of the Sexual Offences Act 2003 (SOA 2003)[43] of which the defendant has been found guilty (or not guilty of an offence listed in Schedule 3 or 5 by reason of insanity) and the court is satisfied that it is necessary to make a sexual harm prevention order, for the purpose of:

(i) protecting the public or any particular members of the public from sexual harm from the defendant, or

41 Sch. 5 inserts s. 103A into the SOA 2003.

42 A long list of sexual offences is set out in that Schedule, which should be consulted for the specific offences concerned. It covers the most serious sexual offences.

43 These provisions relating to sexual harm prevention orders and sexual risk orders were brought into force on 8 March 2015, by Art 2. the Anti-social Behaviour, Crime and Policing Act 2014 (Commencement No. 8, Saving and Transitional Provisions) Order 2015.

(ii) protecting children or vulnerable adults generally, or any particular children or vulnerable adults, from sexual harm from the defendant outside the United Kingdom.

A reference to proceedings for an offence under the SOA as amended includes a reference to proceedings for an offence under section 42 of the Armed Forces Act 2006 (a criminal offence) as respects which the corresponding offence under the law of England and Wales (within the meaning given by that section) is that offence (s. 137(1)(d)).

Section 137(5) and (6), as amended, provides that a service court includes a reference to the CMAC, the Supreme Court on an appeal brought from the CMAC, the court martial and a Service Civilian Court. It follows that an appeal lies to the CMAC in respect of a SHPO made by the court martial.[44] For the purposes of the Court Martial (Appeals) Act 1968, the CMAC treats the order as a sentence passed on the defendant in respect of the offence. Appeal also lies where a variation was sought and refused by the court martial.

A person within subsection (2) may apply to the appropriate court for an order varying, renewing or discharging a sexual harm prevention order.

An application under section 103E for variation, renewal or discharge of a SHPO may be made only by the defendant who at the time of the application is a person subject to service law or a civilian subject to service discipline, or by a Provost Martial, and must be made to the court martial. The requisite consent must be that of the defendant and a Provost Martial.

Where, at the time of the application the defendant is neither a person subject to service law nor a civilian subject to service discipline, the application must be made to the Crown court in England and Wales. An appeal against the making of an order under section 103E, or the refusal to make such an order, must be made to the Court of Appeal in England and Wales.

The Sentencing Guide points out that the court may impose a SHPO in the appropriate terms.[45]

---

44 See: s. 137(3)(c)(iii) SOA 2003, as amended by the ASBCPA 2014, Sch. 5 para. 7 (in force on 8 March2015 by S.I. 2015/373, Art. 2(e).

45 The Sentencing Guide, para. 5.10.4.

Recognizance by Service Parent or Guardian

Where a person aged under 18 is convicted of an offence by the court martial or the Service Civilian Court, is a civilian subject to service discipline, and has a service parent or service guardian, the court may order the parent or guardian to enter a recognizance not exceeding level 3 on the standard scale. However, where the offender is under 16 when convicted, and the court is satisfied, having regard to the circumstances of the case, that the exercise of its powers would be desirable in the interests of preventing the commission by them of further offences, it must order the service parent or service guardian to enter into recognizance (s. 233). The order must not exceed three years or, where the offender will reach the age of 18 in a period shorter than three years, for a period exceeding that shorter period.

There is a right to appeal the recognizance from the court ordering it (s. 235).

Variation Proceedings

Main Provisions: section 163 AFA 2006; rules 29A, 118–124 AF(CM)R 2009; AF(CM)(A)R 2013

Variation proceedings under rule 118 are the court martial version of the Crown court 'slip rule', statutorily provided for in the rule making provision of section 163(3)(h) of the 2006 Act. A period of 56 days[46] is allowed for any necessary correction to the sentence passed. It is generally the court that may vary or discharge the sentence.

The Manual of service law (MSL) points out[47] that where lay members participated in the sentencing or appellate proceedings at which the sentence to be varied was awarded, they should be reassembled to exercise the power. The power can only be exercised by the entire original lay board and a judge advocate or, in exceptional circumstances, by a judge advocate sitting alone. In the case of an equality of votes on whether the sentence should be varied, and if so how, the judge advocate has a casting vote.[48] When announcing the decision to vary sentence, the court must give

46 AF(CM) R 2009, r. 118(1).

47 At para. 133, JSP 830 MSL Version 2.0, p. 2-29-40. See also para. 2.25 of the Sentencing Guide, Version 3.

48 Rule 122.

reasons and explain the sentence, as required by section 252 in every case, by stating in open court the effect of the sentence.

However, in recognition of the exigencies of service life, and the fact that lay members may be called away on duty, the military justice system permits members to attend the proceedings by live link.[49] All the same, when the judge advocate is satisfied that one or more of the original lay members is unavailable and no live link attendance is possible, and that to delay proceedings until such time as the court could be assembled would mean that any sentence passed would have been served, thereby unfairly prejudicing the offender, they may make a direction under rule 120 that variation proceedings be held without lay members, thereby sitting alone.

In relation to courts requiring five lay members, rule 29 requires all members to be present for the purpose of sentencing for a Schedule 2 offence, or where the defendant falls to be sentenced to more than seven years' imprisonment or more than seven years' detention under section 209.[50] This was modified by rule 29A of the 2009 Rules by the AF(CM)(A)R 2013, which disapplies rule 29 (minimum of five lay members) in all variation proceedings relating to sentencing proceedings where the sole defendant has entered a guilty plea before the commencement of trial or, where there are two or more co-defendants, and each co-defendant has entered a guilty plea, before the commencement of trial.

Variation proceedings are not confined to application of the slip rule. Under rule 99, a judge advocate may vary or discharge a special measures direction if it appears to them to be in the interests of justice to do so, either on their own motion or on application by a party to the proceedings. An application is usually made in writing, to the CAO, although leave may be granted to hear it orally. When a written application is made, the judge advocate may grant it without a hearing.

Costs

Main Provisions: sections 26, 27 and 28 AFA 2001; section 192 AFA 2006; AF(P)(C)R 2009

49 See rr. 18 and 121 AF(CM)R 2009.

50 Offenders under 18 convicted of certain serious offences: power to detain for specified period.

Sections 26, 27 and 28 of the Armed Forces Act 2001, as amended by the AFA 2006, set out the circumstances when the court may award costs against the prosecution or defence. The Armed Forces (Proceedings) (Costs) Regulations 2009 are made under those provisions.[51] The 2009 rules revoke the Armed Forces Proceedings (Costs) Regulations 2005.

The combined provisions enable the court martial, the SAC, the CMAC and the SCC to make orders for the payment of costs which have been incurred by one of the parties to proceedings before the court as a result of an unnecessary or improper act or omission by another party (reg. 3). This is wider than the advocates in the case, as it extends to the incurring of costs 'in respect of the proceedings by one of the parties as a result of an unnecessary or improper act or omission by, or on behalf of, another party to the proceedings'. However, no order shall be made by the SCC which requires a person under the age of 18 who has been convicted of an offence to pay an amount by way of costs which exceeds the amount of any fine imposed on them by that court.

Wasted Costs

The rules also make provision in relation to the determination by those courts of the proportion of wasted costs to be disallowed or met by legal or other representatives (reg. 4). Before making a wasted costs order the court shall allow the legal or other representative and any party to the proceedings to make representations. The primary responsibility for the consideration of a wasted costs order lies with the judge. It is their responsibility to keep the question of costs in the forefront of their mind at every stage of the case and in this instance they ought to be prepared to take the initiative even without any prompting from a party. The judge should bear in mind the guidance given by the Court of Appeal in *Re A Barrister (Wasted Costs Order) (No.1 of 1991)*.[52]

The rules provide for appeals against costs orders (reg. 5) and for recovery of certain of Her Majesty's forces' costs as if they were incurred by the Director of Service Prosecutions (DSP) (reg. 6). The categories of costs are set out in Schedule 1 AF(P)(C)R 2009. The recipient of such costs incurred by the DSP is specified to be the court administration officer (reg. 7).

51 Taken from the EN to the Regulations.

52 [1993] QB 293 and Practice Memoranda 2011, paras 8.3.2 (i) and (iii).

Civilian Courts Sentencing for Service Offences

Main Provisions: sections 271(2) and 304H AFA 2006; section 12 AFA 2016; Chapter 1 of Part 12 CJA 2003; AF(CCDSO)(MCJA2003) R 2003

Seriousness

Section 271(2) AFA 2006 provides that the Secretary of State may by regulations modify any provision of Chapter 1 of Part 12 of CJA 2003 (the 2003 Act), and any other enactment that confers functions on sentencing courts, in its application to a *civilian court* dealing with an offender for a service offence. The Armed Forces (Civilian Courts Dealing with Service Offences) (Modification of the Criminal Justice Act 2003) Regulations 2009 contain the relevant provisions on seriousness.

Regulation 4 modifies section 143 of the 2003 Act (determining the seriousness of an offence) so that, when dealing with an offender for a service offence committed while the offender was on bail or charged with a service offence and released from service custody, in considering the seriousness of that offence the court must treat the fact that it was committed while the offender was charged with a service offence and released from service custody as an aggravating factor.

Crediting of Periods of Remand in Custody

Regulations 5 to 9 modified section 240 CJA 2003. This section was repealed by section 108 of LASPOA 2012 which inserted section 240ZA, entitled: 'Time remanded in custody to count as time served: terms of imprisonment and detention.' Schedule 13 to LASPOA also modified section 246 AFA 2006.[53] Subject to what follows, the position is that the number of days for which the offender was kept in service custody in connection with the offence in question or any related offence since being so charged is to count as time served by the offender as part of the sentence. However, if, on any day on which the offender was kept in service custody, the offender was also detained in connection with any other matter, that day is not to count as time served. A day counts as time served only in relation to only one

53 Crediting of time in service custody: terms of imprisonment and detention. See also, on p. 450.

sentence, and only once in relation to that sentence. A day is not to count as time served as part of any period of 28 days served by the offender before automatic release (see s. 255B(1) of the 2003 Act).

15 CIVILIANS IN THE DISCIPLINARY SYSTEM

Jurisdiction in the UK

Main Provisions: section 51 and Part 11 AFA 2006; section 18 AFA 2011

Civilians subject to service discipline (service civilians) may be tried by a service court for offences committed under the Act. Under section 51, the Service Civilian Court (SCC) has jurisdiction to try any service offence committed outside the British Islands by a civilian, except an offence within subsection (3)[1] or an offence in relation to which subsection (6) applies (namely, an offence if the defendant is for the time being a member of the regular or reserve forces; or liable to recall).

Since section 18 of the AFA 2011[2] modified section 277(2) of the 2006 Act, the SCC may sit in any place, whether within or outside the United Kingdom. This means it could sit within UK to deal with an offence committed overseas by a service civilian.[3] Its constitution, proceedings and powers in respect of finding and sentence are set out in Part 11 of the Act. Every defendant has the right to elect trial by court martial, and to appeal to the court martial against a finding or sentence of the SCC.

---

1 Indictable only offences: an offence under s, 266 committed in respect of a financial statement order (requiring the person to give to the court, within such period as may be specified in the order, such a statement of their financial circumstances) made by a court other than the Service Civilian Court; any service offence under regulations under ss 328 (enlistment) or 343 (service inquiries) an offence within section 50(2)(h) (An offence under any of ss 95 to 97 of the Reserve Forces Act 1996, namely, offences against orders and regulations; failure to attend for service on call out or recall; failure to attend for duty or training; or (i) Reserve Forces Act offences – Giving a false answer during enlistment in a reserve force.

2 In force from 2 April 2012: Armed Forces Act 2011 (Commencement No. 1, Transitional and Transitory Provisions) Order 2012.

3 The first such case was heard at MCS Catterick. A civilian wife of a Cpl was tried on a s. 18 wounding charge (Offences Against The Person Act 1861). She was acquitted of the s. 18 charge, but convicted of malicious wounding under s. 20.

Jurisdiction Overseas

Main Provisions: sections 51 and 58 AFA 2006; Schedule 15 AFA 2006 (Appendix J); AF(CSSD)O 2009; Article VII(4) NATO SOFA.

While the SCC has jurisdiction to try any service offence committed outside the British Islands by a civilian, there are exceptions in relation to indictable offences under section 42, any service offence under regulations under section 328 (concerning enlistment) or section 343 (service inquiries), or an offence within section 50(2)(h) or (i) (Reserve Forces Act offences). There is no jurisdiction in relation to ex-service personnel who commit offences at a time when they are liable to recall for service or when they have been served with a recall notice. Section 58 AFA 2006[4] sets out the time limit for charging a civilian formerly subject to service discipline. See on page 577 below for powers of the SCC to try young offenders (under 18) charged with certain indictable offences.

Certain persons are, for the purposes of the AFA 2006, designated as 'civilians subject to Service discipline'[5] which term refers to a civilian who

a. is not subject to service law; and
b. is within any paragraph of Part 1 of Schedule 15, which lists the categories.[6]

Part 2 of Schedule 15[7] excludes a person who is not a United Kingdom national when they are in a country

a. of which they are a national; or
b. in which they are ordinarily resident.

As noted above, on page 35, this excludes spouses who are nationals of the country in which they reside with the service spouse. They remain subject to their own domestic law.

A 'designated area' is defined in paragraph 12 of Schedule 15 (see Appendix J) as an area which is outside the British Islands and designated

4 As amended by Sch. 4(5) AFA 2011, in force from 14 December 2012 by Art. 3 AFA 2011(CO2) O 2012.
5 Herein referred to also as 'service civilians'.
6 See also above, on p. 51 in relation to appointment of CO.
7 See Appendix J, Civilians Subject to Service Discipline, where Sch. 15 is set out.

for the purposes of the Schedule by an order made by the Secretary of State. A designated area may consist of two or more areas (whether or not contiguous, e.g. Germany and Belgium). The current designated areas are to be found in Schedule 3 AF(CSSD)O 2009. Part 1 to Schedule 3 designates the areas for the purposes of paragraphs 4, 6 and 9 of Schedule 15[8] to the Act. Part 2 designates the areas for the purposes of paragraph 8 of Schedule 15 to the Act, which relates to persons residing or staying with a person subject to service law in a designated area. Visiting family and friends would fall into this category.

These categories and the designated areas applicable are detailed in Schedule 15 to the Act, as modified by the AFA 2011[9] and the Armed Forces (Civilians Subject to Service Discipline) Order 2009. These include the families of Armed Forces personnel living with them abroad and certain contractors working with the Armed Forces, even though actually employed at other military headquarters, such as NATO. Although rapidly receding in importance and significance, in view of the withdrawal of British Forces from many overseas bases, most notably those in Germany, it remains important that those civilians accompanying the force are dealt with under UK domestic law, rather than the law of the host nation.

The Categories

Schedule 15 has been modified by section 22 AFA 2011, as explained below. Paragraph 4 of Schedule 15, as amended, provides that a Crown servant who works solely or mainly in support of any of the Armed Forces in a designated[10] area is only a civilian subject to service discipline in two circumstances:

a. if he is in the designated area in which he usually works; or
b. if he is in a designated area but came there wholly or partly to work in support of the armed forces.

8 Crown servants in a designated area working in support of Her Majesty's forces, persons in a designated area who are members or employees of other specified organisations, and persons residing or staying with a person falling within para. 4 or 6 in a designated area.
9 See, Armed Forces Act 2011 (Commencement No. 2) Order 2012.
10 Areas are currently designated for the purposes of each of a number of paragraphs of Sch. 15 by the Armed Forces (Civilians Subject to Service Discipline) Order 2009 (S.I. 2009/836).

Paragraph 5, as modified, applies to a person whenever employed in a specified naval, military or air force organisation by reason of the UK's membership of the organisation and they are either

(i) in a country outside the British Islands, and his normal place of work under that employment is in that country; or
(ii) he is in a country outside the British Islands and he came there wholly or partly for the purposes of his work under that employment.

The only organisation currently specified by the Armed Forces (Civilians Subject to Service Discipline) Order 2009 (AF(CSSD)O 2009) is NATO. The amended paragraph 5 parallels the amendment to paragraph 4. It limits paragraph 5, so that the employee is only a subject to service discipline in the two specified circumstances.

Paragraph 6 applies to any member or employee of specified organisations whenever they are in a designated area and their normal place of work for that organisation is in that designated area, or who does not usually work there but is in that designated area wholly or partly for the purposes of work for that organisation. The organisations currently specified include the Navy, Army and Air Force Institutes and the Soldiers, Sailors, Airmen and Families Association (Forces Help). The same areas are designated as for paragraph 4 (Crown servants).

Paragraph 7 refers to persons designated by or on behalf of the Defence Council only if it appears to the Defence Council or the authorised officer that it is desirable to do so because it is

a. in the interests of the person;
b. for the protection of other persons (whether or not members of any of Her Majesty's forces); or
c. for the purpose of maintaining good order and discipline.

This would cover visiting legal defence representatives to service personnel charged with a service offence.

Paragraph 8 applies to persons residing or staying with a person subject to service law in a designated area, while they are in that area. So, if a service family employ a nanny for their children, who lives with them, they will be subject to service law. Similarly, if the parents, family members or friends of a service person stay with the service family, they too are subject to service discipline. Paragraph 13 importantly points out that references to a person residing or staying with another person include references to the person being about to reside or stay with the other person, and departing after residing or staying with them. This also covers transit arrangements while in the designated area. For example, parents visiting their son, a serviceman and his wife, in Paderborn, Germany, will fall under paragraphs 8 and 13 from the moment they arrive at the German airport or their car or train crosses the international border and enters Germany.

Paragraph 9 applies to persons residing or staying with a civilian of the kind covered by paragraph 4 or 6 in a designated area, while they are in that area.

Paragraph 10 of Schedule 15, as amended,[11] applies to a person who stays with someone within paragraph 5 (an employee of a specified military organisation) in two circumstances. One is if the person stays with the employee in the country or territory in which the employee normally works. The other is if they stay with the employee in a country or territory to which the employee came for the purpose of their work (EN, paras 87–92). The amendment in section 22 AFA 2011 introduces the terms 'qualifying person', defined as a person within paragraph 5 or someone who would fall within paragraph 5 were it not for the fact they are a national of, or ordinarily resident in that relevant country. 'Relevant country' is defined as being either outside the British Isles where that person's normal place of work under that employment is situated, or a country to which that person came wholly or partly for the purposes of their work under that employment.

The rule governing admissibility of evidence of enlistment is to be found in rule 68 Armed Forces (Service Civilian Court) Rules 2009 (AF(SCC) R 2009). Rule 69 refers to admissibility of evidence as to service. Entries in service records are admissible in accordance with rule 70 AF(SCC)R 2009.

In relation to the admissibility of standing or routine orders, rule 72 AF(SCC)R 2009 provides that a certificate purporting to be signed by a person's CO or an officer authorised by the CO, shall be evidence of the matters stated in the certificate.

---

11 By AFA 2011, s. 22.

A similar provision is to be found in rule 71 AF(SCC)R 2009 concerning Defence Council instructions, regulations and certificates relating to decorations, badges or emblems. Any certificate given in the prescribed manner shall be evidence of the matters stated in it.

Jurisdiction in Germany

See the section on NATO Jurisdiction above, on page 30 *et seq*.

Trial of Civilians by Court Martial

Main Provisions: sections 154–157 AFA 2006; Part 4 AF(CM)R 2009

More detailed information about trial by court martial is to be found on Chapter 13. Most service civilians charged with an offence will be tried by the SCC. Only the more serious offences will be tried by court martial. See the sentences available below, on page 503 *et seq*. The practical effect of this section, section 279 and Part 5 is that the SCC will try a charge only where the DSP, the court and the defendant are content that it should do so (EN).

A person is a 'civilian offender' for the purposes of Part 1 of Schedule 3 if they have committed a service offence and

(a) was a civilian when (s)he committed the offence;
(b) is a civilian when sentenced for the offence; and
(c) remained a civilian throughout the period between committing the offence and being sentenced.

A civilian is defined as a person who is

(a) not a member of the regular forces;
(b) not a member of the reserve forces; and
(c) not liable to recall.

Aside from changes to the powers of sentencing, the court martial is constituted slightly differently and the provisions of rule 33 AF(CM)R 2009 apply. As mentioned above, on page 409, this provides that for the trial of a civilian by court martial, each of the lay members must be either,

a. a person not subject to Service law (i.e. a civilian member of the force) who is qualified for membership[12] and not ineligible by virtue of rule 32,[13] or
b. an officer or warrant officer who would be qualified for membership under section 156,[14] and not ineligible by virtue of section 157[15] or rule 32,[16] if this rule did not apply.

There is no need to comply with section 155(3) which, in the case of a person subject to service law, stipulates that a prescribed number of the lay members must be officers or warrant officers. The requirement in rule 33 that the member 'must not be subject to Service law'[17] means that they must not be members of any of the services. As will be recalled, civilians 'subject to Service discipline' fall within the jurisdiction of the military justice system. See Appendix J for civilians 'subject to Service discipline'.

The trial procedure is the same as for trial by court martial. See Appendix H. Service of documents, witness summonsing, listing, etc. will follow the same procedures as set out in the AF(CM)R 2009.

The trial of civilians by military courts remains a contentious issue, which was raised in the case of *Martin v United Kingdom* and subsequently in *Mustafa v Bulgaria* (see above, page 27 *et seq.*). The ECtHR is prepared to countenance such trials, but only if there are 'compelling reasons justifying such a situation', which must be demonstrated for each case.

12 See r. 33(2). This essentially requires members to be over 18, a UK national, not suffering a mental disorder, not disqualified for jury service, not a member of the Military Court Service or of the Service Prosecuting Authority

13 E.g., where they belong to the same unit as the defendant, or were not a member of the court that sentenced the defendant, etc.

14 That is, an officer who has held a commission for at least three years, or for periods amounting in the aggregate to at least three years, or immediately before receiving their commission, they were a warrant officer in any of the forces. A list of criteria disqualifying membership is set out in s. 156(4) AFA 2006.

15 'He (a) was the commanding officer of the defendant at any time from commission of the offence to arraignment, or (b) has taken part in an investigation into the subject matter of the charge, or (c) has conducted an inquiry into the subject matter of any charge against the defendant. In the case of a Warrant Officer, (b) and (c) apply.'

16 This sets out the persons ineligible for membership in particular circumstances, e.g. where they worked together in the same unit.

17 The importance of the distinction is adverted to on p. 35.

Sentence of Civilians Tried by Court-Martial

Main Provisions: section 164 and Schedule 3 AFA 2006

The general provisions of section 164 of the Act apply, as modified by Schedule 3. The Schedule has effect as if subsections (4), (6) and (7) of section 164 were omitted and substituted with the provisions of the table contained in the Schedule. The table is applicable to those civilians who have been tried by court martial for a section 42 offence and found guilty. The following punishments are available.

Civilian Offenders

Section 259 AFA 2006 requires a service court to have regard to any guidelines that are relevant to the offender's case and, in exercising any other function relating to the sentencing of offenders for service offences, have regard to any guidelines which are relevant to the exercise of the function. Nevertheless, it may depart from those guidelines if, in its opinion, the departure is justified by any features of service life or of the service disciplinary system that are relevant to the case.

Election for Trial

As is the case of an election from magistrates' court to the Crown court, whenever the offender elects trial by court martial, sentencing is whatever the court considers appropriate for the offence, within the law.

Ex-Service Personnel

Where the civilian offender is an ex-service person who is not subject to service law when sentenced, is not a member of a volunteer reserve force or an ex-regular reserve force who is subject to an additional duties commitment, and is not a civilian offender as defined above, the following punishments can be awarded under Part 2 of Schedule 3 AFA 2006.

For the purposes of determining the court martial's powers when sentencing an ex-service person, references in the following provisions to punishments under section 164 are modified as follows:

(a) section 25(2) (penalty for misapplying public property, etc.) has effect as if the reference to rows 2 to 12 were to rows 2 to 10 above;

Row #	Punishment	Limitation
1.	Imprisonment	Any sentence of imprisonment imposed in respect of an offence under this section must not exceed— (a) if the corresponding offence under the law of England and Wales is a summary offence, the maximum term of imprisonment that could be imposed by a magistrates' court on summary conviction; (b) if that corresponding offence is an indictable offence, the maximum sentence of imprisonment that could be imposed by the Crown Court on conviction on indictment. (Limitations in respect of young offenders: detention in a Young Offender Institution for offenders aged over 18 and under 21, and detention for serious offences for offenders under 18). [s. 42(4)]
2.	A fine	Any fine imposed in respect of an offence under this section must not exceed— (a) if the corresponding offence is a summary offence, the maximum fine that could be imposed by a magistrates' court on summary conviction; (b) if that corresponding offence is an indictable offence, the maximum fine that could be imposed by the Crown Court on conviction on indictment. [s. 42(4)]
3.	A service community order (defined by s. 178)	Only if s. 164(5) permits, namely (a) the offender is aged 18 or over when convicted; and (b) it appears to the court that they will reside in the United Kingdom when the order is in force
4.	An overseas community order (defined by s. 182)	Only if it appears to the court that the offender will reside outside the United Kingdom when the order is in force.
5.	A conditional discharge (defined by s. 185)	
6.	An absolute discharge (defined by s. 185)	
7.	A service compensation order (defined by s. 175)	

Table 15.1 Sentence of Civilians Tried by Court Martial

Row #	Punishment	Limitation
1.	Imprisonment	Only in respect of persons 18 and over. Maximum of 12 months in respect of any one offence or where two or more terms of imprisonment to run consecutively their aggregate must not exceed 65 weeks (s. 183).
2.	Dismissal with disgrace from Her Majesty's service	Only if the person being sentenced is an officer or a member of the reserve forces
3.	Dismissal from Her Majesty's service	Only if the person being sentenced is an officer or a member of the reserve forces
4.	Detention for a term not exceeding two years	Not if the person being sentenced is an officer, or was an officer when last a member of Her Majesty's forces
5.	Reduction in rank, or disrating	Only if the person being sentenced is a warrant officer or non-commissioned officer, and not to an extent prohibited by regulations under subsection (4)
6.	Fine	
7.	A service community order (defined by s. 178)	Only if s. 164(5) permits, namely (a) the offender is aged 18 or over when convicted; and (b) it appears to the court that he will reside in the United Kingdom when the order is in force
8.	A severe reprimand or a reprimand	Only if the person being sentenced is, or was at the time of the offence, an officer, warrant officer or non-commissioned officer
9.	An absolute discharge (defined by s. 185)	
10.	A service compensation order (defined by s. 175)	Being an order that requires the offender to pay compensation for any personal injury, loss or damage resulting from the offence they have been convicted of or which was taken into consideration).

Table 15.2 Ex-Service Personnel Punishments

(b) section 35(3) (penalty for annoyance by flying) has effect as if the reference to rows 3 to 12 were to rows 3 to 10;
(c) section 42(3) (penalties for criminal conduct offences) has effect as if the reference to rows 5 to 12 were to rows 5 to 10;
(d) any provision of regulations under section 328 (enlistment) which provides that an offence is punishable by any punishment mentioned in rows 2 to 12 of the Table in section 164 has effect as if the reference to rows 2 to 12 were to rows 2 to 10;
(e) any provision of regulations under section 343 (service inquiries) which provides that an offence is punishable by any punishment mentioned in rows 7 to 12 of the Table in section 164 has effect as if the reference to rows 7 to 12 were to rows 6 to 10.

Service Civilian Court

Main Provisions: section 51, Part 11 and Schedule 10 AFA 2006; AF(SCC)R 2009; section 18 AFA 2011

Jurisdiction

Jurisdiction is dealt with above, on page 497.

General

The SCC is the service version of the magistrates' court, which deals with offences committed by persons subject to service discipline. The procedure before it should be familiar to anyone who has appeared in the magistrates' court, with the signal difference that the SCC is presided over by a judge advocate sitting alone.

Mode of Address

The, judge is addressed as 'sir' or 'madam'. Sittings of the court are to be in open court. Advocates are not robed.

Mode of Trial

A requirement arises under section 279 of the Act and rule 42 AF(SCC)R 2009 where the judge advocate must determine if the interests of justice require that the charge is tried by the court martial. The judge advocate must also give the defendant the opportunity to elect trial by court martial. If the defendant does elect trial, they may, at any time before arraignment by the court martial, give notice that they withdraws their election. If this is done after the matter has been adjourned, it must be in writing and shall be served on:

a. the court administration officer;
b. the Director SPA;
c. any other defendant.

In the event that there is any issue arising as to whether the defendant is fit to stand trial, or where it appears that the defendant did the act or made the omission charged against them as the offence, whether they were insane at the time of the act or omission, the court must refer the charge to the court martial for trial by that court.

The provisions applicable to vulnerable witnesses are those discussed above, on page 420 and in Appendix O. The substantive provisions are in Chapter 6 of the AF(SCC)R 2009.

The ability of the SCC to deal with circumstances not provided for, under rule 25 AF(SCC)R 2009, are set out above on page 428.

Appointment of CO

Main Provisions: Schedule 15 AFA 2006; AF(MCO)R 2009

See page 50 above for the position under the Armed Forces (Meaning of 'Commanding Officer') Regulations 2009. While a CO will be appointed for a civilian, for the purposes of the disciplinary process, they have no power to deal summarily with a service civilian as the CO's powers relate to persons subject to service law (s. 52(4)(a) and ss. 367–369). The service police will therefore refer all cases against service civilians directly to the SPA for advice and charging. It is clear that a CO has no initial powers under section 120, in the case of a civilian and that it is for the DSPA to charge (see s. 121A).

Statutory Powers of the Director

There are clear statutory obligations placed on the DSPA when a case is allocated for trial before the SCC. These powers are set out in section 126 and mirror the powers and obligations placed upon the Director in relation to trial by court martial, set out in the preceding section (see page 508 *et seq*).

Duties of the Court Administration Officer

The Court Administration Officer (CAO) will appoint the time and place when proceedings in the SCC shall commence. Once those proceedings have commenced, it is a matter for the court to direct such times and periods each day as it may sit (r. 16). The qualification to this is where an offender has been convicted in trial proceedings, the sentencing proceedings in respect of them shall commence immediately after the conclusion of the trial proceedings, unless the judge advocate *for those proceedings* appoints some later time.

The CAO must serve notice of any time and place appointed by them for the commencement or resumption of any proceedings on each person to whom the proceedings relate. In addition, such notice should be served on any legal representative of each such person, the CO of each such person, the Director and any such other person as the Judge Advocate General may direct.

Disclosure by the Director

The regime applicable to the Director in the case of the SCC is the same as for court martial. See above, on page 360, as the Criminal Procedure and Investigations Act 1996 (Application to Armed Forces) Order 2009 (CPIA1996(AAF)O 2009) applies to all service courts, including an appeal from the CO to the SAC.[18]

Time Limits

Main Provisions: section 58 and Schedule 15 AFA 2006; section 22 and Schedule 4 AFA 2011

18 See Criminal Procedure and Investigations Act 1996 (Application to the Armed Forces) Order 2009, Article 2

For offences charging criminal conduct, under section 42 of the Act, the trial must start within the time limit applicable to that offence under English domestic law.

An interesting point was raised in the 1998 unreported case of *Colohan*, concerning a civilian tried before the Standing Civilian Court on a charge of drink driving. The defence submitted that the prosecution was time-barred under section 127 of the Magistrates' Court Act 1980 (laying of information within 6 months). The Magistrate (Mr Berkson, DJAG) held, as a matter of law, that section 127 of the Magistrates' Court Act did not apply to the SCC. In the subsequent case of *Buchan*,[19] the CMAC decided that section 127(1) of the Magistrates Courts Act 1980 did not apply to courts martial.[20] The appellant, a serviceman subject to military law at all material times, had been tried and convicted by court martial in Germany of driving without due care and attention, contrary to section 3 of the Road Traffic Act 1988 and had argued that the charge had been laid out of time and that the court martial accordingly had no jurisdiction to deal with the matter as no 'information' had been laid within the six-month time limit imposed for summary offences. The court held:

> The first and most fundamental [problem confronting the appellant] is that section 127 simply does not apply to Courts Martial.

That decision was followed in the case of *McCosh* [2011] EWCA Crim 1466, when leave to appeal a conviction for driving with excess alcohol on this point was refused. Both *McCosh* and *Buchan* were cited with approval in the case of *R v S* (see further on page 394).

19 [2007] EWCA Crim 716 CMAC. He was charged under s. 70 of the Army Act 1955. The court went on to say, 'under section 70 of the Army Act and there is no time limit directly applicable to an offence under section 70, save for such time limit as can be gleaned from section 7 of the Armed Forces Act 1986 and that does not relate to any time limit which could affect this charge.' (AFA 1986, s. 7 removed the three-year time limit for prosecution). There is no reason to think that these cases would be decided differently today.

20 'Except as otherwise expressly provided by any enactment and subject to subsection (2) below, a Magistrates' Court shall not try an information or hear a complaint unless the information was laid or the complaint made, within six months from the time when the offence was committed, or the matter of complaint arose.'

Ex-Service Personnel or Service Civilians

For any offender (including a former regular) who has ceased to be subject to service discipline, a charge must be issued within six months from the date on which the offender ceased to be subject to service discipline,[21] unless consent is forthcoming from the Attorney General.[22]

There are two exceptions. First, if the person became subject to service law[23] at the same time as ceasing to be a civilian subject to service discipline (e.g. they enlist overseas), the six-month period does not begin to run until they cease to be subject to service law (when s. 57 applies instead).[24] Second, under Schedule 15, certain civilians are subject to service discipline only while in certain designated areas or while in any area outside the British Islands. If they cease to be a civilian subject to service discipline only because they left such an area but are still residing or staying in that area (e.g. they went on holiday), the six-month period does not begin to run.[25]

Service of Documents

Main Provisions: Part 2, rules 4–14AF(SCC)R 2009

The rules and procedure for service of documents cover service

a. on a person to whom proceedings relate
b. the court administration officer
c. other individuals
d. on a corporation
e. by another method
f. by CO
g. by fax or electronic means.

---

21 See AFA 2006, s. 58.

22 *Ibid.*, s. 61(2).

23 A person is 'subject to service law' if they fall within ss 367–369 of the Act and it essentially means every member of the regular forces, every member of the reserve forces on call-out or training or serving on the permanent staff of a regular force, reserve forces and Members of British overseas territories' forces serving with UK forces.

24 This sets out the time limit for charging person formerly subject to service law, who may not, after the end of six months beginning with the date they ceased to be subject to service law, be charged with a service offence committed while they was so subject.

25 EN, para. 147. See AFA 2011, Sch. 4(5), amending.

Service of any document under the SCC Rules on a person to whom proceedings relate may be effected in any of a number of ways:

a. on them personally;
b. by leaving it at their usual or last known place of abode; or
c. by post in a letter addressed to their usual or last known place of abode;
d. by post in a letter addressed to their legal representative's place of business; or
e. by DX, fax, electronic mail or other electronic means to their legal representative, where their legal representative
f. has given a DX box number, fax number or electronic mail or other electronic means address; and
g. has not refused to accept service by that means.

References to a person's legal representative mean any person whose name and address has been given to the CAO, under rule 26(4).

Service on the Director

The Director may be served by post, DX, fax, electronic mail or other electronic means to either the principal office of the Service Prosecuting Authority or on a prosecuting officer personally, with his or her agreement.

Service on the CAO

The CAO may be served by post, DX, fax, electronic mail or other electronic means to any office of the Military Court Service, or personally on any member of that service, with their agreement.

Presumptions as to the Date of Service

Presumptions as to the date of service of a document on a person (otherwise than personally) are in rule 13, which states that service by DX is on the fifth day after the day on which it was left at the addressee's DX box number or despatched.

Where a document is served by fax, electronic mail or other electronic means, the person serving it need not provide a paper copy as well. Where a document is sent by fax, electronic mail or other electronic means, service is assumed to occur on the day after it was transmitted; and in any other case, on the day on which the addressee responds to it if that is earlier.[26] Proof of service is by certificate of service signed by the person serving it.[27]

Where a document to be served on a person is sent or delivered to their CO, the CO must arrange for the document to be served on them personally as soon as is reasonably practicable. The date of service within the UK is the day on which the document was despatched. In respect of a document sent by post from the UK or Germany to an address within Germany, or from Germany to an address within the United Kingdom, the date of service is on the tenth day after the day on which it was despatched.

In the case of any other document sent by post, date of service is on the tenth day after the day on which it was despatched.

Young Persons

Where proceedings relate to a 'young person' under the age of 18 years at the commencement of the proceedings, any document required by the SCC Rules to be served on such 'young person' must also be served on the young person's parent or guardian. Where the young person has not appointed a legal representative, any right conferred on a party to proceedings by the Rules may be exercised, and any duty imposed on them by these Rules (except pleading to a charge) discharged, by their parent or guardian on their behalf. The court may in fact give leave for the parent or guardian to represent the 'young person' in any proceedings (r. 27).

For other methods of service, such as on corporations or other individuals, see SAC Rules, rules 8–10.

Witness Summonsing

In the case of service witnesses, the defence will notify the CAO who, in turn, will notify the individual, with the place, date and time they are required to attend. This will generally entail also notifying the unit to which the witness belongs. See rule 48 of the SCC Rules.

Non-service witnesses are required to attend by notice from the CAO. However, where there is the concern that a material witness is unwilling to attend, application can be made to the judge for the issue of a witness summons, under rule 49. Provided the judge advocate is satisfied, they shall

26 SCC Rules, r. 13.
27 *Ibid.*, r. 14.

issue a witness summons directed to the person concerned and requiring them to attend and give the evidence or produce the document or thing set out in the summons. The application for a witness summons must be in writing and must be served on the CAO and as directed by the judge advocate. The witness summons issued under this rule must state that failure to comply with the summons may result in the issue of a warrant for the arrest of the person to whom the summons is addressed.[28] Nothing prevents the judge advocate from issuing a witness summons of their own motion should they consider it necessary.[29]

Where there are concerns about the confidentiality of a document required to be produced, rule 50 empowers the judge advocate to make an assessment of its relevance and confidentiality.

A witness summons can be withdrawn on application to the judge advocate under rule 52 by the party who applied for it, on the ground that it is no longer needed, or by the witness himself on the grounds that:

(a) he was not aware of any application for it; and
(b) either
    (i) he cannot give or produce evidence likely to be material evidence; or
    (ii) even if he can, his duties or rights (including rights of confidentiality) or those of any person to whom the evidence relates outweigh the reasons for the issue of the witness summons.

The jurisdictional factors of the NATO SOFA in respect of witnesses for SCC courts in Germany should be noted. This is explored on page 23 *et seq*.

Interpreter

The AF(SCC)R 2009 have been amended by the Armed Forces (Service Courts Rules) (Amendment) Rules 2015 (AF(SCR)(A)R 2015), which insert a new rule 21A into the 2009 rules, concerning the provision of interpretation and translation for persons other than a person to whom proceedings relate, that is the complainant and witnesses.

28 SCC Rules, r. 49(10).
29 *Ibid.*, r. 51.

Procedure at Court

Main Provisions: sections 278, 279 and Schedule 10 AFA 2006; Part 13 of the SCC Rules

The proceedings in the SCC take place before a single judge advocate. There is no supporting cast of lay members. The judge advocate for any proceedings is to be specified by or on behalf of the Judge Advocate General.

Before arraignment, the SCC must decide whether the charge should be tried by that court or the court martial. Before making its decision, it must give the Director an opportunity to inform the court of the defendant's previous convictions (if any) and permit both the Director and the defendant an opportunity to make representations as to whether the SCC or the court martial should try the charge. In coming to its decision on the appropriate forum, the court will consider, inter alia:

a. the nature of the case;
b. the seriousness of the offence;
c. whether its powers of punishment in respect of the offence would be adequate;
d. any other circumstances it considers to be relevant; and
e. any representations made by the Director and the defendant.

It must still give the defendant, before arraignment, the opportunity of electing court martial trial of the charge.[30] A defendant who has indicated that they wish to elect trial by court martial may, at any time before *arraignment by the court martial,* give notice they withdraw their election (r. 42(6)).

However, an election for court martial trial by a defendant or (if more than one person is jointly charged) any of the defendants under section 280 on one charge will mean all charges are to be tried by court martial.[31] Similarly, where the SCC is to try together two or more charges against the defendant, an election for court martial trial in respect of any of the charges takes effect as an election in respect of all of them.

30 See s. 280 and r. 42 of The Armed Forces (Service Civilian Court) Rules 2009.
31 AFA 2006, s. 280(5).

If the court itself decides that the charge should be tried by court martial then it must refer the charge to that court and the charge is to be regarded as allocated for court martial trial.

If the court decides, under section 279 to hear the case itself rather than send it for trial by court martial, it must be nevertheless remembered that where a charge is regarded as allocated for SCC trial, the DSP may exercise any of the powers under section 126(2), to:

(a) amend the charge
(b) substitute for the charge another charge
(c) bring an additional charge
(d) discontinue proceedings on the charge;
(e) allocate the charge for trial by the Court Martial;
(f) make a direction under section 127(1) or (2) (directions barring further proceedings) in respect of the offence charged or any offence that could be charged under paragraph (c) above.

If the DSP makes any amendment, substitution and addition of charges before the arraignment of an accused, a charge sheet must be prepared in accordance with the Schedule; the DSP must sign the charge sheet and the DSP must ensure that a copy of the signed charge sheet is served by hand on the accused. As soon as practicable after service of the charge sheet on the accused, the DSP must provide a copy of the charge sheet to the CAO. Where the earliest time at which it will be practicable to provide a copy of the charge sheet to the CAO is less than 24 hours before the time appointed for a hearing in relation to the accused, the DSP must also provide as soon as practicable a copy of the charge sheet to the judge advocate appointed to conduct the hearing. (r. 14 AF(P5AFA2006)R 2009). This provision applies to both the SCC and the court martial.

Where proceedings are discontinued under section 127, it is treated for the purposes of section 63 as a direction that the person is to be treated as acquitted of the service offence so specified.

If the defendant (or if more than one person is jointly charged, any of the defendants) does not elect trial by court martial, the court may arraign the defendant. If the case is to be dealt in the SCC, the arraignment of the defendant then takes place.

The rules of evidence applicable are governed by the SCC Rules 2009. Rule 58 stipulates that insofar as they are capable of doing so, the rules of evidence applicable in a trial on indictment in England and Wales shall apply, subject to any modifications by any other statute or subordinate legislation.

The Crown will open its case, call its witnesses who may be cross-examined by the defence. At the close of the Crown's case, the defence may make such submissions as appropriate, for example that there is no case to answer. If the judge rules there is a case to answer, the defence proceeds to call the defendant and any witnesses. Closing addresses are made in the usual order, whereafter the judge gives their decision. A separate decision is given for each offence charged.[32] Under section 281 of the Act, it is also open to the court to convict of an offence other than that charged in circumstances where the SCC acquits a person of an offence specifically charged in the charge sheet, but the allegations in the charge sheet amount to or include (expressly or by implication) an allegation of another service offence. This provision applies section 161 (power of court martial to convict of offence other than that charged) to the SCC as it applies in relation to the court martial. See above, page 434.

Joinder, Severance and Amendment

Main Provisions: rule 36 and Part 8 SCC Rules 2009

These provisions deal with both joinder/severance of charges and of defendants. They broadly follow civilian practice. The issue should be raised at preliminary proceedings to enable the court to give such directions as appear to it to be necessary to secure the proper and efficient management of the case. The court may make an order or ruling on any question as to the joinder or severance of charges and any other question of law, practice or procedure relating to the case. It is important to understand that, any direction given in preliminary proceedings shall have effect throughout any related proceedings unless varied or discharged by the judge advocate who gave it or the judge advocate for any related proceedings (r. 36). The order or ruling made in preliminary proceedings remains binding unless subsequently varied or discharged by the court. This caters for an

32 It applies this power under s. 161, relating to court martial, to the SCC.

application where circumstances change and counsel feels that it is proper for an application to be made to vary the order made at a preliminary hearing.

Where a defendant is charged with more than one offence in the same charge sheet, the court may direct, *before* the commencement of trial proceedings in relation to the charge sheet, that the charges be divided between two or more charge sheets.[33] Where two or more defendants are charged in a single charge, the court may direct, before the commencement of trial proceedings in relation to the charge, that the charge be replaced with charges against the defendants separately and in separate charge sheets.[34]

If, as a result of a successful application it is necessary to re-arraign a defendant or adjourn the trial, the court will exercise its power under rule 41.

Where a case is allocated to court martial trial and it is subsequently determined by the DSP that a further defendant should be added to the charge sheet, the matter of joinder of defendants (and related arguments of severance) will be a matter for the court martial to determine, as a preliminary matter. Counsel who wish to make representations should, therefore, and in the first instance, do so at a preliminary hearing.

Bail/Custody During Proceedings Of Service Civilian Court
Main Provisions: sections 105, 108 and 109 AFA 2009

During the period of the trial, the retention of the defendant in custody is subject to the provisions of section 109 of the Act. Essentially, this includes circumstances where the judge advocate is satisfied that the defendant might abscond, interfere with witnesses, or where the case has been adjourned for inquiries or a report and it appears to the judge advocate that it would be impracticable to complete those inquiries or make the report without keeping the accused in service custody. The provisions applicable to the SCC are the same as those in the court martial. See also above, on page 132, 141 and page 436.

33 *Ibid.*, r. 38.
34 *Ibid.*, r. 39.

Sentence – The Service Civilian Court

Main Provisions: sections 164, 282–284 and Schedule 3 and 6 AFA 2006; Chapter 7 AFA 2006; section 17 AFA 2011; section 44 and Schedule 16 CCA 2013; ORA 2014; ORA2014(C2)O 2015

The Armed Forces Act 2006 was amended by Schedule 6 of the Offender Rehabilitation Act 2014, which fully came into force on 1 February 2015.[35] Schedule 6 sets out the requirements that may be included in overseas community orders made in respect of young offenders. Section 164 and Schedule 3 (punishments available to court martial) apply in relation to the SCC as they apply in relation to the court martial subject to the following limitations under:

a. sections 282–284 (concerning the maximum imprisonment, fine or compensation order that may be awarded by SCC); and
b. 282 (3), where the Service Civilian Court sentences an offender to whom Part 2 of Schedule 3 applies (an ex-serviceman), it may not award a punishment in any of rows 2 to 5 or 8 of the Table in section 164.[36]

The sentences therefore available are the following.

Imprisonment

Imprisonment for not more than 12 months for any one offence, may be imposed, subject to a maximum of 65 weeks where the defendant is convicted of two or more offences, the sentence for each of which is ordered to run concurrently (s. 283).

Detention and Training Order[37]

Where the defendant is not older than 21 years of age on the date of conviction. See also pages 582 *et seq*.

35 The Offender Rehabilitation Act 2014 (Commencement No. 2) Order 2015.

36 Dismissal, detention, forfeiture of seniority, service community order made on dismissal.

37 See s. 211 AFA 2006. See also on p. 582 below.

Fine

Where the conviction relates to an offence under section 42 (criminal conduct), the court may not impose a fine which a magistrates' court could not impose on the offender for the corresponding offence under the law of England and Wales (s. 284(2)). Otherwise, the maximum sum of the fine is the 'prescribed sum' as defined by section 32(9) of the Magistrates' Courts Act 1980.[38]

Service Community Order

This is defined by section 178 as an order imposing on the offender one or more of the requirements mentioned in section 177(1) of the CJA 2003, for example unpaid work, curfew, exclusion, residence, prohibited activity, supervision, etc. The service community order (SCO) can only be made in respect of civilians and dismissed service personnel and only if there is no requirement for mental health treatment. It specifies the local justice area in England and Wales, the locality in Scotland or the petty sessions district in Northern Ireland where the offender resides or will reside. It is broadly equivalent to a community order made under section 177 of the CJA 2003 and most of the provisions that apply to such orders are extended to SCOs. The court may not make a service community order unless:

a. the offender is aged 18 or over when convicted; and
b. it appears to the court that they will reside in the United Kingdom when the order is in force (see, section 164(5) AFA 2006). Where there is a drug rehabilitation requirement, the SCO is subject to periodic review, as set out in section 179 AFA 2006 and section 210 of the Criminal Justice Act 2003.

When the court makes a community order, unless there are exceptional circumstances[39] it must include in the order at least one requirement

38 This is set at £5,000 as at December 2020. See s. 374 AFA 2006 (definition of 'prescribed sum').

39 Exceptional circumstances must (a) relate to the offence or to the offender, (b) make it unjust in all the circumstances for the court to impose a requirement for the purpose of punishment in the particular case, and (c) make it unjust in all the circumstances for the court to impose a fine for the offence concerned.

imposed for the purpose of punishment or impose a fine for the offence in respect of which the community order is made or do both.[40]

Overseas Community Order

The Overseas Community Order (ss 182[41] and 183[42]), applies only to service civilian offenders living overseas and is an order imposing on the civilian offender one or more of the requirements mentioned in section 177(1) of the 2003 Act (community orders under that Act) and not specifying anywhere as an area where the offender resides or will reside. The duty on courts to include a punitive element in a community order or to impose a fine also applies to service courts imposing an overseas community order on an offender who is aged 18 or over when convicted.[43] The conditions which may be imposed are set out in Schedule 6 AFA 2006.

As it applies only to service civilian offenders, it is imposed only if it appears to the court that the offender will reside outside the United Kingdom when the order is in force. It is subject to a number of caveats, among which is a provision which prevents a court from imposing a mental health requirement where the offender's mental condition[44] would warrant the making of a hospital order or guardianship order (which are not available on conviction by a service court), those relating to periodic review of a drug rehabilitation requirement, and the power to impose an electronic monitoring requirement. Neither can it impose a foreign travel prohibition requirement or an alcohol abstinence and monitoring requirement (s. 182(1)(a)).

A copy of the order is to be provided without delay to the offender or, if under the age of 14,[45] to their parent or guardian or, if the order imposes an education requirement under Schedule 6 to this Act, to Service

40 See s. 177(2A) of CJA 2003, inserted by s. 44 and Sch. 16(2) of the CCA 2013, in force 11 December 2013.

41 See Crime and Courts Act 2013, Sch. 16(33), in force from 11 December 2013, which inserts a new ss. 3A into s. 182. It states, '"community order" includes an overseas community order if the offender is aged 18 or over when convicted of the offence in respect of which the overseas community order is made'.

42 Several amendments have been made by the Offender Rehabilitation Act 2014, Sch. 6. See the Act for all the details.

43 See Sch. 16 Part 8, CCA 2013 and EN.

44 *Ibid.*, Sch. 16 Part 8 and EN.

45 Guidance on Sentencing in the Court Martial, para. 4.3.4.

Children's Education (SCE). An education requirement in an overseas community order should not be included unless the court has consulted SCE and is satisfied that, in the view of SCE, arrangements exist for the offender to receive efficient full-time education suitable to their age, ability and aptitude and to any special educational need they may have; and that, having regard to the circumstances of the case, the inclusion of the requirement is necessary for securing the good conduct of the offender or for preventing the commission of further offences (Sch. 6(7)).

Breach, revocation or amendment of an overseas community order is dealt with in section 184 and Part 2 of Schedule 5 (application of Schedule 8 to the Criminal Justice Act 2003 to overseas community orders), so that any breach of such an order may only be dealt with by the civilian courts of the UK.

Service Compensation Order

A service compensation order may be made requiring the offender to pay compensation for any personal injury, loss or damage resulting from their offending or any offence so taken into consideration. They closely resemble the compensation orders available to civilian courts in England and Wales (section 175)[46] and are enforceable in the same way as a fine.[47] It follows that no service compensation order may be made for bereavement, funeral expenses, or loss of any other kind suffered by the dependants of a person in consequence of their death (s. 175(4)). Nor will a service compensation order be made in respect of injury, loss or damage due to an accident arising out of the presence of a motor vehicle on a road unless:

a. it is in respect of damage resulting from an offence of unlawfully obtaining any property; or
b. it is in respect of injury, loss or damage as respects which
   (1) the offender is uninsured in relation to the use of the vehicle; and
   (2) compensation is not payable under any arrangements to which the Secretary of State is a party.

46 Exclusions include: bereavement, funeral expenses, losses by dependents.

47 See s. 177 Criminal Justice Act 2003, similar to s. 130 of the Powers of Criminal Courts (Sentencing) Act 2000 (PCC(S)A).

Where a service compensation order is made in respect of injury, loss or damage due to an accident arising out of the presence of a motor vehicle on a road, the amount to be paid may include an amount representing the whole or part of any loss of or reduction in preferential rates of insurance attributable to the accident.

The service compensation order must be of such amount as the court considers appropriate, having regard to any evidence and to any representations that are made by or on behalf of the offender or the prosecutor.[48] Where the offender is under 18, compensation must not exceed the amount for the time being mentioned in section 131(1) of the 'Sentencing Act'[49] (that is, they must not exceed the limits of orders imposed by a magistrates' court). As at December 2020, this sum shall not exceed £5,000. Part 8 of the Crime and Courts Act 2013 (CCA 2013) amends this provision of the AFA 2006 in respect of adults, by removing limits on compensation orders made against them. This is achieved by the insertion of section 284(2A) which makes it clear that the limit under section 131 only applies in respect of those under the age of 18.[50] In other words, there is no limit on the amount of compensation a service court may order an adult offender to pay. An may permit payment by instalments where the court or officer awarding it makes an order under section 251.

A service compensation order may be imposed at a summary hearing, as it is authorised by section 137 and listed in the Table to section 132. There are limitations on powers of subordinate commanders to whom delegation has been made to make SCOs. These are set out in rule 3(5) AF(SHASSSD) R 2009. See above, on page 324.

Conditional or Absolute Discharge

Under sections 185–187, these sentences are available only for civilian offenders, according to the heading to the section. The condition is that the defendant does not commit an offence during the operational period of the order. Section 185(4) prohibits a conditional or absolute discharge from being combined with any other punishment, except a service compensation

48 S. 175(2).

49 PCC(S)A.

50 '(2A) The following subsections apply if (but only if) the Service Civilian Court has convicted a person aged under 18 ("the offender") of an offence or offences.'

order. Section 186 explains the consequences of a breach of a conditional discharge. If the convicting court is the court martial, it may deal with the offender for the offence for which they were conditionally discharged in any way in which the court that conditionally discharged them could deal with them if it had just convicted them of that offence. It is restricted to the powers that the imposing court had. If, on the other hand, the convicting court is the SCC, it may deal with the offender for the offence for which they were conditionally discharged in any way in which it could deal with them if it had just convicted them of that offence.

A conviction of an offence for which an order under section 185 discharging the offender absolutely or conditionally is made shall be deemed not to be a conviction (s. 187).

In relation to persons under 18, the SCC has the same powers as the Youth Court in England and Wales. See Young Offenders in Chapter 17.

See also Chapter 7 AFA 2006; section 17 AFA 2011, referred to on page 489, for court orders other than sentences (service restraining orders and recognisances).

Slip Rule

For detail about how the slip rule operates in the court martial, see Chapter 13, at page 452.

As indicated above,[51] after a trial by court martial, where an error is detected in the sentence handed down, it is possible for a correction to be made. This is achieved by means of variation proceedings which specifically cater for this in the AF(CM)R 2009. It is interesting that, although this slip rule power is available in the court martial, there is no equivalent in the AF(SCC)R 2009. In section 288, the rule making power of the Secretary of State makes no reference to variation proceedings for the SCC.

There is a power under rule 25 that permits the judge advocate at the SCC to ensure that, in circumstances not provided for in the rules, proceedings are conducted in such a way as appears to them to be in the interests of justice. However, the issue is whether, after a trial has concluded and sentence has been handed down, there are still 'proceedings' extant. On a plain construction, it can be said that the proceedings have ended once the

51 See above, p. 452.

final sentence has been handed down or a defendant is acquitted. It is also doubtful that viewing proceedings as still extant is the correct approach as, otherwise, there would have been little need to have specifically included the provision in rule 27(4)(a) AF(CM)R 2009. Furthermore, a similar power is available under section 152 to the Reviewing Authority, on conclusion of a summary hearing by the CO.[52]

This omission in the case of the SCC seems illogical.

52 See Review of Summary Awards, above on p. 323.

16 APPEALS

Bail Pending Appeal

This is dealt with under the heading for each respective court where the proceedings take place. For appeals by way of case stated from the Summary Appeal Court (SAC) to the High Court, see page 540; for appeals from the Service Civilian Court (SCC) to the court martial, see page 544; for appeals from the court martial to the Court Martial Appeal Court (CMAC) see page 549; for appeals from the CMAC to the Supreme Court, see page 564.

Pre-Sentence Report

Apart from an appeal to the SAC against a CO's decision,[1] any court on appeal against a custodial sentence in respect of a service offence, a sentence of dismissal or dismissal with disgrace, a sentence of service detention or a community punishment, must obtain a pre-sentence report (PSR) unless it is of the opinion that the court below was justified in forming an opinion that it was unnecessary to obtain one, or that, although the court below was not justified in forming that opinion, in the circumstances of the case at the time it is before the court it is unnecessary to obtain a PSR.[2]

In relation to the SAC, where a sentence of service detention is passed or confirmed by the SAC, the court must consider the issues of whether a sentence of detention was serious enough to be warranted and the length of sentence commensurate with seriousness of the offence section. See section 256(9) AFA 2006, as modified by sections 242(4) and 243(3) which places a requirement on the SAC to seek a PSR if, after hearing the appeal, it considers imposing service detention. However, section 256(2) permits the

1 See s. 256(8) AFA 2006.
2 See s. 256(5) and (6) AFA 2006.

court to dispense with a PSR if, in the circumstances of the case, it is of the opinion that it is unnecessary to obtain such a report.

In relation to offenders under 18, the service court cannot form the opinion that a PSR is unnecessary unless a previous PSR exists for the offender, and the court has had regard to the information contained in that report, or, if there is more than one such report, the most recent report (s.256(3)).

Judicial Review

There is no power to judicially review a sentence of a court martial. Section 23 AFA 2001 amended section 29 of the Supreme Court Act 1981[3] to make it clear that:

> The High Court shall have no jurisdiction to make orders of mandamus, prohibition or certiorari in relation to the jurisdiction of a court-martial in matters relating to
> (a) trial by court-martial for an offence, or
> (b) appeals from a Standing Civilian Court.

To Summary Appeal Court

Main Provisions: AFDA 2000; sections 140–151 AFA 2006; AF(SAC)R 2009; AF(SCR)(A)R 2015, rule 6; Appendix G Legal Aid Flowchart; Appendix H Procedure at SAC

Prior to the creation of the SAC the incompatibility of the UK summary hearing process before the CO with fair trial guarantees under the European Convention on Human Rights and Fundamental Freedoms was the subject of several legal challenges before the ECtHR. In consequence of the decision in *Hood v United Kingdom*,[4] the right of election for trial was moved from the end of the summary dealing to the beginning of the process, so that the accused was able to elect whether or not to be tried by a compliant court

3 Now to be cited as the 'Senior Courts Act 1981'. See, Constitutional Reform Act 2005, Sch. 11, Part 1, s. 1.
4 [1999] 18 February, ECHR.

martial at the outset of the hearing. However, the concerns had not been fully allayed.

In *Thompson v United Kingdom*[5] the ECtHR did not consider that any choice by the applicant of a summary trial as opposed to electing trial by court martial would have constituted a valid waiver of his rights under Article 6, paragraphs 1 and 3 of the Convention. A waiver of a right guaranteed by the Convention – in so far as it is permissible – must not run counter to any important public interest, must be established in an unequivocal manner and requires minimum guarantees commensurate to the waiver's importance.

Bell v United Kingdom[6] confirmed that a waiver of a right guaranteed by the Convention – in so far as it is permissible – must not run counter to any important public interest, must be established in an unequivocal manner and requires minimum guarantees commensurate to the waiver's importance. It is important to recall that the ECtHR noted that the applicant was directly subordinate, and in close structural proximity, to his CO, a factor which undoubtedly would have affected the free and unambiguous nature of any choice between a summary trial and a court martial. The applicant would also have been influenced by the fact that a summary procedure involved a maximum sentence of 28 days (and 60 days (as it was at that time) only if extended powers were granted) whereas trial by District Court Martial could have in theory led to a sentence of up to two years' imprisonment. Moreover, the fact that the option was presented to him at all meant that his CO considered him to be guilty as charged and, further, that he warranted more than a minor punishment. The ECtHR considered that, even supposing that the Article 6 rights in question could in principle have been be waived, the circumstances were such that there was no valid waiver by the applicant of his rights under that Article. The Court recalled that, in the *Thompson* case, it concluded that the summary procedure presented even clearer structural independence and impartiality problems than those established in the *Findlay* case and, further, that those proceedings were, consequently, unfair. In all the circumstances, the summary procedure before the CO violated the independence, impartiality and, consequently, fairness aspects of Article 6, § 1 and that the exclusion of

5 [2004] 15 June 2004, ECtHR, Application N° 00036256/97.

6 [2007] Application no. 41534/98, 16 January 2007.

legal representation from the applicant's summary trial also constituted a violation of Article 6, § 3(c) of the Convention.

The SAC was created to address the concerns over non-compliance in the summary process and attempts to heal concerns by ensuring that no person exercising the choice to be dealt with by a CO is deprived of the right to challenge any finding or punishment made by them before an ECHR-compliant court may do so without fear of any higher penalty being imposed.

The SAC is an ECHR-compliant court created by the Armed Forces Discipline Act 2000 (AFDA 2000). The AFA 2006 now consolidates its functions in sections 140–151, together with The Armed Forces (Summary Appeal Court) Rules 2009. It can sit anywhere, within or outside the UK. The mode of dress for advocates and form of address are set out at Appendix G. A suggested format for the appeal (Notice of Appeal) is set out below, on page 540.

At the Summary Hearing[7]

The CO records a finding for each charge proved. There is only one sentence handed down by a CO, irrespective of the number of charges and the number of findings made. This is confirmed in section 131(5) that the award the CO must make is a single award (consisting of one or more of those punishments authorised by section 132) in respect of the charges taken together.

The CO has power to amend, substitute, or bring additional charges against the accused at any time before or after the start of any summary hearing of the charge. The DSP may also direct the CO to bring a charge against the accused in a case referred back to the CO under section 121(1B)(4). This seems to leave open whether the power extends to charges appealed to the SAC. Bearing in mind the background to the creation of the SAC, explained above, and its purpose to ensure that the matter about which the accused was dealt with could be reheard before a compliant court, coupled with the absence of any express power in the Act for either the CO or DPS to make changes or substitutions, the SAC will deal with the charge as ultimately determined by the CO, and with the limitation that

7 For more specific information about the procedure at summary hearing, see Summary Hearing.

the punishment it imposes must be no more severe than the punishment originally awarded (s. 147(3)(b)(ii)). However, this would not preclude the SAC from finding proved a different charge to that alleged on the charge sheet and substituting a finding to that effect, under section 147(1)(b).

Commencing the Appeal

The appellant will need to complete Form of Notice of Appeal to the Summary Appeal Court against the finding or against the punishment awarded by a CO to be found in JSP 830 Volume 1, Chapter 15, Annex C. A sample form is available on page 540. Rule 15(1) requires that the Notice of Appeal is served on the CO. The CO then serves on the Court Administration Officer (CAO) and the Director the documents required by rule 15(3), namely:

(a) the written record of the summary hearing;
(b) a copy of the disciplinary record of the appellant;
(c) a copy of any witness statement prepared for the purposes of the summary hearing;
(d) details of all exhibits that form part of the evidence (whether used or unused at the summary hearing) and where and when they can be inspected;
(e) a list of persons whose evidence was not adduced at the summary hearing but who appear to the commanding officer to be potential witnesses in the proceedings before the court;
(f) any material in the possession of the commanding officer which is not annexed or referred to in the written record but which in his opinion may be material to the proceedings before the court;
(g) if the commanding officer was granted authority to hear the charge summarily, a copy of any document conveying that authority;
(h) if the commanding officer had extended powers for the purposes of section 133(1)(a) or (2), 134(1), 135(1) or 136(1)(b), a copy of any document conveying those powers; and
(i) a document specifying the appellant's age, his rank or rate and any recognised acts of gallantry.

If necessary, an application to extend the initial period or for leave to appeal out of time may be made under rule 16 (1).

Aside from the ability of the defendant to bring an appeal to challenge finding and punishment, it is open to the reviewing authority to refer a finding or punishment to the court under:

(a) section 152(4), with the leave of the SAC (as on appeal), or
(b) section 152(7), after the SAC has completed the hearing of the appeal, when a referral is made of new matters arising.

The reviewing authority must make the application in writing to the CAO.[8] See further, on page 323 (Review of Summary Awards).

Whenever an appeal is entered, it must comply with the time limit (see below). Once the written notice is served, the DSP must inform the CAO in writing, within 28 days, whether or not they intend to contest the appeal, in accordance with rule 19(1).

Where the DSP gives notice under rule 19(1) that it does not intend to contest an appeal, the SAC shall quash the finding against which the appeal is brought, along with any punishment awarded which relates to one or more findings each of which is the subject of an appeal and in respect of each of which the DSP has given notice under rule 19(1) that they do not intend to contest the appeal. The power of the court to quash the finding(s) under section 147(1)(a) AFA may be exercised by the Judge Advocate General without a hearing, and any decision of the Judge Advocate General in exercise of those powers shall be recorded in writing and dated and signed by him or her.[9]

It should also be noted that it is open to the SAC to hear more than one appeal at the same time where it appears to be in the interests of justice to do so. This decision may be made by the SAC of its own motion, or on the application of the Director, or by the appellant in one or more of the appeals to which the application relates.[10] Before refusing an application or making a determination, the judge advocate shall afford the Director and any person who is the appellant in any of the appeals to which the application or determination relates the opportunity of making representations to him.

8 AF(SAC)R 2009, r. 17.
9 AF(SAC)R 2009, r. 20.
10 See AF(SAC)R 2009, r. 48(1).

Appeal – Time Limit

Main Provision: sections 141 and 290(2) AFA 2006; Part 3 Chapter 1 AF(SAC)R 2009

Appeals from CO's decisions must be lodged within 14 days[11] beginning with the date on which the punishment was awarded ('the initial period'), unless leave is given to extend that period. The CO must inform the accused of this right and the accused's right to seek independent legal advice on whether to appeal. An interesting feature of the legislation is that, by virtue of section 290(2), the sentence can be postponed from taking effect for 14 days, otherwise the award takes effect from the beginning of the day on which the award is made. This enables the accused to take legal advice at private expense on the question of appeal. If no election is made under section 209(2), the award takes effect from the end of the appeal period (unless an appeal is brought within that period). If an appeal is brought within that period, the award takes effect from the beginning of the day when the appeal is abandoned or determined (s. 290(3)(b)). Where an appellant fails to appear before the SAC at the time appointed for the commencement or resumption of the appellate proceedings and the judge advocate considers that there is no reasonable explanation for the failure to appear, the judge advocate may direct under rule 24 that the appeal be treated as abandoned.

Bail

Because of the power to defer the effect of any sentence awarded at summary hearing, discussed above on page 298, the question of 'bail' pending the appeal to SAC does not arise. Indeed, the AF(CRFC)O 2009 only applies to appeals from the SAC to the High Court under section 149(2) of the 2006 Act and appeals from SCC to the court martial.

Legal Aid

Whereas there is no right to legal representation or financial assistance for representation at a summary hearing, the situation is different on an appeal to the SAC. Legal aid is available to any service person appealing before the SAC. See Appendix F Legal Aid Flowchart and Legal Aid.

11 AFA 2006, s. 141(2).

Advance Information
Main Provision: rule 42 AF(SAC)R 2009

For an appeal against finding, 'advance information' means: (a) copies of the statements of those witnesses on whom the Director intends to rely against the appellant; (b) a list of all exhibits which the Director intends to adduce in evidence against the appellant, and a statement of where any non-documentary exhibits are held; and (c) a transcript of any sound recording of an interview with the appellant.

Rule 42 requires the Director, as soon as is practicable, to serve advance information on the CAO, the appellant and the appellant's legal representative (if any).

Where the appeal is against punishment only, 'advance information' means a statement of the information that the Director proposes to present to the court in pursuance of rule 86(2)[12] and the written record. If the Director intends subsequently to adduce any evidence not included in the advance information, they must as soon as is practicable serve on the appellant and the court administration officer such documents as they would have been required to include in the advance information if they had had that intention at the time when they served advance information. If proceedings have already commenced, the supplementary documents must be served on the judge advocate instead of the CAO.[13]

Preliminary and Ancillary Proceedings

Whereas the SAC will ordinarily be constituted in accordance with section 142, as indicated below, it is possible in preliminary proceedings for the judge advocate to sit alone. As in civilian courts, such proceedings will deal with case management matters, such as special measures, discussed below (r. 34 SACR 2009).

Special Measures
Main Provisions: YJCEA 1999; AF(SAC)R 2009; AF(SCR)(A)R 2015; SAC(YJCEA1999)R 2015

12 Information before punishment, setting out appellant's background details, service record, other offences, etc.
13 AF(SAC)R 2009, r. 42(4) and (5)

The AF(SCR)(A)R 2015, rule 6, introduced an important amendment to the 2009 rules, pursuant to Directive 2012/29/EU of the European Parliament and of the Council of 23 October 2012. After rule 74, a new Chapter 7 is inserted, entitled 'Special Measures Directions', set out in rules 74D to 74O, to provide protections to vulnerable and young witnesses (that is, persons under the age of 18 years at the time when the application is considered). These are detailed in Appendix N Special Measures Directions.

Additionally, the SAC(YJCEA1999)R 2015 also apply, which make provision for a range of measures designed to help young, disabled, vulnerable or intimidated witnesses give evidence in proceedings before the SAC[14] so as to give effect to Directive 2012/29/EU of the European Parliament and of the Council of 25 October 2012 establishing minimum standards on the rights, support and protection of victims of crime and replacing Council Framework Decision 2001/220/JHA (OJ L 315, 14.11.2012, p. 57–73), Articles 18 and 21. These measures are contained in Part 2 of the Youth Justice and Criminal Evidence Act 1999 (YJCEA 1999), certain provisions of which are applied with modifications to, or modified in their effect for the purposes of, proceedings before service courts by the Youth Justice and Criminal Evidence Act 1999 (Application to Service Courts) Order 2009, as amended by the Youth Justice and Criminal Evidence Act 1999 (Application to Service courts) (Amendment) (No. 2) Order 2015.

Part 2 of the Rules deals with restrictions on cross-examination of a witness by an appellant acting in person. Rule 4 makes provision for the Director of Service Prosecutions to make an application to the judge advocate to give a direction that the appellant be prevented from cross-examining a witness. Rule 5 sets out the procedure for the judge advocate to explain to the appellant that they are prevented from cross-examining the witness and the procedure required for the appellant to appoint a legal representative to conduct the cross-examination. Rule 6 deals with the appointment of a legal representative by the judge advocate and rule 7 deals with the appointment of a legal representative by the appellant.

Part 3 of these Rules contain provisions dealing with restrictions on reporting. Rules 8 to 11 govern applications under sections 25, 45A and 46 of the 1999 Act for directions restricting the reporting of proceedings and

14 See: EN to the SI.

directions dispensing with such restrictions. Rule 12 contains provisions relating to applications for the revocation of reporting directions or the variation or revocation of an excepting direction and rule 13 contains provision for applications for an extension of time. Rule 14 sets out the procedure for determining applications for reporting directions.

The SAC(YJCEA1999)R 2015 came into force on 15 November 2015 and can be found at Appendix O Young, Disabled, Vulnerable, Etc. Witnesses.

Constitution of the Court

The SAC is presided over by a specially appointed judge advocate and two other persons: an officer and either another officer or a warrant officer. It hears appeals from COs' decisions in relation to both findings and awards (sentence) made against persons subject to service law. Thus, it does not apply to service civilians. Any decision of the SAC must be determined by a majority of the votes of the members of the court.[15] However, rulings and directions on questions of law, procedure or practice are to be given by the judge advocate and are binding on the court.[16]

Appeal Against Punishment

Main Provisions: sections 140–150 AFA 2009; Part 12, AF(SAC)R 2009; Appendix H Procedure at SAC

An appeal against punishment takes the form of a rehearing, by virtue of section 146(1)(b) or (2), when the SAC may

(a) confirm the punishment awarded; or
(b) quash that punishment and award in substitution for it any punishment which
    (i) it would have been within the powers of the officer who conducted the summary hearing to award; and
    (ii) in the opinion of the court, is no more severe than the punishment originally awarded (r. 147).

It is clear under section 147(3)(b)(ii) that the SAC has no power to increase the punishment awarded by the CO. This is to ensure that there is no

15 Section 149(1).
16 AFA 2006, s. 146(3) and (4).

disincentive to a service person appealing to a compliant court. Any confirmation or substitution of punishment can only be awarded by the full court.[17] The result is that, unless the SAC directs otherwise, any substituted punishment has effect as if awarded on the day on which the original punishment was awarded, but this is subject to section 148(2), which deals with sentences of detention and provides that, when the SAC awards a punishment of detention it may either direct that the punishment is to take effect at the end of any other period of service detention that has been awarded to the accused on a previous occasion, or it takes effect from the day on which the SAC awards it.

Where the SAC quashes the finding or (if there is more than one finding) every finding made in respect of the appellant, it must quash the punishment which relates to that finding or those findings (s. 147(2)).

On an appeal against punishment, the specific information that the DSP must provide to the court (the antecedent history) is to be found in rule 86(2), and include the details of any activation order made by the CO. The appellant may call witnesses in mitigation of punishment or as to character, produce any document and address the court in mitigation. The decisions of the SAC are by simple majority[18] and include the judge advocate,[19] save that rulings and directions on any questions of law, procedure or practice are to be given by the judge advocate only. The judge advocate pronounces punishment and makes the statement of reasons, and gives the explanation, required by section 252(1).[20]

Substitution of Punishment

Unless it otherwise directs, any punishment substituted by the SAC has effect as if awarded on the day on which the original punishment was awarded. If the court substitutes a term of immediate detention, then the court may direct that the award shall take effect from the end of any other sentence of service detention that has been passed on the offender on a previous occasion, or from the beginning of the day on which the punishment is substituted.[21]

17 AF(CM)R 2009, r. 35(b).
18 AFA 2006, s. 149(1).
19 *Ibid.*, s. 142(1) and s. 146(3).
20 This places a duty to give reasons and explain sentence.
21 AFA 2006, s. 148.

Appeal Against Finding

On an appeal against finding, the DSP and the appellant may each address the court once with respect to the case against the appellant on the charges being heard by the court (s. 75(1)). The address by the DSP is before they adduce any evidence, while the address by the appellant is immediately after the close of their own case or, where there is more than one appellant, after the close of the case for all of the appellants. With the court's leave, the DSP or appellant may address it at any time during the hearing on any matter relating to the appeal or the charges that are being heard by the court. As stated above, the SAC shall not exercise its powers under section 147(1)(b) to substitute another finding without affording the appellant and the Director the opportunity to address it on the exercise of those powers in the particular case.

Where there are two or more appellants, and the court has, in accordance with rule 48, decided to hear the appeals at the same time, rule 76 applies. The procedure is that the DSP puts their case on each of the charges, but they may only address the court once. Before the appellants open their cases, the judge advocate decides the order they are to put their case and to address the court. Where two or more appellants are represented by the same advocate, that advocate may only address the court once (r. 76(4)). At the close of the case for the respondent an appellant may submit, in respect of any charge, that the DSP has failed to establish a case for them to answer. The judge advocate will determine the submission in the absence of the lay members and, if they allow it, they will direct the court to quash the finding (r. 7)]. Moreover, at any time after the close of the case for the respondent, the court may find an appellant not guilty on a charge.

If the appellant wishes to give evidence, they must do so before calling any witnesses. A unique feature of the SAC, and in complete contrast to the other service courts, is that the judge advocate participates in determining the facts and the decision whether the finding should be upheld or quashed, as they are a member of the court.[22] The determination of the court is by simple majority of the votes of the members of the court.[23] It is the judge advocate who announces the finding of the court as to whether the appeal against the CO's finding succeeds or not.

---

22 See, s. 142 AFA 2006.

23 *Ibid.*, s. 149 and r. 3 AF(CM)R 2009.

Substituted Finding
The SAC may not alter the charge nor may it exercise its powers under section 147(1)(b) to substitute another finding without affording the appellant and the Director the opportunity to address it on the exercise of those powers in the particular case.[24] On a simple reading of section 147(1)(b) the court could, for example, instead of a charge of fraud substitute a charge of making a false entry under section 18 founded on the same facts but, where a charge of theft includes a monetary amount, it may not, ostensibly, substitute a lesser amount. This may be an occasion when the judge advocate employs rule 33 in such a way as appears to them to be in the interests of justice. However, substituting a higher amount would seem to increase the severity of the charge, which the court is not able to do.

Pre-Sentence Report
In relation to the requirements for a pre-sentence report, see above, on page 525. Where the CAO has arranged for a PSR, they must serve a copy on the Director and the appellant prior to the proceedings.[25] Before the proceedings, the DSP is to serve on the appellant and the CAO a copy of any convictions the appellant may have.

The AF(SAC)R 2009 cater for Newton hearings in order to determine the factual basis for sentencing. See in particular, rule 84.

Procedure
The procedure for the hearing is broadly similar to the court martial except that the judge and members enter and exit the courtroom together. A detailed explanation of the order of procedure is explained in the very helpful free publication entitled The Court Martial And The Summary Appeal Court Guidance, Volume 1.[26] An extract is produced at Appendix G.

Activation of Suspended Sentence[27]
When a CO finds proved a charge against the offender in respect of an offence committed during the operational period of a suspended sentence,

24 AF(CM)R 2009, r. 75(4).
25 See r. 85(1) AF(SAC)R 2009.
26 At Annex D, Order of Procedure in the Summary Appeal Court.
27 See also, above p. 307.

or the offender is convicted of an offence under British law[28] which was committed during that operational period, and subsequently appears before their CO, the CO may activate the sentence either for the original term unaltered, or with the substitution of a lesser term for the original term. If in addition they award detention for the activating offence they found proved, the aggregate must not exceed 28 days, unless they have extended powers in respect of the activating offence, when the aggregate must not exceed 90 days.[29] See more generally above, on page 307.

Any appeal to the SAC, or application for leave to appeal, against the finding or the punishment awarded in respect of the CO's activation order is to be treated as also being an appeal or application for leave to appeal against the order and punishment.[30] The SAC has the power to confirm the order, quash the order, or make, in substitution for the order, any order that the CO could have made.[31] But the court cannot exercise its powers in such a way that, taking the case as a whole, the appellant is dealt with more severely on appeal 'than he was dealt with by the' CO (s .195(7)).

Specimen Form of Notice

Form of Notice of Appeal to the Summary Appeal Court against the finding or against the punishment awarded by a Commanding Officer.
Ref: JSP 830: Volume 1, Chapter 15, Annex C
Appeal to the Summary Appeal Court is under s. 141 AFA 2006.

* Notes: Unless extended by the judge, any appeal under this section must be brought within 14 days beginning with the date on which the punishment was awarded ('the initial period')

The Commanding Officer must serve the notice on the court administration officer and a copy of the notice and the documents required under paragraph 15(3) SAC(R) 2009 on the Director Service Prosecuting Authority, who is the Respondent.

Figure 16.1 on pages 540–541 is not a statutory form but is the one included in the *Manual of service law* (Crown Copyright). The form can

28 'An offence under the law of any part of the British Islands.' Amendment inserted by Sch. 3(14) AFA 2011 and applies to ss 190(1)(a), 191(2)(b) and 193(2)(b)AFA 2006.
29 AFA 2006, ss 193, 194 and 194A.
30 *Ibid.*, s. 195(1) and (2).
31 *Ibid.*, s. 195(4).

be completed online. See https://www.gov.uk/government/uploads/system/uploads/attachment_data/file/262732/MSLchapter15annexC.pdf.

To the High Court, from the SAC

Main Provisions: section 149(2) AFA 2006; rules 92–94 AF(SAC)R 2009

The appellant or respondent may question any decision of the SAC on the ground that it is wrong in law or is in excess of jurisdiction, by applying to the SAC to have a case stated for the opinion of the High Court in England and Wales.

Time Limit

Main Provision: rule 92 AF(SAC)R 2009

Although there is legislation to specify the time limit in relation to other jurisdictions, that is, the magistrates' court,[32] no specific statutory provision is to be found in the 2006 Act in relation to caseX stated by the SAC. Instead, rule 92 of the AF(SAC)R 2009 makes the necessary provision, stipulating the application must be made in writing and shall be served on the CAO and the appellant or the Director, as the case may be, within 21 days of the date of the decision in respect of which the application is made. It must state the grounds on which the decision of the court is questioned. A suggested format is set out below, on page 543.

The procedure on stating a case is contained in AF(SAC)R 2009, rule 93, and requires the applicant, within 21 days of receiving the notice from the Court Martial Administration Office (CMAO) confirming that the judge advocate has stated a case, to draft a case and serve a copy of it on the CAO and the appellant or the Director, as the case may be. Thereafter, within 21 days of receiving a copy of the draft case from the CMAO, the Director or the appellant must give notice in writing to the applicant and the CAO that they do not intend to take part in the proceedings before the High Court, or indicate in writing on the copy of the draft case that they agree with it and send the copy to the CAO, or draft an alternative case and

32 See s. 111(2) Magistrates Courts Act 1980 which states that the application is to be made within 21 days after the day on which the decision of the magistrates' court was given.

MINISTRY OF DEFENCE

PROTECT - PERSONAL DATA (WHEN COMPLETED)

ANNEX C TO
VOL1 CH 15
JSP 830 MSL
revised 11/09

Save as | Print Form

NOTICE OF APPEAL T-SL-SAC01

To: COMMANDING OFFICER (for onward transmission to the CAO)

I

First name ____________ Family name ____________

Rank/rate ____________ Service number ____________

Ship/unit/establishment/civilian address* ____________

* For ex-service personnel only

Do HEREBY GIVE NOTICE that in respect of:

Charge ____________

I appeal against both finding and punishment ☐ / ☐ punishment only;

Charge ____________

I appeal against both finding and punishment ☐ / ☐ punishment only;

Charge ____________

I appeal against both finding and punishment ☐ / ☐ punishment only;

Charge ____________

I appeal against both finding and punishment ☐ / ☐ punishment only; 1-15-C-1

PROTECT - PERSONAL DATA (WHEN COMPLETED)

Figure 16.1 Appeal Form

Punishment(s) awarded

The above mentioned finding(s) and punishment were imposed by way of summary hearing/activation hearing

on date

at

Officer who conducted the summary hearing/activation hearing, and the capacity in which he conducted it :

First name

Family name

Rank

☐ Commanding Officer
☐ Subordinate Commander

Appellant's signature

Date

1-15-C-2

serve it, together with a copy of the applicant's case, on the CAO. The judge advocate will then consider the applicant's draft case and any alternative draft case served on the CMAO. The next stage is that the judge advocate must state and sign a case within 14 days after either the date on which they receive all the documents required to be served on the CMAO under rule 93(3), or the expiration of the period of 21 days, whichever is the sooner.

A case stated by the court shall state the facts found by the court, the submissions of the parties (including any authorities relied on by the parties during the course of those submissions), the decision of the court in respect of which the application is made and the question on which the opinion of the High Court is sought (r. 94).

If the judge advocate refuses to state a case, they may be made to do so by mandatory order of the High Court (r. 94(3)).

See Figure 16.3 Case Stated, Suggested Form.

Bail Pending Appeal

Main Provisions: Article 3 and Schedule to AF(CRFC)O 2009; SCSRSR 2009; JSP 837

Prior to the 2006 Act, there was no statutory right to bail. Ultimately, the applicant would have to seek *habeas corpus* from the High Court.[33] Since the introduction of the 2006 Act, the position is now regulated by law in the AF(CRFC)O 2009 and more generally referred to in the section on appeals from the SCC to court martial on page 544. A judge advocate may grant bail to a person pending the determination of an appeal to the High Court, under the authority of Article 3 AF(CRFC)O 2009.

The Service Custody and Service of Relevant Sentences Rules 2009, rule 5, provides for a person serving a sentence of detention to be retained in 'Service custody premises' unless

33 *M v Judge Advocate General*, Divisional Court, 29 July 2003 (unreported), where the Claimant sought judicial review of the failure to grant bail and a declaration that a High Court judge has an inherent power to grant bail. Although in the instant case there was no arguable case before the High Court judge, so that permission was refused, all the same, the judge was troubled by the ECHR implications and said that the requirements of the Human Rights Act and Art. 5 of the ECHR indicate a strong case for the existence of such a power for a grant of bail, although the position of a convicted person may be different to that of a person accused. In his mind there was much to be said for the convicting court to have that power because it had immediate knowledge of the facts. But there should also be an additional power to appeal further on the matter of bail. See now, the AF(CRFC)O 2009.

(a) he is released temporarily, for a circumstance set out in rule 33 (e.g., proceedings before any court, tribunal or inquiry, to consult his legal representative in circumstances where it is not reasonably practicable for the consultation to take place in Service custody premises, compassionate or medical);
(b) by an order of a court of competent jurisdiction; or
(c) on the expiry of his sentence, less any remission provided for in these Rules.

Joint Service Publication 837 sets out the policies, practices and procedures that apply to all categories of detainees held in service custody other than on operations, and amplifies the SCSRSR 2009. Although not subordinate legislation, they interpret 'Service custody' as including detainees serving sentences of service detention and those held in service custody without charge. See also section 300 AFA 2006 which permits the Secretary of State to make rules about service custody and the service of relevant sentences, including the 'the provision, classification, regulation and management of Service custody premises'.

As mentioned above, 'Service custody premises' is defined in section 300(7) AFA as 'premises under the control of the Secretary of State for the keeping of persons in Service custody'. This would include a unit guardroom. This is confirmed by the distinction drawn in JSP 837 between the Commandant of MCTC and the CO of Service Custody Premises, imposing on both the requirement to ensure compliance with the JSP by all those under their command and the need to ensure that unit standing orders reflect who is responsible at any time for the custody and care of detainees.[34]

It is important to note that time spent on bail does not count as time served (see Art. 5(1) AF(CRC)R 2009).

Form – Bail Application (SAC and SCC)

Main Provisions: AF(CRFC)O 2009, Articles 6(2), 8(3) and 8(5), and the Schedule

See Figure 16.2 Bail Application SAC and SCC.

---

34 *Ibid.*, paragraph 101 *et seq.*

Form 1
To : THE COURT ADMINISTRATION OFFICER

Trenchard Lines
Upavon
Wiltshire SN9 6BE

NOTICE OF APPLICATION RELATING TO BAIL

Take Notice that an application relating to bail will be made to a judge advocate at:

on .... at............ am/pm

on behalf of *(here insert description of applicant i.e. the appellant, DSP or surety).*

Particulars of applicant:
Name:

Address:

Particulars of convicted person/appellant:

Forenames: Surname:
Rank/Title: Service No (if any):

Sentenced by *Service Civilian Court/Summary Appeal Court* held at:
....
* *delete as appropriate*
Offences of which convicted:

Sentence:

Date when sentence passed:
Address: (give address where detained)
Address if granted bail:
Nature and grounds of application *(State fully facts relied on. Give details of any proposed sureties and answer any objection raised previously)*:

The court administration officer should be consulted about the time and place of the hearing before this notice is sent to the other party to the application. A copy of this notice should be sent to the Office of the Judge Advocate General.

Figure 16.2 Bail Application SAC and SCC

To:
Court Administration Officer
Appellant or the Director

APPLICATION TO THE SUMMARY APPEAL COURT TO STATE A CASE FOR THE OPINION OF THE HIGH COURT
(section 149(2) Armed Forces Act 2006 and Part 15 Summary Appeal Court Rules 2009)

Case details

No, rank, name, unit of defendant:

Charge(s):

This is an application by [...................................... (name of defendant)**]**

[the prosecutor]

---

for the court to state a case for the opinion of the High Court on an appeal on a question of law or jurisdiction.

Use this form ONLY for an application to the court to state a case for the opinion of the High Court on a question of law or jurisdiction, under Part 15 Summary Appeal Court Rules 2009.

1. Complete the boxes above and give the details required in the boxes below. You may attach extra sheets.
2. Sign and date the completed form.
3. Send a copy of the completed form to:
   (a) the Court Administration Officer
   (b) each other party to the case.

You must send this form so as to reach the recipients **not more than 21 days after the decision about which you want to appeal to the High Court. The judge advocate may extend any time limit referred to in Part 15 of the Rules either before or after it expires.**

---

1) Decision under appeal. Give brief details of the decision about which you want to appeal to the High Court (including the date of that decision).

Figure 16.3 Case Stated

2) Question(s) for the High Court. What question(s) of law or jurisdiction do you want the court to state for the opinion of the High Court ?

3) Grounds of appeal. Explain briefly why you think the decision against which you want to appeal was wrong, and how that decision depended on the question(s) specified in box 2 above. Set out the basis on which the Summary Appeal Court decision is said to be wrong in law or in excess of jurisdiction

4) Other applications. I am also applying for:

- ☐ **an extension of time for asking the court to state a case for the High Court.**
- ☐ **pending my appeal, bail.**
  You can only apply for bail pending appeal in accordance with art 3 Armed Forces (Conditional Release from Custody) Order 2009, if the court sentenced you to detention.

Give reasons for any of these applications you are making:

Signed: .... [defendant / defendant's solicitor]

[prosecutor]

Date: ....

To Court Martial from Service Civilian Court

Main Provisions: sections 285–287 AFA 2006

Appeals from the SCC lie to the court martial in relation to both conviction and sentence.[35] In relation to an appeal against conviction, it is by way of a rehearing of the charge (including, where the appellant is convicted, a rehearing as respects sentence).[36] An appeal against sentence only is to be by way of a rehearing as respects sentence.[37] When dealing with an appeal from SCC, the court martial may only pass a sentence that the SCC had power to pass in respect of the offence.[38] In other words, the court martial may impose a more severe sentence than the SCC actually imposed, provided it does not exceed the punishment that the SCC could have imposed.

Any finding made, or sentence passed, by the court martial on an appeal replaces the finding or sentence of the SCC. Unless the court martial otherwise directs, any sentence passed on an appeal takes effect from the beginning of the day on which the SCC passed sentence. If the appellant wishes to appeal further, they are to be treated, for the purpose of enabling them to appeal under the Court Martial (Appeals) Act 1968 against the sentence, as if they had been convicted by the court martial of the offence in respect of which the sentence was passed.

The respondent to an appeal under this section is the DSP. For the purposes of the section, 'sentence' includes any order made by a court when dealing with an offender.

Where a sentence (including an order) is passed on a person on an appeal under section 285 against sentence, the person is to be treated, for the purpose of enabling them to appeal under the Court Martial (Appeals) Act 1968 against the sentence, as if they had been convicted by the court martial of the offence in respect of which the sentence was passed.

The court martial may impose a suspended sentence order, under section 200 AFA 2006, in respect of any sentence of imprisonment. The court is able to suspend the sentence for an operational period of up to 24 months (s. 190 AFA 2006, amended by s. 6 (2)(b) AFA 2016).

---

35 AFA 2006, s. 285.
36 *Ibid,*, s. 286(2).
37 *Ibid.*, s. 286(3).
38 *Ibid.*, s. 286(6).

The trial judge advocate who passed the sentence is not to be a member of the court hearing the appeal nor any other judge advocate before whom there was (in the proceedings under appeal) a hearing (s. 286(5)).

Procedure

Main Provisions: section 285 AFA 2006; rule 125 AF(CM)R 2009

The right to bring an appeal from the SCC to court martial is contained in section 285 of the AFA 2006. AF(CM)R 2009, rule 125(1), stipulates that the appeal shall be brought by serving a written notice of appeal on the CAO who shall then serve the notice on the DSP. Unfortunately, there is no official form for the written notice required. However, a suggested form appears on page 552.

Time Limit

Main Provisions: section 285 AFA 2006; rules 126, 127 AF(CM)R 2009

Any appeal to a court martial must be brought[39]

(a) within the period of 28 days beginning with the date on which the person was sentenced ('the initial period'); or
(b) within such longer period as the Court Martial may allow by leave given before the end of the initial period. Nevertheless, the Court Martial may at any later time give leave for an appeal to be brought within such period as it may allow.

An appeal lies from here to the CMAC and the appellant is to be treated, for the purpose of enabling them to appeal under the Court Martial (Appeals) Act 1968 against the sentence, as if they had been convicted by the court martial of the offence in respect of which the sentence was passed.

Where two or more defendants are convicted in the same proceedings by the SCC and both or all of them appeal against conviction, their appeals shall be heard together.[40] The court martial may only pass a sentence that the SCC had power to pass in respect of the offence. In other words, it is

39 *Ibid.*, s. 285(3).
40 AF(CM)R 2009, r. 128(5).

not restricted to the sentence actually passed but must be limited to such sentence as was open to the SCC to pass.

Bail Pending Appeal

Main Provisions: section 42 AFA 2006; section 25 CJPOA 1994; Articles 3–6 and Schedule 1 AF(CRFC)O 2009

Where the SAC or SCC has imposed a custodial sentence, and the defendant wishes to appeal to the court martial, the question of release from custody pending appeal to the court martial is dealt with in the Armed Forces (Conditional Release from Custody) Order 2009. A person seeking bail will make their application to a judge advocate under Article 3(1)(b), using the Form in Schedule 1 (see above, on page 543). While the rules seem to cater for such an application being entertained by the trial judge who has just passed sentence, it is probably advisable to have it listed before another judge, so that a fresh mind looks at the merits of the application.

The judge advocate may grant the appellant bail pending the determination of their appeal. The judge advocate may also revoke bail which has been already granted to an appellant or vary conditions of bail granted. Such an application for revocation or variation would be made by the DSP or a surety, or on a reference to a judge advocate by the court administration officer within 28 days from:

a. the date of the conviction which is the subject of the appeal;
b. in the case of appeal against sentence, the date on which the sentence was passed.

No bail is available for defendants convicted of certain offences after a relevant previous conviction of such offences, that is civil offences under section 42 of the 2006 Act for which the corresponding offence under the law of England and Wales, is an offence specified in section 25(2) of the Criminal Justice and Public Order Act 1994.

'Relevant previous conviction' means a previous conviction for:

a. manslaughter;
b. an offence under section 70 of either of the 1955 Acts or section 42 of the 1957 Act for which the corresponding civil offence, within the meaning of those Acts, is an offence of manslaughter;

c. an offence under section 42 of the 2006 Act for which the corresponding offence under the law of England and Wales, within the meaning of that section, is an offence of manslaughter; or
d. culpable homicide.

Any time spent on release on bail does not count as part of the term of imprisonment or detention (r. 5). The procedure for making an application to the judge advocate about bail pending appeal is set out in Article 6. Relevant forms are in Schedule 1 to the Order. The applicant must serve an application in the form set out as form 1 in the Schedule on the CAO and on each other person referred to in Article 3(3)(a), namely, the DSP or a surety.

Article 10 provides for two offences, namely the offence of failing without reasonable cause to surrender to custody and the offence of failing to surrender as soon as reasonably practicable after having failed to surrender because of reasonable cause.

Article 11 gives the judge advocate the power to issue a warrant for the arrest of a person or who has absconded following their surrender. The Article also allows the arrest of a person in the absence of a warrant in certain specified circumstances.

Article 12 applies where the grant of bail is conditional on the provision of sureties to secure the person's surrender to custody. It sets out the factors which may be considered by the judge advocate in deciding on the suitability of a surety, together with the procedure for the surety to enter recognizances.

Article 13 sets out the procedure for the judge advocate to order forfeiture of a recognizance in the event of the appellant failing to surrender to custody. Finally, Article 14 creates the offence of agreeing to indemnify sureties.[41]

Special Measures

The applicable rules are those which apply to all cases before the court martial. See above, on pages 420 and 424 and also Appendix N Special Measures Directions.

41 Extracted from the EN.

Form of Notice – Appeal from SCC to Court Martial
Suggested written notice of appeal under section 285 Armed Forces Act 2006 from the Service Civilian Court to the Court Martial: see Figure 16.4 Appeal from SCC to CM on page 552.

Form of Notice – Appeal from SCC to Court Martial – Extension of Time
Suggested written notice of under section 285 Armed Forces Act 2006 from the Service Civilian Court to the Court Martial: see Figure 16.5 Appeal from SCC to CM Extension on page 553.

To the Court Martial Appeal Court

Main Provisions: CM(A)A 1968; section 272 and Schedule 8 AFA 2006; CMACR 2009; CMAC(E)O 2009; CMAC(B)O 2009; YJCEA 1999(ACMAC)O 2006

It was not until the advent of the Armed Forces Act 2001 that a convicted service person had any right to appeal their sentence to the CMAC. Until then, an appeal only lay against conviction. Section 8 of the Courts Martial (Appeals) Act 1968 and the relevant provisions of the three Service Discipline Acts were amended to set out a procedure for such an appeal to take place, provided a petition was first submitted to the Defence Council. This was later consolidated in section 272 and Schedule 8 of the AFA 2006 (which amended the Courts Martial (Appeals) Act 1968 and also renamed the court as the Court Martial Appeal Court.) The right of appeal under section 8 CM(A)A 1968, as amended, now extends to an appeal against conviction and against any sentence (not being a sentence fixed by law) passed for the offence for which the appellant was convicted (whether passed on conviction or in subsequent proceedings).[42]

The CM(A)A 1968, section 5, provides that the Appeal Court shall be duly constituted if it consists of an uneven number of judges not less than three, when it

(a) determines an appeal against conviction, or a finding of not guilty by reason of insanity; or a finding of unfitness to stand trial;

42 The words in brackets were added by Sch. 8(7) AFA 2006.

To:
Court Administration Officer

NOTICE OF APPEAL TO THE COURT MARTIAL
In accordance with s. 285 Armed Forces Act 2006 and rule 125 Armed Forces (Court Martial) Rules 2009

I, *(Full name of appellant)*

Having been found guilty by the **Service Civilian Court** sitting at *(place)* on *(date of SCC decision)* of the following offence(s)

(List the charges)

And sentenced on *(specify the date)*** as follows:

(detail the sentence awarded)

Intend to appeal against my Conviction/Sentence *(delete as appropriate)**

On the following grounds: *(Set out the grounds on an accompanying sheet on which your appeal from the Service Civilian Court is based. Provide sufficient information to enable matters relied upon to be clearly identified. Wording such as "the conviction is unsafe and unsatisfactory" or "the sentence is in all the circumstances too severe" will be ineffective as grounds and an extension of time may have to be applied for.)*

Signed.. Date: ....

Notes:
* Where you pleaded guilty at the SCC trial, you may appeal against your sentence. If you did not plead guilty, but were found guilty, you may appeal against your conviction or sentence
** Any appeal must be brought within the period of 28 days beginning with the date on which the person was sentenced

Figure 16.4 Appeal from SCC to CM

To:
Court Administration Officer

NOTICE OF APPLICATION FOR PERMISSION TO APPEAL TO THE COURT MARTIAL
In accordance with s. 285 Armed Forces Act 2006 and rule 125 Armed Forces (Court Martial) Rules 2009

I, *(Full name of appellant)*

Having been found guilty by the **Service Civilian Court** sitting at *(place)* on *(date of SCC decision)* of the following offence(s)

(List the charges)

And sentenced on *(specify the date)*** as follows:

(detail the sentence awarded)

Give notice that I apply to the Court Martial for the following

Extension of time within which to apply for permission to appeal:

Yes/No?

Leave to appeal out of time: *(the application must state why you did not (appeal within the initial period ** and be (accompanied by the proposed notice of appeal.*

Yes/No?

Signed...*** Date: ....

Notes:
* Where you pleaded guilty at the SCC trial, you may appeal against your sentence. If you did not plead guilty, but were found guilty, you may appeal against your conviction or sentence.
** An application for extension of time must be brought within the initial period of 28 days beginning with the date on which you were sentenced.
*** This notice must be signed by the appellant or by his legal representative.

Figure 16.5 Appeal from SCC to CM Extension

(b) determining an application for leave to appeal to the Supreme Court; and
(c) refusing an application for leave to appeal to the Appeal Court against conviction or any finding of insanity or unfitness, other than an application which has been refused by a single judge.

However, if one of the appointed judges is unable to continue, the court may continue to hear the appeal.

At least one of the judges of which the Appeal Court consists at any sitting must be a judge of the court (i.e. the CMAC).

With the leave of the court,[43] appeals may be brought against conviction and against any sentence (not being a sentence fixed by law) passed on the appellant for the offence for which they were convicted whether passed on conviction or in subsequent proceedings.[44] In addition, an appeal may also be brought, with the leave of the Appeal Court, by a person on whom a fine is imposed or against whom a compensation order is made under paragraph 13 of any of those Schedules (parents and guardians subject to service jurisdiction).[45]

The appellant must provide the information and documentation required by rule 46 of the CMACR 2009. The Rules provide for service of documents, oaths, evidence to be given through live link, interpreters, qualifications of legal representatives and rights of audience. Public interest considerations are found in rule 23. Nothing 'shall affect any rule of law which authorises or requires the withholding of any document or the refusal to answer any question on the grounds that the disclosure of the document or the answering of the question would be injurious to the public interest.'

An amendment was introduced by the AF(SCR)(A)R 2015 which introduced a new rule 29A into the AF(SAC)R 2009, providing for interpretation and translation for persons other than the appellant, namely, the complainant and any other witness who is required to give evidence at a hearing. Provided the court administration officer is satisfied that the person needs interpretation, they shall appoint an interpreter to act at the hearing.

43 See s. 8(1)CM(A)A 1968 and CMACR 2009, r. 46.
44 S. 8, CM(A)A 1968, as amended by Sch. 8(7)(a)(ii) AFA 2006.
45 CM(A)A 1968, s. 8(1)(a).

Time Limits

Main Provisions: section 9 CM(A)A 1968; rule 47 Part 7 CMACR 2009; CM(RCMS)(SP) R 2009

Anyone wishing to appeal to the CMAC may only do so with leave of that court. They must lodge their application in accordance with the Court Martial Appeal Court Rules 2009. The period within which an appeal notice must be lodged with the registrar is the period of 28 days after the date of the

a. conviction, verdict, or finding,
b. sentence, or
c. order,

in relation to which the appellant wants to appeal.[46]

A party on whom the registrar serves an appeal notice may serve a respondent's notice, and must do so if that person wants to make representations to the court, or the registrar or a judge of the court so directs,[47] not more than 14 days after the registrar serves the appeal notice or the direction to do so. It is also possible for a judge of the court, or the registrar, to extend the service period of the respondent's notice under rule 49(4), even after it has expired.[48]

Bail Pending Appeal

Main Provisions: section 25(2) CJPOA 1994; Articles 3–10 CMAC(B)O 2009; Schedule 8 AFA 2006

Save for those defendants convicted of certain offences after a previous conviction of such offences,[49] the CMAC may grant bail to an appellant pending their appeal, revoke bail already granted or vary conditions of

46 CMACR 2009, r. 47.
47 *Ibid.*, r. 49.
48 *Ibid.*, r. 50(1).
49 These are offences specified in s. 25(2) Criminal Justice and Public Order Act 1994, as amended by s. 139 and Sch. 6, para. 32(1) and (2) of the Sexual Offences Act 2003, namely, defendants charged with or convicted by a UK court of homicide or rape after previous conviction of such offences and, in the case of a previous conviction of manslaughter or of culpable homicide, for which they were sentenced to a term of imprisonment or, if they were then a child or young person, to long-term detention under any of the relevant enactments.

their bail.[50] It is also open to the Director SPA to make application to the CMAC in order to seek a revocation or variation of the appellant's bail.

The procedure for making an application for bail pending appeal is set out in Article 6 of the CMAC(B)O, which requires the service of an application (in the form set out as form 1a in the Schedule to that Order) on the registrar, the Director SPA and on any other person referred to in Article 3(2)(a).

It is important to note that Article 5 provides that any time spent on release on bail does not count as part of the term of imprisonment or detention.

If the court grants bail, it may require a surety or sureties, or a security, under rule 7(4), or impose conditions in accordance with Article 7(5):

(a) to secure that he surrenders to custody at an appointed time and place;
(b) to secure that he does not commit an offence while on bail;
(c) to secure that he does not obstruct the course of justice whether in relation to himself or any other person;
(d) for his own protection or, if he is a child or young person, for his own welfare or in his own interests;
(e) to secure that he makes himself available for the purpose of enabling inquiries or a report to be made to assist a court in dealing with him;
(f) to secure that before the time appointed for him to surrender to custody, he attends an interview with a legal adviser.

Article 8 applies where the court grants the appellant bail subject to a condition that must be met before they are released, when the court may direct how that condition is to be met, for example entering into a recognizance.

Where a person has given security in pursuance of Article 7(4)(b), and the court is satisfied that they failed to surrender to custody then, unless it appears that they had reasonable cause for their failure, the court may order the forfeiture of the security under Article 9(3) or a lesser amount under Article 9(4).

---

50 CMAC(B)O2009, Art. 3.

Article 10 creates an offence of absconding while on bail which is committed when (a) a person fails without reasonable cause to surrender to custody, or (b) having failed to surrender then fails to do so at the appointed place as soon after the appointed time as is reasonably practicable.

Schedule 8 AFA 2006 amended section 43 CM(A)A 1968 by substituting a new section 43(1) and adding two new subsections, as follows:

> (1) The Appeal Court may make an order under this section where
>   (a) but for the decision of the Appeal Court, the accused would be liable to be detained; and
>   (b) immediately after that decision, the Director of Service Prosecutions is granted leave to appeal or gives notice that he intends to apply for leave to appeal.
>
> (1A) An order under this section is
>   (a) an order providing for the detention of the accused so long as any appeal to the Supreme Court is pending; or
>   (b) an order directing that, so long as any appeal to the Supreme Court is pending, the accused is not to be released except on bail.
>
> (1B) Where an order within subsection (1A)(b) is made, the Appeal Court may grant the accused bail pending the appeal.

An appellant may apply to the CMAC for legal aid for an appeal. See JSP 838 (The Armed Forces Legal Aid Scheme) and at Appendix F.

Powers of the CMAC

Main Provisions: section 12–14A and 16–25 CM(A)A 1968; rules 61 and 162 AF(CM)R 2009; CMACR 2009; CMAC(IT)R 2013

The CMAC may confirm, vary or quash any conviction or sentence. It may also activate any suspended sentence of detention. Schedule 8(10) AFA 2006 amended section 12 CM(A)A 1968 to insert the provision that, where the CMAC quash a conviction, the appellant is to be treated as if they had been acquitted by the court martial; but this does not apply if an order under section 19 authorising the appellant to be retried is made.

In adjusting the sentence under section 13, in the case of conviction on two or more charges (multiple relevant offences, where the court has passed a single sentence) when some but not all convictions are successfully appealed and the appellant remains convicted of two or more of the joined offences, those offences are to be treated for the purposes of section 13(2) of the 1968 Act as a single offence.[51]

On an appeal against conviction for an offence to which the appellant did not plead guilty, although the court martial could lawfully have found them guilty of some other offence, if it appears to the CMAC on an appeal against conviction that the court martial must have been satisfied of facts which prove them guilty of that other offence, then instead of allowing or dismissing the appeal, the CMAC can substitute for the finding of the court martial a finding of guilty of the other offence, and may pass on the appellant, in substitution for the sentence passed on them by the court martial, any sentence that they think appropriate, and which is a sentence that the court martial would have had power to pass in respect of that other offence and which is not more severe than the sentence passed by the court martial.[52]

Section 15 CM(A)A 1968, which sets out the power to vary a conviction so as to attract a different sentence, was repealed by Schedule 8(14) AFA 2006.

In an appeal against sentence, section 16A (inserted by Sch. 8(16) AFA 2006) provides that, where the court martial passes two or more sentences on a person, an appeal or application for leave to appeal against any of those sentences is to be treated as an appeal or application in respect of both or all of them. The CMAC may quash the sentence passed by the court martial and pass in substitution for it any sentence that

a. they think appropriate; and
b. is a sentence that the court martial had power to pass in respect of the offence.

But the CMAC may not exercise those powers in such a way that, taking the case as a whole, the appellant is dealt with more severely on appeal than they were dealt with by the court martial. While there is a power in section

51 This amendment to s. 13 CMAA 1968 was made by Sch. 1 AFA 2011 which inserted a new Sch. 3A into the 2006 Act. See r. 162 AF(CM)R 2009, and also Sch. 3A(15) set out on p. 465 *et seq*.
52 CM(A)A 1968, s. 14, as amended by Sch. 8(12) AFA 2006 and Sch. 1 AFA 2011.

29 of the Criminal Appeal Act 1968 for the Court of Appeal to make a loss of time order, relating to forfeiture of time during which an appellant is in custody pending the determination of their appeal, there is no equivalent provision in the CM(A)A.

Interpreter

The Court Martial Appeal Court (Interpretation and Translation) Rules 2013 amend rule 16 of the CMACR 2009 to provide that in the case where a person to whom proceedings relate and who is due to attend a hearing, the registrar must appoint an interpreter to act at the hearing where satisfied that the person needs interpretation. On application or on its own initiative, the court may require a written translation of any document or part of a document to be provided for a person to whom proceedings relate.

Retrial

See also, page 376, above.

There is no automatic right to retrial. The provisions of section 19 CM(A)A 1968, as amended by Schedule 8 of the 2006 Act, give the CMAC the power to order a re-trial for the offence of which the appellant was convicted by the original court martial and in respect of which their appeal was allowed – in other words, where the conviction is overturned on appeal, and it appears to the court that the interests of justice require that an order under section 19 CM(A)A should be made.[53] Where the CMAC directs that a fresh charge or charges may be brought, section 19(4), as amended, stipulates that the DSP must bring the charge or charges so specified, and they are to be regarded as allocated for court martial trial for the purposes of Part 5 of the 2006 Act (relating to investigation, charging and mode of trial). Such charges must be in respect of the case as respects which the original charge was brought (s. 125(5) AFA 2006). See also rule 61 of the AF(CM)R 2009.

Where the CMAC authorise the retrial of a person, they may by order authorise, under the power in the amended section 20(1E), the keeping of

53 CM(A)A 1968, s. 19(1): 'The Appeal Court shall have the power, on quashing a conviction, to make an order authorising the appellant to be retried by the Court Martial], but shall only exercise this power when it appears to the Court that the interests of justice require that an order under this section should be made.'

that person in service custody for up to eight days or, if the person is legally represented and consents, for such period, not exceeding twentyeight days, as the court think appropriate. Alternatively, if they propose to release the person from custody, they may require them to comply with such requirements as seem necessary and for a purpose mentioned in section 107(3) of the 2006 Act (relating to conditions imposed on release from custody after charge – see on page 141 above).

Any arraignment of the defendant must take place within the period of three months of the date of the order authorising re-trial, unless extended by the CMAC. These provisions are contained in section 20(1) CM(A)A 1968, as amended by Schedule 8(21) AFA 2006.

Most importantly, where a person authorised to be retried is again convicted on the retrial, the court martial may not pass a sentence that is (or sentences that, taken together, are) more severe than the sentence (or the sentences, taken together) passed at the original trial.[54]

Prosecution Appeal

Main Provisions: section 58 CJA 2003; section 273 AFA 2006; rule 50 AF(CM) R 2009; rules 4–6 CM(PA)O 2009; Schedule AF(ROCMS) O 2009; AF(ROCMS)(SP)R 2009

Prosecution appeals fall into two categories: appeals during the course of court martial proceedings which are terminated by a ruling by the judge advocate (terminating rulings) and those where the Attorney General refers an unduly lenient sentence to the court. There is, in addition, an equivalent right to that found under Part 10 of the CJA 2003 for a prosecutor to appeal an acquittal where there is new, compelling, evidence of a 'qualifying offence'.[55] In fact, the double jeopardy provisions do not apply, for the reasons explained on page 375.

54 See Sch. 1 CM(A)A 1998 as amended by Sch. 8(54) AFA 2006.

55 Section 75 makes clear that its provisions apply only to qualifying offences, which are listed in Part 1 to Sch. 5 CJA 2003. These are murder, attempted murder, soliciting murder, sexual offences, etc.

Terminating Ruling

The right to appeal a terminating ruling under section 58 of the Criminal Justice Act 2003 also applies to court martial and is governed by rule 4 of the Court Martial (Prosecution Appeals) Order 2009,[56] which is in terms materially identical to section 58 Criminal Justice Act 2003. Rule 4 CM(PA) O 2009 applies where a judge advocate makes a ruling in relation to trial proceedings at an 'applicable time'[57] and the ruling relates to one or more of the charges being tried. The DSP may not appeal unless, *following the making of the ruling*, they inform the court that they intend to appeal, or request an adjournment to consider whether to appeal and, if such an adjournment is granted, inform the court following the adjournment that they intend to appeal. It is vital that the technical provisions are followed. So, 'legislation ... requires that the court be informed of the intention to appeal, or, alternatively, that an adjournment should be requested for the question to be considered. Postponement of both these alternatives is not an option. In other words, unless the prosecution informs the court of its intention to appeal immediately following the making of the ruling, or immediately requests an adjournment to consider whether to appeal, this first pre-condition to an appeal is not fulfilled.'[58]

Furthermore, prosecutors appealing against rulings must give the Article 4(7)/section 58(8) undertaking ('acquittal agreement') in open court at the time of invoking the right of appeal. This acquittal agreement required under rule 4(7) places on the prosecutor the mandatory duty to give an undertaking to the effect that, in respect of the charge or each charge which is the subject of the appeal, the accused should be acquitted of that charge if either of the conditions mentioned in paragraph (8) is fulfilled following the making of the ruling, namely, either leave to appeal to the CMAC is not obtained or the appeal is abandoned before it is determined by the CMAC.

56 This Order, made under powers conferred by s. 323 AFA 2006, replaces and revokes two statutory instruments of 2006 (1786 and 1788) which contained similar provisions, in order to align the provisions with the 2006 Act. Rule 9 amends ss 39(1) (right of appeal) and 42(1) (bail) of the 1968 Court Martial (Appeals) Act.

57 Being any time after the commencement of those trial proceedings and before the time when the judge advocate starts their summing-up to the lay members of the court, *ibid.*, r. 4(12).

58 *R v M* [2012] EWCA Crim 792, citing *R v T (N)* [2010] 2 Cr App R 12; *LSA v The Queen* [2008] EWCA Crim 1034; see also, r. 4 AF(PA)O 2009.

In *R v M*[56] it was pointed out that the acquittal agreement must also be given by the time that notice of the intention to appeal needs to be given, whether that is 'immediately' following the making of the terminating ruling, or by the time allowed by way of adjournment, immediately sought.[59]

See *The Queen v LSA* [2008] EWCA Crim 1034,[60] an interlocutory appeal by the prosecution in a court martial,[61] where the appeal to the CMAC was dismissed for want of jurisdiction.

> The prosecution may not inform the court ... that it intends to appeal, unless, at or before that time, it informs the court that it agrees that, in respect of the charge or each charge which is the subject of the appeal the accused in relation to that charge should be acquitted of that charge if either of the conditions mentioned in paragraph (9) is fulfilled.[62]

The judge advocate's ruling is to have no effect while the DSP is able to take any steps to appeal it.

Where the ruling relates to two or more charges, the DSP must inform the court not only of the intention to appeal but, at the same time, of the charge or charges which are the subject of the appeal.[63]

If the DSP intends to appeal a ruling of 'no case to answer', proceedings may be continued in respect of any charge which is not the subject of the appeal (r. 6).

Where the DSP notifies their intention to appeal, the judge advocate must decide whether or not the appeal should be expedited. If they decide it should not be expedited, they may order an adjournment or discharge

---

59 *R v M*, above p. 561, at para. 9, citing Lord Judge in *R v T(N)*.

60 Mentioned above on p. 554 on a different point, concerning two charges arising out of the same facts.

61 See also, *Brown (Kirsty Elizabeth) R v* [2007] EWCA Crim 2632, discussed above at p. 197, which was the first appeal to be brought under the CM(PA)O 2006. At the outset, the CMAC dealt with a conundrum as to its jurisdiction. How was Article 1, which provides that the Order only applies to trials which commence after 5 July 2006, to be reconciled with Arts 4(1) and 4(13) that purport to make Art. 4 applicable to a ruling that is made before the trial commences? The CMAC stated that Art. 1(2) must be read as if it provides 'this Order shall apply only to trials by courts martial which commence or are to commence on or after 5 Jul 06'. See also the decision in *Armstrong*, [2012] EWCA Crim 83, at p. 162, FN 27; p. 406, FN 34.

62 *The Queen v LSA*, para. 23.

63 CM(PA)O 2009, r. 4(5).

the lay members of the court (r. 5). It would seem an appropriate order to make in serious cases, or multi-handed cases, or where delay impacts on key witness availability.

Unduly Lenient Sentence

Main Provisions: SOA 1956; section 273 AFA 2006;[64] Schedule AF(ROCMS) O 2009; AF(ROCMS)(SP)R 2009; PFA2012(CA)O 2013; Schedule 25 CJIA 2008; section 269 CJA 2003; AF(MA)R 2019

The offences specified as within section 273 AFA 2006, for the purpose of a prosecutor's appeal are listed in the schedule to AF(ROCMS)O) 2009. The AF(MA)R 2019 added three offences to the schedule, namely, section 6 SOA 1956 (intercourse with girl between 13 and 16); section 14 or 15 SOA 1956 (indecent assault on a woman or a man); and section 1 of the Indecency with Children Act 1960 (indecent conduct towards young child). These were offences under section 70 of the AA and AFA 1955 and section 42 NDA 1957, which existed prior to the AFA 2006 coming into force.

This power of referral is reserved to the Attorney General, under section 273 AFA 2006 where they consider

(a) that a sentence passed by the Court Martial in respect of an offence under section 42 (criminal conduct) is unduly lenient, and
(b) that condition A or B is satisfied.

Condition A is that the corresponding offence under the law of England and Wales is under that law an offence which, if committed by an adult, is triable only on indictment.

Condition B is that the case is of a description specified for the purposes of this subsection in an order made by the Secretary of State. The Schedule to the AF(ROCMS)O 2009, made under such authority, lists more than 30 cases which are specified for the purposes of section 273(3) of the Act. This Order is supplemented by the Armed Forces (Review of

64 Sch. 25(28) CJIA 2008 amended the section references in s. 273(6)(b), substituted a new s. 273(7) (determination of minimum term in relation to mandatory life sentence), and inserted a new s. 273(7A) (orders for determination of minimum term in relation to mandatory life sentence/discretionary life sentences and certain other sentences).

Court Martial Sentence) (Supplementary Provisions) Regulations 2009, the provisions of which are broadly equivalent to those set out in Schedule 3 Criminal Justice Act 1988. The Schedule was further supplemented by the AF(MA)R 2019, which added an offence under one of the following sections of the Sexual Offences Act 1956: section 6 (intercourse with girl between 13 and 16); section 14 or 15 (indecent assault on a woman or a man); an offence under section 1 of the Indecency with Children Act 1960 (indecent conduct towards young child).

The statutory provisions are clarified in the AF(ROCMS)O 2009 and the AF(ROCMS)(SP)R 2009. There is a provision in rule 8 AF(ROCMS)(SP) R 2009 that

> [the] term of any sentence passed by the Court Martial Appeal Court under section 273(5)(b) of the Act or the Supreme Court under section 274(4)(b) of the Act shall, unless they otherwise direct, begin to run from the time when it would have begun to run if passed in the proceedings in relation to which the reference was made.

Article 3 AF(ROCMS)O 2009 has been amended by the Protection of Freedoms Act 2012 (Consequential Amendments) Order 2013 to include section 59A[65] and section 61[66] of the Sexual Offences Act 2003 (as amended) as offences in respect of which an appeal lies against an unduly lenient sentence. It came into force on 13 May 2013.

In *Attorney General's Reference No. 32 of 2015*, the defendant, a 28-year-old woman, married to a serviceman serving in Germany, pleaded guilty to four offences of sexual activity with a child – a boy aged 14 – contrary to section 9(1) SOA 2003. She was sentenced to imprisonment for two years on each charge, suspended for two years. The Attorney General successfully applied for the sentence to be reviewed as unduly lenient, pursuant to section 273 AFA 2006. A concurrent sentence of three years' imprisonment was substituted. During the court martial proceedings, the defence had requested a *Goodyear* indication on sentence. The prosecutor, in an endeavour to adopt a pragmatic approach, gave an indication that,

65 S. 59A creates the offence of trafficking people for sexual exploitation, and replaces ss 57–59 of the SOA 2003. See the Protection of Freedoms Act 2012, s. 109(2).
66 Administering a substance with intent.

in the event of a suspended sentence, he would not refer the case to the Attorney General for review. The CMAC underlined, at paragraph 27, that *Goodyear* required prosecuting counsel to remind the judge that the position of the Attorney General to refer any eventual sentencing decision as unduly lenient is not affected by any assistance the prosecuting counsel provides the judge, and that prosecuting counsel should not say anything that may create the impression that the sentence indication has the support or approval of the Crown.

Notice of an application for leave to refer a case to the CMAC must be given within 28 days from the day on which the sentence in the case was passed.[67]

An application to the CMAC for leave to refer a case to the Supreme Court must be made within the period of 14 days beginning with the date on which the CMAC concluded its review of the case referred to it under section 273(1) of the Act.[68]

Regulation 6 AF(ROCMS)(SP)R 2009 provides that time spent in custody by a person whose case is pending its review and pending any reference to the Supreme Court shall be reckoned as part of the term of his sentence.

Regulation 7 explains the entitlement of a person whose sentence is the subject of a reference to the CMAC or the Supreme Court to be present on the hearing of the reference or related proceedings. It also provides that the CMAC may, on a reference under section 273(1) of the Act, pass sentence on a person although they are not present.

Reference of a Point of Law

Main Provisions: section 34 CM A)A 1968; Schedule 25 CJIA 2008; section 274 AFA 2006

As mentioned above, on page 12, the Act makes no provision for a general reference for the purpose of clarifying the law, for example after an acquittal. Section 36 of the Criminal Justice Act 1972 will not apply, as it is concerned with a reference of a point of law following acquittal on indictment. Section 36 permits the prosecutor refer a point of law to Court of Appeal, following

67 AF(ROCMS)(SP)R 2009 rule 3 and CMACR, r. 47.
68 *Ibid.*, r. 5.

acquittal on indictment, for that court's opinion on the point referred. The purpose is to clarify the law. The CMAC may also refer the matter to the Supreme Court. A reference under section 36 does not affect the trial in relation to which the reference is made or any acquittal in that trial. In other words, the defendant is not in jeopardy from the ruling (s. 36(7)).

However, section 34 of the Court Martial (Appeals) Act 1968[69] permits a reference only by the Judge Advocate General[70] or the Secretary of State for Defence. So, the anomaly exists that an independent prosecuting authority is unable to refer the matter, even via the Attorney General, for clarification of a point of law,[71] yet the Secretary of State for Defence, from whom the Director SPA is independent, can. This is a very unsatisfactory provision and directly impinges upon the independence from the chain of command of both the prosecutor and the military justice system as a whole.

Sentence – Point of Law

Where the CMAC has reviewed an unduly lenient sentence, under section 273(1) AFA 2006, the Attorney General or the offender may refer to the Supreme Court a point of law involved in any sentence passed on the offender in the proceedings. That is the only prosecution power of reference on a point of law. The reference will be made where the CMAC has certified that the point of law is of general public importance and it appears to the CMAC or the Supreme Court (as the case may be) that the point is one which should be considered by the Supreme Court. See section 274(1). An error of law on any other ruling cannot be corrected by an appeal by the prosecutor.

Attendance of Witnesses

The Court Martial Appeal Court (Evidence) Order 2009 provides for section 32(1) and (3) of the Criminal Evidence Act 1988, and Chapter 1A of Part 2 (and sections 63 and 65 so far as having effect for the purposes of that Chapter) of the Youth Justice and Criminal Evidence Act 1999, to have effect, with a modification, in relation to proceedings before the CMAC,

69 The title was amended by Sch. 8 AFA 2006, from Courts Martial (Appeals) Act 1968.
70 See also r. 51 Court Martial Appeal Court Rules 2009. The effect of s. 34 is that the appeal by JAG is to be treated as an appeal by the person convicted against their conviction: s. 34(2).
71 This may be corrected in a future Armed Forces Bill.

allowing a witness who is not in the country where the court is sitting to give evidence through a live television link.

Criminal Cases Review Commission

Main Provisions: section 321 and Schedule 11(2), AFA 2006; section 25C and 29A CMAA 1968; section 12A CAA 1995; CJIA 2008; CMACR 2009

Section 12A of the Criminal Appeal Act 1995 provides for a referral by the Commission to the CMAC in respect of any conviction by the court martial (including on an appeal brought from the SCC). Even if they do not refer the conviction they may at any time refer to the CMAC any sentence (other than one fixed by law) imposed by the court martial on, or in subsequent proceedings relating to, the conviction. All the same, the Appeal Court may dismiss the appeal against conviction if

a. the only ground for allowing it would be that there has been a development in the law since the date of the conviction or finding that is the subject of the appeal, and
b. the condition in subsection (3) is met.

The condition referred to stipulates that if the reference had not been made, but the appellant had made (and had been entitled to make) an application for an extension of time within which to seek leave to appeal on the ground of the development in the law, the court would have refused such an extension.[72]

Section 29A was inserted into the Court Martial (Appeals) Act 1968,[73] which provides a power to the CMAC to order an investigation by the Criminal Cases Review Commission for them to report to the court on any matter if it appears to the court that

(a) in the case of an appeal, the matter is relevant to the determination of the appeal and ought, if possible, to be resolved before the appeal is determined;

72 S. 25(3)(c) inserted by Sch. 25 Part 1 CJIA 2008.

73 Inserted by Armed Forces Act 2006, Sch. 11, Powers of the Criminal Cases Review Commission.

(b) in the case of an application for leave to appeal, the matter is relevant to the determination of the application and ought, if possible, to be resolved before the application is determined;
(c) an investigation of the matter by the Commission is likely to result in the Court's being able to resolve it; and
(d) the matter cannot be resolved by the Court without an investigation by the Commission.

A reference by the CCRC of a person's conviction shall be treated for all purposes as an appeal by the person under section 8 of the Court Martial Appeals Act against the conviction. The Commission may give notice to the CMAC that any related conviction specified in the notice is to be treated as referred to that court.

If the CMAC does order an investigation, and the Commission believe that inquiries should be made to assist them in the exercise of any of their functions in relation to any case, they may require the appointment of an investigating officer to carry out the inquiries. That requirement may be imposed on any Provost Marshal to appoint a person serving in a service police force selected by them, or in another service police force selected by them.[74]

In relation to cases dealt with by the SCC, section 12B permits provides that the Commission

(a) may at any time refer the conviction to the Court Martial; and (whether or not they refer the conviction)
(b) may at any time refer to the Court Martial any sentence imposed by the Service Civilian Court on, or in subsequent proceedings relating to, the conviction.

On a reference by the Commission, rule 51 CMACR 2009 stipulates on whom a copy of the reference must be served by the Registrar, namely, the individual concerned and any other person directly affected by the reference.

---

74 Criminal Appeal Act 1995, s. 19, amended by ss 8 and 9 Sch. 11 to AFA 2006.

Notice of Appeal to CMAC

The form of notice of application for permission to appeal is found in Schedule 2 to the Court Martial Appeal Court Rules 2009. It is replicated below, followed by the Notice of Requisition under section 36(2) AFA 2006 (where leave to appeal has been denied and the applicant wishes to make their application to the full court).

Form – Notice of Appeal to CMAC

See Figure 16.6 Notice of Appeal to CMAC on page 570.

The period for appeal against conviction is 28 days from the date of conviction, verdict or order regardless of whether the applicant has been sentenced.

The period for appeal against sentence is 28 days from the date of sentence.

Form – Notice of Requisition

See Figure 16.7 Notice of Requisition on page 571.

Form – Notice of Application Relating to Bail

See Figure 16.8 Notice of Application Relating to Bail on page 572.

Form – Suggested Skeleton Appeal Notice, Completed

See Figure 16.9 Suggested Skeleton Appeal Notice on page 573.

Appeal to the Supreme Court

Main Provisions: section 39 CM(A)A 1968; Schedule 8 AFA 2006; Supreme Court Practice Directions

Section 39 of Court Martial (Appeals) Act 1968, provides that an appeal lies to the Supreme Court, at the instance of the accused or the DSP,[75] from any decision of the CMAC on an appeal to them under Part II of this Act. The appeal may only proceed with leave of the CMAC or the Supreme Court

75 This provision was inserted by Sch. 8, s. 42, AFA 2006, amending the CM(A)A.

NOTICE OF APPLICATION FOR PERMISSION TO APPEAL Rule 46(1) TO THE REGISTRAR OF THE COURT MARTIAL APPEAL COURT Royal Courts of Justice Strand London, WC2A 2LL Email: courtmartialappealoffice@hmcts.x.gsi.gov.uk	
Details required	*Notes*
Details of appellant: Name of appellant: Number: Unit or ship: Address: Rank or rate:	
Details of application Convicted by the Court Martial at: Offence(s) of which convicted: Sentence: Date when sentence passed: Name of place of detention/address:	If not in custody set out the appellant's address in full
Details required	*Notes*
I the above name applicant give notice that I apply to the Court Martial Appeal Court for the following: Extension of time within which to apply for permission to appeal: Yes/No? Permission to appeal against conviction: Yes/No? Permission to appeal against sentence: Yes/No? Leave to call a witness: Yes/No? Leave to be present: Yes/No? If you are applying for legal aid is any solicitor, solicitor advocate or Counsel now acting for you? Yes/No?	The reasons why you consider that your conviction should be quashed and/or that your sentence should be reduced should be set out in a separate document(s) and attached to this form. Where grounds of appeal have been settled by counsel they must be signed by counsel and attached to this form. There is no obligation to include a copy of counsel's advice although in some cases it may be helpful to do so. Grounds must be settled with sufficient detail to enable matters relied upon to be clearly identified. Wording such as "the conviction is unsafe and unsatisfactory" or "the sentence is in all the circumstances too severe" will be ineffective as grounds and an extension of time may have to be applied for. If this form is lodged outside of the prescribed period it must be accompanied by the reasons why the application was delayed. If yes, please give his name and address
Signed: Dated:	This notice must be signed by the appellant or by his legal representative.

Figure 16.6 Notice of Appeal to CMAC

TO THE REGISTRAR OF THE COURT MARTLAL APPEAL COURT

Royal Courts of Justice
Strand
London, WC2A 2LL

NOTICE OF REQUISITION UNDER SECTION 36(2)

Details required	*Notes*
Details of appellant: Name of appellant: Number: Unit or ship: Address: Rank or rate:	If not in custody set out the appellant's address in full
Details of Application	
I the above named applicant having received on –	Insert date of receipt.
your notification that my application for – has/have been refused by a judge of the court. give you notice that I desire that the said application(s) shall be considered and determined by the full court.	Insert details of application.
State any reasons, in addition to those set out by you in your original application, that you wish give –	You must not repeat reasons that you have already stated in previous applications.
Signed: Dated:	This notice must be signed by the appellant or by his legal representative.

Figure 16.7 Notice of Requisition

Form Ia[a]

To: THE REGISTRAR OF THE COURT MARTIAL APPEAL COURT
Royal Courts of Justice
Strand, London WC2A 2LL

NOTICE OF APPLICATION RELATING TO BAIL
Take notice that an application relating to bail will be made to the Court Martial Appeal Court
at:

on ....at.........am/pm
on behalf of [description of applicant]

Particulars of applicant:

Name:

Address:

Particulars of appellant:

Forenames: Surname:

Rank/Title: Service No (if any):

Sentenced by the Court Martial held at:

Offence(s) of which convicted:
Sentence:
Date when sentence passed:
Address (give address where detained):
Address if granted bail:
Nature and grounds of application *(State fully facts relied on. Give details of any proposed sureties and answer any objections raised previously):*

The Registrar should be consulted about the time and place of the hearing before this notice is sent to the other party to the application. A copy of this notice should be sent to the Office of the Judge Advocate General
a Article 6(2), 8(3) and 8(5) and Schedule to CMACB(O) 2009,

Figure 16.8 Notice of Application Relating to Bail
Note: a. Article 6(2), 8(3) and 8(5) and Schedule to CMACB(O) 2009

IN THE COURT MARTIAL APPEAL COURT Case No:
BETWEEN:
LAC JOHN JONES

Applicant

-and-
THE QUEEN

Respondent.

APPEAL NOTICE[a]
Pursuant to Rule 46 of the Court Martial Appeal Court Rules 2009
The Appellant appeals to the Court Martial Appeal Court against conviction [and sentence] by the Court Martial sitting at Colfordrick on the 1st January 2015.

1. The Judge Advocate in the proceedings was HH JA Smith.
2. The Appellant appeals against his conviction [and sentence] on the following charges:

[as in Indictment, stating in each case, the sentence imposed]

THE CASE FOR THE PROSECUTION.

3. [Summarise the prosecution's case on the facts as opened to the Court, subject to any qualifications in the course of the trial.]

THE CASE FOR THE DEFENCE.

4. [Summarise the defence case.]

GROUNDS OF APPEAL AGAINST CONVICTION
[Issue of Fitness to Plead]

5.1 The learned Judge Advocate erred in law by holding that the Appellant was fit to stand trial. (Day 2, 7th December 2014 16.00hrs[b])
5.2 The learned Judge misdirected himself in the way that he analysed the evidence of the two expert doctors called by the prosecution and the two expert doctors called by the defence on the issue of PTSD arising out of conflict in Afghanistan.
5.3 *and ff.*[c]
[Summarise the relevant evidence on both sides]
5. **and ff*
[The relevant law is set out in X Act as interpreted in the case of R v Bloggs, etc.]
5* *and ff*
The learned Judge Advocate in his Ruling stated that [Summarise, or if given in writing at time of delivery, quote passages complained of[d]]
5* *and ff*
It is submitted that:
(a)
(b)
(c)

a. Preferably, the Appeal Notice should be settled by the trial advocate unless his or her conduct of the case is the subject of criticism in the appeal.
b. When transcripts are available, the Grounds of Appeal should be perfected by the advocate who is to conduct the appeal, with appropriate references to the volume, page and line numbers of the transcripts set out in a marginal note.
c. Number each paragraph consecutively.
d. Quote from transcripts when perfecting Grounds with appropriate references.

Figure 16.9 Suggested Skeleton Appeal Notice

[Issue of intent]

6.1 On count 1, The learned Judge Advocate misdirected the Panel on the issue of intent as follows:
[Summarise and give reference: e.g. Summing Up: Day 8 15th December 2014 11.30 hrs'[e]]

6.2 It is submitted that the directions were wrong in law in the following respects:
(a)
(b)
(c)

[Issue of Self-defence]

7.1 On Counts 3 and 4, the learned Judge Advocate failed to remind the Panel of the significant evidence of witness X, to the effect that the Appellant was the subject of a serious attack.

7.2 *and ff*
[Summarise evidence of Witness X and give reference – e.g. Day 5, 11th December 15.00 hrs[f].]

GROUNDS OF APPEAL AGAINST SENTENCE.

8.1 The sentences passed by the Court Martial in respect of each count were excessive in that:
(a) no or insufficient account was taken of the unblemished service of the Appellant in Her Majesty's Forces.
(b) [Details of all mitigation – if advanced at the hearing, include references to the Transcript when available].[g]

8.2 The relevant sentencing powers of the Court Martial in respect of the charges in the Indictment were as follows:
(a) *and ff* [Set out the powers in respect of the offences charged].

APPLICATION UNDER RULE 46 (2). [if required]

9.1 The Appellant applies to the Court for the attendance of the following witness to give evidence at the hearing:
[Give name and address of witness if known]

9.2 The witness was [not] examined at the Court Martial [for the following reasons:
(a) [His identity was unknown to the Appellant at the time of the hearing;]
(b) [Any other reason: eg witness sick or abroad]

9.3 The following is a summary of the evidence the witness is expected to give:
[Summarise evidence].

NOTICE OF APPLICATION FOR PERMISSION TO APPEAL

10.1 The Appellant's Application for Permission to Appeal [or for extension of time for service; for adducing fresh evidence; for an order requiring a witness to attend court] is attached pursuant to Rule 46 (1) (j).

[DONALD MACTHOMAS (Counsel)][h]

Signed: *John Jones* (Appellant)
or *Bloggs & Co* (Solicitors for the Appellant)

5th December 2020

---

e. The Summing Up will be transcribed. Give appropriate references when perfecting Grounds.
f. Quote from transcripts when perfecting Grounds with appropriate references.
g. All matters relied upon must be clearly identified.
h. Where Grounds of Appeal are settled by Counsel, Counsel must sign the document.

itself. Leave shall not be granted unless it is certified by the CMAC that a point of law of general public importance is involved.

The Form of application is set out in rule 10 the Supreme Court Rules 2009:

a. Every application to the Court for permission to appeal shall be made in the appropriate form.
b. An application for permission to appeal must be made first to the court below, and an application may be made to the Supreme Court only after the court below has refused to grant permission to appeal.

Forms

The actual forms are to be found in the Supreme Court Practice Directions.[76] In particular, see PD 4 (Notice of Appeal), PD 5 (Papers for the Appeal Hearing), PD 7 (Applications, documents, forms and orders) and PD 12 (Criminal Proceedings).

An application to the CMAC for leave to appeal to the Supreme Court shall be made within the period of 28 days[77] beginning with the relevant date, defined as the date of the CMAC's decision, or if later, the date on which the CMAC gives reasons for its decision. Either court has the power to extend the time limit. The practice is that where permission to appeal is granted by the Supreme Court, the application for permission to appeal will stand as the notice of appeal and the grounds of appeal are limited to those on which permission has been granted: rule 18(1). The appeal is to be determined by not less than three Supreme Court justices.

Bail Pending Appeal

Main Provisions: Part III CM(A)A1968; Schedule 8 AFA 2006; CMAC(B)O 2009; Article 9 CM(PA)O 2009

76 Available online at http://www.supremecourt.gov.uk/procedures/practice-directions.html.

77 CM(A)A1968, s. 40, as amended, by s. 91(2) Courts Act 2003 (from 14 days).

Prosecution Appeal

Article 9 CM(PA)O 2009 amends provisions of the CMAA 1968 so as to provide for a right of appeal to lie to the Supreme Court against any decision of the CMAC in relation to an appeal against terminating rulings. It mirrors the provisions of the Criminal Appeal Act 1968 in relation to the granting of bail to an accused who has appealed to the Supreme Court.

Where, but for the decision of the Appeal Court, the accused would be liable to be detained; and immediately after that decision, the DSP is granted leave to appeal or gives notice that they intend to apply for leave to appeal, the CMAC[78] can make:

a. an order providing for the detention of the accused so long as any appeal to the Supreme Court is pending; or
b. an order directing that, so long as any appeal to the Supreme Court is pending, the accused is not to be released except on bail.

Generally

While the appeal is pending, a person not subject to service law may be granted bail by the Supreme Court pending their appeal.[79]

78 CM(A)A, s. 43, as amended by Sch. 8 AFA 2006.
79 CM(A)A 1968, s. 43.

17 YOUNG OFFENDERS

Main Provisions: section 61 CJCSA 2000; Schedule 38 CJA 2003; section 51, Part 8, Chapter 5 (sections 208–214) AFA 2006; AF(CM)R 2009; AF(SCC)R 2009; Schedule 2 AFA2006(TP)O 2009

The age at which a person can join HM forces is 16, with full-time adult service from the age of 18. Consequently, there are personnel not yet of adult age who would be classified as juveniles. Where the service court sentences a young offender, that is, a person under the age of 18, there are certain limitations applicable. It is important to note that Schedule 2(5) AFA(TP)O 2009 substitutes '21' for '18' in respect of certain provisions relating to the age when prison is an available sentencing option. The range of sentencing options is laid out in Part 8, Chapter 5 AFA, in particular sections 208–214.[1]

The Service Civilian Court's (SCC) jurisdiction over juveniles is explained in section 51 AFA 2006 as including any service offence committed outside the British Islands by a civilian. Where the young offender is charged with an offence under section 42 which is 'indictable-only', being an offence where the corresponding offence under the law of England and Wales is murder, manslaughter or an offence under section 5 of the Domestic Violence, Crime and Victims Act 2004 (causing or allowing death of child, etc.), or section 227 (firearms offences) would apply, the SCC has no jurisdiction (s. 51(3)(5)). Otherwise the SCC must first consider

1 When s. 61 of the Criminal Justice and Court Services Act 2000 eventually comes into force, no court will be able to pass a sentence of detention in a young offender institution or a sentence of custody for life, and no court will be able to make a custodial order except in relation to a person who is aged at least 17 but under 18. Nor is any court to commit a person to be detained under s. 108 of the Powers of Criminal Courts (Sentencing) Act 2000 (detention of persons aged at least 18 but under 21 for default or contempt) or make an order fixing a term of detention under that section.

whether trial by court martial is more appropriate. This takes place before arraignment. The factors determining the appropriate forum are

a. the nature of the case;
b. the seriousness of the offence;
c. whether its powers of punishment in respect of the offence would be adequate;
d. any other circumstances it considers to be relevant; and
e. any representations made by the Director and the defendant.

If the court decides that the charge should be tried by the court martial, it must refer the charge to that court. The charge is to be regarded for the purposes of Part 5 as allocated for court martial trial.

A young civilian offender subject to court martial proceedings who is under the age of 18 at the commencement of the proceedings (a 'young person'), and who has not appointed a legal representative, may with the leave of the judge advocate be represented in any court martial proceedings by their parent or guardian, in accordance with rule 40(3)(b) AF(CM)R 2009. That person may exercise all rights and duties imposed on the defendant on their behalf, with the exception of pleading to a charge, which the young defendant can do themself.[2] Where there is a requirement to serve a document on the young person, it must be served on the parent or guardian of the young person, in accordance with rule 40(2).

In the SCC, rule 27 AF(SCC)R 2009 makes provision for a parent or guardian of a young person to exercise on their behalf any right they have (save for entering a plea) and, with leave of the court, to represent them in any proceedings.

Prohibition on Imprisonment

Main Provisions: Schedule 2(5) AFA 2006(TP)O 2009; section 61 of CJCSA 2000; sections 208, 209, 211, 212, 223 AFA 2006

Neither the court martial nor the SCC can impose a sentence of imprisonment on a young offender.[3] Indeed, no custodial sentence may be passed unless

2 AF(CM)R 2009, r. 40(3)(a).
3 AFA 2006, s. 208.

the court considers the offence was so serious that no less severe a sentence can be justified for the offence.[4] The provisions contained in the Criminal Justice Act 2003 reducing the minimum age for imprisonment from 21 to 18 years are not in force. In the meantime, Schedule 2(5) AFA 2006(TP)O 2009 provides that, until the coming into force of section 61 of the Criminal Justice and Court Services Act 2000 (CJCSA 2000) (abolition of certain sentences for offenders aged between 18 and 21), '21' shall be treated as substituted for '18' in each of the following provisions of AFA 2006:[5]

(a) section 208 (prohibition on imposing imprisonment on persons under 18);
(b) section 209(2)(a), in relation to an offence punishable in the case of an offender aged 18 or over with imprisonment for 14 years or more, and 209(6) in respect of a sentence of detention that the offender be detained for such period (not exceeding the maximum term of imprisonment with which the offence under section 42 is punishable in the case of a person aged 18 or over) as may be specified in the sentence;
(c) the reference in section 211(1)(a)[6] to an offence punishable with imprisonment if convicted by the Court Martial or the Service Civilian Court in the case of a person aged 18 or over;
(d) section 212(1)(b) in relation to a detention and training order, which may not exceed the maximum term of imprisonment with which the offence is punishable in the case of a person aged 18 or over; and 212(2)(c) concerning the maximum term of imprisonment with which a section 42 offence is punishable in the case of a person aged 18 or over is 51 weeks.

4 S. 260(2).

5 S. 222 contained provisions relating to certain violent or sexual offences. This was removed by LASPOA 20212, Sch. 22, para. 10.

6 Maximum term of imprisonment where the corresponding offence under the law of England and Wales is under that law a summary offence.

Young Offenders Who Commit Murder

Main Provisions: section 51 and Schedule 2(6) AFA 2006(TP)O 2009; sections 217 and 218 AFA 2006; section 93 PCC(S)A 2000; Schedule 2(5) AFA 2006(TP)O 2009

Following conviction by court martial of an offence of murder or any other offence the sentence for which is fixed by law as imprisonment for life, the court must sentence an offender who is aged less than 21 to custody for life.[7]

Schedule 2(6) AFA(TP)O 2009 makes some key amendments in relation to a mandatory sentence of custody for life. The provisions apply until the coming into force of section 61 of CJCSA 2000, where

(a) section 217 of AFA 2006 (conviction of offence for which sentence fixed by law as life imprisonment) applies;
(b) the offender is aged under 21 on conviction; and
(c) section 218(1) of that Act (detention at Her Majesty's pleasure) does not apply.

Life Sentence for Dangerous Offenders Aged Under 18

Section 221 applies where a person aged under 18 is convicted by the court martial of an offence under section 42 (criminal conduct), and the corresponding offence under the law of England and Wales is a serious offence, and the court is of the required opinion (defined by section 223) that there is a significant risk to members of the public of serious harm occasioned by the commission by the offender of further specified offences, or further acts or omissions that would be specified offences if committed in England or Wales.

'Serious harm' means death or serious personal injury, whether physical or psychological and 'specified offence' is a specified violent offence, a specified sexual offence or a specified terrorism offence within section 224 and Schedule 15 of the 2003 Act.

---

7 AFA 2006, s. 218. as amended by Sch. 2(6)AFA(TP)O 2009 and s. 93 PCC(S)A 2000.

Detention

Main Provisions: section 89(1) PCC(S)A 2000; sections 209, 210 AFA 2006; section 221 AFA 2006; Schedule 2(5)(b) and Schedule 2(6) AFA 2006(TP)O 2009; Part 1 Schedule 22 LASPOA 2012 (Dangerous Offenders subject to Service law).

These provisions are distinct from those relating to the power of a CO or service court to impose service detention. Instead of imprisonment, the court martial has other options for those under 21, where the offender is convicted of certain serious criminal offences. There is a power to detain for a specified period. The seriousness of the case is within this provision if the corresponding offence under the law of England and Wales is an offence punishable in the case of an offender aged 21 or over with imprisonment for 14 years or more, and not an offence where the sentence is fixed by law. Cases falling within section 208 are fixed by law.

In relation to certain serious offences, the court has power to detain for a specified period if it is of the opinion that none of the other methods by which the offender may legally be dealt with is suitable. The power under section 209 to pass sentence of detention applies where the offence is

(a) a criminal offence under section 42 of the AFA,[8] and
(b) the case falls under any of the following:
   i. The corresponding offence under the law of England and Wales is an offence punishable in the case of an adult offender with imprisonment for 14 years or more; and is not an offence the sentence for which is fixed by law.[9]
   ii. The corresponding offence under the law of England and Wales is an offence under
      a. section 3 of the Sexual Offences Act 2003 (c. 42) (sexual assault);
      b. section 13 of that Act (child sex offences committed by children or young persons);

8 *Ibid.*, s. 209(1)(a).
9 *Ibid.*, s. 209(2).

c. section 25 of that Act (sexual activity with a child family member); or
d. section 26 of that Act (inciting a child family member to engage in sexual activity). Or
e. The case falls within section 227(1) (certain firearms offences).

The offender can be detained for any period which does not exceed the maximum term of imprisonment with which the offence under section 42 is punishable in the case of a person aged 21 or over.[10]

It is important to remember that there are certain required custodial sentences in respect of dangerous young offenders,[11] young offenders convicted of certain violent or sexual offences,[12] or any offence which falls within section 51A of the Firearms Act 1968, requiring a minimum sentence for certain offences under section 5 of that Act.[13]

Section 210 states that a person sentenced to be detained under section 209 is liable to be detained in such place, and under such conditions, as may be determined by the Secretary of State or by such other person as may be authorised by them for the purpose and that such detention is to be deemed 'legal custody'.

Detention and Training Order

Main Provisions: Schedule 2(5) and (6) AFA 2006(TP)O 2009; sections 211–216 AFA 2005

10 *Ibid.*, s. 209(6), as amended by Sch. 2(5)(b) Armed Forces Act 2006 (Transitional Provisions, etc.) Order 2009. Until the coming into force of s. 61 CJCSA 2000 (abolition of certain sentences for offenders aged between 18 and 21), '21' shall be treated as substituted for '18' in each of the following provisions of AFA 2006— (a) section 208 (prohibition on imposing imprisonment on persons under 18); (b)section 209(2)(a) and (6); (c)the reference in section 211(1)(a) to an offence punishable with imprisonment in the case of a person aged 18 or over; (d) section 212(1)(b) and (2)(c); (e) section 222(3)(d).

11 See s. 221 AFA 2006 and Part 1 Sch. 22 Legal Aid, Sentencing and Punishment of Offenders Act 2012 which deals with and amends the provisions relating to dangerous offenders subject to service law, etc. New sections are inserted into the AFA 2006, namely, s. 218A (life sentence for second listed offence); s. 219A (extended sentence for certain violent or sexual offenders aged 18 or over); and s. 221A (extended sentence for certain violent or sexual offenders aged under 18).

12 *Ibid.*, s. 222.

13 *Ibid.*, s. 227.

The court has[14] the option to pass a detention and training order (DTO) on a person *under* 21 where the accused is convicted of an offence which is punishable with imprisonment in the case of a person aged 21[15] *or over* and the court is of the opinion that the offence, or the combination of the offence and one or more offences associated with it, was so serious that no less severe sentence can be justified for the offence.

Any DTO will consist of a period of detention and training followed by a period of supervision. It replaces the 'custodial order' that was available under the Service Discipline Acts for offenders aged 17.

The term of an order made under section 211 for a detention and training order shall be 4, 6, 8, 10, 12, 18 or 24 months, but may not exceed the maximum term of imprisonment with which the offence is punishable in the case of a person aged 18 or over.[16] However, where the corresponding civil offence is a summary only offence, and the maximum term of imprisonment with which that offence is punishable in the case of a person aged 18 or over is 51 weeks, then the detention order may not exceed six months.[17]

Where a person is convicted by a *civilian court* in England and Wales of an offence punishable with imprisonment which they committed during the supervision period of a detention and training order made by a *civilian court*, section 105 of the Sentencing Act and section 213 AFA enable the court convicting the defendant to make an order for their detention for a further period, up to the period of supervision that remained outstanding at the date of the new offence. One effect of section 213 AFA 2006 is that a civilian court in England and Wales (but not a service court)[18] has the same powers in the case of a person subject to a detention and training order made by a service court.

14 Ss 211(1)(b) and 260(3), as amended by Sch. 22 LASPOA 2012.
15 See Sch. 2(5) AFA 2006(TP)O 2009.
16 S. 212(1).
17 S. 212(2).
18 EN para. 436: 'This section applies to orders made under section 211 many of the provisions governing detention and training orders made by civilian courts under section 100 of the Sentencing Act, so that both kinds of order work in much the same way. In particular, the section enables a civilian court in the UK (but not a service court) to deal with an offender who, following his release from custody, fails to comply with the supervision requirements imposed by the order.'

Section 214 confers similar powers on the court martial and the SCC where they convict a person of a service offence punishable with imprisonment and committed during the supervision period of a detention and training order made by a service court.[19] This means that when a person subject to a DTO is released from detention before the term of the order ends and is then convicted by a service court of a service offence, or an offence under the law of any part of the British Islands, punishable with imprisonment, the service court may make an order under section 214(3) that the person to be detained in secure accommodation[20] in addition to the sentence for the new offence.

The court martial can also exercise these powers if the offender was convicted of the new offence by a civilian court anywhere in the British Islands, or at a summary hearing.[21] In this case the court can issue a summons or a warrant for the offender's arrest, so that it can consider whether to exercise its powers (s. 214(7)). Section 216 provides for an appeal in such a case.

Offender Under 15

In relation to any offender under 15 years of age when convicted, the court cannot make a DTO under section 211, unless it is of the opinion that they are a persistent offender.

Offender Under 12

If the offender is under 12 when convicted, before the court can make an order it must be of the opinion that they are a persistent offender and only a custodial sentence would be adequate to protect the public from the offender. However, at present no order can be made until provision has been made under section 100(2) of the Sentencing Act enabling civilian courts to make orders for offenders of that age (s. 211(3)(c)). As at December 2020, the Secretary of State has not yet made an order appointing a date for this provision to come into force.[22]

---

19 EN para. 437.

20 S. 214(3)(m) as amended, specifies 'youth detention accommodation'.

21 '... otherwise than by the Court Martial or the Service Civilian Court', s. 214(5).

22 See, 'Sentencing Children and Young People Definitive Guideline', para. 6.4, which makes no reference to DTOs for persons under 12. More importantly, the table on p. 23 confirms

Overseas Community Orders

Main Provisions: section 105 PCC(S)A 2000; CJA 2003; sections 182–184 and Schedule 6 AFA 2006; section 78 LASPOA 2012; Chapter 7 AFA2006(TO)O 2009

These orders are available to service courts in respect of young offenders. An overseas community order is an order that imposes on the offender one or more of the requirements mentioned in section 177(1) of the 2003 Act (community orders under that Act) and which does not specify an area where the offender resides or will reside. The order is designed to apply to those who are residing outside the UK and who will continue to do so throughout its duration. It is for this reason that there are some restrictions on the conditions that may be imposed.

The CJA provisions are modified by Schedule 6 AFA. The order may not include a foreign travel prohibition requirement or an alcohol abstinence and monitoring requirement. Section 182[23] and Schedule 6 to the AFA 2006 set out the general requirements that may be specified. These are:

1. **Unpaid work requirement:** This requirement may not be included in an overseas community order if the offender is aged under 16 on conviction.[24] Otherwise, the offender must perform such unpaid work for the number of hours specified in the order and at such times as they may be instructed by the responsible officer.[25] The work required to be performed under an unpaid work requirement of a community order or a suspended sentence order must be performed during a period of 12 months, and remains in force until the offender has worked under it for the number of hours specified in it.

Where an unpaid work requirement is imposed by a suspended sentence order, the supervision period as defined by section 189(1)(a) CJA 2003 continues until the offender has worked under the order for the number of hours specified in the order, but does not continue beyond the

---

that DTOs are not available for children under 12, https://www.sentencingcouncil.org.uk/wp-content/uploads/Sentencing-Children-and-Young-People-definitive-guideline-Web.pdf.

23 Amended by s. 78 LASPOA 2012.

24 Sch. 6(1) AFA 2006.

25 Sch. 6 AFA 2006, s. 199 and s. 200 CJA 2003.

end of the operational period as defined by section 189(1)(b)(ii) of the 2003 Act.[26]

2. **Exclusion requirement:** This means a provision prohibiting the offender from entering a place specified in the order for a period so specified. In the case of a community order, the period specified must not exceed two years. The exclusion requirement may be made to operate only during the periods specified in the order, and may specify different places for different periods or days.[27] Where an overseas community order made in respect of an offender aged under 18 on conviction contains an exclusion requirement prohibiting the offender from entering a place specified in the order for a specified period, such specified period must not exceed three months.[28]

3. **Residence requirement:** In relation to an offender aged under 18 on conviction, the effect of Schedule 6(3) on section 206(1) of the 2003 Act (residence requirement) has effect as if after the words 'at a place specified in the order' there were added 'or with an individual so specified'. In other words, the offender may be required to reside with a specified individual at a specified place. There is an additional requirement that the individual has consented to the requirement, namely that the offender resides with them. However, the individual may not permit the offender to reside at a place other than that specified.

However, Schedule 6(3)(4) provides that no residence requirement can be imposed in relation to an offender under 16 years of age at the date of conviction.

Before making a community order or suspended sentence order containing a residence requirement, the court must consider the home surroundings of the offender.[29]

4. **Mental health requirement:** This is any requirement that the offender must submit, during a period or periods specified in the order, to treatment by or under the direction of a registered medical practitioner

26 Amended by s. 20 LASPO 2012 which, pending the coming into force of s. 61 of the Criminal Justice and Court Services Act 2000, alters terms used for certain punishments in s. 189, e.g. 'in subsection (1), after the first "imprisonment' insert" or, in the case of a person aged at least 18 but under 21, a sentence of detention in a young offender institution'.

27 CJA 2003, s. 205.

28 AFA 2006, Sch. 6(2).

29 CJA 2003, s. 206(3) .

and/or a registered psychologist. It requires the consent of the offender. However, that requirement for consent does not apply if the offender is aged under 14.[30] Moreover, in section 208(1) of the 2003 Act as it applies in relation to an overseas community order, if a medical practitioner is of the opinion that part of the treatment can be better or more conveniently given in or at an institution or place which is not specified in the relevant order they may, with the consent of the offender, make arrangements for them to be treated accordingly. Once again, where the offender is under 14, their consent is not required.

5. **Drug rehabilitation requirement:** In relation to an overseas community order made in respect of an offender aged 14 or over but under 18 on conviction, the drug rehabilitation requirement under section 209(1) of the 2003 Act means that the order requires that the offender must submit to treatment by or under the direction of a specified person having the necessary qualifications or experience, with a view to the reduction or elimination of the offender's dependency on or propensity to misuse drugs. However, the effect of Schedule 6(5) on section 209(1) is to prohibit a drug rehabilitation requirement from including a *requirement* to provide samples for testing unless the offender agrees to do so (as well as agreeing to undergo treatment).

An offender aged under 14 can be required to undergo treatment without their consent,[31] but cannot be required to provide samples at all.

The conditions for imposing the order are those set out in section 209 CJA 2003:

The court must be satisfied

(i) that the offender is dependent on, or has a propensity to misuse, drugs, and
(ii) that their dependency or propensity is such as requires and may be susceptible to treatment,
(iii) that arrangements have been or can be made for the treatment intended to be specified in the order (including arrangements for the reception of the offender where he is to be required to submit to treatment as a resident),

30 Sch. 6(4) AFA 2006 and ss 207 and 208 CJA 2003.
31 Sch. 6(5)(3).

(iv) the requirement has been recommended to the court as being suitable for the offender by an officer of a local probation board[32] or an officer of a provider of probation services, and

(v) the offender expresses his willingness to comply with the requirement. However, this does not apply if the offender is aged under 14.

Formerly, the six-month minimum treatment and testing period specified in section 209(3) did not apply in relation to an overseas community order made in respect of an offender aged under 18 on conviction. However, section 78(10) LASPOA 2012 repealed Schedule 6(5)(4) AFA 2006.

6. **Alcohol treatment requirement:** An alcohol treatment requirement, in relation to a community order or suspended sentence order, means a requirement that the offender must submit, during a period specified in the order, to treatment by or under the direction of a specified person having the necessary qualifications or experience with a view to the reduction or elimination of the offender's dependency on alcohol. The court must be satisfied of the offender's dependency on alcohol. Schedule 6(6) provides that the requirement, as defined by section 212 of the 2003 Act, *may not be included in an overseas community order if the offender is aged under 18 on conviction*.

7. **Education requirement:** In respect of an offender aged under 18 on conviction, the order may include a requirement requiring the offender to comply, during a period or periods specified in the order, with arrangements for their education

(a) made for the time being by their parent or guardian; and
(b) approved by Service Children's Education ('SCE').

The court cannot make the education requirement unless it has consulted SCE and is satisfied that arrangements exist for the offender to receive efficient full-time education suitable to their age, ability and aptitude and to

32 S. 327 substitutes a new s. 5A of the Criminal Justice and Court Services Act 2000 to enable the Probation Board to carry out activities anywhere in the world in relation to persons who are or have been subject to proceedings before the court martial, the SAC or the SCC. Such activities must correspond to activities which the board is required or authorised to carry out in relation to persons who have been charged with or convicted of criminal offences.

any special educational need they may have and that, having regard to the circumstances of the case, the inclusion of the requirement is necessary for securing the good conduct of the offender or for preventing the commission of further offences. The order cannot extend beyond compulsory school age (within the meaning of s. 8 of the Education Act 1996.

Conditional or Absolute Discharge

Main Provision: sections 185–187 and Schedule 3(3)(1) AFA 2006; Article 161 AFA2006(TP)O 2009

This is an order that is only available in respect of a service civilian. There is no equivalent provision in the CO's powers under section 132, or the court martial under section 164, when sentencing a person subject to service law.

A conditional discharge is defined as an order by a service court discharging the offender on condition that they commit no further service offence during a specified period. The maximum period that can be specified is the same as that for which a civilian court in England and Wales can impose a conditional discharge (as at December 2020, three years).

An absolute discharge is an order by a service court discharging the offender absolutely.

These sentences are available under section 185 AFA 2006. A conditional discharge is available only for civilian offenders.[33] Absolute discharge is available for civilians and also for persons *previously* subject to service law (under Sch. 3, para. 3(1)).

Where a court sentences an offender by conditionally or absolutely discharging them, the sentence must not include any other punishment except a service compensation order.

Section 187(1) provides that a conviction of an offence for which an absolute or conditional discharge is ordered shall be deemed not to be a conviction for any purpose other than the purposes of the proceedings in which the order is made and of any subsequent proceedings which may be taken against the offender under section 186 for breach of the order by committing further offences during the period of conditional discharge. If

33 See EN paras 383–386 and Sch. 3, para. 1(1) for the sentencing powers of the court martial in relation to civilians, and s. 282 for those of the SCC.

the convicting court is the *court martial,* it may deal with the offender for the offence for which they were conditionally discharged in *any way in which the court that conditionally discharged them could deal with them* if it had just convicted them of that offence.

If the convicting court is the *SCC,* it may deal with them for the offence for which they were conditionally discharged in *any way in which it could deal with them* if it had just convicted them of that offence.

18 VICTIMS OF AND WITNESSES TO CRIME

Main Provisions: JSP 839; Council Framework Decision 15 March 2001;[1] CMCHA 2007; CJ(AFCPVC)R 2015 (Code of Practice for Victims of Crime)

The military justice system (MJS) has lagged behind its civilian counterpart in dealing with victims of crime, but is trying to catch up, assisted by some independent charitable organisations providing support to service persons and their families. While the CO, even at the summary hearing level, can make a service compensation order (SCO) in favour of a victim of crime, it is the pastoral care of, protection of and support to victims which is perhaps even more important. However, SCOs are not without their limitations.[2] Moreover, they are inadequate to deal with certain types of violent conduct, for example, domestic abuse. In these cases, the court martial can issue service restraining orders, whether it convicts or even acquits the defendant of an offence. This is a very useful tool in tackling harassment. See page 602 below.

Several publications are available to assist victims of crime, some of which are set out hereafter.

Code of Practice on Victims of Crime

JSP 839[3] contains the guidance in support of the code of practice on services to be provided by the Armed Forces to victims of crime. The

1 On the standing of victims in criminal proceedings. It provides that victims' needs should be considered and addressed in a comprehensive, coordinated manner, avoiding partial or inconsistent solutions which may give rise to secondary victimisation.

2 See above, on p. 320 *et seq*.

3 This important document has been made publicly available by MoD and can be accessed online, https://www.gov.uk/government/publications/jsp-839-code-of-practice-on-services-to-be-provided-by-the-armed-forces-to-victims-of-crime.

previous, non-statutory code, was modelled on the Ministry of Justice Code which was originally issued by the Home Office under section 32 of the Domestic Violence, Crime and Victims Act 2004, adjusted for the circumstances and requirements of the Armed Forces. It is mandatory and governs the minimum level of service to be provided. The code was put on a legal footing on 16 November 2015, following the coming into force of the Criminal Justice (Armed Forces Code of Practice for Victims of Crime) Regulations 2015 (CJ(AFCPVC)R 2015) and requires among other things, for service police, COs, victim liaison officers (VLO), the Military Court Service (MCS) and the Service Prosecuting Authority (SPA) to provide certain services in addition to professionally trained victim support. The code is set out in the Schedule to the CJ(AFCPVC)R 2015. For those cases not covered by the code, while service providers are not obligated to deliver the entitlements set out by the code to victims in such instances, the needs of the victim should always be a key consideration in any investigation.[4]

The code explains that a victim is entitled to receive written information on what to expect from the service justice system, such as the victim and witness information leaflets, or the details of a website which contains that information. Depending on the type of crime, the personal circumstances of the complainant, or its relevance to the particular stage of the investigation or of the proceedings, the code specifies the information that must be offered without unnecessary delay from first contact with the service police or CO exercising powers of criminal investigation. This includes information about where and how to get advice or support, including access to medical support, any specialist support (such as psychological support) and alternative accommodation.

All Armed Forces personnel who are victims of relevant criminal conduct wherever it occurs (United Kingdom or overseas) are entitled to and are offered, if practical, immediate unfettered access to victim support services, regardless of whether they consent to their CO being notified. The service police will ask for consent to pass their details to *victim support services*, who are a charity, completely independent of the military chain of command. They will also provide the victim with a leaflet containing the contact details of such support services, should the victim wish to

4 JSP 839, at para. 1.4.

make independent contact at a later date. In the United Kingdom, serving personnel who are victims can access the same services as their civilian counterparts.

For those serving (and their dependants) overseas, the Soldiers, Sailors, Airmen and Families Association (SSAFA) run an accredited independent victim support scheme for the military community. Staffed by skilled volunteers, SSAFA provide information, emotional support and practical help to people affected by a range of crimes, including assault, sexual violence, burglary, racism and domestic abuse. SSAFA Forces Help operates in partnership with the UK national charity Victim Support to provide this service to forces based in Germany and Cyprus. SSAFA recognises that women with a military connection face particular issues, as in cases of domestic violence, as women often do not have the freedom, confidence or independent financial means to find somewhere to go. The service provided is confidential and outside the military chain of command. It does not make a difference whether or not the crime was reported to the police or when or where it happened.[5] Confidential helpline phone numbers are detailed on their website. Also, a confidential email can be sent to Forcesline. SSAFA also provides secure sheltered housing for women and children who may find themselves without somewhere to live or require a place of safety. Additional help is available from the Armed Forces Federation.

The VLO is a person appointed by the suspect's CO and is responsible for keeping the victim informed of events in respect of relevant conduct concerning the suspect or accused. In cases where the suspect is unknown, the VLO will be appointed by the CO of the victim, or if the victim is a civilian the role is likely to be taken by the service police. While it is not the VLO's role to provide emotional support to the victim, they should be aware of the victim support agencies available and be able to provide advice to victims on how to contact them. In the event that the suspect and the victim are from within the same unit, and the victim does not wish the CO to be informed, the unit welfare services are, again, empowered to appoint a VLO while maintaining the confidentiality of the victim.[6]

---

5 http://www.ssafa.org.uk/.

6 Much helpful information has been provided by the response by Mark Francois MP to a Parliamentary question posed by Madeleine Moon MP on 14 January 2013. See HC Deb, 14 January 2013, c593W.

Section 9 of the JSP contains a very useful list of points of contact, support services and resources, along with relevant addresses, email details and telephone numbers.

The SPA has its own policy, 'Victims Right To Review',[7] which is available from their website.[8] The current version is dated November 2015. It sets out how victims can give effect to their right to seek a review of certain decisions taken by the SPA and applies in all qualifying cases from 5 June 2013. The following are 'qualifying decisions', where the SPA:

a. makes the decision not to bring proceedings (i.e. at the pre-charge stage); or
b. decides to discontinue all charges involving the victim, thereby entirely ending all proceedings relating to them;
c. offers no evidence in all proceedings relating to the victim; or
d. decides to leave all charges in the proceedings to 'lie on file'.

The code places obligations upon the SPA to deliver services to victims in cases which have been referred to the SPA by the service police or CO. The SPA is independent of the chain of command and has its own policy which complies with the obligations of the code. Section 6 JSP 839 provides a brief description of these services in order to assist eligible victims to better understand the minimum level of service the SPA provides:

a. charge (including directing the bringing of a charge), allocate the charge for trial by a Service court or refer the case to the Commanding Officer;
b. substitute for the charge another charge against the accused;
c. substantially alter a charge;
d. bring an additional charge against the accused;
e. discontinue proceedings on a charge;
f. offer no evidence on a charge;
g. not bring proceedings (i.e., at the pre-charge stage);

---

7 In *R (FNM) v DPP* [2020] EWHC 870 (Admin), the Court of Appeal held that 'The VRR Scheme gives a complainant dissatisfied with a decision not to prosecute the right to seek an independent review which is carried out by the ARU ... Paragraph 42 of the Guidance states that where the complainant has given reasons for requesting the review, the issues raised will be addressed in the decision letter, where appropriate … There is no duty on the DPP positively to invite representations from complainants seeking a review.

8 http://spa.independent.gov.uk/test/about_us/publication_scheme.htm.

h. leave all charges in the proceedings to lie on the file; and
i. make a direction barring further proceedings.

These decisions are known as qualifying decisions, which are decisions made on after 5 June 2013. Where a qualifying decision is made, the victim will be provided with the reason for the decision (such as that there is insufficient evidence, or that it is not in the public and service interest to prosecute), details of how the victim can obtain further information regarding this decision and how the victim can request a review of the decision in accordance with the SPA Victims Right to Review policy.

The SPA website contains a copy of their policy, entitled, 'Victim's Right To Review', which is a downloadable document explaining, inter alia, which decisions are subject to the scheme; who can apply; how victims can exercise the right to review; what the SPA does with a request for a review; principles for reconsidering a prosecuting decision; who will carry out the review; and the complaints procedure.

Domestic Abuse

'Domestic Abuse: Guidance and Support for the Armed Forces Community – A Handbook for Civilian Support Services', which is available online,[9] provides information about welfare support within the Armed Forces community. It includes a section on children and young people affected by domestic abuse, who will often display signs of anxiety, anger or loss of self-esteem as a result of the abuse they hear and witness.

Also available – although difficult to locate on the government website – is JSP 913 'Tri-Service Policy on Domestic Abuse And Sexual Violence'. It is under review. The extant JSP is structured in two parts: Part 1 is a directive, which provides the direction that must be followed, in accordance with statute or policy mandated by Defence or on Defence by Central Government. Part 2 provides guidance and best practice to assist

9 https://assets.publishing.service.gov.uk/government/uploads/system/uploads/attachment_data/file/642108/Practitioners_Handbook_-_Aug_17_update.pdf. For an interesting research paper see, 'Domestic Violence and Abuse (DVA) in Military Families: Improving Signposting to Specialist Support', by Emma Williamson and Andrea Matolcsi, with Kurda Yar, *Centre for Gender and Violence Research, University of Bristol With Women's Aid (England)*, https://www.fim-trust.org/wp-content/uploads/2019/09/Bristol-Report.pdf.

the user in complying with the directive in Part 1. It includes definitions of relevant terms, legislation and information about support generally available.

Domestic violence is an issue that causes real concern within the service community and is something which the services will not tolerate under any circumstances. The major problem in understanding just how serious an issue it is, is that COs may be dealing with these matters summarily and thereby bypassing the SPA. It is an unfortunate consequence of the system that retains the CO in the disciplinary process for the purpose of criminal offences. Domestic violence is unquestioningly a criminal matter. While policy may advise that such cases be referred to the SPA, policy cannot override the CO's statutory powers set out in the legislation. In cases of sustained abuse, regulation 3 AF(P5AFA2006)R 2009 defines 'prescribed circumstances'[10] where there are circumstances which would indicate to a reasonable person, that a person subject to service law has or may have been the victim of a course of conduct by a person subject to service law (excludes civilian spouses) or an offence under section 42 as respects which the corresponding offence under the law of England and Wales is the offence under section 4 of the Protection from Harassment Act 1997 (putting people in fear of violence on at least two occasions), committed by a person subject to service law (covers civilian and service spouses).

This would require the CO to notify the service police of the matter. However, if the offence subsequently reported by the police, after investigation, is one of assault (up to and including assault occasioning actual bodily harm), and is a first offence, the CO may choose to deal with it themselves and not report it to the SPA. In this way, potential serious offences may not be spotted and dealt with effectively.[11]

10 See the list on p. 113.

11 The civilian case of Rachel Williams illustrates the dangers of failing to act properly and in time. She was shot at point blank range by her estranged husband Darren, who was on bail for beating her up. Remarkably she survived the attack. She had called police seven times between July 9 and August 18, 2011, visited her local station twice and was 'at high risk of harm'. The police arrested Darren Williams twice in a week for threatening to kill his wife and assaulting her – but he was released on bail to live near her. After the attack on Mrs Williams, he hanged himself. Mrs Williams' agony was compounded when her son was found hanged while she was recovering in hospital. Reported in the *Daily Mail*, 28 November 2012. See also the case of Cardiff woman, Joanna Michael, who made two emergency calls at about 2.30am and 2.45am on Wednesday 5 August 2009, while an unnamed witness also twice dialled 999. Ms Michael was stabbed to death by her ex-boyfriend, Cyron Williams, who inflicted 72 separate wounds

Once a domestic violence incident occurs, the accused may be arrested under section 67 and held, without charge, for a maximum of 48 hours, unless extended by a judge advocate to the maximum period of 96 hours. Pre-charge custody before the CO only exists to charge or release when someone has been placed in custody, for example by the service police on arrest. That person's CO must be notified of the fact and the grounds for keeping that person in custody without charge. However, even on release without charge, the absence of a form of police bail means that there is little the CO can do to prevent a violent spouse from returning to the scene of the abuse, other than to give them a clear military order. Some may regard that as an effective and pragmatic way of dealing with the matter, pending the involvement of the SSAFA and the Families' Officer, or a decision by the dependant spouse to be returned to the United Kingdom. After all, it is the CO's responsibility to maintain good order and military discipline within their unit and to protect all those under their command, which, overseas, includes civilians and dependants subject to service discipline. The order would be in place for such time as was necessary to resolve an investigation of the circumstances – even after release from custody without charge, where the investigation was incomplete – and the determination of any decision whether or not to charge or, in the absence of evidence, to repatriate the dependant spouse.[12] As pointed out in the section above on Disobedience to a Lawful Command, on page 196 *et seq*, the order must be a lawful order, that is, it must not be contrary to domestic or international law, and it must have a clear military purpose. Maintaining good order and discipline within the unit is the clearest example of a military purpose.

It therefore would be utterly irresponsible for a CO to allow service persons under their command to return to the matrimonial home, or to contact the victim of their domestic abuse – without it being established there was no risk to the victim and/or any children. At present, once the investigation is complete, if there is insufficient evidence to charge, then

---

in the knife attack. Williams was jailed for life in March 2010. Ms Michael's family attempted to sue the police, but their claim was rejected by the UK Supreme Court by a majority of 5–2 (Lady Hale and Lord Kerr dissenting): *Michael and others v The Chief Constable of South Wales Police and another* [2015] UKSC 2.

12 *Cf.* a standing order of a continuing nature which can prevent servicemen from visiting specified locations where there is a proper service reason for placing them 'out of bounds'.

there would, technically, be nothing to prevent the service spouse from returning home. The CO has no power to make a restraining order, as they can only be issued by a service court. Therefore, once again in order to maintain good order and discipline, even after a charge has been dealt with, the CO can issue a lawful order to the service offender to stay away from the victim.

Experience has shown that failure to deal properly with instances of abuse can lead to a tragic loss of life. Failure on the part of the CO to act diligently according to the risk posed could even lead to disciplinary and/or criminal liability in the event of the death of a victim, including, in an extreme case, gross negligence manslaughter against the CO or corporate manslaughter against the MoD. The CO, as 'senior management' would owe a duty of care to the victim in such case.

Section 1 of the Corporate Manslaughter and Corporate Homicide Act 2007 states:

> (1) An organisation to which this section applies is guilty of an offence if the way in which its activities are managed or organised
> (a) causes a person's death, and
> (b) amounts to a gross breach of a relevant duty of care owed by the organisation to the deceased.
>
> (2) The organisations to which this section applies are
> (a) a corporation;
> (b) a department or other body listed in Schedule 1;
> (c) a police force;
> (d) a partnership, or a trade union or employers' association, that is an employer.
>
> (3) An organisation is guilty of an offence under this section only if the way in which its activities are managed or organised by its senior management is a substantial element in the breach referred to in subsection (1).
>
> (4) For the purposes of this Act
> (a) 'relevant duty of care' has the meaning given by section 2, read with sections 3 to 7;
> (b) a breach of a duty of care by an organisation is a 'gross' breach if the conduct alleged to amount to a breach

of that duty falls far below what can reasonably be expected of the organisation in the circumstances;

(c) 'senior management', in relation to an organisation, means the persons who play significant roles in
   (i) the making of decisions about how the whole or a substantial part of its activities are to be managed or organised, or
   (ii) the actual managing or organising of the whole or a substantial part of those activities.

Under section 2, a 'relevant duty of care' in relation to an organisation[13] includes a duty owed to its employees or to other persons working for the organisation or performing services for it and also *a duty owed as occupier of premises*. The duty is owed to someone for whose safety the organisation is responsible. In Germany, all buildings and barracks are owned by the German government and occupied by the MoD and its personnel. But service spouses and families are within subsection (2) and are owed a duty as persons 'for whose safety the organisation is responsible' (s. 2(1)(d)). The exemptions from liability of the military under section 4 do not apply to an organisation's duties as employer or occupier. The exemption only applies to the conduct, preparation and support of military operations as well as other hazardous and unpredictable circumstances, including peacekeeping operations and operations dealing with terrorism or serious public disorder.[14]

Section 17 stipulates that the consent of the Attorney General is required to prosecute offences of corporate manslaughter.[15] However, section 326 AFA 2006 makes it clear that such consent is not required, save in cases where section 61(2) applies (time limits laid down by any of sections 55 to 58).

Of course, should the CO charge the service person, then they may decide to place them in custody, and that detention will then be subject to judicial scrutiny – it is only a judge advocate who can impose conditions – although it would still be possible, where the judge advocate ordered the

---

13 The Ministry of Defence is an organisation listed in Sch. 1 to the Act.

14 *Ibid.*, s. 4 and see also CPS guidance: 'Corporate Manslaughter'.

15 The SPA and Attorney General would doubtless discuss this in accordance with the agreement set out in the Prosecutors' Convention.

release of the individual, for a CO to give an order to, for example, restrict the movement of the individual if there were an imminent danger of a repeat incident, for example. Banning a spouse from returning to the family married quarter is relatively straightforward. Alternative accommodation must be found for the allegedly abusive spouse. This will usually mean moving them into barracks/single accommodation. However, if the parties reconcile, it is difficult for a CO to prevent this in the absence of a clear and obvious risk of repetition. One can have some sympathy for a CO in such circumstances where the parties to an abusive relationship wish to re-unite.[16] It is a much wider welfare and safety issue, especially where children are concerned; and here a multi-agency approach is advisable.

Prosecutors

Under the JSP 839 guidance, the victims of domestic abuse (DA) are automatically classified as intimidated victims, and prosecutors are trained to consider their specific needs. COs, on the other hand, are neither equipped nor resourced to deal with it, even if they in fact appreciate the point. They may choose not to seek or follow legal advice. Moreover, that legal adviser may not have the same level of expertise in domestic abuse/violence as a SPA specialist lawyer.

In its review of how to assist victims of crime, the Crown Prosecution Service permit victims a right of review of decisions not to prosecute. This was introduced in 2013 and is consonant with the view of the Court of Appeal in *Killick*,[17] where it was held:

> 51. Although in our judgment a complainant has a right to a review, it is for the independent prosecutor to reach a decision impartially, subject only to review by the courts under well-established principles.

16 The unreported General Court Martial case of *C* (1992), involved a charge of manslaughter, prosecuted by the author, where the parties were married with two small children and were involved in a loving, though abusive relationship, especially at weekends when they would engage in an almost ritual of heavy drinking. Each spouse was violent towards the other, but there was little doubt about the depth of affection they felt for each other, and their children. Tragically, on one evening the violence went too far and resulted in a serious injury to the wife, who sustained a broken rib during a fall from a blow delivered by her husband. The rib penetrated the lung and, due to a failure of proper medical diagnosis and treatment, resulted in her very painful death due to infection.

17 See FN 725.

In concluding this chapter, it is worth pointing out the existence of a new charity, the Centre for Military Justice, which provides free, independent, expert legal services to serving or former members of the Armed Forces or their bereaved families. They will assist with inquests, service complaints, advice to victims of crime and other casework, strategic litigation and psychological support. They run a free helpline for initial advice and information on a wide variety of topics. If they cannot help the inquirer, they aim to direct them to someone who can.

19 OTHER IMPORTANT PROVISIONS

Service Restraining Order

Main Provisions: sections 50 and 229 and 232 AFA 2006

Where the defendant is subject to service law or is a civilian subject to service discipline, a service court may make an order under section 229 where it convicts or acquits a defendant of an offence. Such an order prohibits the defendant from doing anything described in the order, and has effect for a fixed period specified in the order or until further order. An order under this section may be made, and a prohibition may be included in the order, only for the purpose of protecting a person mentioned in the order from conduct that amounts to harassment. The section adopts the definition in section 7 of the Protection from Harassment Act 1997 that a 'course of conduct' must involve

a. in the case of conduct in relation to a single person … , conduct on at least two occasions in relation to that person, or
b. in the case of conduct in relation to two or more persons … , conduct on at least one occasion in relation to each of those persons.

It is an offence to breach the order without reasonable excuse, rendering a person who does so guilty of an offence under section 229 AFA 2006 and liable to any punishment mentioned in the table in section 164, but any sentence of imprisonment imposed in respect of the offence must not exceed five years.

In proceedings for an order under this section, the Director of Service Prosecutions and the defence may lead (as further evidence) any evidence that would be admissible in proceedings in the High Court in England and Wales for an injunction under section 3 of the Protection from Harassment Act 1997.

Sexual Harm Prevention Orders and Sexual Risk Orders

Main Provision: Anti-social Behaviour, Crime and Policing Act 2014 Schedule 11, Part 2

The former provisions relating to the service court's power to make a Service Sexual Offences Prevention Order were repealed by the ASBCPA 2014. See above, page 489 for sexual harm prevention orders (SHPOs).

Recognizance Order for Service Parent or Service Guardian

Main Provisions: sections 233–236, 251, 285 to 287 AFA 2006

Two situations are considered by section 233 in respect of an offender subject to service discipline: the offender aged under 16 when convicted and the offender aged under 18 when convicted.

Offender Under 16

Where the offender is aged under 16 when convicted and the court is satisfied, having regard to the circumstances of the case, that it would be desirable in the interests of preventing the commission by them of further offences, it may exercise the power to order that a parent or guardian enter into a recognizance to take proper care of the offender and exercise proper control over them. However, it is essential that this is done with the consent of the offender's service parent or service guardian. If the parent or guardian refuses consent, and the court considers the refusal unreasonable, it may order that parent or guardian pays a fine not exceeding level 3 on the standard scale. If the court decides not to exercise its powers under section 233 to make an order, and the offender is under 16 when convicted, the court must state in open court that it is not satisfied, having regard to the circumstances of the case, that the exercise of those powers would be desirable in the interests of preventing the commission by them of further offences.

An order must not require the parent or guardian to enter into a recognizance for a period exceeding three years; or where the offender will reach the age of 18 in a period shorter than three years, for a period exceeding that shorter period. The financial circumstances of the parent or guardian must be taken into account when fixing the amount of a recognizance. A recognizance under section 233 may, where the court has passed an overseas community order on the offender, include a provision that the service parent or service guardian ensure that the offender complies with the requirements of that order (s. 234(4)).

Offender Under 18

Where the offender is under 18 when convicted, the court has a discretion whether to order the parent or guardian to enter into a recognizance. If it determines to do so, then the same obligations apply as explained above as, too, the consequences of failing to consent to a recognizance (s. 233(1)).

A recognizance may be forfeit under section 236 if the offender commits a service offence during the period of its operation. The court will then order the parent or guardian to pay the sum to which they are bound, or order them to pay part of that sum, or it may remit that sum. Where payment is ordered, the court may make an order under section 251 to allow payment by instalments.

There is a right of appeal, under section 235, to either the court martial, where the order was made by the Service Civilian Court (SCC), or the Court Martial Appeal Court (CMAC), where it was made by the court martial.

Remission – Service Detention

Main Provisions: SCSRSR 2009

Background

The Service Custody and Service of Relevant Sentences Rules 2009 (SCSRSR 2009) make provision concerning service custody and the service of custodial sentences imposed in respect of service offences under the Armed Forces Act 2006, including committal of persons to civil prison or service custody premises, the management of service custody premises, the treatment of persons in service custody and the inspection of service custody premises. They revoke and replace the Imprisonment and Detention (Army) Rules 1979, the Imprisonment and Detention (Air Force) Rules 1980 and the Naval Detention Quarters Rules 1973.

The SCSRSR broadly follow the earlier rules made under Service Discipline Acts preceding the AFA 2006. The main new provisions in the Rules include a new system of disciplinary powers at the Military Corrective Training Centre (MCTC) (under Part 4 of the Rules). This includes power for a Judge Advocate (acting as an adjudicator under the rules) to award up to 28 additional days of service detention for prescribed breaches of discipline by a detainee (rule 51). A punishment imposed by an adjudicator is subject to review under rule 52.

Automatic Remission of Sentence

Under rule 8, a detainee who has been sentenced to 25 or more days of service detention shall be entitled to a period of remission. Where the sentence does not exceed 28 days, they shall be entitled to a period of remission equal to the number of days by which the sentence exceeds 24 days.

If the sentence exceeds 28 days of service detention, the detainee shall be entitled to a period of remission equal to one-third of the period of the sentence, except that if this would result in the detainee serving fewer than 24 days, the period of remission shall be such as to require the detainee to serve 24 days.

If the detainee is serving two or more periods of service detention concurrently, remission shall be calculated under this rule in respect of each of those periods, and the date of final release shall be the later or last date calculated. For the purposes of calculating a period of remission under this rule, a fraction of a day shall be ignored.

Earned Remission of Sentence

Importantly, rule 70 provides that a detainee serving more than 90 days detention may earn extra remission for good conduct (in addition to automatic remission under rule 8) up to a maximum of one-sixth of the period of their sentence in excess of 90 days. This is at the grant of the Commandant of the MCTC.[1] Although the Commandant can no longer order forfeiture of remission, it is possible for a Judge Advocate (sitting as an adjudicator, as mentioned above) to impose additional days of detention at a compliant hearing. The Memorandum explains that the additional days are not strictly forfeiture of remission, although the effect is similar.[2]

Defence legal representatives should bear in mind the difference in remission applicable to a sentence of detention.[3] This is, subject to what is

1 SCSRSR 2009, r. 70.

2 SCSRSR 2009 r. 51, and the Practice in the Court Martial: Collected Memoranda, section 6.

3 See §3.4.2 of the Practice Memoranda.

said above, one third, rather than the half that one would expect for a similar term of imprisonment (a determinate sentence, under the Criminal Justice and Courts Act 2015).[4] In the cases of *R v Ball* and *R v Rugg*,[5] decided pre-2006 AFA, the two appellants were each found guilty of common assault and assault occasioning actual bodily harm on another soldier and were each sentenced to imprisonment for a year and dismissed the service. In the exercise of its discretion, the Reviewing Authority substituted a punishment of a year in detention, being of the view that this was a less severe punishment than imprisonment. Both appealed to the CMAC where the disparity in remission rules was pointed out, which showed that the appellants would serve six months' imprisonment, but eight months in detention. The CMAC was of the view that however benign service in a military corrective training centre may seem to the outsider, the essence of a sentence of detention was loss of liberty. The sentence was varied to nine months' detention.

Rehabilitation Periods

Main Provisions: section 5, section 6 and Sch. 1(7) of ROA 1974; section 65, section 209, section 218, section 378, section 383, Sch. 16 para. 65(2)(a)(i); section 139 LASPOA 2012; LASPOA2012(C9SPSCD)O 2014;[6] PA1997(CRCRM)(A)(EW)O2020[7]

---

4 As at 1 December 2020, the government is considering scrapping automatic early release at the halfway point of a sentence so that those given sentences of four years or more for serious violent crimes would be made to serve a minimum two-thirds of that time in prison before being released.

5 (1998) CMAC, No: 9706363/S2-9706386/S2.

6 Helpful guidance can be obtained from the NACRO website: https://www.nacro.org.uk/resettlement-advice-service/support-for-individuals/disclosing-criminal-records/rehabilitation-offenders-act/ and the Government website, 'Guidance on the Rehabilitation of Offenders Act 1974' https://assets.publishing.service.gov.uk/government/uploads/system/uploads/attachment_data/file/299916/rehabilitation-of-offenders-guidance.pdf. Both were consulted, in addition to the legislation, in the preparation of this section.

7 The Police Act 1997 (Criminal Record Certificates: Relevant Matters) (Amendment) (England and Wales) Order 2020, which amends the rules governing disclosure of convictions to employers through the Disclosure and Barring Service (DBS) certificates by removing the requirement to reveal automatic disclosure of, e.g. youth cautions, warnings or reprimands, or disclosure of multiple convictions. See the very helpful analysis by Samuel March, 'New criminal record disclosure rules take effect', UK Human Rights Blog, 1 December 2020, https://ukhumanrightsblog.com/2020/12/01/new-criminal-record-disclosure-rules-take-effect/.

The provisions of the Rehabilitation of Offenders Act 1974 (ROA 1974) were amended by section 139 of the Legal Aid, Sentencing and Punishment of Offenders Act 2012 (LASPOA 2012), which came into force on 10 March 2014.

Section 5 ROA sets out the rehabilitation periods for particular sentences, while section 6 covers the rehabilitation period applicable to a conviction.

Certain sentences are excluded from rehabilitation under section 5(1) of ROA 1974. These are:

(a) a sentence of imprisonment for life [see also s. 218 AFA 2006];
(b) a sentence of imprisonment, youth custody, detention in a young offender institution or corrective training for a term exceeding forty-eight months;
(c) a sentence of preventive detention;
(d) a sentence of detention during Her Majesty's pleasure or for life under section 90 or 91 of the Powers of Criminal Courts (Sentencing) Act 2000 or under section 209[8] or 218[9]of the Armed Forces Act 2006 or under section 205(2) or (3) of the Criminal Procedure (Scotland) Act 1975, or a sentence of detention for a term exceeding forty-eight months passed under section 91 of the said Act of 2000 (young offenders convicted of grave crimes) or section 209 of the said Act of 2006 or under section 206 of the said Act of 1975 (detention of children convicted on indictment) [see also s. 218 AFA 2006 concerning conviction by court martial of offenders who commit murder, or any other offence the sentence for which is fixed by law as imprisonment for life, when under 18. In these circumstances, the court must

8 Offenders under 18 convicted of certain serious offences: power to detain for specified period. Section 5(1A)(a) of ROA states that references to section 209 of the AFA 2006 include references to section 71A(4) of the Army Act 1955 or Air Force Act 1955 or section 43A(4) of the Naval Discipline Act 1957.

9 Offenders who commit murder, etc., when under 18: mandatory detention at Her Majesty's pleasure. Section 5(1A)(b) of ROA states that the reference to section 218 of the Armed Forces Act 2006 includes a reference to section 71A(3) of the Army Act 1955 or Air Force Act 1955 or section 43A(3) of the Naval Discipline Act 1957.

impose a mandatory sentence of detention at Her Majesty's pleasure] and

(e) a sentence of custody for life;

(f) a sentence of imprisonment for public protection under section 225 of the Criminal Justice Act 2003, a sentence of detention for public protection under section 226 of that Act[10] or an extended sentence under section 227 or 228 of that Act (including any sentence within this paragraph passed as a result of any of sections 219 to 222 of the Armed Forces Act 2006).

The effect is that sentences of over four years' imprisonment are never spent.

The rehabilitation period for a sentence is the period beginning with the date of the conviction in respect of which the sentence is imposed, and ending at the time listed in the Table (set out below as Table 19.1), as amended by section 139(4) of LASPOA 2012, in relation to that sentence. Rehabilitation periods are subject to reduction by half for persons aged under 18.

> There is no rehabilitation period for an order discharging a person absolutely for an offence, or any other sentence in respect of a conviction where the sentence is not dealt with in the Table or under subsection (3), and, in such cases, references in the Act to any rehabilitation period are to be read as if the period of time were nil.
>
> Consecutive terms of imprisonment or other custodial sentences are to be treated as a single term. Imprisonment or other custodial sentences which are wholly or partly concurrent (that is terms of imprisonment or other custodial sentences imposed in respect of offences of which a person was convicted in the same proceedings) are to be treated as a single term. [s. 139(4)]

10 They were abolished by the coalition government and have not been available to those convicted on or after 3 December 2012. However, this was not retrospective, and existing prisoners remained on this sentence. See House of Commons Briefing Paper Number 6086, 6 June 2019, by Georgina Sturge and Jacqueline Beard, https://commonslibrary.parliament.uk/research-briefings/sn06086/.

Sentence	End of rehabilitation period for adult offenders	End of rehabilitation period for offenders under 18 at date of conviction
A custodial sentence of more than 30 months and up to, or consisting of, 48 months	The end of the period of 7 years beginning with the day on which the sentence (including any licence period) is completed	The end of the period of 42 months beginning with the day on which the sentence (including any licence period) is completed
A custodial sentence of more than 6 months and up to, or consisting of, 30 months	The end of the period of 48 months beginning with the day on which the sentence (including any licence period) is completed	The end of the period of 24 months beginning with the day on which the sentence (including any licence period) is completed
A custodial sentence of 6 months or less	The end of the period of 24 months beginning with the day on which the sentence (including any licence period) is completed	The end of the period of 18 months beginning with the day on which the sentence (including any licence period) is completed
Removal from Her Majesty's service	The end of the period of 12 months beginning with the date of the conviction in respect of which the sentence is imposed	The end of the period of 6 months beginning with the date of the conviction in respect of which the sentence is imposed
A sentence of service detention	The end of the period of 12 months beginning with the day on which the sentence is completed	The end of the period of 6 months beginning with the day on which the sentence is completed
A fine	The end of the period of 12 months beginning with the date of the conviction in respect of which the sentence is imposed	The end of the period of 6 months beginning with the date of the conviction in respect of which the sentence is imposed
A compensation order	The date on which the payment is made in full	The date on which the payment is made in full
A community or youth rehabilitation order (Overseas Community Order)	The end of the period of 12 months beginning with the day provided for by or under the order as the last day on which the order is to have effect	The end of the period of 6 months beginning with the day provided for by or under the order as the last day on which the order is to have effect
A relevant order (see below)	The day provided for by or under the order as the last day on which the order is to have effect	The day provided for by or under the order as the last day on which the order is to have effect

Table 19.1 Rehabilitation Periods
Source: ROA as amended by section139(4) of LASPOA 2012

A community or youth rehabilitation order means a service community order or overseas community order under the Armed Forces Act 2006. [s. 139(8)(b)]

Where in respect of a conviction a detention and training order was made under section 100 of the Powers of Criminal Courts (Sentencing) Act 2000, or an order under section 211 of the Armed Forces Act 2006 was made, the rehabilitation period applicable to the sentence shall be —

(a) in the case of a person aged fifteen years or over at the date of his conviction, five years if the order was, and three and a half years if the order was not, for a term exceeding six months;

(b) in the case of a person aged under fifteen years at the date of his conviction, a period beginning with that date and ending one year after the date on which the order ceases to have effect.

A 'custodial sentence' means, *inter alia,* a sentence of imprisonment; a sentence of detention under section 209 for offenders under 18 convicted of certain serious offences, detained for a specified period; and a detention and training order under section 211 of the Armed Forces Act 2006.[11]

'relevant order' means—

(a) an order discharging a person conditionally for an offence,
(b) an order binding a person over to keep the peace or be of good behaviour,
(c) an order under section 1(2A) of the Street Offences Act 1959,
(d) a hospital order under Part 3 of the Mental Health Act 1983 (with or without a restriction order),
(e) a referral order under section 16 of the Powers of Criminal Courts (Sentencing) Act 2000,
(f) an earlier statutory order, or
(g) any order which imposes a disqualification, disability, prohibition or other penalty and is not otherwise dealt with in the Table or under subsection (3),

11 S. 139(8) LASPOA 2012.

but does not include a reparation order under section 73 of the PCC(S)A 2000,

'removal from Her Majesty's service' means a sentence of dismissal with disgrace from Her Majesty's service, a sentence of dismissal from Her Majesty's service or a sentence of cashiering or discharge with ignominy,

'sentence of imprisonment' includes a sentence of penal servitude (and 'term of imprisonment' is to be read accordingly),

'sentence of Service detention' means—

(a) a sentence of service detention (within the meaning given by section 374 of the Armed Forces Act 2006), or a sentence of detention corresponding to such a sentence, in respect of a conviction in service disciplinary proceedings, or

(b) any sentence of a kind superseded (whether directly or indirectly) by a sentence mentioned in paragraph (a).

Section 139 applies in relation to convictions or (as the case may be) cautions before the commencement date (as well as in relation to convictions or cautions on or after that date).[12]

The LASPOA reforms change the way some rehabilitation periods are set so that they are fairer and reflect better the seriousness of the sentences imposed.[13] As a consequence, rehabilitation periods for community orders and custodial sentences will comprise the period of the sentence plus an additional specified period, rather than all rehabilitation periods starting from the date of conviction as it is under the current regime. So, for an example, an adult offender sentenced to two and a half years custody, who would previously have had to declare their criminal conviction for ten years from the date of conviction, will now have to disclose their conviction for the period of the sentence plus a further four years (giving a total rehabilitation period of 6 and a half years). See the NACRO site for more detail (referenced on page 616, FN 6 above).

12 *Ibid.*, s. 141 LASPOA 2012 (transitional provisions).

13 See Ministry of Justice announcement http://bit.ly/1f1gGPR.

Where no provision is made by or under a community or youth rehabilitation order or a relevant order for the last day on which the order is to have effect, the rehabilitation period for the order is to be the period of 24 months beginning with the date of conviction.

Effect of Rehabilitation

A person who has become a rehabilitated person for the purposes of the ROA 1974 in respect of a conviction is to be treated, under section 4, for all purposes in law as a person who has not committed or been charged with or prosecuted for, or convicted of, or sentenced for, the offence or offences which were the subject of that conviction; and, notwithstanding the provisions of any other enactment or rule of law to the contrary, no evidence about that conviction may be adduced in evidence in proceedings, nor may the person be asked about it in proceedings. Section 5 excludes questions being asked seeking information with respect to a person's previous convictions, offences, conduct or circumstances from them or any other person otherwise than in proceedings before a judicial authority.

Subsequent Offending

Offences committed during the a rehabilitation period for an earlier offence have the effect of extending that period until the rehabilitation period for the subsequent offence expires. If the subsequent offence is, for example, an offence of murder, for which there is no rehabilitation period, then any earlier rehabilitation period during which the murder offence was committed will also never expire. For example, on 1 January 2018, D was sentenced to 4 months imprisonment for theft. That conviction would become spent on 1 May 2020. However, on 1 January 2020 he was convicted of murder. This means that the conviction for theft will never be spent. However, a service disciplinary conviction listed in Schedule 1 to the 1974 Act ranks as a conviction referred to in section 6(6)(bb) of that Act and is to be disregarded for the purposes of extending a period of rehabilitation following subsequent conviction.

The Rehabilitation of Offenders Act 1974 (Exceptions) Order 1975 makes exceptions in the operation of the ROA 1974 which came into force on 1 July 1975. The exceptions are concerned with various types of work (Schedule 1), licences, certificates and permits (Schedule 2) and proceedings

(Schedule 3). Exceptions are also made in relation to action taken for the purpose of safeguarding national security (Articles 3(b) and 4(c)). A person applying for employment in the service of the Crown is not exempted from being asked any question in order to assess, for the purpose of safeguarding national security, their suitability where the person questioned is informed that, by virtue of this Order, spent convictions are to be disclosed for the purpose of safeguarding national security.

Motoring Offences

Section 5(8) of ROA 1974 provides that, where in respect of a conviction an order was made imposing on the person convicted any disqualification, disability, prohibition or other penalty, the rehabilitation period applicable to the sentence shall be a period beginning with the date of conviction and ending on the date on which the disqualification, disability, prohibition or penalty ceases or ceased to have effect.

Forfeiture and Deductions from Pay

Main Provisions: sections 341 and 342 AFA 2006; AF(FD)R 2009; JSP 754.

Fines and other payments imposed by court order can be paid through deductions from pay. This has to have a statutory basis, which is to be found in section 341. This expressly states that there can be no forfeiture of or deduction from a service person's pay unless authorised by statute. The following section sets out the circumstances when forfeiture and deductions may be made under regulations made by the Secretary of State enabling the Defence Council, or officers authorised by them, to make orders authorising deductions from pay for the matters listed in section 342(1). The relevant regulations are the Armed Forces (Forfeitures and Deductions) Regulations 2009. The EN to the Regulations summarise the relevant provisions:

Regulation 3 – the Defence Council, or an officer authorised by them, may make an order authorising a day's pay of such a person to be forfeited for certain days of *absence from duty*.

Regulation 4 – the Defence Council, or an officer authorised by them, may by order authorise deductions from the pay of such a person to satisfy amounts paid by or on behalf of the Armed Forces to meet an amount the person has been ordered to pay by a *court of ordinary criminal jurisdiction*.

Regulation 5 – the Defence Council, or an officer authorised by them, may in certain circumstances by order authorise a deduction from the pay of such a person, to be used as *compensation* for any loss of, or damage to *public or service property* caused by that person's wrongful act or negligence.

Regulation 6 – the Defence Council, or an officer authorised by them, may by order authorise a deduction from the pay of such a person to satisfy a payment which he is required to make in respect of certain *financial penalties imposed* under the Act.

Regulation 7 – the Defence Council, or an officer authorised by them, may by order authorise a deduction from the pay of such a person to satisfy a payment which he is required to make under a *maintenance order* of a court *in the United Kingdom or the Sovereign Base Areas*.

Regulation 8 – the Defence Council, or an officer authorised by them, may by order authorise in certain circumstances a deduction from the pay of such a person to satisfy a payment which he is required to make under a maintenance order of a court, tribunal or person *outside the United Kingdom or the Sovereign Base Areas*.

Regulation 9 – defines certain expressions for the purposes of regulations 7 and 8.

Regulation 10 – the Defence Council, or an officer authorised by them, may by order authorise a deduction from the pay of such a person to meet an obligation (s)he has to make a payment in respect of a *child* in accordance with a *maintenance* calculation or maintenance assessment made under the Child Support Act 1991 or the Child Support (Northern Ireland) Order 1991.

Regulation 11 – permits the Defence Council, or an officer authorised by them, to make an order authorising a deduction from the pay of a service person to satisfy any amount the person is required pay by virtue of any *judgment* or order enforceable by a court in the United Kingdom.

Regulations 12 and 13 make transitional provisions.

Minimum Pay

Section 341 importantly provides that a person subject to service law shall nevertheless continue to be paid at not less than such minimum rate as may be prescribed by the Defence Council. Joint Service Publication (JSP) 754 specifies a Minimum Drawing Rate (MDR) which is the minimum rate of pay beyond which no further compulsory deductions should be made from an individual's pay account in respect of the items listed in 02.0406. The MDR mechanism is designed to protect service individuals by guaranteeing that certain otherwise compulsory deductions will not be made from their pay once the MDR threshold is reached. A specific formula is provided to work out the MDR.[14] Any amount authorised to be deducted from the pay of a person subject to service law may also be deducted from any bounty, allowance or grant that may be due to them.

Minor Punishments

Main Provisions: AF(MPLPRR)R 2009

The Armed Forces (Minor Punishments and Limitation on Power to Reduce in Rank) Regulations 2009 (AF(MPLPRR)R 2009) authorise minor punishments that may be imposed on offenders by the court martial and by commanding officers (COs) who have heard charges summarily.

The minor punishments authorised are:

a. stoppage of leave orders – only where the offender is below the rank or rate of warrant officer. A stoppage of leave order is an order that the offender shall not leave a relevant place on a specified number of days without the permission of their commanding officer;
b. restriction of privileges orders – only if the offender is an able rate, marine, soldier, airman or military officer cadet. A restriction of privileges order is an order that on a specified number of days the offender shall perform such extra duties as their CO shall decide; and
c. admonition.

Regulations 3 and 4 make more detailed provision in relation to stoppage of leave orders and restriction of privilege orders respectively. Relevant

14 See JSP 754, Tri-Service Regulations for Pay and Charges, Ch 2, s 4.

place is defined as a naval ship or establishment, a military establishment or an air force station. Restriction of Privileges (ROPs) requires the service person to perform 'extra duties', which means work, training or any other duty performed by the offender at times when they would not otherwise be required to perform any duty. ROPS may not exceed 14 days.

Stoppage of leave and ROPs (but not admonition) may be deferred for up to 28 days.[15] So, for example, the captain of a ship at sea can defer the award until the ship docks.

Regulation 5 prohibits the combination of a minor punishment with certain other punishments (minor or otherwise). Stoppage of leave may be combined with an award of a fine, severe reprimand, reprimand, ROPs or a service compensation order. ROPs may only be combined with a fine, stoppage of leave or a service compensation order. If the CO awards an admonition, the only additional punishment they may award is a service compensation order.

Regulation 6 permits the court martial, or the CO who has heard a charge summarily, after awarding a minor punishment other than an admonition, to defer its effect to a date not later than 28 days after it was awarded. In other words, the punishment is to be treated as if it were awarded at a time not later than 28 days after it was awarded. In that event, any reference in these Regulations to when an order is awarded shall be construed as a reference to when, in accordance with this regulation, it is to be treated as awarded.

Regulation 7 provides for the delegation by COs of the functions conferred on them by these Regulations, that is, to any person subject to service law who is of or above the rank of naval lieutenant, military or marine captain or flight lieutenant.

Regulation 8 prevents the court martial from reducing a member of any of Her Majesty's air forces below the highest rank they have held as an airman (ranks held as an 'airman' are ranks below that of corporal for these purposes). A similar limitation is placed by section 135(3) of the Armed Forces Act 2006 on COs powers to impose reduction in rank.[16]

---

15 *Ibid.*, r. 6.

16 Explanatory Note to AF(MPLPRR)R 2009.

Limitations on Combinations of Punishment

There are limitations on the combinations of punishments that may be awarded. If the CO awards a ROPs order, the only additional punishments they may award are:

a. a fine
b. a stoppage of leave order
c. a service compensation order.

If the CO awards an admonition, the only additional punishment they may award is a service compensation order, as set out in regulation 5(4) AF(MPLPRR)R 2009.

Admissibility of Evidence Before a Civilian Court

Main Provisions: AF(EPCC)R 2009; section 372 AFA 2006; regulation 29 AF(SCC)R 2009; Supra, pp 72 *et seq*

The Armed Forces (Evidence in Proceedings before Civilian Courts) Regulations 2009 (AF(SCC)R 2009) concern the admissibility of evidence in proceedings for an offence created by or under the Act before a court of ordinary criminal jurisdiction in the United Kingdom, the Isle of Man or a British overseas territory. The Regulations are made under the authority of section 372 of the AFA 2006.

The Regulations make provision in respect of evidence of enlistment, evidence as to service, evidence of standing or routine orders and the admissibility of service records, Defence Council instructions or regulations and Defence Council certificates.

Regulation 4 relates to a document purporting to be an enlistment paper as evidence of enlistment of the individual and anything recorded in the document as the answer given by them.

Regulation 5 relates to the admissibility of certified true copies of the enlistment document in Regulation 4.

Regulation 6 provides for evidence as to service, or discharge, or rank, or posting and medals and honours awarded, etc. of an individual in HM forces. Such a document shall, if it purports to be issued by or on behalf of the Defence Council or by a person authorised by them, be evidence of the matters stated in the document.

Regulation 7 deals with a record purporting to be a service record made in pursuance of any Act or of Queen's Regulations, or otherwise in pursuance of naval, military or air force duty. It details the conditions for admissibility (it must be signed by the CO of the person to whom the record relates or by a person whose duty it was to make or keep the record). The accused's service record will hold details such as their career, postings, posts held, promotions and honours and awards.

Regulation 8 deals with certified copies of the service record mentioned in Regulation 7.

Regulation 9 is concerned with the admissibility of Defence Council instructions or regulations.

Regulation 10 deals with Defence Council certificates, namely that any certificate purporting to be issued by or on behalf of the Defence Council or by a person authorised by them and stating that a decoration of a description specified in, or as annexed to, the certificate is or is not a naval, military or air force decoration, or that a badge or emblem of a description specified in, or as annexed to, the certificate is or is not one supplied or authorised by the Defence Council, shall be evidence of the matters stated in the certificate.

Regulation 11 covers standing or routine orders, stating that a certificate purporting to be signed by a person's CO or an officer authorised by the CO to give the certificate, and stating the contents of, or of any part of, standing orders, or other routine orders of a continuing nature, of any of HM forces shall, in proceedings against the person concerned, be evidence of the matters stated in the certificate. An example of how these provisions are relevant would be the case of a person who was AWOL or a deserter and is arrested by, or surrenders to, the civilian police. Relevant Part 2 Orders from the unit, posting a person as being absent from their unit on a particular day, would be sufficient evidence to prove the point of commencement of their absence from the unit and would be admissible in a civilian court, along with a supporting certificate from the CO or an officer authorised by the CO. See above, pp 72 *et seq*.

Case Law – Evolution of CMAC Sentencing Appeals

A substantial body of case law from the military justice system is now available. The CMAC's jurisdiction was extended by the amendment of section 8 of the Court Martial (Appeals) Act 1968 by section 17 of the Armed Forces Act 1996 that introduced a right of appeal against a sentence imposed at a court martial. What follows is a small snapshot of cases that summarise the CMAC's approach to a jurisdiction and ethos few service personnel had experience of. When the first appeals against sentence were heard, it appeared to service personnel that the court struggled to understand the particular requirements of the military environment and the context of offending. This led to what was perceived as some unfortunate decisions that displayed how different the military system is, in some key respects, from the civilian, and how its ethos was perhaps not fully understood.

Love (1997) This misunderstood difference was demonstrated first in *Love*,[17] a case concerning a Royal Military Policeman who sent obscene material through the diplomatic courier system to a contact in Southampton, whom he had never met. He was dismissed from the service after conviction at his court martial. On his appeal to the CMAC, the sentence of dismissal was quashed. The court stated that,

> Court martial sentences are concerned at one and the same time to achieve two things. First to punish service personnel for the criminality of their conduct; second, to deal with them also on a disciplinary basis. In that they are unique. Members of other professions and occupations who transgress the law of the land are dealt with quite separately, (a) by the civilian criminal courts, followed (b) if appropriate by disciplinary proceedings before their own professional bodies. This would be so, for example, in the case of lawyers, doctors, nurses, architects and police officers. These considerations seem to us to be of some importance when it comes to determining what should be this court's approach to these appeals. Hitherto this court has been concerned exclusively with an Appellant's criminality and whether his sentence is wrong in principle or manifestly excessive. By

17 1 Cr App 458, CMAC (1997).

> contrast, statutory appeals from professional disciplinary bodies have gone traditionally to other courts, generally either the Divisional Court or the Privy Council, where a quite different approach is brought to bear upon them ... In the present class of appeal therefore it seems to us that this Court is exercising a somewhat hybrid jurisdiction and that, whilst we are free and clearly intended by Parliament to correct any injustice which we perceive in a court martial sentence, we must nevertheless be mindful that those imposing and confirming such sentence are, generally speaking, better placed than we are when it comes to assessing the seriousness of offending in the context of service life, and deciding upon what particular penalty is required to maintain the discipline and efficiency of the armed forces.

Although the court remarked that it, 'had to assess the seriousness of the offending in the context of service life and the need to maintain the discipline and efficiency of the Armed Forces', in quashing the sentence it showed it did not really appreciate the serious practical difficulties of employing a military policeman in another role in a small service community. Military police are not infantrymen, or engineers, or signallers, each of whom is trained in their role through a series of training courses. A military police officer is similarly specialist trained. Consequently, a service person from one branch or specialisation cannot be transferred easily to another.

Andrews (1998) The case was followed by *Andrews.*[18] A was charged with unlawful possession of 55 milligrams of class B controlled drug (cannabis). Approving *R v Love,* the court held '... court-martial sentences are concerned at one and the same time to achieve two things. First, to punish service personnel for the criminality of their conduct; second, to deal with them also on a disciplinary basis. In that they are unique ...' The service community is not the same as a civilian community and offences which may be at the margin of criminality in the civilian community may be far more serious when committed by service personnel in a service environment. (The sentence was varied on other grounds.)

In earlier times, 'financial loss' was often exploited by defence counsel to the advantage of the service defendant. It seemed that every appellant

18 (1998) Case No 98/0600/S2, decided 23 February 1998. See p. 350, FN 37 above.

was able to produce a financial forecast that showed that they would rise to the highest ranks and that their potential earnings were substantial. Using that as a basis for pension calculation, some eye-watering amounts were being cited as potential losses. Of course, what was ignored was the reality that many appellants never would have a full career. Moreover, little regard was had to the fact that, for similar offending in civilian life, similar financial loss might follow.

Cooney, Allam and Wood (1999) A year after *Love*, the court heard the appeals in *Cooney, Allam and Wood*.[19] The issue common to all three cases was the financial consequence of the sentence on the appellant. In quashing the sentences of dismissal, in each case the court agreed that it 'seems to us clear that, in a closely-knit and inherently hierarchical organisation such as the armed services, there may be a particular need to mark criminal conduct that is inimical to trust or to discipline with a substantial form of punishment'.

McEnhill (1999) This case[20] was concerned with an appellant who was at the time of the commission of the offences a colour sergeant in the Scots Guards, with an unblemished record of 18 years' service. He was married, with three dependent children. He had been sentenced to 112 days imprisonment, dismissal from Her Majesty's service and reduction to the ranks. Dismissal carried with it the loss not only of his salary for the remaining four years of his service, but of his terminal grant and his accrued pension right to the age of 60. The total value of loss was calculated as exceeding £240,000.[21] In addition, the appellant, lacking civilian skills,

19 31 March 1999, [1999] EW Misc 1 (CM). The cases also dealt with the difference in periods of remission for sentences of service detention and sentences of imprisonment. It pointed out that it 'would be 50% in relation to a term of imprisonment up to 2 years, in relation to service detention, as appears in Rule 28 of the Detention Rules, and as described in paragraph 82 of the Sentencing Guide, the maximum period of remission generally available is one-third of the sentence'. This period of remission for detention remains at one third, according to para. 3.4.2 of the Sentencing Guide, Version 3 – Oct 2011.

20 CMAC, 22 January 1999.

21 It is suggested there is a considerable bias in favour of the servicemen in such calculations, as they do not take into account properly the salary that may be earned in civilian life and assume that the dismissed serviceman will never work again and will never again earn pension.

had been unable to find any except menial employment and his and his family's future was uncertain.

In its *dicta,* the court distinguished the approach in *Love* from that in *Andrews*. It held that *Love* was a case where trust and discipline within the appellant's unit were not threatened, whereas in *Andrews*, an offence of possession of Class B drugs, the breach of discipline was a serious element. The court looked at the differences with civilian life and held that:

> The differences – and they are important – are the practical ones (a) that an offence involving indiscipline will in most cases be a serious aggravating feature, and (b) that the range of prescribed penalties differs significantly from those in the civilian courts.

The court acknowledged that the 'service community is not the same as a civilian community and offences which may be at the margin of criminality in the civilian community may be far more serious when committed by servicemen and personnel in a service environment'.

Hanley (2001) Hanley[22] concerned a regimental accountant responsible for the maintenance of the unit's Central Bank Account, which was held on trust for the benefit of service personnel and comprised assets received from such personnel via the shop or mess bar. In October 1999, the appellant was due to take a holiday in the United States, so he handed over the account to a Staff Sergeant colleague and left on holiday. These offences came to light when the SSgt and the bar manager were discussing what action to take over an accounting deficiency on the takings of the mess bar. It became apparent that the bar had a total surplus of roughly £900. On each of the three occasions the bar manager had taken money to the appellant for payment into the bank, the appellant had not paid it in. He was convicted of three offences of theft and sentenced to 120 days' imprisonment, to be dismissed from HM service and to be reduced to the ranks. On appeal, he argued that there would be substantial financial loss to be suffered as a consequence of his discharge from the army, which was in turn the automatic consequence of the sentence of imprisonment imposed. That loss would be in the order of £360,000. CMAC quashed the sentence

22 [2001] EWCA Crim 87.

of imprisonment, quashed the dismissal from the service and substituted a sentence of detention for a period of 90 days. The sentence that he be reduced to the ranks remained.[23]

The court does have regard to the overall criminality and the likely effect on a civilian in employment having committed a similar crime.

Begley (2010) In *Begley*,[24] the appellant was convicted by court martial of two charges of battery. He was sentenced to be reduced from the rank of Bombardier to Lance Bombardier. In refusing his application for leave to appeal, the court remarked that the fact that he was reduced in rank, even though that had obviously severe financial consequences to him in the region about £6,000 a year, and a significant impact on his promotion prospects, was actually a less serious consequence than would have happened in civilian life where he would not only have lost his job, but no doubt been subject to some form of community order.

Since then, things have moved on considerably.

Downing (2010) In the case of *Downing*,[25] heard before the Lord Chief Justice, the appellant pleaded guilty at court martial to one offence of fraud contrary to section 1 of the Fraud Act 2006, in that between 31 January 2007 and 12 December 2008 he submitted 47 claims for payment, with a total value of £7,959.05. He was sentenced to military detention for 12 months and was ordered to be dismissed from Her Majesty's service. He appealed on the basis the sentence of military detention was manifestly excessive or wrong in principle. The CMAC held that this was not a tiny 'expenses fiddle'. There was no basis for concluding that the decision to impose dismissal was manifestly excessive or wrong in principle. Accordingly, the application was dismissed.

R v I (2010) In *R v I*,[26] the appellant pleaded guilty at court martial to making a false entry in an official document contrary to section 62(1)(b) of the Army Act 1955 (see s. 18 AFA 2006 for a similar offence) and to committing the

23 The CMAC emphasised that this case turned upon its particular facts.
24 [2010] EWCA Crim 1758.
25 [2010] EWCA Crim 739.
26 [2010] EWCA Crim 1645.

civil offence of forgery, the forgery being a contravention of section 1 of the Forgery and Counterfeiting Act 1981. For the second offence he was sentenced to a term of 6 months' imprisonment, the operation of which was suspended for 2 years. For the first offence, he was dismissed from Her Majesty's service. On appeal against his dismissal from the service, he argued that his financial loss was just short of £1 million. Referring to the Guidance on Sentencing in the Court Martial, the Appeal Court held that the financial loss flowed from the loss of his career. They said that in order to succeed in an appeal against sentence it is not sufficient to cherry-pick the single offence to which the sentence has been attached even though the Act requires a sentence for each offence. Rather, it is necessary to look at the overall picture and determine whether the sentence for the offending was, in the round, wrong in principle or manifestly excessive. The court cited the *dicta* of the Lord Chief Justice in the cases of *Downing*[27] and *Glenton*,[28] where he held:

> The question whether the criminal activities of a member of the military require dismissal from the Service is pre-eminently, although not exclusively, a decision for the Court Martial. For this purpose, for the assessment of the impact of the applicant's convictions on his ability to continue to serve in the relevant force, the Court Marital must be regarded as an expert tribunal, entitled to the same level of respect to which any such tribunal is entitled when an appeal court is considering its decision. (*Downing*, para. 13)
>
> In addressing the submissions ... we have reminded ourselves that the Court Martial is a specialist criminal court. That does not mean that we accept blindly the decision of the Court Martial, but we must attach due respect to a court which is designed to deal with service issues. (*Glenton*, para. 19)

This approach of looking at the overall criminality was consistent with that taken in *Begley*. The court in *R v I* went on to approve the text of

27 [2010] EWCA Crim 739.
28 [2010] EWCA Crim 930. Also referred to on p. 176, *et seq*.

paragraph 3.3.6 of the Guidance on Sentencing in the Court Martial, and stated that

> the issue of loss of pension and gratuity has been exaggerated and misinterpreted by those seeking to use this fact to argue against dismissal. In civilian life a professional person convicted of serious offences would inevitably lose his employment. He would retain contributions conferring entitlement to a pension which he would be due to receive when reaching pensionable age (normally 65), and he would most likely suffer financial loss by not being able to earn the same amount of money in any subsequent employment. In the Service context, the individual retains a preserved pension when dismissed which reflects the years already served, but he loses the opportunity to qualify for an immediate pension (AFPS75) or Early Departure Payment (AFPSO5) — in effect that future loss is no different from future loss of earnings in the civilian context. It is, therefore, a flawed argument to suggest that someone who has committed an offence meriting dismissal should be retained in the Service solely on the basis that he otherwise would lose the opportunity to qualify for an immediate pension. Nevertheless, any potential loss should be one of the factors which is relevant when considering dismissal.

In spite of an occasional blip,[29] some balance has, it is submitted, at last been introduced into the consideration of dismissal and its related financial consequences.

The Sentencing Guide issued by the Judge Advocate General's Office 'contains judicial guidance on the approach to sentencing and appropriate sentencing tariffs while taking account of the Armed Forces discipline policy. It is intended to assist all concerned in the sentencing of offenders in the Court Martial and other Service courts. Sentencing is a complex and difficult exercise and whilst it must not be reduced to a rigid or mechanistic

29 For example, the struggle to understand the quintessentially military assessment of a sentence of dismissal with disgrace, a sentence specific to the services and without rival in (civilian) criminal sentencing.

process, consistency of approach is essential to maintain public confidence.' It is a very useful document and is available online.[30]

Claims Against Visiting Forces: Transfer of Liability

Main Provisions: VFA 1952; section 25 AFA 2011, AFA 2011(C3) 2013; NATO SOFA

The Explanatory Notes to section 25 helpfully clarify the background to the provision under section 25 of AFA 2011.

The NATO Status of Forces Agreement (SOFA), Article VIII(5), concerns the handling of claims arising from the activities of the armed forces of a sending state while they are in the territory of a receiving state. The Article requires claims in tort arising out of certain acts or omissions of members of a visiting force or of the civilian component of such a force to be dealt with by the receiving state. These are acts done in the performance of official duty and anything else for which a visiting force or civilian component is legally responsible and which causes damage in the territory of the receiving state to anyone but the receiving state itself.

Under the Article, claims must be filed, considered and settled or adjudicated in accordance with the laws and regulations of the receiving state with respect to claims arising from the activities of its own armed forces. The Article also provides for the amount of the costs incurred and any award to be shared by the sending state and the receiving state in specified proportions. For example, where one sending state is responsible, the amount is to be borne 75% by the sending state and 25% by the receiving state.

Section 9 of the Visiting Forces Act 1952 provides for the Secretary of State to make arrangements to handle claims and to settle them but not to defend proceedings for a claim. This is different from the practice of other NATO member states. If the Secretary of State does not succeed in settling the claim and the matter is to be decided by the courts, the sending state has to act for itself. This is not necessarily easy for the sending state, which finds itself in unfamiliar proceedings.

---

30 https://www.judiciary.uk/about-the-judiciary/who-are-the-judiciary/judicial-roles/judges/judge-advocate-general/. This is version 5, from January 2018.

To remedy this, section 25 of the 2011 Act provides for the insertion of a new section 9A into the Visiting Forces Act 1952. This enables the Secretary of State (SoS), if a sending state requests it, to transfer any liability in tort in respect of a relevant claim to the Ministry of Defence. The SoS does this by a written declaration specifying the claim and the time from which any liability in tort is transferred. This will enable them to be substituted as a party in the proceedings in place of the sending state. It does not prevent the SoS defending the claim. They will have only the same liability, and accordingly any defence, that the sending state would have had in the proceedings.

20 SERVICE INQUIRIES AND CORONERS INQUESTS

While these two legal processes are not, strictly speaking, part of the disciplinary system of the services, they are important areas about which practitioners should be aware. The two processes perform separate functions, albeit there is some overlap, for example, in identifying systemic flaws.

The Armed Forces Act (AFA 2006) replaced the single Service Board of Inquiry with the service inquiry, which is an internal process that looks at identifying, for example, flaws in procedures or training, while the coroner's inquest is specific to inquiring into the circumstances of the death of a service person. The coroner may not name who was legally responsible for causing the death. However, they do have the power to investigate not just the main cause of death, but also any acts or omissions that directly led to the cause of death.

This is a complex and specialist area, so what follows is just a summary of the two systems. Further detail should be sought from the legislation and guidance available.

Service Inquiries

Main Provisions: section 343 AFA 2006; AF(SI)R 2008; JSP 832

A 'service inquiry' means an inquiry referred to in section 343 of the AFA 2006. That section provides a power for the Secretary of State to make regulations for causing inquiries to be held in prescribed circumstances in relation to matters connected with any of Her Majesty's forces. Those regulations are the Armed Forces (Service Inquiries) Regulations 2008, which contain the rules governing service inquiries. In other words, if the service authorities require circumstances to be investigated, for example, to understand why equipment failed or malfunctioned, or how training errors could be improved, a service inquiry could be convened. The strict rules of evidence do not apply as this is not a formal court,[1] but the general principle of fairness to those affected is recognised in rule 18.

Joint Service Publication (JSP) 382 provides policy and guidance on the procedures to be followed when convening and conducting service inquiries, and for their subsequent staffing and disclosure. Although a service inquiry is an internal fact-finding investigation convened to establish the facts concerning an accident or incident and to make recommendations to prevent recurrence, the findings and recommendations of an inquiry may be of value to external authorities such as the HM Coroner (or Procurator Fiscal) and to the next of kin where the matter involves death or serious injury (where the serious injury is such that the individual cannot act for themselves). Matters that are of a high-profile nature (see *ibid.* Chapter 1 para 1.18) may also be of general public interest.

The functions of a service inquiry panel are to investigate and report on the facts relating to the matters specified in the terms of reference provided to it by the convening authority and to comply with those terms of reference. After considering the provisional report, the convening authority may either declare the provisional report as the final report or, by notice in writing to the president, amend the terms of reference at any time until it has either declared the inquiry report as final or has required the panel to undertake such further work as they consider appropriate, under regulation 19(4)(b).

Pursuant to rule 4, matters for referral to a service inquiry include the death on or after 1 October 2008 of a person while subject to service law. The convening authority must cause a service inquiry to be held if they consider that anything of consequence to any of the regular or reserve forces, which is not in their opinion apparent from the death, may be learned by any of those forces by means of such an inquiry.

This requirement, however, shall not apply if a service, UK, British overseas territory or overseas police force is conducting, has conducted or informs the convening authority that it will conduct an investigation into the events that caused the death. In other words, a service inquiry will not

1 JSP 832, para. 4.1.

take place until the conclusion of a police investigation and a determination by the prosecuting authority whether to institute criminal or disciplinary proceedings before a service or civilian court, as appropriate.

However, it remains open to the Defence Council to cause a service inquiry to be held in relation to any matter connected with any of Her Majesty's forces, and for that purpose may carry out, or appoint a person to carry out, any of a convening authority's functions under these Regulations. In other words, this is not limited to the death of a service person.

As service inquiries are not open to the public, regulation 18 provides that anyone who, in the president's opinion, whose character or professional reputation may be affected by the inquiry's findings is entitled to attend the inquiry. Attendance is subject to such conditions as the president, after consulting the convening authority, may reasonably impose.

Where a person is required to appear before a service inquiry, as a person affected, (also known as a 'Reg. 18 witness') they are entitled to legal representation or, with the consent of the president, they may be represented by a person other than a legal representative. A person affected has the right to give evidence and ask questions of witnesses or produce any witness to give evidence. Where they are represented, their representative may question witnesses and may, with the permission of the president, address the panel; and the president shall provide 'a potentially affected person' with a copy of any part of the record of the proceedings of the panel, if the president considers it appropriate to do so (reg. 18).

Evidence given by a person to a service inquiry panel shall not be admissible against a person at a summary hearing or in proceedings before a civilian court or a service court unless it is for an offence against section 42 of the Act for an offence under section 2 or 5 of the Perjury Act 1911(3).[2]

Suspension

Where a Service Inquiry is convened, a suspension notice to the president to suspend or defer the proceedings of the panel may be issued by the convening authority for such period as the convening authority considers necessary to allow for[3]

---

2 Armed Forces (Service Inquiries) Regulations 2008, r. 12.

3 *Ibid.*, r. 8.

a. the carrying out of any other investigation, whether in the United Kingdom or abroad, relating to any of the matters to which the Service Inquiry relates or would relate, or
b. the determination of any
   (1) civil proceedings whether in the United Kingdom or abroad;
   (2) proceedings for any criminal offence in any court whether in the United Kingdom or abroad;
   (3) proceedings in a service court; or
   (4) a summary hearing.

This provision covers investigations by a Home Department Police Force, Procurator Fiscal or Director of Public Prosecutions. Their view will generally prevail in determining whether or not a service inquiry should continue concurrently. This is very important to avoid any prejudice to the formal legal process. What is particularly interesting is that the power of the convening authority to issue a suspension order extends to summary proceedings before the Commanding Officer (CO) (reg. 8(1)).

Inquests

Main Provisions: JSP 751 Volume 2 (Management of the Deceased); Boards of Inquiry and Coroners' Inquests; Guidance No. 7, A Cadre of Coroners for Service Deaths (Chief Coroner); MOU between ACPO and the MoD & Service Police 2008 / Home Office circular 028 / 2008; The United Kingdom Forces (Jurisdiction Of Colonial Courts) Order 1965; QRs §J7.147; Coroners and Justice Act 2009; Guide to Coroner Services[4]

Deaths Within the United Kingdom

Where a service person dies in the United Kingdom, the coroner for the area in which the death occurs must, as soon as practicable, conduct an investigation into the person's death if they have reason to suspect that[5]

4 S. 10.3, https://www.coronersociety.org.uk/_img/pics/pdf_1503323588.pdf.

5 Coroners and Justice Act 2009, s. 1. Where a death occurs in or near the Coroner's area and the circumstances of the death are such that there should be an investigation into it, the Senior Coroner has a duty to report it to the Chief Coroner (*ibid.*, s. 1(4)).

(a) the deceased died a violent or unnatural death,
(b) the cause of death is unknown, or
(c) the deceased died while in custody or otherwise in state detention.

Certain circumstances, specified in section 7 of the 2009 Act, require a jury. The responsibility for investigation of the death will be a matter resolved by the protocol[6] between the service police and the local police authority. At any incident involving death or serious injury likely to lead to death or at any incident likely to lead to the investigation of terrorism, murder or manslaughter in the UK and in national security cases, the Ministry of Defence Police (MDP) and the service police will take any action immediately necessary at the scene. They will simultaneously inform the local Home Office Police Force who will lead the investigation.

In the case arising out of the deaths of army recruits at Deepcut, the Surrey police carried out the investigations. Under the Protocol existing at the time, primacy for conducting the investigation of all deaths (on MoD premises) rested with the Chief Officer of the Home Department Police Force under whose jurisdiction the deaths occurred.[7] This jurisdiction was repeated in the updated Home Office Circular 17/1999 dated 25 March 1999.

The Report found that

> [t]here is no doubt that in the case of each of the four deaths at Deepcut the notification [to the civilian police] was given as the civilian police were called to the scene, as was the Coroner's officer.[8]

Deaths Overseas

A The Ministry of Justice and Ministry of Defence produced a useful booklet entitled Boards of Inquiry and Coroners' Inquests, Information for

---

6 MOU between ACPO and the MOD & Service Police 2008: http://library.college.police.uk/docs/hocirc/ho-circ-2008-028.htm and A protocol between police forces and the MoD police Home Office circular 028/2008: https://webarchive.nationalarchives.gov.uk/20130125152335/http://www.homeoffice.gov.uk/about-us/corporate-publications-strategy/home-office-circulars/circulars-2008/028-2008/.

7 The Deepcut Review, Nicholas Blake QC, para. 3.1, https://assets.publishing.service.gov.uk/government/uploads/system/uploads/attachment_data/file/228930/0795.pdf.

8 *Ibid.*, para. 3.5

Bereaved Families, to provide a general overview of the coroner's inquest that might take place after a death and to help explain the purpose. It is available online.[9] Although it refers to Boards of Inquiry, they no longer exist having been replaced in section 343 the 2006 Act by the service inquiry (see above, Service Inquiries).

The Chief Coroner's Guidance No. 7[10] explains that the Chief Coroner has created a specialist cadre of coroners in England and Wales. Members of the cadre, who will be specially trained, conduct investigations and inquests into the deaths of service personnel on active service. Active service also includes training and preparation for active service. Guidance is also contained in the respective QRs for each service. The use of the cadre for investigations and inquests will extend only to 'service deaths' as defined by section 17(2) of the 2009 Act, relating principally therefore to death on active service. Although many of these deaths will occur overseas, the jurisdiction of the cadre will also extend to such deaths in England and Wales.

QRs J7.147 provide:

> When, following a death aboard a ship or aircraft outside territorial waters or in an aircraft while airborne over the United Kingdom (in which case the death is registerable under the Service Departments Registers Orders), the body is landed for burial:
>
> a. In England or Wales – it is necessary to apply to the Registrar for the sub-district in which burial is to take place for a certificate of no liability to register; this certificate must be produced for the purpose of burial; but if (under para. J7.024) an inquest into the death is to be held, the authority for burial will be the disposal certificate issued by the coroner.
> b. In Scotland or Northern Ireland – no such certificate is required but the keeper of the burial-ground must

9 https://webarchive.nationalarchives.gov.uk/20121109062132/http://www.mod.uk/NR/rdonlyres/0FADF3F8-A820-4081-8E33-5F170FFC4B0C/0/BoardsofInquiryandCoronersInquests.pdf.

10 https://www.judiciary.uk/wp-content/uploads/2013/09/guidance-no-7-a-cadre-of-coroners-for-service-deaths.pdf.

> be satisfied that death occurred outside the United Kingdom; the authorities in para. J7.123b (Marriage or Civil Partnership office) should be consulted if cremation is proposed or any difficulty arises.

Since the decision of the Court of Appeal in *R v West Yorkshire Coroner, ex parte Smith* [1983] QB 335 (the case of Helen Smith), coroners in England and Wales must hold an inquest into a death overseas if the body is returned to the coroner's district and the circumstances are such that an inquest would have been held if the death had occurred in England and Wales.

In 1983 Home Office Circular No.79/1983[11] advised that in the case of a single service death overseas the coroner local to where the funeral is to be held should normally conduct the inquest. But in the case of more than one death, the inquest should be held in the jurisdiction where the bodies first arrive (currently usually RAF Brize Norton in Oxfordshire, previously until 1 September 2011 at RAF Lyneham in Wiltshire).

Where the next of kin wish the investigation and inquest to be conducted in their coroner area and the investigation and inquest can take place there, the investigation (including the inquest) will be transferred to that area. The senior coroner of the receiving area will be expected to accept the transfer (under section 2, Coroners and Justice Act 2009).[12]

Defence Inquests Unit

In 2008 the MoD established the Defence Inquests Unit (DIU)[13] to coordinate and manage all defence-related inquests into the deaths of service and MoD personnel who die on, or as a result of, injuries sustained while on operations, and those who die as a result of training activity.

In contrast to a service inquiry, an inquest is an independent judicial inquiry into the facts surrounding a death, and in England and Wales is conducted by a coroner. The inquest may not name who was legally

---

11 Affirmed by the then Minister of State at the Department for Constitutional Affairs, Harriet Harman MP, in the House of Commons on 23 October 2006, see Chief Coroner's Guidance No. 7, para. 5.

12 Guidance No. 7, para. 6, page 3.

13 As the majority of the fatalities were soldiers, the Army set up Project Ajax to improve the situation. In 2008 it became tri-service, and so the DIU was formed: 'Defence Inquests Unit: helping to find the answers', https://www.gov.uk/government/news/defence-inquests-unit-helping-to-find-the-answers.

responsible for causing the death. However, the coroner does have the power to investigate not just the main cause of death, but also any acts or omissions that directly led to the cause of death.

The purpose of the inquest is to determine:

- Who the deceased was; and
- When, where and in what circumstances the deceased came by his or her death.[14]

The inquest will establish the medical cause of death and by what means it arose and may draw attention to the existence of circumstances which, if nothing is done, might lead to further deaths in the same manner.

Inquests are conducted in England and Wales by a coroner who is an independent judicial officer who follows all laws. There are separate arrangements in Scotland and Northern Ireland. An inquest is an inquiry to determine a number of facts surrounding a death. The coroner will conduct an inquest into the death of a service person that occurred overseas and whose body is repatriated to England or Wales where an inquest would be held had the person died in England or Wales. The coroner does not work for the MoD.[15]

Where the death of a service person occurs overseas, whether on duty or not, the service police will invariably represent the HM Coroner in collating evidence to support the subsequent inquest held in the United Kingdom. They will recover/seize evidence that supports their own independent investigation into the death (whether of a criminal nature or not) and may also assist other policing agencies in the recovery of evidential material.

Consequently, the guiding principle pertaining to the property of the deceased is that no items are to be removed, altered, damaged or destroyed until their evidential value has been assessed and the service police have sanctioned the release of the property to the Standing/Unit Committee of Adjustment[16] representative (Army/RAF) or Logistics Officer/Imprest Holder (Royal Navy). To expedite this process, a senior service police officer may allocate resources to assist in the collection recording, retention

14 Coroners and Justice Act 2009, s. 5.
15 Boards of Inquiry and Coroners' Inquests, Information for Bereaved Families.
16 Debts (Deceased Servicemen etc) Act 1893.

and distribution of public/service property or items that are identified as the personal effects of the deceased.[17]

Scotland

The Chief Coroner's remit does not extend to Scotland, but a new law is now in force under sections 12 and 50 of the Coroners and Justice Act 2009 whereby the Chief Coroner may recommend to the Lord Advocate that it may be appropriate for the circumstances of a service death to be investigated in Scotland under the Fatal Accidents and Sudden Deaths Inquiry (Scotland) Act 1976 (as amended). This allows for possible transfers to Scotland of investigations into the deaths of service personnel from Scotland who die on active service abroad, particularly where the family of the deceased wish the transfer to take place. It will not allow a fatal accident inquiry to be carried out where the person was injured abroad and subsequently died on return to England or Wales.[18] If the family do not wish to transfer the investigation to Scotland, a coroner's inquest will be held in England or Wales, usually in the district to which the body is repatriated.[19]

Northern Ireland

A Northern Ireland coroner will only investigate a death that occurred in Northern Ireland and will not hold an inquest into a death that occurred outside the Province. Coroners' inquests into the death of Northern Irish service personnel that occur outside Northern Ireland will be held in England or Wales.[20]

Criminal Prosecution Possible

The question may arise whether the inquest should proceed prior to any prosecution.[21] Historically, Health and Safety Executive (HSE) prosecutions generally took place after the conclusion of the inquest. It was HSE policy that a final decision on enforcement action would not normally be made

17 Annex A to Volume 2, Part 1, Chapter 6 of JSP 751, para. 1.
18 Guidance No. 7, para. 16.
19 JSP J751, §02.01.0503 at p. 2-1-5-1.
20 *Ibid.*, para. 02.01.0504.
21 This paragraph is based upon the HSE 'Coroners Inquests': https://www.hse.gov.uk/enforce/enforcementguide/wrdeaths/chronology.htm#hse-prosecution.

until after the inquest was held.[22] However, consideration must be given in each case to the potential risk of a prosecution for an offence under health and safety offences legislation amounting to a bar to any subsequent criminal proceedings being brought for gross negligence manslaughter and corporate manslaughter. HSE's policy of not proceeding with a prosecution was designed to prevent a repeat of the circumstances that occurred in *R v Beedie*.[23] In that case, a manslaughter conviction was quashed, as the accused had already been prosecuted for a health and safety offence relating to the defective state of a gas installation, which had resulted in the death. Consequently, the second (manslaughter) prosecution should have been stayed as an abuse of process. To minimise the risk of a repeat of *R v Beedie*, commencing legal proceedings before an inquest should only be considered for those cases where the risk of an unlawful killing verdict at a subsequent inquest is minimal. HSE has provided guidance OC 165/10, 'Work-related deaths: liaison with police, prosecuting authorities, local authorities, and other interested authorities including consideration of individual and corporate manslaughter/homicide',[24] and the 'Work-related Deaths Protocol: Practical guide'.[25] This latter guidance covers the issues to consider before a decision is made to commence legal proceedings before an inquest. It also recommends that the following parties should be consulted as to their view to commence legal proceedings before inquest:

- the police and/or where appropriate the CPS, that is where the CPS has given the police early investigative advice or carried out a full evidential review;
- the Coroner, any decision should take into account what delay there might be in the hearing of any inquest, and whether the Coroner will suspend his/her investigation if criminal proceedings for health and safety offences are commenced;
- the bereaved family members.

---

22 *Ibid.*, para. 24 *et seq.*
23 [1997] 3 WLR 758.
24 https://www.hse.gov.uk/foi/internalops/ocs/100-199/165_10.htm.
25 https://www.hse.gov.uk/pubns/wrdp2.pdf.

The Chief Coroner's Guidance No. 33[26] explains that the law on suspending an investigation and adjourning an inquest is based upon the premise that investigations into a death should not be duplicated. Schedule 1 to the Coroners and Justice Act 2009 sets out when a coroner can or must suspend and resume investigations. There are three specific circumstances in which the coroner is obliged to suspend the investigation and to adjourn any inquest being held as part of that investigation.[27] There also exists a general power to suspend and adjourn, and a discretion to suspend and adjourn. The inquest must be compulsory suspended where criminal proceedings are pending or contemporaneous (s. 4).

Paragraph 5 of Schedule 1 to the Act provides a general power to a coroner to suspend an investigation 'if it appears to the coroner that it would be appropriate to do so'. This may be used, for example, if another investigation is being conducted into the death by the Independent Office for Police Conduct, the HSE or an Accident Investigation Branch, or if an investigation is being conducted in another jurisdiction, for example, if the death occurred abroad. Where this power is exercised, the coroner may resume the investigation at any time with sufficient reason (paragraph 10 of the Schedule). In addition, under rule 25(1) of the Coroners (Inquests) Rules 2013, a coroner may adjourn the inquest where it is deemed 'reasonable to do so'.

Schedule 1 provides that a senior coroner must suspend an investigation into a person's death where a Provost Marshal or the Director of Service Prosecutions requests the coroner to do so on the ground that a person may be charged with

a. the service equivalent of a homicide offence involving the death of the deceased, or
b. a service offence that is alleged to be a related offence.

Where an investigation was suspended following such a request, it may be resumed once a specified period has elapsed, namely, a period of 28 days

26 http://library.college.police.uk/docs/appref/Guidance-No-33-Suspension-Adjournment-and-Resumption-of-Investigations-and-Inquests.pdf.
27 Where certain criminal charges may be brought; where certain criminal proceedings are brought; suspension pending inquiry under the Inquiries Act 2005.

beginning with the day on which the suspension first takes effect, or whatever longer period (beginning with that day) the coroner specifies.[28]

Legal Aid

This is a highly contentious matter. Legal aid is available for early advice and assistance (legal help) for all inquests. Legal aid for representation at an inquest may also be available via the Exceptional Case Funding (ECF) scheme.[29] However, legal aid for representation at an inquest is means tested.

The Joint Committee on Human Rights recommended that if inquests are to remain inquisitorial, families must be given non-means-tested funding for legal representation at inquests where the state has separate representation for one or more interested persons.[30] Despite a number of reports[31] supporting the need for legal aid to be made available to bereaved families, and the obvious issue of equality of arms in Article 2 European Convention on Human Rights and Fundamental Freedoms (ECHR) matters, the Ministry of Justice, in its final report of 2019,[32] decided that they 'will not be introducing non-means-tested legal aid for inquests where the state has representation', although legal aid is likely to be awarded for representation of the bereaved at an inquest following the non-natural death or suicide of a person detained by police, in prison or in a mental health unit, as set out in the Ministry of Justice report at paragraph 13.[33]

Currently, ECF[34] is available to provide legal aid for representation at inquests, subject to means and merits tests, where there would be a breach or a risk of a breach of the ECHR or an enforceable EU law right if legal aid

28 Coroners and Justice Act 2009, Sch. 1 Part 1, s. 1(4).

29 Ministry of Justice Final report: Review of legal aid for inquests, February 2019, presented to Parliament by the Lord Chancellor and Secretary of State for Justice, at para. 81, https://assets.publishing.service.gov.uk/government/uploads/system/uploads/attachment_data/file/777034/review-of-legal-aid-for-inquests.pdf.

30 Enforcing human rights, Tenth Report of Session 2017–19, para. 74, https://publications.parliament.uk/pa/jt201719/jtselect/jtrights/669/669.pdf.

31 Legal Aid for Inquests Timeline, https://www.inquest.org.uk/legal-aid-for-inquests-timeline.

32 Ministry of Justice Final report, at para. 131.

33 See above, p. 631.

34 Para. 37, Ministry of Justice Final report, 2019.

was not made available, or where conditions are met for a 'wider public interest determination'. When such an application is received, Legal Aid Agency caseworkers will look at the inquest case and, in relation to legally aided representation at an inquest, make a decision as to whether, in their view:

a. legal aid for representation at an inquest is required by a Human Rights Act Convention right (most commonly Article 2 of ECHR) or it is considered appropriate to grant legal aid because of a risk to the rights of the applicant; and/or
b. the conditions are met for a 'wider public interest determination' that, in the particular circumstances of the case, the provision of representation for the individual at the inquest is likely to produce significant benefits for a class of person, other than the applicant and the members of the applicant's family.

The Ministry of Justice remains unmoved by suggestions that the legal aid scheme should remove the requirement for means testing. Their report states, at paragraph 45:

> we are mindful that a significant expansion of legal aid could have the unintended consequence of undermining the inquisitorial nature of the inquest system. It could also reinforce the commonly held misconception that an inquest's role is to apportion blame, as opposed to finding fact and learning lessons.

APPENDIX A RANK STRUCTURE

Rank Structure**

Royal Navy	Army / Royal Marines	Royal Air Force
	Officer Ranks	
NATO OF10 Admiral of the Fleet 	NATO OF10 Field Marshal 	NATO OF10 Marshall of the Royal Air Force 
NATO OF9 Admiral 	NATO OF9 General 	NATO OF9 Air Chief Marshall 
NATO OF8 Vice-Admiral 	NATO OF8 Lieutenant General 	NATO OF8 Air Marshall 
NATO OF7 Rear Admiral 	NATO OF7 Major General 	NATO OF7 Air Vice Marshall 
NATO OF6 Commodore 	NATO OF6 Brigadier 	NATO OF6 Air Commodore 
NATO OF5 Captain 	NATO OF5 Colonel 	NATO OF5 Group Captain 
NATO OF4 Commander 	NATO OF4 Lieutenant Colonel 	NATO OF4 Wing Commander 
NATO OF3 Lieutenant Commander 	NATO OF3 Major 	NATO OF3 Squadron Leader 

Royal Navy	Army / Royal Marines	Royal Air Force
NATO OF2 Lieutenant 	NATO OF2 Captain 	NATO OF2 Flight Lieutenant 
NATO OF1 Sub-Lieutenant 	NATO OF1 Lieutenant 	NATO OF1 Flying Officer 
NATO OF1 Midshipman (Lapel) 	NATO OF1 Second Lieutenant or Cornet (Royal Horseguards/ Dragoons) 	NATO OF1 Pilot Officer 
Other Ranks		
NATO OR-9 Warrant Officer Class 1 	NATO OR-9 Warrant Officer Class 1 (Regimental Sergeant Major)/ Regimental  Corporal Major (RCM) Conductor	NATO OR-9 RAF Master Aircrew  Warrant Officer 
NATO OR-8 Warrant Officer Class 2 (WO2) 	NATO OR-8 Warrant Officer Class 2 British Army (Company/ Squadron Sergeant Major) Squadron Corporal Major — SCM – Household Division only Regimental Quartermaster Corporal (RQMC) - an appointment  Regimental Quarter Master Sergeant (RQMS) 	NATO OR-8

Royal Navy	Army / Royal Marines	Royal Air Force
NATO OR-7 Chief Petty Officer (CPO) (Buttons on cuff) 	NATO OR-7 Staff / Colour Sergeant / Or Staff Corporal - SCpl (Household Division) 	NATO OR-7 RAF Flight Sergeant Aircrew  Flight Sergeant (FS)  Chief Technician 
NATO OR-6 Petty Officer (PO) 	NATO OR-6 Sergeant Corporal of Horse – CoH (3 chevrons and a metal crown Household Division) 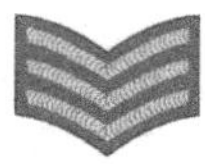	NATO OR-6 Sergeant Aircrew  Sergeant 
NATO OR-4 Leading Rate (LH) 	NATO OR-4 Corporal / Bombardier Lance Sergeant (Three white chevrons) Lance Corporal of Horse — LCoH (Household Division; 3 chevrons & cloth crown) 	NATO OR-4 Corporal 
NATO OR-3	NATO OR-3 Lance Corporal / Lance Bombardier Lance Corporal – LCpl (Household Division) 	NATO OR-3 LCpl - RAF Regt 

Royal Navy	Army / Royal Marines	Royal Air Force
NATO OR-2 Able Rate 		NATO OR-2 Senior Aircraftman (Technician)  Senior Aircraftman  Leading Aircraftman 
	NATO OR-1 Private (Pte)/ Trooper (Tpr)/ Gunner (Gnr) / Signaller (Sig)/ Sapper (Spr)/ Guardsman (Gdsm)/ Rifleman (Rfn)/ Kingsman (Kgn) (No insignia)	NATO OR-1 Aircraftman (No insignia)

** Rank structure courtesy of MoD Service websites.
The NATO rank OR 5 does not exist in the UK armed forces.
NATO grades as per STANAG 2116 (unclassified).

Further Notes:
A reference to 'Star' rank is to the rank of Brigadier (and equivalents) and higher. So a 1* is a Brigadier / Commodore / Air Commodore. A 2* is Major General / Air Vice Marshall / Vice Admiral and so on up to 5*. Star rankings are shown on vehicle pennants.

In the Royal Navy, all rear-admirals and above are officers of flag rank, but only those officers of flag rank who are authorised to fly a flag are formally called 'flag officers'. Therefore, a Commodore is not the holder of Flag rank.

The Five Star ranks of Admiral of the Fleet, Field Marshal, and Marshall of the Royal Air Force no longer exist as serving ranks. Field marshal is the highest rank in the British Army. Throughout the 20th century it was reserved for army and army group commanders in wartime, and retiring Chiefs of the General Staff. Peacetime promotions to the rank of field marshal have now been discontinued. The last officer to achieve the rank of field marshal is Sir Peter Inge when he was appointed Chief of the Defence Staff (CDS) in 1992. The holder of the rank never retires. In June 2012, Lord Guthrie of Craigiebank was promoted to honorary field marshal. In 2014, Lord Walker of Aldringham was also appointed to the honorary rank.

The current Marshall of the Royal Air Force is HRH The Prince of Wales who holds this as an honorary rank. He also holds the honorary rank of Admiral of the Fleet.

NATO grades have been included for officers (OF) and enlisted personnel (OR). It will be noticed that there is no OR-5 grade in the UK armed forces. So, an OR-4 is a Leading Rate/ Corporal /Bombardier/ Corporal (RAF). Similarly, OF-4 grade is held by a Commander RN/ Lieutenant Colonel/Wing Commander.

It will be noted that the officer grade of OF-1 encompasses two rank bands (e.g. Lieutenant and 2nd Lieutenant).

In the Royal Navy, Good Conduct badges are gold chevrons that look very much like the Army's rank symbols for Lance Corporal, Corporal and Sergeant and are worn on the left arm by entitled ratings below the rate of Chief Petty Officer. When worn without a rate badge, the centre of the device is to be midway between the point of the shoulder and the point of the elbow. See, BR-3, volume 1, chapter 39, paragraph 3921.

In 2015 the Army created the role of Army Sergeant Major whose rank badge is the royal coat of arms within a wreath, worn on the arm, just below the shoulder, on dress uniform. The Army Sergeant Major is the most senior non-commissioned appointment in the British Army.

APPENDIX B UNIT CHAIN OF COMMAND

Basic command structures for the three services.

Royal Navy

Structure varies according to on-shore or at sea and, when at sea, the size of ship commanded. Military command is vested in Officers of the Royal Navy and Royal Marines (except the RM Band Service) and Officers of the Maritime Reserve.[1] Sea command is vested in Officers of flag rank when holding certain appointments which will be promulgated from time to time by the Ministry of Defence (Naval Secretary) and Officers of the RN and RNR who are borne for seaman duties in sea-going ships. In sea-going ships, command descends first to the officer appointed as Executive Officer and then by seniority through those officers entitled to exercise sea command. In small ships, the Captain, following any directions given by the senior officer of the flotilla, squadron or administrative authority, is to prescribe in their standing instructions the order in which officers and ratings are to succeed to the command of the ship and exercise sea command.[2]

For example, the Type 26 (City Class) Frigate has a crew of 157 personnel. The organisation of a typical RN frigate is as follows:

CoC Navy

Commanding Officer	Usually a Commander RN Known as "The Captain" Supported by an Executive Department (Commanded by a Lieutenant Commander) Master At Arms and Leading Regulator 5 Specialist Departments
Department	Commanded by Lieutenant Commander 1 Welfare Dept

2 Supply Dept
3 Flight Dept
4 Marine Engineering Dept
5 Weapons Engineering Dept

The complement of a frigate relates to the requirement to man the ship for battle. A Type 23 has a complement of 181 (including 13 Officers).

CoC Army

Division	Commanded by Major-General Contains 2 or more Brigades Up to 40,0000 personnel
Brigade	Commanded by a Brigadier Contains 3–5 Battalions/Regiments Over 2,500 troops
Battalion/Regiment	Commanded by a Lieutenant Colonel Regimental Sergeant Major HQ Company 4 or 5 Companies 650 soldiers
Company	Commanded by a Major Company Sergeant Major 3 Platoons and an HQ (approximately 80 personnel)
Platoon	Commanded by a Lieutenant Platoon Sergeant Consists of 3 Sections (24 other ranks)
Section	Commanded by a Corporal 2 Fire Teams
Fire Team	Commanded by a Lance Corporal 3 private soldiers

The Royal Air Force[3]

The RAF Regiment has a different organisational structure to the rest of the RAF. The Regiment adopted two distinct types of operational establishments, the field squadron and the anti-aircraft flight. All members of the Regiment, however, were initially trained as combat infantrymen, an arrangement that continues to the present day.[4] The RAF Regiment comes under command of 2 Group, Air Command and has seven regular and six reserve squadrons organised into seven wings.[5]

A field squadron is divided into flights, which are equivalent in size to an army platoon. Each squadron contains several rifle flights, whose task is to engage the enemy at close range, and a support weapons flight, which provides fire support to the rifle flights using machine guns, mortars and snipers. The field squadrons are 171 strong.[6]

CoC RAF

Station	Commanded by a Group Captain Approximately 5000 personnel
Wing	Commanded by a Wing Commander Approx 300 personnel Wings are also the administrative units of an RAF station Usually consist of three Wings, Operations/ Flying, Administrative and Engineering
Squadron	Most flying squadrons are commanded by a wing commander who oversees around 200 personnel & between 12 and 15 aircraft. A squadron is further divided into flights, under the command of a squadron leader

3 RAF website, https://www.rafweb.org/Organsation/Org-index.htm and http://www.armedforces.co.uk/raf/listings/l0011.html.
4 RAF Regiment Association, https://www.rafregt.org.uk/regiment-info/raf-regiment-history/.
5 https://www.raf.mod.uk/our-organisation/force-protection/raf-regiment/.
6 RAF Regiment, https://military.wikia.org/wiki/RAF_Regiment.

APPENDIX C DISCIPLINARY SYSTEM

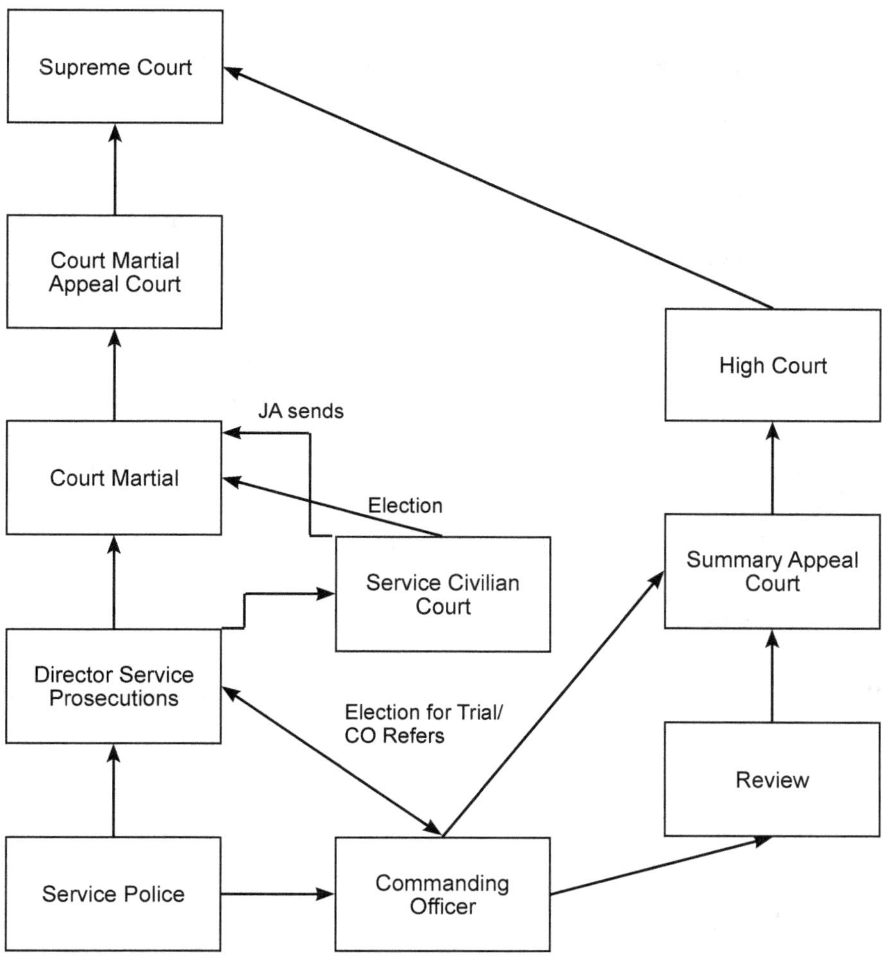

APPENDIX D SCHEDULE 1 OFFENCES AFA 2006

Part 1 Offences that may be dealt with without permission

1. An offence under section 1 of the Theft Act 1968 (c. 60) (theft).
2. An offence under section 12 of that Act (taking vehicle, etc. without consent).
3. An offence under section 5(2) of the Misuse of Drugs Act 1971 (c. 38) (possession of controlled drug).
4. An offence under section 1(1) of the Criminal Damage Act 1971 (c. 48) (criminal damage).
5. An offence under section 3 of the Theft Act 1978 (c. 31) (making off without payment) where the payment required or expected did not exceed £100.
6. An offence under section 9 of the Criminal Attempts Act 1981 (c. 47) (interference with vehicles).
7. An offence under section 39 of the Criminal Justice Act 1988 (c. 33) (assault and battery).
8. An offence under section 3 of the Road Traffic Act 1988 (c. 52) (careless driving, etc.).
9. An offence under section 5 of that Act (driving a vehicle where driver has consumed excessive amount of alcohol, etc.).

9A. An offence under section 5A of that Act (driving a vehicle with concentration of specified controlled drug above specified limit) [Inserted by the Crime and Courts Act 2013, Sch. 22, para. 17].

10. An offence under section 25 of that Act (tampering with vehicles, etc.) where the vehicle was on a road.
11. An offence under section 28 of that Act (dangerous cycling).
12. An offence under section 29 of that Act (careless cycling, etc.).

Part 2 Offences that may be dealt with only with permission

13. An offence under section 47 of the Offences against the Person Act 1861 (assault occasioning actual bodily harm).
14. An offence under section 1 of the Prevention of Crime Act 1953 (c. 14) (possession in public place of offensive weapon).
15. An offence under section 13 of the Theft Act 1968 (c. 60) (abstracting of electricity).
16. An offence under section 139 of the Criminal Justice Act 1988 (c. 33) (possession in public place of point or blade).
17. An offence under section 125 of the Communications Act 2003 (c. 21) (dishonestly obtaining electronic communications services).
18. An offence under section 126 of that Act (possession or supply of apparatus, etc. for contravening s. 125 of that Act).
19. An offence under section 1 of the Fraud Act 2006 (fraud).
20. An offence under section 11 of that Act (dishonestly obtaining services).

Section 54(2)(b) of the 2006 Act stipulates that in addition to an offence listed in Part 2 of Schedule 1, an offence under section 1 of the Criminal Attempts Act 1981 of attempting to commit an (indictable) offence so listed also requires permission from higher authority.

APPENDIX E SCHEDULE 2 OFFENCES AFA 2006

1. An offence under section 1 (assisting an enemy).
2. An offence under section 2(1) (misconduct on operations).
3. An offence under section 3 (obstructing operations) which relates to an action or operation against an enemy.
4. An offence under section 4(1) or (2) (looting).
5. An offence under section 6 (mutiny).
6. An offence under section 7 (failure to suppress mutiny).
7. An offence under section 8 (desertion) where the accused intended to avoid a period of active service (within the meaning of that section).
8. An offence under section 31(1) (hazarding of ship).
9. An offence under section 33(1) (dangerous flying, etc.).
10. An offence under section 39 of attempting to commit an offence within any of paragraphs 1 to 9.
11. An offence under section 40 of encouraging or assisting the commission of an offence within any of paragraphs 1 to 9.
12. An offence under section 42 (criminal conduct) as respects which the corresponding offence under the law of England and Wales is—
    (a) murder;
    (b) manslaughter;
    (c) kidnapping;
    (d) high treason;
    (e) piracy;
    (f) cheating the public revenue;
    (g) an offence under section 2 of the Treason Act 1842 (c. 51) (attempt to injure or alarm the Sovereign);
    (h) an offence under section 3 of the Treason Felony Act 1848 (c. 12) (compassing the deposition of the Sovereign, etc.);
    (i) an offence under section 4, 18, 22, 23, 28 or 29 of the Offences against the Person Act 1861 (c. 100) (soliciting murder, wounding with intent, using chloroform, etc. to commit indictable offence,

administering poison, causing injury by explosives, using explosives, etc. with intent);

(j) an offence under section 20 of that Act of inflicting grievous bodily harm;

(k) an offence under section 2 or 3 of the Explosive Substances Act 1883 (c. 3) (causing explosion likely to endanger life or property etc);

(l) & (m) repealed (1.7.2011) by Bribery Act 2010 (c. 23), s. 19(1), Sch. 2 (with ss. 16, 18(3), 19(5)); S.I. 2011/1418, Art. 2

(n) an offence under section 1 or 2 of the Perjury Act 1911 (c. 6) (perjury or false statements on oath);

(o) an offence under section 1 or 7 of the Official Secrets Act 1911 (c. 28) (spying or harbouring spies);

(p) an offence under section 1 of the Infant Life (Preservation) Act 1929 (c. 34) (child destruction);

(q) an offence under section 1 of the Children and Young Persons Act 1933 (c. 12) (cruelty to children);

(r) an offence under section 1 of the Infanticide Act 1938 (c. 36) (infanticide);

(ra) an offence under section 1A of the Prevention of Crime Act 1953 (threatening with offensive weapon in public) ;

(s) an offence under section 33 or 33A of the Sexual Offences Act 1956 (c. 69) (keeping a brothel, etc.);

(t) an offence under section 1 of the Geneva Conventions Act 1957 (c. 52) (grave breaches of conventions);

(u) an offence under section 2 of the Suicide Act 1961 (c. 60) (assisting suicide, etc.);

(v) an offence under section 5, 16, 16A, 17, 18 or 20 of the Firearms Act 1968 (c. 27) (unlawful possession or use of firearm, etc.);

(w) an offence under section 8, 10 or 21 of the Theft Act 1968 (c. 60) (robbery, aggravated burglary, blackmail);

(x) an offence under section 12A of that Act (aggravated vehicle taking) involving an accident which caused the death of any person;

(y) an offence under section 4, 5(3) or 8 of the Misuse of Drugs Act 1971 (c. 38) (production and supply of controlled drugs, possession of such drugs with intent to supply, permitting production of such drugs);
(z) an offence under section 1(2) of the Criminal Damage Act 1971 (c. 48) (destroying or damaging property with intent to endanger life);
(aa) an offence under section 1 of the Biological Weapons Act 1974 (c. 6) (developing biological agents, etc.);
(ab) an offence under section 51 of the Criminal Law Act 1977 (c. 45) (bomb hoaxes);
(ac) an offence under section 1 of the Protection of Children Act 1978 (c. 37) (indecent photographs of children);
(ad) an offence under section 170 of the Customs and Excise Management Act 1979 (c. 2) (fraudulent evasion of duty, etc.);
(ae) an offence under section 1 of the Taking of Hostages Act 1982 (c. 28) (hostage-taking);
(af) an offence under any of sections 1 to 4 of the Aviation Security Act 1982 (c. 36) (hijacking, destroying, damaging or endangering safety of aircraft, etc.);
(ag) an offence under section 1 or 2 of the Child Abduction Act 1984 (c. 37) (abduction of child);
(ah) an offence under any of sections 1, 18 to 23 and 29B to 29G of the Public Order Act 1986 (c. 64) (riot, stirring up racial or religious hatred, possession of inflammatory material);
(ai) an offence under section 134, 139AA or 160 of the Criminal Justice Act 1988 (c. 33) (torture, threatening with article with blade or point or offensive weapon, possession of indecent photograph of child);
(aj) an offence under section 1, 3A or 22A of the Road Traffic Act 1988 (c. 52) (causing death by dangerous driving, causing serious injury by dangerous driving, causing death by careless driving when under the influence of drink or drugs, causing danger to road-users);

(ak) an offence under any of sections 1 to 6 or 8(6) of the Official Secrets Act 1989 (c. 6) (disclosure of information relating to security, intelligence, defence, international relations, etc.);

(aka) an offence under section 3ZA of the Computer Misuse Act 1990 (unauthorised acts causing, or creating risk of, serious damage);

(al) an offence under any of sections 1 or 9 to 13 of the Aviation and Maritime Security Act 1990 (c. 31) (endangering safety at aerodromes, offences against the safety of ships and fixed platforms);

(am) an offence under section 72 of the Value Added Tax Act 1994 (c. 23) (evasion of VAT);

(an) an offence under Part II of the Channel Tunnel (Security) Order 1994 (S.I. 1994/570) (offences relating to Channel Tunnel trains and the tunnel system);

(ao) an offence under section 2 of the Chemical Weapons Act 1996 (c. 6) (use, etc. of chemical weapons);

(ap) an offence under section 11, 12, 15, 16, 17, 18, 38B, 39, 54, 56, 57 or 58 of the Terrorism Act 2000 (c. 11);

(aq) an offence under section 51 or 52 of the International Criminal Court Act 2001 (c. 17) (genocide, crimes against humanity, war crimes, etc.);

(ar) an offence under section 47, 79, 80, 113 or 114 of the Anti-terrorism, Crime and Security Act 2001 (c. 24);

(as) an offence under section 1 of the Dealing in Cultural Objects (Offences) Act 2003 (c. 27) (dealing in tainted cultural objects);

(at) any offence under Part 1 of the Sexual Offences Act 2003 (c. 42) except one under section 71;[1]

(au) an offence under any of sections 1, 2, 5, 6 or 8 to 11 of the Terrorism Act 2006;

(av) an offence under section 62(1) of the Coroners and Justice Act 2009 (possession of prohibited images of children);

(aw) an offence under section 1, 2 or 6 of the Bribery Act 2010;

1 Amended by the AFA2006(AOS2)O 2018, which removed from the exception offences under sections 3 (sexual assault), 66 (exposure) and 67 (voyeurism). Sexual activity in a public lavatory remains a Sch. 2 offence.

(ax) an offence under section 69 of the Serious Crime Act 2015 (possession of paedophile manual);

(ay) an offence under section 1 or 2 of the Modern Slavery Act 2015 (slavery, servitude and forced or compulsory labour and human trafficking).

13. An offence under section 42 as respects which the corresponding offence under the law of England and Wales is—

14. (a) an offence under section 1 of the Criminal Attempts Act 1981 (c. 47) of attempting to commit an offence within a sub-paragraph of paragraph 12;

(b) an offence under section 1 of the Criminal Law Act 1977 (c. 45) of conspiracy to commit such an offence;

(c) an offence under Part 2 of the Serious Crime Act 2007 of encouraging or assisting the commission of such an offence.

15. An offence under paragraph 4 of Schedule 2A (lay member of the Court Martial engaging in prohibited conduct).

16. An offence under paragraph 5 of Schedule 2A (disclosing information about the deliberations of members of the Court Martial) committed by a person described in sub-paragraph (2) of that paragraph.

APPENDIX F LEGAL AID FLOWCHART

Flow charts appear on the following pages.

Flow Chart – Legal Aid Processes: Trial (including referral to DSP)[1]

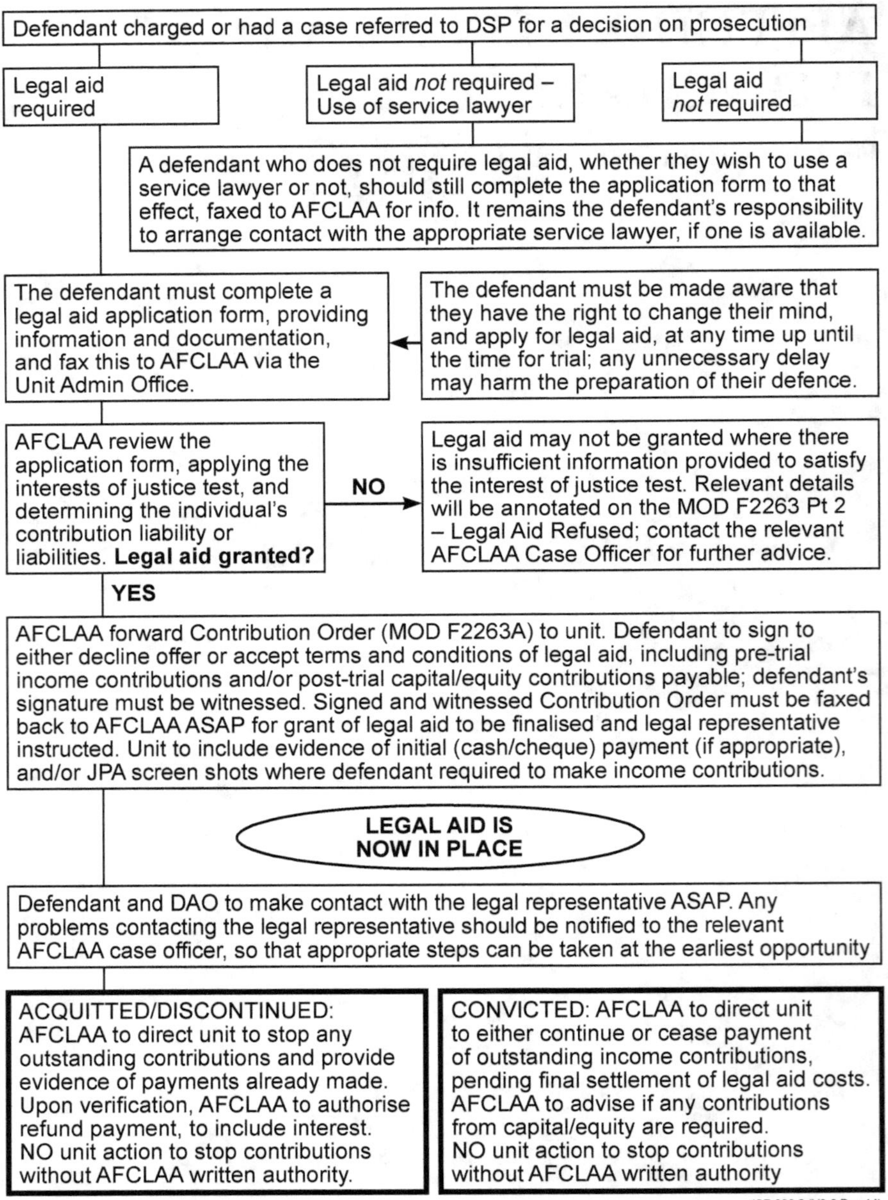

1 Includes Court Martial (CM), Service Civilian Court (SCC) and overseas criminal court trials.

Flow Chart – Legal Aid Processes: Summary Appeal or Election

Appellant lodges an appeal against summary dealing (including appeals lodged on their behalf by the Review Authority) or elects for CM trial

Legal aid required

Legal aid *not* required – Use of service lawyer

Legal aid *not* required

The appellant/defendant who does not require legal aid, whether they wish to use a service lawyer or not, should still complete the application form to that effect, faxed to AFCLAA for info.
It remains the appellant/defendant's responsibility to arrange contact with the appropriate service lawyer, if one is available.

The appellant/defendant must complete a legal aid application form, providing information and documentation, and fax this to AFCLAA via the Unit Admin Office.

The appellant/defendant must be made aware that they have the right to change their mind, and apply for legal aid, at any time up until the time for trial; any unnecessary delay may harm the preparation of their case.

AFCLAA review the application form to determine whether the appellant/defendant may be liable for a contribution (dependent upon outcome), before making a formal offer of legal aid.

AFCLAA forward a completed Contribution Order (MOD F2263B) to unit. Appellant/defendant must sign to decline offer or accept terms and conditions of legal aid, including liability for post-proceedings income contribution (determined by outcome of appeal/trial, and type of appeal); defendant's signature must be witnessed. Signed and witnessed Contribution Order must be faxed back to AFCLAA ASP for grant of legal aid to be finalised and legal representative instructed; no contributions are required in advance of proceedings.

LEGAL AID IS NOW IN PLACE

Appellant/defendant and DAO to make contact with the legal representative ASAP. Any problems contacting the legal representative should be notified to the relevant AFCLAA case officer, so that appropriate steps can be taken at the earliest opportunity

Summary Appeals: AFCLAA issue formal notification to appellant with contribution liability (identified on MOD F2263B), whose appeal was wholly unsuccessful (i.e. appeal failed/denied), or partially unsuccessful (i.e. finding upheld, but sentence varied); notification identifies level of contribution payable, and the payment schedule. Appellants whose appeal was upheld, or whose appeal was lodged by the Reviewing Authority, or who are below the income threshold, will NOT be required to contribute towards their legal aid costs.
Elections: AFCLAA issue formal notification to convicted offender with contribution liability, detailing contribution payment plan. Where actual legal aid costs are less than the contribution originally notified to the convicted offender, AFCLAA will authorise refund of the excess.
Defendants acquitted at trial, or who were below the income threshold, will NOT be required to contribute.

JSP 838 2 (V3.0 Dec 14)

Armed Forces Criminal Legal Aid Authority (AFCLAA)

Contact Details:
Armed Forces Criminal Legal Aid Authority
Building 398c Room 10
Trenchard Lines
Upavon
Pewsey
Wiltshire SN9 6BE

Telephone: +44 (0) 1980 61 5973 (Head)
+44 (0) 1980 61 8915 (Dep Head)
+44 (0) 1980 61 8008 (Office Manager)

Email:	mcs-afclaa-group@mod.gov.uk
Email address (internal):	LF-MCS-AFCLAA-&Group
Email address (external):	LF-MCS-AFCLAA-&Group@mod.uk

Phone Numbers:

External dialling code:	(44) (0)1980 61 – followed by extension
Internal dialling code:	94344 – followed by extension
Head:	5973
Dep Head:	8915
Team 1 Case Officer	8035
Assistant Case Officer	8814
Team 2 Case Officer	5974
Assistant Case Officer	8041
Team 3 Case Officer	8028
Assistant Case Officer	8982
Office Manager:	8008
Fax:	5691
Out of Hours:	+44 (0) 7766 511314

JSP 838

APPENDIX G PROCEDURE AT SUMMARY APPEAL COURT

The mode of dress for advocates is as for a court martial. Service personnel (including the defendant if still serving, and including members of the Special Investigation Branch) normally wear formal dress uniform in the court martial, including preliminary and arraignment proceedings. As in the court martial, the Judge Advocate is addressed as 'Your Honour'. When a High Court Judge presides they are addressed as 'My Lord'/'My Lady'.[1]

Order of Procedure in the Summary Appeal Court[2]

Serial	Activity	Remarks
1.	No later than 5 minutes before the appointed time Court Usher opens the court.	Appellant must enter the court. Representatives for the Appellant and Respondent and Verbatim Court Recorder may enter the court as required. Public and press enter and take their places. Service personnel wear headdress
2.	At the appointed time Court Usher invites the Judge Advocate and Members to enter the court.	All persons in court rise. Counsel bow, all service personnel salute. Service Members return salute.
3.	Judge Advocate and Members take their places.	Members sit on the bench at each side of the Judge Advocate.
4.	Judge Advocate reads out SAC notice.	Judge Advocate invites all in the court to remove headdress. As the name of each Member is read out, they identify themselves.
5.	Judge Advocate asks Appellant and Respondent if there are any objections.	Judge Advocate deals with any objections.

1 See Para. 3.2 The Court Martial And The Summary Appeal Court Guidance

Serial	Activity	Remarks
6.	Judge Advocate directs the waiting Member to withdraw.	
7.	Judge Advocate reads out the charge and confirms with Appellant (or their Counsel / legal representative) whether appeal is against both finding and punishment, or punishment only.	If the appeal is against both finding and punishment this will form a complete rehearing of all the evidence. Serials 7 to 24 apply. If the appeal is against punishment, go to serial 21 onwards.
8.	Trial commences.	Appellant sits.
9.	Respondent opens its case.	
10.	Respondent's witnesses called to give evidence.	Court Usher calls in each witness as instructed. Witnesses entering the courtroom salute/ bow as appropriate. Court Usher administers the oath/affirmation. If appropriate, Judge Advocate invites the witness to sit.
11.	Examination in chief; Cross-examination; Re-examination.	
12.	Questions from the Members.	The Judge Advocate and the Members may ask questions of the witness.
13.	Release of witnesses.	See guide paragraph 4.6. Judge Advocate gives the witness an appropriate witness warning. Witness stands, replaces headdress, salutes/bows as appropriate, and leaves.
14.	Appellant's case. Opening address, calls witnesses.	Same procedure as for Respondent.
15.	Appellant may give evidence.	Appellant puts on headdress, salutes the court and takes the witness stand in same way as every other witness. On completion, returns to their seat in the court.
16.	Closing speeches.	Advocates address the court, with the appellant having the last word.

Serial	Activity	Remarks
17.	Deliberation on finding.	All except the Judge Advocate, Members and Court Usher leave the court room. Court Usher receives any further instructions from the Judge Advocate before also leaving. The Judge Advocate and Members deliberate the finding alone.
18.	Judge Advocate calls for Court Usher to reopen the court.	All persons involved in SAC return to respective places in the court room.
19.	Judge Advocate announces the finding.	Finding either: confirms the original finding, or quashes the original finding, or substitutes a new finding that a different offence is proved.
20.	If original finding of guilt is quashed.	The conviction and punishment are reversed. Judge Advocate signs TRN2 form confirming the outcome.
21.	If appeal is against punishment only.	Respondent reads out the undisputed facts of the case to the court.
22.	If finding of guilt is not quashed, or was not appealed: Respondent and Appellant address the court on punishment and mitigation.	Deals with reports on the appellant, antecedents, any time already spent in custody, evidence of character witnesses.
23.	Judge Advocate closes the court to allow for deliberation on punishment.	All except the Judge Advocate, Members and Court Usher leave the court room. Court Usher receives any further instructions from the Judge Advocate before also leaving. The Judge Advocate and Members deliberate the sentence alone.
24.	Judge Advocate calls for Court Usher to reopen the court.	All persons involved in SAC return to respective places in the court room.

Serial	Activity	Remarks
25.	Judge Advocate formally announces the decision on punishment.	Judge Advocate gives reasons for the court's decision. Judge Advocate signs the TRN2 Record of Proceedings form.
26.	March out.	An offender who is in custody, or whose custodial sentence still stands, is always marched out of court. In other cases it is at the discretion of the senior service Member of the court.
27.	Judge Advocate concludes the case.	Judge Advocate announces that the appeal hearing is concluded, directs the Court Usher to clear the court. Uniformed personnel replace headdress and salute, civilians bow. Judge Advocate and Members leave.

APPENDIX H ORDER OF PROCEDURE IN COURT MARTIAL

Extracted from The Court Martial And The Summary Appeal Court Guidance, Volume 1, Table B.

Form of Dress for Advocates: Legal representatives such as barristers and solicitors appear in the Court Martial robed as if appearing in the Crown Court, as appropriate for their qualification. Otherwise civilians dress as they choose, provided styles adopted are not likely to cause offence. If asked, the Court Officer advises civilian attendees to wear clothing which is neat, tidy, restrained and commensurate with the dignity of the court.

Form of Address: In court the Judge Advocate is addressed as 'Your Honour' and referred to as 'the Judge Advocate'. On the rare occasion when a High Court Judge presides he is addressed as 'My Lord' and referred to as 'His Lordship'.

Order of Procedure in the Court Martial[1]

Serial	Activity	Remarks
1.	No later than 5 minutes before the appointed time Court Usher opens the court.	Defendant must enter the court. Defendant's Assisting Officer, legal representatives and Verbatim Court Recorder may enter the court as required. Public and press enter and take their places. Service personnel wear headdress.
2.	Judge Advocate enters.	Counsel bow, Judge Advocate bows and sits down. All sit.
3.	Judge Advocate deals with any preliminary matters.	
4.	Defendant(s) stand.	

Serial	Activity	Remarks
5.	If the defendant has not been arraigned previously at a plea & case management hearing: **Arraignment**.	Arraignment takes place in the absence of the Board. Judge Advocate reads the heading of the charge sheet. Judge Advocate deals with any negative responses.
6.	Arraignment.	Judge Advocate reads charges and asks each defendant how he pleads to each charge. Judge Advocate records the pleas on the Record of Proceedings (TRN1 form).
7.	President of the Board and Board members enter.	All stand except Judge Advocate, who remains seated. Counsel bow, Service personnel salute. President of the Board returns salute. If there has already been a guilty plea, the Judge Advocate directs the Board to join them on the bench; otherwise they are directed to the members' box to the side of the courtroom.
8.	On direction from Judge Advocate all sit except the Defendant(s) who remains standing.	Judge Advocate may ask the Defendant(s) to identify themselves.
9.	Judge Advocate reads out the Court Martial notice.	As the name of each officer or warrant officer (or civilian) member of the Board is read out they identify themselves.
10.	Judge Advocate asks if there are any objections.	Judge Advocate deals with any objection.
11.	Administration of oaths.	All stand including Judge Advocate who administers oath to President of the Board and members in order of seniority. For sentencing proceedings (where there has already been a guilty plea or finding) members are sworn once only, not for each case.
12.	Judge Advocate directs the waiting member to withdraw.	All sit.

Serial	Activity	Remarks
13.	*Morris* Direction.	Judge Advocate addresses the members on their duties, reminds them that they will not be reported on for their performance in court, and that they must report to the Judge Advocate any attempts to interfere with or pressurise them before or during the trial.
14.	Judge Advocate directs that members be provided with copies of the charge sheet.	Unless any objections raised by counsel.
15.	The defendant(s) has been arraigned previously and has entered a plea(s).	Judge Advocate tells the Board how the defendant(s) pleaded.
16.	No charge to be tried (plea of guilty to all charges, or prosecution accepts pleas to sufficient charges for justice to be done).	Omit serials 17 to 34. Turn to serial 35.
17.	Charges to be tried; plea of not guilty.	
18.	Trial proceedings commence.	Defendant sits.
19.	Prosecution opens its case.	
20.	Prosecution witnesses called to give evidence.	See guide paragraph 4.6. Court Usher administers the oath and invites witness to sit.
21.	Examination in chief; Cross-examination; Re-examination.	
22.	Questions from the Board.	President of the Board or member writes question on paper and passes to Judge Advocate, who decides whether to ask it.
23.	Release of witnesses.	See guide paragraph 4.6.
24.	Legal submissions during the trial.	President of the Board and members retire while Judge Advocate adjudicates. If at any stage the Judge Advocate decides to withdraw the matter from the Board and directs the Board to find the defendant not guilty, turn to serial 31.

Serial	Activity	Remarks
25.	Defence case.	Same procedure as for prosecution opening and witnesses. See serials 19–23.
26.	Defendant may give evidence.	Defendant puts on headdress, salutes the court and takes the stand in the same way as any other witness. On completion, they replace headdress, salute the court and return to their seat.
27.	Closing speeches.	Advocates address the court, with the defence having the last word.
28.	Judge Advocate sums up the case.	On completion, Board retires to consider findings (verdicts). Judge Advocate leaves. Court Usher clears the court.
29.	President of the Board informs Court Usher that Board has reached finding(s).	Court Usher informs Judge Advocate.
30.	Court Usher re-opens the court.	See serials 1–3 & 7.
31.	Finding: unanimity.	In respect of each charge for each defendant, Judge Advocate asks the President of the Board whether the Board has reached a finding upon which they **are all agreed**: response is **YES** or **NO**. If response was **NO**, Judge Advocate asks the President of the Board whether the Board has reached a finding upon which a majority are agreed: response is **YES** or **NO**.
32.	Finding: announcement.	As soon as the response **YES** has been given, Judge Advocate asks the President of the Board to announce the finding on that charge for that defendant: **GUILTY** or **NOT GUILTY**. Judge Advocate records the findings on the Record of Proceedings (TRN1 form).

Serial	Activity	Remarks
33.	In the case of findings of **NOT GUILTY** to all charges.	Judge Advocate and President of the Board sign the Record of Proceedings (TRN1 form). Prosecutor hands up diary of events Judge Advocate declares the proceedings concluded.
34.	Board and Judge Advocate leave the court.	Service personnel salute. Board and Judge Advocate leave. Court Usher clears the court.
35.	**Sentence Stage.** In the event of finding of **GUILTY** to any charges; or plea of **GUILTY** to any charges.	Members move from the members' box to sit alongside the Judge Advocate. After they are seated, all in court sit.
36.	**If plea(s) of guilty:** Prosecution opens the facts of the case and lists antecedents of offender. **If finding of guilty:** Prosecution lists antecedents of offender if relevant.	If there is a dispute about the facts following a guilty plea, court will consider a ***Newton* Hearing**.
37.	In the event of a dispute about facts affecting sentence after a guilty plea.	Procedure for *Newton* hearing follows serials 20–23 & 26, after which Judge Advocate and Board retire together to consider the facts.
38.	Adjournment for Pre-Sentence Reports or other reports.	If, following submissions from defence or on the court's own motion, reports are considered necessary, the court is adjourned until some later date for sentence. Judge Advocate and Board leave.
39.	Court re-opens.	Serials 1–3 & 7.
40.	Plea in mitigation.	Character witnesses may be called and/or written character evidence and other reports are put before the court.
41.	Court retires to consider sentence.	Judge Advocate and Board may retire together or may remain in the courtroom while they deliberate, in which event the court is cleared.

Serial	Activity	Remarks
42.	Judge Advocate informs Court Usher that the court is ready to deliver sentence.	Court Usher reopens the court. Serials 1–3 & 7.
43.	Judge Advocate and Board enter together if they retired to deliberate. If court has remained in the courtroom, all other parties are readmitted.	Counsel bow, Service personnel salute. All sit except offender who remains standing.
44.	Judge Advocate gives reasons for sentence.	Headdress remains on.
45.	President of the Board formally announces sentence.	
46.	With leave of the Judge Advocate, the President of the Board may address the offender.	Offender is addressed about the effects of their offending on the Service, and the effects of the sentence on their service career.
47.	March out.	An offender sentenced to custody is always marched out whatever their rank. If non-custodial sentence, it is at the discretion of the President of the Board depending on the rank of the offender and the sentence. Court Usher orders offender and escort to attention, to salute, and then to march out of the court smartly.
48.	Miscellaneous rulings and Diary of Events	Judge Advocate deals with costs if appropriate. They may do so before the offender is marched out. Prosecutor hands up Diary of Events. Judge Advocate and President of the Board sign the completed Record of Proceedings / Trial Result Notification (TRN1 form).
49.	Judge Advocate declares the proceedings concluded.	Judge Advocate and Board leave. Court cleared.

APPENDIX I NATO SOFA

NATO Status of Forces Agreement – Extract

Article VII

1. Subject to the provisions of this Article,
   a. the military authorities of the sending State shall have the right to exercise within the receiving State all criminal and disciplinary jurisdiction conferred on them by the law of the sending State over all persons subject to the military law of that State;
   b. the authorities of the receiving State shall have jurisdiction over the members of a force or civilian component and their dependents with respect to offences committed within the territory of the receiving State and punishable by the law of that State.

2.
   a. The military authorities of the sending State shall have the right to exercise exclusive jurisdiction over persons subject to the military law of that State with respect to offences, including offences relating to its security, punishable by the law of the sending State, but not by the law of the receiving State.
   b. The authorities of the receiving State shall have the right to exercise exclusive jurisdiction over members of a force or civilian component and their dependents with respect to offences, including offences relating to the security of that State, punishable by its law but not by the law of the sending state.
   c. For the purposes of this paragraph and of paragraph 3 of this Article a security offence against a State shall include:
      (i) treason against the State;
      (ii) sabotage, espionage or violation of any law relating to official secrets of that State, or secrets relating to the national defence of that State

3. In case where the right to exercise jurisdiction is concurrent the following rules shall apply:
   a. The military authorities of the sending State shall have the primary right to exercise jurisdiction over a member of a force or of a civilian component in relation to
      (i) offences solely against the property or security of that State, or offences solely against the person or property of another member of the force or civilian component of that State or of a dependent;
      (ii) offences arising out of any act or omission done in the performance of official duty.
   b. In the case of any other offence the authorities of the receiving State shall have the primary right to exercise jurisdiction.
   c. If the State having the primary right decides not to exercise jurisdiction, it shall notify the authorities of the other State as soon as practicable. The authorities of the State having the primary right shall give sympathetic consideration to a request from the authorities of the other State for a waiver of its right in cases where that other state considers such waiver to be of particular importance.
4. The foregoing provisions of this Article shall not imply any right for the military authorities of the sending State to exercise jurisdiction over persons who are nationals of or ordinarily resident in the receiving State, unless they are members of the force of the sending State.
5.
   a. The authorities of the receiving and sending states shall assist each other in the arrest of members of a force or civilian component or their dependents in the territory of the receiving State and in handing them over to the authority which is to exercise jurisdiction in accordance with the above provisions.
   b. The authorities of the receiving State shall notify promptly the military authorities of the sending State of the arrest of any member of a force or civilian component or a dependent.
   c. The custody of an accused member of a force or civilian component over whom the receiving state is to exercise jurisdiction

shall, if he is in the hands of the sending State, remain with that State until he is charged by the receiving State.

6.

a. The authorities of the receiving and sending States shall assist each other in the carrying out of all necessary investigations into offences, and in the collection and production of evidence, including the seizure and, in proper cases, the handing over of objects connected with an offence. The handing over of such objects may, however, be made subject to their return within the time specified by the authority delivering them.

b. The authorities of the Contracting parties shall notify one another of the disposition of all cases in which there are concurrent rights to exercise jurisdiction.

7.

a. A death sentence shall not be carried out in the receiving State by the authorities of the sending State if the legislation of the receiving state does not provide for such punishment in a similar case.

b. The authorities of the receiving State shall give sympathetic consideration to a request from the authorities of the sending State for assistance in carrying out a sentence of imprisonment pronounced by the authorities of the sending State under the provision of this Article within the territory of the receiving State.

8. Where an accused has been tried in accordance with the provisions of this Article by the authorities of one Contracting Party and has been acquitted, or has been convicted and is serving, or has served, his sentence or has been pardoned, he may not be tried again for the same offence within the same territory by the authorities of another Contracting Party. However, nothing in this paragraph shall prevent the military authorities of the sending State from trying a member of its force for any violation of rules of discipline arising from an act or omission which constituted an offence for which he was tried by the authorities of another Contracting Party.

9. Whenever a member of a force or civilian component of a dependent is prosecuted under the jurisdiction of a receiving State he shall be entitled:

a. to a prompt and speedy trial;

b. to be informed, in advance of trial, of the specific charge or charges made against him;
c. to be confronted with the witnesses against him;
d. to have compulsory process for obtaining witnesses in his favour, if they are within the jurisdiction of the receiving State;
e. to have legal representation of his own choice for his defence or to have free or assisted legal representation under the conditions prevailing for the time being in the receiving State;
f. if he considers it necessary, to have the services of a competent interpreter; and
g. to communicate with a representative of the Government of the sending State and when the rules of the court permit, to have such a representative present at his trial.

10.

a. Regularly constituted military units or formations of a force shall have the right to police any camps, establishment or other premises which they occupy as the result of an agreement with the receiving State. The military police of the force may take all appropriate measures to ensure the maintenance of order and security on such premises.
b. Outside these premises, such military police shall be employed only subject to arrangements with the authorities of the receiving State and in liaison with those authorities, and in so far as such employment is necessary to maintain discipline and order among the members of the force.

11. Each Contracting Party shall seek such legislation as it deems necessary to ensure the adequate security and protection within its territory of installations, equipment, property, records and official information of other Contracting Parties, and the punishment of persons who may contravene laws enacted for that purpose.

APPENDIX J CIVILIANS SUBJECT TO SERVICE DISCIPLINE

AFA Sch. 15

(As amended by s. 22 AFA 2011)

Part 1
Civilians subject to Service discipline

Persons in one of Her Majesty's aircraft in flight

1. (1) A person is within this paragraph if he is in one of Her Majesty's aircraft in flight.
   (2) For the purposes of sub-paragraph (1) the period during which an aircraft is in flight includes—
      (a) any period from the moment when power is applied for the purpose of the aircraft taking off on a flight until the moment when the landing run (if any) at the termination of that flight ends;
      (b) any period when the aircraft is on the surface of the sea or navigable waters.
   (3) In sub-paragraph (2)(a) a 'flight' means a journey by air beginning when the aircraft takes off and ending when it next lands.
   (4) In this paragraph 'Her Majesty's aircraft' means all aircraft belonging to or used for the purposes of any of Her Majesty's forces.

Persons in one of Her Majesty's ships afloat

2. (1) A person is within this paragraph if he is in one of Her Majesty's ships afloat.
   (2) In this paragraph 'Her Majesty's ships' means all ships belonging to or used for the purposes of any of Her Majesty's forces.
   (3) For the purposes of this paragraph 'afloat' means not on shore.

Persons in Service custody, etc.

3. (1) A person is within this paragraph if—
   (a) he is in Service custody; and
   (b) his being in Service custody is lawful by virtue of any provision of or made under this Act.

   (2) A person is also within this paragraph if he is in the course of being arrested, or of having an attempted arrest made of him, by a person who has a duty under Service law to apprehend him.

Crown servants in designated area working in support of Her Majesty's forces

4. (1) A person is within this paragraph (subject to paragraph 11) if—
   (a) he is a Crown servant;
   (b) his sole or main role is to work in support of any of Her Majesty's forces; and
   (c) either—
      (i) he is in a designated area, and his normal place of work in that role is in that designated area; or
      (ii) sub-paragraph (i) does not apply, but he is in a designated area and he came there wholly or partly for the purposes of his work in that role.

   (2) In this paragraph 'Crown servant' means a person employed by or in the service of the Government of the United Kingdom.

Persons working for specified military organisations

5. (1) A person is within this paragraph (subject to paragraph 11) if—
   (a) he is employed by or in the service of a specified naval, military or air-force organisation of which the United Kingdom is a member;
   (b) he is so employed by reason of the United Kingdom's membership of that organisation; and
   (c) either—
      (i) he is in a country outside the British Islands, and his normal place of work under that employment is in that country; or
      (ii) sub-paragraph (i) does not apply, but he is in a country outside the British Islands and he came there wholly or partly for the purposes of his work under that employment.

(2) In this paragraph 'specified' means specified by order of the Secretary of State under this paragraph.

Persons in designated area who are members or employees of other specified organisations

6. (1) A person is within this paragraph (subject to paragraph 11) if—
   (a) he belongs to or is employed by a specified organisation; and
   (b) either—
      (i) he is in a designated area, and his normal place of work for that organisation is in that designated area; or
      (ii) sub-paragraph (i) does not apply, but he is in a designated area and he came there wholly or partly for the purposes of work for that organisation.

(2) In this paragraph 'specified organisation' means an organisation which—
   (a) does not fall within paragraph 5; and
   (b) is specified by order of the Secretary of State under this paragraph.

Persons designated by or on behalf of Defence Council

7. (1) A person is within this paragraph (subject to paragraph 11) if—
   (a) he is designated for the purposes of this paragraph by or on behalf of the Defence Council or by an officer authorised by the Defence Council; and
   (b) he is outside the British Islands.

(2) A person may be designated for the purposes of this paragraph only if it appears to the Defence Council or the authorised officer that it is desirable to do so—
   (a) in the interests of the person;
   (b) for the protection of other persons (whether or not members of any of Her Majesty's forces); or
   (c) for the purpose of maintaining good order and discipline.

(3) In deciding whether to designate a person for the purposes of this paragraph, the Defence Council or the authorised officer must have regard in particular to—

(a) the characteristics of the justice system (if any) in any country or territory where the person is or is likely to be;

(b) the terms of any treaty, agreement or arrangement relating to the legal status, or the treatment, of visiting forces to which the United Kingdom and any such country or territory are parties;

(c) the likelihood of the person's being subject to the law applicable to the armed forces of any country or territory outside the British Islands.

(4) A designation under this paragraph—

(a) may designate persons by name or by description;

(b) may provide, in relation to any person designated by it, that it applies to him only for a specified period or in specified circumstances;

(c) may be withdrawn by any person entitled to make designations under this paragraph.

(5) In sub-paragraph (4) 'specified' means specified by the designation.

Persons residing or staying with person subject to Service law in designated area

8. A person is within this paragraph (subject to paragraph 11) if—

(1) he resides or is staying with a person subject to Service law in a designated area; and

(2) he is in that designated area.

Persons residing or staying with person falling within paragraph 4 or 6 in designated area

9. (1) A person is within this paragraph (subject to paragraph 11) if—

(a) he resides or is staying with a relevant person in a designated area; and

(b) he is in that designated area.

(2) In this paragraph a 'relevant person' means a person who—

(a) falls within paragraph 4 or 6; or

(b) would fall within paragraph 4 or 6, but for paragraph 11 or his not being in a designated area. Persons residing or staying with person falling within paragraph 5

10. (1) A person is within this paragraph (subject to paragraph 11) if—

(a) he resides or is staying with a qualifying person in a relevant country; and
(b) he is in that country.

(2) In this paragraph a 'qualifying person' means a person who—
(a) falls within paragraph 5; or
(b) would fall within paragraph 5, but for paragraph 11 or his not being in a relevant country.

(3) For the purposes of this paragraph, a country is 'relevant' in relation to a person employed as mentioned in paragraph 5 if it is outside the British Islands and—
(a) is where that person's normal place of work under that employment is situated; or
(b) is a country to which that person came wholly or partly for the purposes of his work under that employment.

Part 2

Exclusion and definitions

Exclusion

11. (1) A person who is not a United Kingdom national is not within any of paragraphs 4 to 10 at any time when he is in a country—
(a) of which he is a national; or
(b) in which he is ordinarily resident.

(2) In this paragraph a 'United Kingdom national' means an individual who is—
(a) a British citizen, a British overseas territories citizen, a British National (Overseas) or a British Overseas citizen;
(b) a person who under the British Nationality Act 1981 (c. 61) is a British subject; or
(c) a British protected person within the meaning of that Act.

(3) In determining for the purposes of this paragraph whether a person is ordinarily resident in a country, no account shall be taken of any period during which he has been or intends to be present there while falling (apart from this paragraph) within any of paragraphs 4 to 10.

Definitions

12. (1) In this Schedule 'designated area' means an area which—
    (a) is outside the British Islands; and
    (b) is designated for the purposes of this Schedule by an order made by the Secretary of State.
  (2) An area designated for the purposes of this Schedule may consist of two or more areas (whether or not contiguous).

13. In this Schedule references to a person residing or staying with another person include references to the person—
    (a) being about to reside or stay with the other person; and
    (b) departing after residing or staying with him.

For the purposes of paragraphs 5(1)(c), 10 and 11, a territory that is not within a country is to be treated as a country.

APPENDIX K SERVICE POLICE CODES OF PRACTICE

Summary of Provisions

Section 113(3) of the Police and Criminal Evidence Act 1984 (PACE) (as amended) provides for the Secretary of State to issue a code of practice, or a number of such codes, concerned with

(a) the exercise of powers conferred by or under Part 3 of the Armed Forces Act 2006, relating to Powers of Arrest, Search and Entry; or

(b) investigations of service offences

A failure on the part of any person to comply with any provision of a code shall not of itself render him liable to any criminal or civil proceedings except those to which s. 113(8) PACE applies, namely, proceedings in respect of an offence under a provision of Part 1 of the Armed Forces Act 2006 other than section 42 (criminal conduct).

Any of the codes is admissible in evidence in all criminal and civil proceedings and any provision of a code may be taken into account where it is deemed relevant to any question arising in the proceedings. [See s. 113(10)].

Section 113(12A) makes clear that

'service offence' has the meaning given by section 50 of the Armed Forces Act 2006;

'criminal proceedings' includes service proceedings;

'service proceedings' means proceedings before a court (other than a civilian court) in respect of a service offence; and

'civilian court' has the meaning given by section 374 of the Armed Forces Act 2006.

The following is extracted from the JSP:[1]

Code A – Statutory Powers of Stop and Search and the Recording of Encounters with the Service Community. The exercise by the Service Police of statutory powers to stop and search and requirements for the Service Police to record encounters with members of the Service Community. This Code was introduced on 30 September 2003.

Code B – Searches of Premises by Service Policemen and the Seizure of Property Found by Service Policemen on Persons or Premises. This Code governs the exercise by the Service Police of powers in respect of the searching of premises and the seizure of property found by the Service police on persons or premises. This Code was first introduced on 30 September 2003.

Code C – Code of Practice for the Detention, Treatment and Questioning of Persons by Service Policemen. The purpose of this Code is to ensure that all persons suspected of being involved in offences under the Armed Forces Act 2006 are dealt with fairly and properly in accordance with the law. This Code was first introduced on 1 February 1997.

Code D – Code of Practice for the Identification of Persons by Service Policemen. This Code concerns the principle methods used by the Service Police for identifying persons in connection with the investigation of offences under the Armed Forces Act 2006 and the keeping of accurate and reliable criminal records. This Code was first introduced on 1 February 1997.

Code E – Code of Practice on Audio Recording of Interviews with Suspects. This Code deals with the audio recording of interviews of persons suspected of certain types of offences under the Armed Forces Act 2006 and governs the way in which audio recorded interviews are carried out. This Code was first introduced on 1 February 1997.

Code F – Code of Practice on Visual Recordings With Sound of Interviews With Suspects. This Code deals with the visual recording of interviews of persons suspected of certain types of offences under the Armed Forces Act 2006 and governs the way in which visual recorded interviews are carried out. This Code was introduced for the first time on 31 October 2009.

Code G – Code of Practice for the Statutory Power of Arrest by Service Policemen. This Code sets out the criteria the Service Police must consider when exercising their power of arrest under sections 67, 69, 303, 110 and 111 of the Armed Forces Act 2006. This Code was introduced on 31 December 2006.

The Codes contained in the JSP 397 have been issued by the Secretary of State under section 113(3) of the Police and Criminal Evidence Act 1984 and have been approved by Parliament.

These Codes of Practice came into effect at midnight (GMT) on 31 of October 2009, and apply to all activities referred to in them undertaken on or after that date. The existing Service Police Codes of Practice (JSP 397) dated 31 December 2006 cease to have effect.

APPENDIX L DISCLOSURE CODE

Code of Practice – Disclosure

Under the CPIA1996(CP)(AF)O 2009 (Schedule)

Preamble

This code of practice ('the Code') is made under article 3 of the Criminal Procedure and Investigations Act 1996 (Code of Practice) (Armed Forces) Order 2009. It sets out the manner in which Service policemen are to record, retain and reveal to the Director material obtained in service investigations which may be relevant to the investigation, and related matters.

Introduction

1.—(1) Subject to paragraph 11(2), the Code applies in relation to service investigations which begin on or after 31st October 2009.

(2) Nothing in the Code applies to material intercepted in obedience to a warrant issued under section 5 of the Regulation of Investigatory Powers Act 2000(1), or to any copy, as defined in section 15 of that Act, of that material.

Interpretation

2.—(1) In the Code—

'2006 Act' means the Armed Forces Act 2006;

'accused' includes a person bringing an appeal to the Summary Appeal Court under section 141 of the 2006 Act;

'Director' means the Director of Service Prosecutions appointed under section 364 of the 2006 Act;

'disclosure of material' to an accused includes disclosure of material to his legal representative;

'disclosure officer' means the Service policeman responsible for—

(a) examining material retained by a Service police force during the service investigation;

(b) revealing material to the Director during the service investigation and any proceedings in a service court resulting from it; and

(c) certifying that he has done this;

'investigator' means any Service policeman involved in the conduct of a service investigation. All investigators have a responsibility for carrying out the duties imposed on them under the Code, including in particular recording information and retaining records of information and other material;

'judge advocate' has the same meaning as in the 2006 Act;

'legal representative' means, in relation to an accused, a person appointed by him as his legal representative, who—

(d) has a general qualification within the meaning of section 71 of the Courts and Legal Services Act 1990(2),

(e) is an advocate or solicitor in Scotland,

(f) is a member of the bar of Northern Ireland or a solicitor of the Court of Judicature of Northern Ireland, or

(g) a person who has in any of the Channel Islands, the Isle of Man, a Commonwealth country or a British overseas territory, rights and duties similar to those of a barrister or a solicitor in England and Wales and is subject to punishment or disability for breaching professional rules,

and in this definition the reference to the Court of Judicature of Northern Ireland is to be read as a reference to the Supreme Court of Northern Ireland until paragraph 5 of Schedule 11 to the Constitutional Reform Act 2005(3) comes into force;

'material' means material of any kind, including information and objects; this includes not only material coming into the possession of the investigator, such as documents seized in the course of a search of premises, but also material generated by him such as records of interview';

'officer in charge of an investigation' is the Service policeman responsible for directing a service investigation. He is also responsible for ensuring that proper procedures are in place for recording information and retaining records of information and other material in the investigation;

'prosecution' includes the respondent in an appeal under section 141 of the 2006 Act; and references to the prosecution case, prosecution disclosure and prosecution witness shall be understood accordingly;

'prosecution disclosure' refers to any duty of the Director under articles 4 and 13 of the Application Order to disclose material which is in his possession or which he has inspected in pursuance of the Code and which might reasonably be considered capable of undermining the case against the accused or of assisting the case for the accused;

'service court' means the Summary Appeal Court, the Service Civilian Court, the Court Martial or the Court Martial Appeal Court;

'service investigation' means an investigation conducted by a Service policeman with a view to it being ascertained—

(h) whether a person should be charged with a Service offence; or

(i) whether a person charged with a Service offence is guilty of it;

'Service offence', subject to paragraph 11, has the meaning given by section 50 of the 2006 Act;

'Service police force' means the Royal Navy Police, the Royal Military Police or the Royal Air Force Police; and

'the Application Order' means the Criminal Procedure and Investigations Act 1996 (Application to the Armed Forces) Order 2009(4).

(2) In the Code—

(a) subject to paragraphs (b) and (c), 'Service policeman' means a member of a Service police force;

(b) a Provost Marshal is to be taken to be a member of the appropriate Service police force (if he is not a member of that force);

(c) an officer in the Royal Air Force or the Royal Auxiliary Air Force who is appointed to exercise functions conferred by or under the 2006 Act on Service policemen is to be taken to be a member of the Royal Air Force Police.

(3) References in the Code to material which 'may be relevant to a service investigation' are references to where, in relation to any material, it appears to an investigator or the officer in charge of the investigation or to the disclosure officer that it has some bearing on any Service offence under investigation or any person being investigated, or on the surrounding circumstances of the case, unless it is incapable of having any impact on the case.

(4) In the Code material is 'sensitive' if the disclosure officer believes that its disclosure would give rise to a real risk of serious prejudice to an important public interest.

General responsibilities

3.—(1) The functions of the investigator, the officer in charge of a service investigation and the disclosure officer are separate. Whether they are undertaken by one, two or more persons will depend on the complexity of the case and the administrative arrangements within each Service police force. Where they are undertaken by more than one person, close consultation between them is essential to the effective performance of the duties imposed by the Code.

(2) In any service investigation, one or more deputy disclosure officers may be appointed to assist the disclosure officer, and the deputy disclosure officer may perform any function of a disclosure officer.

(3) The Provost Marshal of each of the Service police forces is responsible for putting in place arrangements to ensure that in every service investigation the identity of the officer in charge of the investigation and the disclosure officer is recorded. The Provost Marshals shall ensure that disclosure

officers and deputy disclosure officers have sufficient skills and authority, commensurate with the complexity of the investigation, to discharge their functions effectively. An individual must not be appointed as disclosure officer, or continue in that role, if that is likely to result in a conflict of interest, for instance, if the disclosure officer is the victim of the alleged offence which is the subject of the service investigation. The advice of a more senior officer must always be sought if there is doubt as to whether a conflict of interest precludes an individual acting as disclosure officer. If thereafter the doubt remains, the advice of the Director should be sought.

(4) The officer in charge of a service investigation may delegate tasks to another investigator, to other persons employed in support of the Service police or to other persons participating in the investigation under arrangements for joint investigations, but he remains responsible for ensuring that these have been carried out and for accounting for any general policies followed in the investigation. In particular, it is an essential part of his duties to ensure that all materials which may be relevant to a service investigation is retained, and either made available to the disclosure officer or (in exceptional circumstances) revealed directly to the Director.

(5) In conducting an investigation, the investigator should pursue all reasonable lines of inquiry, whether these point towards or away from the suspect. What is reasonable in each case will depend on the particular circumstances. For example, where material is held on computer, it is a matter for the investigator to decide how many of the files on the computer it is reasonable to inquire into, and in what manner.

(6) If the officer in charge of a service investigation believes that other persons may be in possession of material that may be relevant to the investigation, and if this has not been obtained, he should ask the disclosure officer to inform them of the existence of the investigation and to invite them to retain the material in case they receive a request for its disclosure. The disclosure officer should inform the Director that they may have such material. However, the officer in charge of a service investigation is not required to make speculative enquiries of other persons; there must be some reason to believe that they may have relevant material. That reason may come from information provided to the police by the accused or from other inquiries made or from some other source.

(7) If, during a service investigation, the officer in charge of the investigation or disclosure officer for any reason no longer has responsibility for the functions falling to him, either his officer commanding or the officer in charge of service investigations for the Service police force concerned must assign someone else to assume that responsibility. That person's identity must be recorded, as with those initially responsible for these functions in each investigation.

Recording of information

4.—(1) If material which may be relevant to the service investigation consists of information which is not recorded in any form, the officer in charge of the investigation must ensure that it is recorded in a durable or retrievable form (whether in writing, on video or audio tape, or on computer disk).

(2) Where it is not practicable to retain the initial record of information because it forms part of a larger record which is to be destroyed, its contents should be transferred as a true record to a durable and more easily-stored form before that happens.

(3) Negative information is often relevant to an investigation. If it may be relevant it must be recorded. An example might be a number of people present in a particular place at a particular time who state that they saw nothing unusual.

(4) Where information which may be relevant is obtained, it must be recorded at the time it is obtained or as soon as practicable after that time. The requirement to record information promptly does not require an investigator to take a statement from a potential witness where it would not otherwise be taken.

Retention of material

5.—(1) The investigator must retain material obtained in a service investigation which may be relevant to the investigation. Material may at any time be photographed, video-recorded, captured digitally or otherwise retained in the form of a copy rather than te original, if—

(a) the original is perishable;

(b) the original was supplied to the investigator rather than generated by him and is to be returned to its owner; or

(c) the retention of a copy rather than the original is reasonable in all the circumstances.

(2) Where material has been seized by a Service policeman in exercise of any power of seizure conferred by or under Part 3 of the 2006 Act, the duty to retain it under the Code is subject to any provisions on the retention of seized material set out in an order made under Part 3 of the 2006 Act dealing with powers of stop and search, search, seizure and retention.

(3) If the officer in charge of a service investigation becomes aware as a result of developments in the case that material previously examined but not retained (because it was not thought to be relevant) may now be relevant to the investigation, he should, wherever practicable, take steps to obtain it or ensure that it is retained for further inspection or for production in court if required.

(4) The duty to retain material includes in particular the duty to retain material falling into the following categories, where it may be relevant to the investigation—

(a) crime reports (including crime report forms, relevant parts of incident report books, including Daily Occurrence Books, or Service policemen's notebooks);

(b) arrest and custody records (including all written records relating to custody and produced by Service policemen or other service personnel);

(c) records which are derived from tapes of telephone messages containing descriptions of an alleged offence or offender;

(d) final versions of witness statements (and draft versions where their content differs from the final version), including any exhibits mentioned (unless these have been returned to their owner on the understanding that they will be produced in court if required);

(e) interview records (written records, or audio or video tapes, of interviews with actual or potential witnesses or suspects);

(f) communications between the police and experts such as forensic scientists, reports of work carried out by experts, and schedules of scientific material prepared by the expert for the investigator, for the purposes of proceedings before a service court;

(g) records of the first description of a suspect by each potential witness who purports to identify or describe the suspect, whether or not the description differs from subsequent descriptions by that witness or other witnesses.

(5) The duty to retain material which may be relevant to the investigation also includes in particular the duty to retain material which may satisfy the test for prosecution disclosure in the Application Order, such as—

(a) information provided by an accused person which indicates an explanation for the offence with which he has been charged;

(b) any material casting doubt on the reliability of a confession;

(c) any material casting doubt on the reliability of a witness.

(6) The duty to retain material falling into these categories does not extend to items which are purely ancillary to such material and possess no independent significance (for example, duplicate copies of records or reports).

(7) All material which may be relevant to the service investigation must be retained until a decision is taken whether to institute proceedings against a person for a Service offence.

(8) If a service investigation results in proceedings being instituted, all material which may be relevant must be retained at least until the accused is acquitted or convicted or the Director decides not to proceed with the case.

(9) Where the accused is convicted, all material which may be relevant must be retained at least until—

(a) the convicted person is released from custody, or discharged from hospital, in cases where the court imposes a custodial sentence or a hospital order;

(b) six months from the date of conviction, in all other cases.

If the service court imposes a custodial sentence or hospital order and the convicted person is released from custody or discharged from hospital earlier than six months from the date of conviction, all material which may be relevant must be retained at least until six months from the date of conviction.

(10) If an appeal against conviction is in progress when the release or discharge occurs or at the end of the period of six months specified in paragraph 5(9), all material which may be relevant must be retained until the appeal is determined. Similarly if the Criminal Cases Review Commission is considering an application at that point in time, all material which may be relevant must be retained at least until the Commission decides not to refer the case to the appropriate court.

(11) In paragraph 5(10) 'appropriate court'—

(a) where an application relates to a conviction by the Service Civilian Court, means the Court Martial; and

(b) where an application relates to a conviction by the Court Martial (including on appeal from the Service Civilian Court), means the Court Martial Appeal Court.

(12) In paragraph 5(8) to 5(10)—

(a) a reference to a person being acquitted includes a reference to the Summary Appeal Court quashing under section 147(1)(a) of the 2006 Act a finding that a charge against that person had been proved; and

(b) a reference to a person being convicted includes a reference to the Summary Appeal Court confirming under section 147(1)(a) of the 2006 Act, or substituting under section 147(1)(b) of that Act, a finding that a charge against that person has been proved.

Preparation of material for Director

6.—(1) The officer in charge of the investigation, the disclosure officer or an investigator may seek advice from the Director about whether any particular item of material may be relevant to the investigation.

(2) Material which may be relevant to an investigation, which has been retained in accordance with the Code, and which the disclosure officer believes will not form part of the prosecution case, must be listed in a schedule.

(3) Material which the disclosure officer does not believe is sensitive must be listed in a schedule of non-sensitive material. The schedule must include a statement that the disclosure officer does not believe the material is sensitive.

(4) Any material which is believed to be sensitive must be either listed in a schedule of sensitive material or, in exceptional circumstances, revealed to the Director separately. If there is no sensitive material, the disclosure officer must record this fact in a schedule of sensitive material.

(5) Paragraphs 6(6) to 6(10) below apply to both sensitive and non-sensitive material. Paragraphs 6(11) to 6(13) apply to sensitive material only.

(6) The disclosure officer must ensure that a schedule is prepared if—

> (a) a person is charged under section 120 or 121 of the 2006 Act, and the charge is allocated for the purposes of Part 5 of the 2006 Act for trial by the Court Martial or the Service Civilian Court; or
>
> (b) a person brings an appeal under section 141 of the 2006 Act to the Summary Appeal Court.

(7) As soon as reasonably practicable after it becomes clear to the disclosure officer that a charge is to be tried by the Court Martial following an election under the 2006 Act for Court Martial trial, he must ensure that a schedule is prepared.

(8) The disclosure officer should ensure that each item of material is listed separately in the schedule, and is numbered consecutively. The description of each item should make clear the nature of the item and should contain sufficient detail to enable the Director to decide whether he needs to inspect the material before deciding whether or not it should be disclosed.

(9) In some enquiries it may not be practicable to list each item of material separately. For example, there may be many items of a similar or repetitive nature. These may be listed in a block and described by quantity and generic title.

(10) Even if some material is listed in a block, the disclosure officer must ensure that any items among that material which might satisfy the test for prosecution disclosure under the Application Order are listed and described individually.

(11) Subject to sub-paragraph (12) below, the disclosure officer must list in the schedule of sensitive material any material the disclosure of which he believes would give rise to a real risk of serious prejudice to an important public interest, and the reason for that belief. The schedule must include a statement that the disclosure officer believes the material is sensitive. Depending on the circumstances, examples of such material may include the following—

(a) material relating to national security;

(b) material received from the intelligence and security agencies;

(c) material relating to intelligence from foreign sources which reveals sensitive intelligence-gathering methods;

(d) material given in confidence;

(e) material relating to the identity or activities of informants, undercover Service police officers, undercover police officers or other persons supplying information to the Service police who may be in danger if their identities are revealed;

(f) material revealing the location of any premises or other place used for Service police or police surveillance, or the identity of any person allowing a Service policeman or a police officer to use them for surveillance;

(g) material revealing, either directly or indirectly, techniques and methods relied upon by a Service policeman or a police officer in the course of an investigation, for example covert surveillance techniques or other methods of detecting offences;

(h) material whose disclosure might facilitate the commission of other offences or hinder the prevention and detection of offences;

(i) material upon the strength of which search warrants were obtained;

(j) material containing details of persons taking part in identification parades;

(k) material supplied to an investigator during a service investigation which has been generated by an official of a body concerned with the regulation or supervision of bodies corporate or of persons engaged in financial activities, or which has been generated by a person retained by such a body;

(l) material supplied to an investigator during a service investigation which relates to a child or young person and which has been generated by a local authority social services department, an Area Child Protection Committee or another person or body contacted by an investigator during the investigation;

(m) material relating to the private life of a witness.

(12) In exceptional circumstances, where an investigator considers that material is so sensitive that its revelation to the Director by means of an entry in the schedule of sensitive material is inappropriate, the existence of the material must be revealed to the Director separately. This will apply only where compromising the material would be likely to lead directly to the loss of life, or directly threaten national security.

(13) In such circumstances, the responsibility for informing the Director lies with the investigator who knows the detail of the sensitive material. The investigator should act as soon as is reasonably practicable after the Service police report on the case is sent to the Director. The investigator must also ensure that the Director is able to inspect the material so that he can assess whether it is disclosable and, if so, whether it needs to be brought before a judge advocate for a ruling on disclosure.

Revelation of material to Director

7.—(1) The disclosure officer must give the schedules prepared in accordance with paragraph 6 to the Director. Wherever practicable this should be at the same time as he gives him the Service police report on the case (or as soon as is reasonably practicable after it becomes clear that a charge is to be tried by the Court Martial, in cases for which paragraph 6(7) applies).

(2) The disclosure officer should draw the attention of the Director to any material an investigator has retained (including material not listed in a schedule and to which paragraph 6(12) applies) which may satisfy the test for prosecution disclosure in the Application Order, and should explain why he has come to that view.

(3) At the same time as complying with the duties in paragraphs 7(1) and 7(2), the disclosure officer must give the Director a copy of any material which falls into any of the following categories (unless such material has already been given to the Director as part of the Service police report on the case)—

(a) information provided by an accused which indicates an explanation for the Service offence with which he has been charged;

(b) any material casting doubt on the reliability of a confession;

(c) any material casting doubt on the reliability of a prosecution witness;

(d) any other material which the investigator believes may satisfy the test for prosecution disclosure in the Application Order.

(4) If the Director asks to inspect material which has not already been copied to him, the disclosure officer must allow him to inspect it. If the Director asks for a copy of material which has not already been copied to him, the disclosure officer must give him a copy. However, this does not apply where the disclosure officer believes, having consulted the officer in charge of the investigation, that the material is too sensitive to be copied and can only be inspected.

(5) If material consists of information which is recorded other than in writing, whether it should be given to the Director in its original form as a whole, or by way of relevant extracts recorded in the same form, or in the form of a transcript, is a matter for agreement between the disclosure officer and the Director.

Subsequent action by disclosure officer8.—(1) At the time a schedule of non-sensitive material is prepared, the disclosure officer may not know exactly what material will form the case against the accused, and the Director may not have given advice about the likely relevance of particular items of material. Once these matters have been determined, the disclosure officer must give the Director, where necessary, an amended schedule listing any additional material—

(a) which may be relevant to the investigation,

(b) which does not form part of the case against the accused,

(c) which is not already listed on the schedule, and

(d) which he believes is not sensitive,

unless he is informed in writing by the Director that the Director intends to disclose the material to the defence.

(2) Article 13 of the Application Order imposes a continuing duty on the Director, for the duration of proceedings against the accused for a Service offence, to disclose material which meets the tests for disclosure (subject to public interest considerations). To enable him to do this, any new material coming to light should be treated in the same way as the earlier material.

(3) In particular, after a defence statement has been given, the disclosure officer must look again at the material which has been retained and must draw the attention of the Director to any material which might reasonably be considered capable of undermining the case for the prosecution against the accused or of assisting the case for the accused; and he must reveal it to him in accordance with paragraphs 7(4) and 7(5).

Certification by disclosure officer

9. The disclosure officer must certify to the Director that to the best of his knowledge and belief, all material which has been retained and made available to him has been revealed to the Director in accordance with the Code. He must sign and date the certificate. It will be necessary to certify, not only when any schedule prepared in accordance with paragraph 6 and accompanying material is submitted to the Director and when material which has been retained is reconsidered in accordance with paragraph 8(3) after the accused has given a defence statement, but also whenever in

accordance with paragraph 8(2) a schedule is otherwise given or material is otherwise revealed to the Director.

Disclosure of material to accused

10.—(1) If material has not already been copied to the Director, and he requests its disclosure to the accused on the ground that it satisfies the test for prosecution disclosure under the Application Order, the disclosure officer must disclose it to the accused.

(2) If material has been copied to the Director and it is to be disclosed, whether it is disclosed by the Director or the disclosure officer is a matter for agreement between the two of them.

(3) The disclosure officer must disclose material to the accused either by giving him a copy or by allowing him to inspect it. If the accused asks for a copy of any material which he has been allowed to inspect, the disclosure officer must give it to him, unless in the opinion of the disclosure officer that is either not practicable (for example because the material consists of an object which cannot be copied, or because the volume of material is so great), or not desirable (for example because the material is a statement by a child witness in relation to a sexual offence).

(4) If material which the accused has been allowed to inspect consists of information which is recorded other than in writing, whether it should be given to the accused in its original form or in the form of a transcript is a matter for the discretion of the disclosure officer. If the material is transcribed, the disclosure officer must ensure that the transcript is certified to the accused as a true record of the material which has been transcribed.

(5) If a judge advocate concludes that an item of sensitive material satisfies the test for prosecution disclosure under the Application Order and that the interests of the defence outweigh the public interest in withholding disclosure, it will be necessary to disclose the material if the case is to proceed. This does not mean that sensitive documents must always be disclosed in their original form: for example, the judge advocate may agree that sensitive details still requiring protection should be blocked out, or that documents may be summarised, or that the Director may make an

admission about the substance of the material under section 10 of the Criminal justice Act 1967(5).

Transitional Provisions

11.—(1) In the Code—

(a) 'commencement' means 31st October 2009.

(b) 'Director' includes, with respect to anything done before commencement in relation to a service investigation to which the Code applies, a prosecutor within the meaning of the old code;

(c) 'judge advocate' includes, with respect to anything done before commencement in relation to a service investigation to which the Code applies, a judicial officer within the meaning of the old code.

(d) 'old code' means the code of practice under the Criminal Procedure and Investigations Act 1996 (Code of Practice) (Armed Forces) Order 2008;

(e) 'SDA offence' means an offence under the Army Act 1955, the Air Force Act 1955, the Naval Discipline Act 1957, the Armed Forces Act 1991 or the Reserve Forces Act 1996 committed before commencement;

(f) 'Service offence' includes an SDA offence.

(2) Where before commencement the old code applied in relation to a service investigation, the Code shall apply in relation to it, but anything done in relation to the investigation before commencement in accordance with the old code shall be treated as having been done in accordance with the Code.

(3) In paragraph 5(2), in relation to material seized before commencement, the reference to Part 3 of the 2006 Act includes a reference to Part 2 of the Armed Forces Act 2001(11).

(4) In paragraph 6(6) the reference to a person charged under section 120 or 121 of the 2006 Act includes a person who before commencement has been

charged under section 83B of the Army Act 1955, section 83B of the Air Force Act 1955 or section 52I of the Naval Discipline Act 1957.

(5) In paragraph 6(7) the reference to the 2006 Act includes a reference to any of the Army Act 1955, the Air Force Act 1955 and the Naval Discipline Act 1957.

APPENDIX M ALCOHOL AND DRUGS TESTING

Testing for alcohol and drugs in connection with a suspected offence or accident

CHAPTER 3A

(Inserted by s. 11 of the AFA 2011, in force 1st November 2013)[1] as amended by s. 2(2)(g) of the AFA 2016.[2]

Testing for alcohol and drugs on suspicion of offence.

Preliminary testing for alcohol and drugs

93A **Commanding officer's power to require preliminary tests**

(A1) This section applies in the situations described in subsections (1) to (3C).

(1) The first situation is where the commanding officer of a person Subject to Service Law has reasonable cause to believe that that person—

(a) is committing a safety-critical duty offence; or

(b) has committed such an offence and still has alcohol or a drug in the body or is still under the influence of a drug.

(1A) In a situation described in section 93A(3A), (3B) or (3C), a preliminary breath test is a procedure administered by a Service policeman under which—

(a) the affected person provides a specimen of breath; and

(b) the specimen is used for the purpose of obtaining, by means of an approved device, an indication of the proportion of alcohol in the person's breath or blood.

(2) Repealed by AFA 2016, s. 2(2)(e)

(3) The second situation is where the commanding officer of a person who is a civilian subject to service discipline has reasonable cause to believe that that person—

(a) is committing an offence under section 42 as respects which the corresponding offence under the law of England and Wales is an offence under section 78, 79, 92 or 93 of the Railways and Transport Safety Act 2003 (maritime and aviation offences); or

(b) has committed such an offence under section 42 and still has alcohol or a drug in the body or is still under the influence of a drug.

(3A) The third situation is where the commanding officer of a person subject to Service law or of a person who is a civilian subject to service discipline has reasonable cause to believe that—

(a) there has been an accident involving an aircraft or a ship; and

(b) at the time of the accident, the person was carrying out an aviation function in relation to the aircraft or a marine function in relation to the ship.

(3B) The fourth situation is where the commanding officer of a person subject to service law or of a person who is a civilian subject to service discipline has reasonable cause to believe that—

(a) there has been an accident involving an aircraft or a ship;

(b) before the accident, the person carried out an aviation function in relation to the aircraft or a marine function in relation to the ship; and

(c) it is possible that the carrying out of the function by the person may have caused or contributed to—

(i) the occurrence of the accident;

(ii) any death, injury to a person, damage to property or environmental harm resulting from the accident; or

(iii) any risk of death or of such injury, damage or harm created by the accident.

(3C) The fifth situation is where the commanding officer of a person subject to Service law or of a person who is a civilian subject to service discipline has reasonable cause to believe that—

(a) there has been an accident which resulted in or created a risk of—

(i) death;

(ii) serious injury to any person;

(iii) serious damage to property; or

(iv) serious environmental harm;

(b) the person—

(i) was carrying out a safety -critical function at the time of the accident; or

(ii) carried out a safety-critical function before the accident; and

(c) it is possible that the carrying out of the safety-critical function by the person may have caused or contributed to—

(i) the occurrence of the accident;

(ii) the death, injury, damage or harm; or

(iii) the risk of death, in jury, damage or harm.

(4) Where this section applies the commanding officer may require the person mentioned in subsection (1), (3), (3A), (3B) or (3C) ('the affected person') to co-operate with any one or more of—

(a) a preliminary breath test (see section 93B);

(b) a preliminary impairment test (see section 93C);

(c) a preliminary drug test (see section 93D).

(5) The Defence Council may by regulations provide for the delegation by a commanding officer of the commanding officer's functions under this section.

(6) A person who, without reasonable excuse, fails to comply with a requirement imposed under subsection (4) commits an offence.

(7) A person guilty of an offence under this section is liable to any punishment mentioned in the Table in section 164, but any sentence of imprisonment imposed in respect of the offence must not exceed two years.

(8) For the purposes of this section, a person does not co-operate with a preliminary test unless the person's co-operation—

(a) is sufficient to enable the test to be carried out; and

(b) is provided in such a way as to enable the objective of the test to be satisfactorily achieved.

93AA **Section 93A: interpretation**

(1) In section 93A(1), 'safety-critical duty offence' means—

(a) an offence under section 20A; or

(b) an offence under section 20(1)(a) in respect of a safety-critical duty (as defined in section 93I).

(2) In section 93A(3A) and (3B)—

'aviation function' means a role or activity in connection with aviation that is specified, or of a description specified, by regulations made by the Defence Council for the purposes of those subsections;

'marine function' means a role or activity in connection with a ship or ships that is specified, or of a description specified, by regulations made by the Defence Council for the purposes of those subsections.

(3) The Defence Council may specify a role or activity (or description of role or activity) under subsection (2) only if carrying it out with ability impaired by alcohol or drugs would result in a risk of—

(a) death,

(b) serious injury to any person,

(c) serious damage to property, or

(d) serious environmental harm,

but this is subject to subsection (4).

(4) The Defence Council's powers under subsection (2) include power to specify a role or activity that is undertaken in preparation for, or in connection with, the carrying out of a role or activity (or description of role or activity) that satisfies the test in subsection (3), either by specifying such a role or activity generally or by specifying a particular role or activity.

(5) For the purposes of section 93A(3A) and (3B), an accident does not involve an aircraft or a ship simply because it takes place on an aircraft or ship.

(6) In section 93A(3C), references to the carrying out of a safety-critical function are to—

(a) the performance by a person subject to Service law of a duty specified, or of a description specified, by regulations under section 20A(2) or of any other safety-critical duty (as defined in section 93I); or

(b) the carrying out by a person who is a civilian subject to service discipline, in the course of the person's employment, of a role or activity which, if it were carried out by a person subject to Service law in the course of his or her duty, would be a safety-critical duty.

(7) References in section 93A and this section to a person carrying out a function include a failure by the person to carry out a function at a time when the person is responsible for carrying it out (and related expressions are to be read accordingly).

93B **Preliminary breath test**

(1) In a situation described in section 93A(1) or (3), a preliminary breath test is a procedure administered by a Service policeman under which—

(a) the affected person provides a specimen of breath; and

(b) the specimen is used for the purpose of obtaining, by means of an approved device, an indication whether the proportion of alcohol in the person's breath or blood is likely to be such as is necessary for the commission of the suspected offence.

(1A) In a situation described in section 93A(3A), (3B) or (3C), a preliminary breath test is a procedure administered by a Service policeman under which—

(a) the affected person provides a specimen of breath; and

(b) the specimen is used for the purpose of obtaining, by means of an approved device, an indication of the proportion of alcohol in the person's breath or blood.

(2) A preliminary breath test may be administered only—

(a) at or near the place where the requirement to co-operate with the test is imposed;

(b) at a Service police establishment determined by the Service policeman; or

(c) at a medical establishment.

(3) In this section 'the suspected offence' means an offence mentioned in section 93A(1)(a) or (3) (a) which the commanding officer has reasonable cause to believe has been committed.

93C **Preliminary impairment test**

(1) A preliminary impairment test is a procedure under which a Service policeman—

(a) observes the affected person performing tasks specified by the Service policeman; and

(b) makes such other observations of the affected person's physical state as the Service policeman thinks expedient.

(2) A preliminary impairment test may be administered only—

(a) at or near the place where the requirement to co-operate with the test is imposed;

(b) at a Service police establishment determined by the Service policeman; or

(c) at a medical establishment.

(3) The Provost Marshals (acting jointly) must issue, and may from time to time revise, a code of practice about—

(a) the kind of task that may be specified for the purposes of a preliminary impairment test;

(b) the kind of observation of physical state that may be made in the course of a preliminary impairment test;

(c) the way in which a preliminary impairment test should be administered; and

(d) the inferences that may be drawn by a Service policeman from observations made in the course of a preliminary impairment test.

(4) In subsection (3) 'the Provost Marshals' means the Provost Marshals of each of the Service police forces.

(5) A Service policeman administering a preliminary impairment test must have regard to the code of practice.

(6) A Service policeman may administer a preliminary impairment test only if the Service policeman is approved for that purpose by a Provost Marshal of a Service police force.

(7) A code of practice under this section may include provision about—

(a) the giving of approval under subsection (6) ; and

(b) in particular, the kind of training that a Service policeman should have undergone, or the kind of qualification that a Service policeman should possess, before being approved under that subsection.

93D **Preliminary drug test**

(1) A preliminary drug test is a procedure administered by a Service policeman under which—

(a) a specimen of sweat or saliva is obtained from the affected person; and

(b) the specimen is used for the purpose of obtaining, by means of an approved device, an indication whether there is a drug in the person's body.

(2) A preliminary drug test may be administered only—

(a) at or near the place where the requirement to co-operate with the test is imposed;

(b) at a Service police establishment determined by the Service policeman; or

(c) at a medical establishment.

Provision of specimens for analysis

93E **Provision of specimens for analysis**

(1) This section applies in relation to an investigation into whether a person has committed—

(a) an offence under section 20A;

(b) an offence under section 20(1) (a) in respect of a safety-critical duty (as defined by section 93I) ; or

(c) an offence under section 42 as respects which the corresponding offence under the law of England and Wales is an offence under section 78, 79, 92 or 93 of the Railways and Transport Safety Act 2003.

(2) In the course of the investigation a Service policeman may require the person—

(a) to provide two specimens of breath for analysis by means of an approved device;

(b) to provide a specimen of blood or urine for a laboratory test.

(3) A requirement under this section may be imposed only at a Service police establishment or a medical establishment.

(4) For the purposes of this section and section 93F, a person does not provide a specimen of breath for analysis unless the specimen—

(a) is sufficient to enable the analysis to be carried out; and

(b) is provided in such a way as to enable the objective of the analysis to be satisfactorily achieved.

(5) For the purposes of this section and section 93F, a person provides a specimen of blood if and only if—

(a) he consents to the taking of such a specimen from him;

(b) the specimen is taken from him by a registered medical practitioner or registered nurse; and

(c) the specimen is of sufficient quantity to enable it to be divided into two parts for the purposes of analysis.

(6) For the purposes of this section and section 93F, a person provides a specimen of urine if and only if the specimen—

(a) is provided within one hour of the requirement for its provision being made and after the provision of a previous specimen of urine; and

(b) is of sufficient quantity to enable it to be divided into two parts for the purposes of analysis.

(7) Where the provision of a specimen may be required under this section, the question of whether it is to be breath, blood or urine, and in the case of blood the question of who is to be asked to take it, is to be decided by the Service policeman imposing the requirement.

(8) But where a Service policeman decides for the purposes of subsection (7) to require the provision of a specimen of blood, there shall be no requirement to provide such a specimen if—

(a) the registered medical practitioner who is asked to take the specimen is of the opinion that, for medical reasons, it cannot or should not be taken; or

(b) the registered nurse who is asked to take it is of that opinion and there is no contrary opinion from a registered medical practitioner;

and where by virtue of this subsection there can be no requirement to provide a specimen of blood, the Service policeman may require a specimen of urine instead.

(9) A Service policeman must, on requiring a person to provide a specimen in pursuance of this section, warn the person that a failure to provide it may render the person liable to proceedings for a service offence.

(10) A person who, without reasonable excuse, fails to provide a specimen when required to do so in pursuance of this section is guilty of an offence.

(11) A person guilty of an offence under this section is liable to any punishment mentioned in the Table in section 164, but any sentence of imprisonment imposed in respect of the offence must not exceed two years.

93F **Further provision about specimens under section 93E**

(1) Where two specimens of breath are provided by a person in pursuance of section 93E, the one with the lower proportion of alcohol in the breath is to be used and the other is to be disregarded.

(2) If the specimen with the lower proportion of alcohol contains no more than a prescribed proportion of alcohol, the person who provided it may claim that it should be replaced by such a specimen of blood or urine as may be required under section 93E.

(3) If the person then provides such a specimen, neither specimen of breath is to be used.

(4) In subsection (2) 'prescribed' means prescribed by regulations made by the Defence Council for the purposes of this section; and the regulations may prescribe different proportions of alcohol in relation to different kinds of offence.

(5) On a request made at the time a specimen of blood or urine is provided under section 93E, the person who provided the specimen must be given a part of the specimen sufficient for the purposes of analysis.

93G **Specimens of blood from persons incapable of consenting**

(1) A Service policeman may request a registered medical practitioner to take a specimen of blood from a person ('the person concerned'), irrespective of whether that person consents, if—

(a) the Service policeman would (in the absence of any incapacity of the person concerned and of any objection under section 93H) be entitled under section 93E to require the person concerned to provide a specimen of blood for a laboratory test;

(b) it appears to the Service policeman that the person concerned has been involved in an accident that constitutes or is comprised in the matter that is under investigation or the circumstances of that matter;

(c) it appears to the Service policeman that the person concerned is or may be incapable of giving a valid consent to the taking of a specimen of blood (whether or not consent has purportedly been given) ; and

(d) it appears to the Service policeman that that person's incapacity is attributable to medical reasons.

(2) It is lawful for a registered medical practitioner to whom a request is made under this section, if that practitioner thinks fit—

(a) to take a specimen of blood from the person concerned irrespective of whether that person consents; and

(b) to provide the specimen to a Service policeman.

(3) The specimen must be of sufficient quantity to enable it to be divided into two parts for the purposes of analysis.

(4) If a specimen is taken in pursuance of a request under this section, it must not be subjected to a laboratory test unless the person concerned—

(a) has been informed that it was taken;

(b) has been required by a Service policeman to give permission for a laboratory test of the specimen; and

(c) has given permission.

(5) A Service policeman, on requiring a person to give permission for the purposes of this section for a laboratory test of a specimen, must warn the person that a failure to give the permission may render the person liable to proceedings for a service offence.

(6) On a request made at the time a person gives permission under this section for a laboratory test of a specimen, that person must be given a part of the specimen sufficient for the purposes of analysis.

(7) A person who, without reasonable excuse, fails to give permission for a laboratory test of a specimen taken from the person under this section is guilty of an offence.

(8) A person guilty of an offence under this section is liable to any punishment mentioned in the Table in section 164, but any sentence of imprisonment imposed in respect of the offence must not exceed two years.

General provisions relating to testing

93H **Patients in medical establishments**

(1) This section applies in relation to a person who is at a medical establishment as a patient.

(2) The person shall not be required to co-operate with a preliminary test or to provide a specimen under section 93E unless the responsible medical professional has been notified of the proposal to impose the requirement.

(3) If the responsible medical professional objects on medical grounds the requirement must not be imposed.

(4) If the responsible medical professional does not object on medical grounds and the requirement is imposed, the requirement must be for co-operation with a preliminary test administered, or for the provision of a specimen, at the medical establishment.

(5) No specimen of blood may be taken from the person under section 93G, and the person may not be required to give permission for a laboratory test of a specimen taken under that section, unless the responsible medical professional—

(a) has been notified of the proposal that the specimen be taken or of the proposal to make the requirement; and

(b) has not objected on medical grounds.

(6) In this section 'the responsible medical professional' means—

(a) the registered medical practitioner in immediate charge of the person's case; or

(b) if there is no such registered medical practitioner, the registered nurse in immediate charge of the person's case.

(7) In this section 'medical grounds' means—

(a) in relation to a requirement to co-operate with a preliminary test or to provide a specimen under section 93E, the ground that the requirement, or compliance with it by the patient, or any warning required by section 93E(9), would be prejudicial to the proper care and treatment of the patient;

(b) in relation to the taking of a specimen under section 93G or a requirement to give permission for a laboratory test of a specimen taken under that section, the ground that the taking of the specimen, the requirement, or any warning required by section 93G(5), would be so prejudicial.

93I **Definitions for purposes of Chapter 3A**

(1) In this Chapter—

'approved', in relation to a device, means approved by the Secretary of State;

'drug' includes any intoxicant other than alcohol;

'medical establishment' means any facility at which medical or surgical treatment for in- or out-patients is provided;

'preliminary test' means—

(a) a preliminary breath test within the meaning of section 93B;

(b) a preliminary impairment test within the meaning of section 93C; or

(c) a preliminary drug test within the meaning of section 93D;

'safety-critical duty' means a duty which the commanding officer of the person mentioned in section 93A(1), 93AA(6)(a) or 93E(1) reasonably believes is such that performing the duty with ability impaired by alcohol or drugs would result in a risk of—

(a) death;

(b) serious injury to any person;

(c) serious damage to property; or

(d) serious environmental harm;

'Service police establishment' means any building or part of a building, any structure, or any room (whether on land or on a ship) which is used by a Service policeman for the performance of his duties.

(2) In this Chapter any reference to a Service policeman includes a Royal Navy coxswain.'

APPENDIX N SPECIAL MEASURES DIRECTIONS

Amendments by AF(SCR)(A)R 2015
SI 2015 No. 1812

Citation and commencement

1. These Rules may be cited as the Armed Forces (Service Courts Rules) (Amendment) Rules 2015 and shall come into force on 16th November 2015.

Summary Appeal Court rules

2. The Armed Forces (Summary Appeal Court) Rules 2009 are amended as follows.

3. In paragraph (2) of rule 3 (interpretation: general), after the definition of 'civilian police force' insert –

> 'the complainant', in relation to any offence (or alleged offence), means a person against or in relation to whom the offence was (or is alleged to have been) committed.

4. After rule 29 (interpreters) insert—

> **'Interpretation, translation for persons other than the appellant**
>
> **29A.**—(1) Where the complainant is due to attend a hearing as a witness, the court administration officer, where satisfied that the person needs interpretation, shall appoint an interpreter to act at the hearing.
>
> (2) The court administration officer may also appoint an interpreter for any other witness who is required to give evidence at a hearing, other than an appellant.

(3) Before an interpreter begins to act at a hearing, an oath shall be administered to the interpreter.

(4) Before an interpreter is sworn, the interpreter's name shall be read out, and any party to the proceedings may object to the interpreter on any reasonable ground.

(5) If the judge advocate upholds any such objection, the interpreter shall not be sworn, and the court administration officer shall appoint another interpreter.

(6) On application or on his own initiative, the judge advocate may require a written translation of any document or part of a document to be provided for a complainant or other person who attends a hearing as a witness (other than an appellant) and who needs interpretation.

(7) In its application by virtue of paragraph (6), nothing in this rule is limited by anything in Chapter 7 of Part 11 (special measures directions).'

5. In rule 30(2) (matters to be included in record of proceedings)—

(a) at the end of rule 30(2)(h) delete 'and';

(b) in rule 30(2)(i), delete the full stop and insert '; and';

(c) after rule 30(2)(i), insert—

'(j) a record of any decision on an application under rule 29A(6).'.

6. After rule 74C insert—

'CHAPTER 7
Special Measures Directions

Interpretation of Chapter 7

74D.—(1) In this Chapter—

'the 1999 Act' means the Youth Justice and Criminal Evidence Act 1999;

'eligible witness' means a witness eligible for assistance by virtue of rule 74E or 74F;

'intermediary' has the same meaning as in section 29 of the 1999 Act;

'special measures direction' means a direction providing for one or more of the special measures available in relation to a witness to apply to evidence given by the witness;

'the special measures provisions' means the provisions of Chapter 1 of Part 2 of the 1999 Act applied by an order under section 61(1) of that Act.

(2) In this Chapter—

(a) references to the quality of a witness's evidence are to its quality in terms of completeness, coherence and accuracy (and for this purpose 'coherence' refers to a witness's ability in giving evidence to give answers which address the questions put to the witness and can be understood both individually and collectively); and

(b) references to the special measures available in relation to a witness are to be construed in accordance with rule 74G.

Witnesses eligible for assistance on grounds of age or incapacity

74E.—(1) A witness is eligible for assistance by virtue of this rule if the witness is under the age of 18 at the time when it falls to the judge advocate to consider whether to give a special measures direction in relation to the witness.

(2) A witness is also eligible for assistance by virtue of this rule if the judge advocate considers that the quality of evidence given by the witness is likely to be diminished because the witness—

(a) suffers from mental disorder within the meaning of the Mental Health Act 1983;

(b) otherwise has a significant impairment of intelligence and social functioning; or

(c) has a physical disability or is suffering from a physical disorder.

Witnesses eligible for assistance on grounds of fear or distress about testifying

74F. A witness (other than an appellant) is eligible for assistance by virtue of this rule if the judge advocate is satisfied that the quality of evidence given by the witness is likely to be diminished by reason of fear or distress on the part of the witness in connection with testifying in the proceedings.

Special measures available

74G.—(1) Where a witness (other than an appellant) is eligible for assistance by virtue of rule 74E, the special measures available in relation to him are those for which provision is made by sections 23, 25 to 27, 29 and 30 of the 1999 Act.

(2) Where a witness is eligible for assistance by virtue of rule 74F, the special measures available in relation to him are those for which provision is made by sections 23 and 25 to 27 of that Act.

(3) Where an appellant is eligible for assistance by virtue of rule 74E, the special measures available in relation to him are those for which provision is made by sections 29 and 30 of that Act.

Special measures direction relating to eligible witness

74H.—(1) Subject to the special measures provisions and this Chapter, a judge advocate may give a special measures direction in relation to a witness if—

(a) the witness is an eligible witness; and

(b) any of the special measures available in relation to the witness (or any combination of them) would, in the judge advocate's opinion, be likely to improve the quality of evidence given by the witness.

(2) A special measures direction must specify particulars of the provision made by the direction in respect of each special measure which is to apply to the witness's evidence.

(3) In determining whether any special measure or measures would be likely to improve the quality of evidence given by the witness, and if so whether to give a direction providing for the measure or measures to apply, a judge advocate must consider all the circumstances of the case, including in particular—

(a) any views expressed by the witness; and

(b) whether the measure or measures might tend to inhibit such evidence being effectively tested by a party to the proceedings.

(4) A special measures direction may provide for one or more special measures to apply in combination with a direction under rule 25 (live links), and for the purposes of this Chapter a measure would be likely to improve the quality of the witness's evidence if, were it combined with such a direction, it would be likely to do so.

(5) A judge advocate may give a special measures direction—

(a) on an application made by a party to the proceedings; or

(b) of the judge advocate's own motion.

(6) A judge advocate who gives, or refuses an application for, a special measures direction must state in open court his reasons for doing so.

(7) Nothing in this Chapter is to be regarded as affecting any power of the court or a judge advocate to make an order or give leave of any description—

(a) in relation to a witness who is not an eligible witness; or

(b) in relation to an eligible witness, where the order is made or the leave is given otherwise than by reason of the fact that the witness is an eligible witness.

Evidence given in private

74I. A special measures direction may not provide for the exclusion of persons under section 25 of the 1999 Act unless it appears to the judge advocate that there are reasonable grounds for believing that any person other than an appellant has sought, or will seek, to intimidate the witness in connection with testifying in the proceedings.

Video recorded evidence in chief

74J.—(1) A special measures direction may not provide for a video recording, or a part of such a recording, to be admitted under section 27 of the 1999 Act if the judge advocate is of the opinion, having regard to all the circumstances of the case, that in the interests of justice the recording, or that part of it, should not be so admitted.

(2) In considering for the purposes of paragraph (1) whether any part of a recording should not be so admitted, the judge advocate must consider whether any prejudice to an appellant which might result from that part being so admitted is outweighed by the desirability of showing the whole, or substantially the whole, of the recorded interview.

(3) Where a special measures direction provides for a recording to be admitted under section 27 of the 1999 Act, the judge advocate may nevertheless subsequently direct that it is not to be so admitted if—

(a) it appears to the judge advocate that—

(i) the witness will not be available for cross-examination (whether conducted in the ordinary way or in accordance with any such direction); and

(ii) the parties to the proceedings have not agreed that there is no need for the witness to be so available; or

(b) rule 74M has not been complied with to the satisfaction of the judge advocate.

(4) Paragraph (3) is without prejudice to rule 74N (power to vary or discharge special measures direction).

(5) Where a recording is admitted under section 27 of the 1999 Act—

(a) the witness must be called by the party tendering it in evidence, unless the parties to the proceedings have agreed that there is no need for the witness to be called; and

(b) the witness may not give evidence in chief otherwise than by means of the recording—

(i) as to any matter which, in the opinion of the judge advocate, has been dealt with adequately in the witness's recorded testimony; or

(ii) without the leave of the judge advocate, as to any other matter which, in the opinion of the judge advocate, is dealt with in that testimony.

(6) Where a special measures direction provides for part only of a recording to be admitted under section 27 of the 1999 Act, references in paragraphs (3) and (4) to the recording or to the witness's recorded testimony are references to the part of the recording or testimony which is to be so admitted.

(7) The judge advocate may give leave for the purposes of paragraph (5)(b)(ii) if it appears to him to be in the interests of justice to do so, and may do so either—

(a) on an application by a party to the proceedings; or

(b) of his own motion.

Examination of witness through intermediary

74K.—(1) Any examination of a witness conducted in pursuance of a provision included in a special measures direction by virtue of section 29(1) of the 1999 Act (examination of witness through intermediary) must take place—

(a) in the presence of such persons as the direction may provide; and

(b) in circumstances in which the members of the court, and legal representatives acting in the proceedings, are able to see and hear the examination of the witness and to communicate with the intermediary (and for this purpose any impairment of eyesight or hearing is to be disregarded).

(2) Where two or more legal representatives are acting for a party to the proceedings, paragraph (1)(b) is to be regarded as satisfied in relation to those representatives if at all material times it is satisfied in relation to at least one of them.

(3) Before an intermediary begins to act, he shall make a declaration in the following form:

'I solemnly, sincerely and truly declare that I will well and faithfully communicate the questions and answers and make true explanation of all matters and things as shall be required of me according to the best of my skill and understanding.'

(4) In this rule 'the intermediary' has the same meaning as in section 29 of the 1999 Act.

Application for special measures direction: general

74L.—(1) An application for a special measures direction must be made in writing to the court administration officer, unless a judge advocate gives leave for it to be made orally.

(2) A written application must specify—

(a) unless the application is made by an appellant and does not relate to evidence in support of an alibi, the name and date of birth of the witness in relation to whom it is made;

(b) the special measure or measures sought;

(c) where the application is for a direction including provision by virtue of section 27 of the 1999 Act (video recorded evidence in chief), the information mentioned in rule 74M(4);

(d) the grounds on which the applicant asserts—

(i) that the witness is an eligible witness; and

(ii) that the measure or measures will improve the quality of the witness's evidence; and

(e) the views of the witness as to the matters specified in accordance with sub-paragraph (d).

(3) In paragraph (2)(a) 'evidence in support of an alibi' has the same meaning as in article 7 of the CPIA Order.

(4) A written application must be made, and a copy served on all other parties to the proceedings—

(a) where the application is made by the Director, not more than 14 days after the Director serves advance information in respect of the charge to which the proposed evidence relates;

(b) where the application is made by an appellant, not more than 14 days after the Director complies or purports to comply with article 4 of the CPIA Order.

(5) Notwithstanding paragraph (4), a judge advocate may at his discretion consider a written application made outside the period of 14 days there mentioned.

(6) Where a written application has been made, a judge advocate may—

(a) grant the application without a hearing; or

(b) direct a hearing.

(7) But the application may not be granted without a hearing unless—

(a) at least 14 days have elapsed since the application was served on each other party to the proceedings; and

(b) no other party has served notice on the court administration officer that he opposes the application.

(8) Any party to the proceedings—

(a) may attend a hearing of the application, and be heard;

(b) may, with leave of the judge advocate, adduce evidence (including expert evidence) at the hearing.

Application for special measures direction permitting admission of video recorded evidence in chief

74M.—(1) This rule applies where an application is made for a special measures direction including provision by virtue of section 27 of the 1999 Act.

(2) The application must be accompanied by a copy of the video recording which (or part of which) it is proposed to tender in evidence.

(3) Where the application is made by the Director, he must at the same time serve on the appellant a copy of that recording.

(4) The application must include the following information—

(a) the date on which the recording was made;

(b) the times at which the recording commenced and finished, including details of any interruptions;

(c) the address of the premises where the recording was made, and the usual function of those premises;

(d) in relation to each person present at any point during, or immediately before, the recording—

(i) the name, age and occupation of the person;

(ii) the time for which he was present; and

(iii) his relationship (if any) to the witness;

(e) in relation to the equipment used for the recording—

(i) a description of the equipment;

(ii) the number of cameras used;

(iii) whether the cameras were fixed or mobile;

(iv) the number and location of the microphones;

(v) the video format used; and

(vi) whether it offered single or multiple recording facilities and, if so, which were used; and

(f) if the recording is a copy—

(i) the location of the master recording; and

(ii) details of when and by whom the copy was made.

(5) Where the applicant is an appellant and the application is granted, the applicant must, not later than the close of the case for the prosecution, serve on each other party to the proceedings a copy of the video recording which (or part of which) it is proposed to tender in evidence under the direction.

Variation or discharge of special measures direction

74N.—(1) A judge advocate may vary or discharge a special measures direction if it appears to him to be in the interests of justice to do so.

(2) A judge advocate may exercise the power conferred by paragraph (1)—

(a) on an application made by a party to the proceedings; or

(b) of the judge advocate's own motion.

(3) An application under this rule must be made in writing to the court administration officer, unless—

(a) a judge advocate gives leave for it to be made orally; or

(b) paragraph (8) applies.

(4) A copy of a written application under this rule must be served on each other party to the proceedings.

(5) Where a written application has been made under this rule, a judge advocate may—

(a) grant the application without a hearing; or

(b) direct a hearing.

(6) But the application may not be granted without a hearing unless—

(a) at least 14 days have elapsed since the application was served on each other party to the proceedings; and

(b) no other party has served notice on the court administration officer that he opposes the application.

(7) Rule 74L(8) applies in relation to a hearing of the application as it applies in relation to a hearing of an application for a special measures direction.

(8) Where the direction was made on the application of an appellant and includes provision for the admission of a video recording which had not been served on the Director, the Director may make an oral application without leave.

(9) A judge advocate who varies or discharges, or refuses an application for the variation or discharge of, a special measures direction must state in open court his reasons for doing so.

(10) In this rule, references to the variation of a special measures direction include the further variation of a direction previously varied.

Warning to lay members

74O. Where in proceedings with lay members evidence has been given in accordance with a special measures direction, the judge advocate must give the lay members such warning (if any) as he considers necessary to ensure that the fact that the direction was given in relation to the witness does not prejudice the appellant.'.

7. In rule 86 (information before punishment) —

(a) at the end of rule 86(2)(g) delete 'and';

(b) in rule 86(2)(h), delete the full stop and insert '; and';

(c) after rule 86(2)(h), insert—

'(i) any statement of the effect of the offence on the victim, the victim's family or others.'.

Court Martial rules

8. The Armed Forces (Court Martial) Rules 2009 are amended as follows.

9. In paragraph (2) of rule 3 (interpretation: general), after the definition of 'civilian police force' insert –

"the complainant', in relation to any offence (or alleged offence), means a person against or in relation to whom the offence was (or is alleged to have been) committed;'.

10. After rule 22 (interpreters) insert—

'Interpretation, translation for persons other than a person to whom proceedings relate

21A.—(1) Where the complainant is due to attend a hearing as a witness, the court administration officer, where satisfied that the person needs interpretation, shall appoint an interpreter to act at the hearing.

(2) The court administration officer may also appoint an interpreter for any other witness who is required to give evidence at a hearing, other than a person to whom any proceedings relate.

(3) Before an interpreter begins to act at a hearing, an oath shall be administered to the interpreter.

(4) Before an interpreter is sworn, the interpreter's name shall be read out, and any party to the proceedings may object to the interpreter on any reasonable ground.

(5) If the judge advocate upholds any such objection, the interpreter shall not be sworn, and the court administration officer shall appoint another interpreter.

(6) On application or on his own initiative, the judge advocate may require a written translation of any document or part of a document to be provided for a complainant or other person who attends a hearing as a witness (other than a person to whom any proceedings relate) and who needs interpretation.'.

11. In rule 23(2) (matters to be included in record of proceedings)—

(a) at the end of rule 23(2)(h) delete 'and';

(b) in rule 23(2)(i), delete the full stop and insert '; and';

(c) after rule 23(2)(i), insert—

'(j)a record of any decision on an application under rule 22A(6).'.

12. In rule 114 (information before sentencing)—

(a) at the end of rule 114(2)(f) delete 'and';

(b) in rule 114(2)(g), delete the full stop and insert '; and';

(c) after rule 114(2)(g), insert—

'(h) any statement of the effect of the offence on the victim, the victim's family or others.'.

Service Civilian Court rules

13. The Armed Forces (Service Civilian Court) Rules 2009(4) are amended as follows.

14. In paragraph (2) of rule 3 (interpretation: general), after the definition of 'civilian police force' insert –

> "the complainant', in relation to any offence (or alleged offence), means a person against or in relation to whom the offence was (or is alleged to have been) committed;'.

15. After rule 21 (interpreters) insert—

> **'Interpretation, translation for persons other than a person to whom proceedings relate**
>
> **21A.**—(1) Where the complainant is due to attend a hearing as a witness, the court administration officer, where satisfied that the person needs interpretation, shall appoint an interpreter to act at the hearing.
>
> (2) The court administration officer may also appoint an interpreter for any other witness who is required to give evidence at a hearing, other than a person to whom any proceedings relate.
>
> (3) Before an interpreter begins to act at a hearing, an oath shall be administered to the interpreter.
>
> (4) Before an interpreter is sworn, the interpreter's name shall be read out, and any party to the proceedings may object to the interpreter on any reasonable ground.
>
> (5) If the judge advocate upholds any such objection, the interpreter shall not be sworn, and the court administration officer shall appoint another interpreter.

(6) On application or on his own initiative, the judge advocate may require a written translation of any document or part of a document to be provided for a complainant or other person who attends a hearing as a witness (other than a person to whom any proceedings relate) and who needs interpretation.'.

16. In rule 22(2) (matters to be included in record of proceedings)—

(a) at the end of rule 30(2)(j) delete 'and';

(b) in rule 30(2)(k), delete the full stop and insert '; and';

(c) after rule 30(2)(k), insert—

'(l) a record of any decision on an application under rule 21A(6).'.

17. In rule 96 (information before sentencing)—

(a) at the end of rule 96(2)(e) delete 'and';

(b) in rule 96(2)(f), delete the full stop and insert '; and';

(c) after rule 96(2)(f), insert—

'(g) any statement of the effect of the offence on the victim, the victim's family or others.'.

APPENDIX O YOUNG, DISABLED, VULNERABLE ETC. WITNESSES

2015 No. 1805

The Youth Justice and Criminal Evidence Act 1999 (Application to Service Courts) (Amendment) (No. 2) Order 2015

Made	*19th October 2015*
Laid before Parliament	*21st October 2015*
Coming into force	*15th November 2015*

The Secretary of State, in exercise of the powers conferred by sections section 61(1) and (2) of the Youth Justice and Criminal Evidence Act 1999(1), makes the following Order:

Citation and Commencement

1. This Order may be cited as the Youth Justice and Criminal Evidence Act 1999 (Application to Service Courts) (Amendment) (No. 2) Order 2015 and shall come into force on 15th November 2015.

Interpretation

2. In this Order the '2009 Order' means the Youth Justice and Criminal Evidence Act 1999 (Application to Service Courts) Order 2009(2).

Amendments to the 2009 Order

3.—(1) Article 2 (interpretation) of the 2009 Order is amended as follows.

(2) In paragraph (1), after the definition of 'CMAC Rules', insert –

"SAC Rules' means rules made under section 151 of the 2006 Act;'.

(3) In the definition of 'Rules of court' –

(a) at the end of sub-paragraph (b), delete 'and';

(b) at the end of sub-paragraph (c), insert –

'and

(d) the Summary Appeal Court;'.

(4) At the end of paragraph (2) (c) –

(a) substitute a semi-colon for the full stop;

(b) after that semi-colon, insert –

'(d) in relation to proceedings before the Summary Appeal Court, as a reference to a special measures direction within the meaning of the SAC Rules.'.

4.—(1) Article 3 (application of special measures provisions) of the 2009 Order is amended as follows.

(2) At the end of paragraph (2) (b), delete 'and'.

(3) At the end of paragraph (2) (c), insert –

'and

(d) the Summary Appeal Court,'.

5.—(1) Article 4 (modification of special measures provisions) of the 2009 Order is amended as follows.

(2) In paragraph (2), for 'or the Court Martial Appeal Court', substitute ', the Court Martial Appeal Court or the Summary Appeal Court'.

(3) At the end of paragraph (5) (c) –

(a) substitute a semi-colon for the full stop;

(b) after that semi-colon, insert –

'(d) in relation to proceedings within the meaning of the SAC Rules, as a reference to the appellant.'.

(4) In paragraph (6), for 'or the Service Civilian Court', substitute ', the Service Civilian Court or the Summary Appeal Court'.

(5) In paragraph (11) (a), after 'Court Martial' insert 'or the Summary Appeal Court'.

(6) In paragraph (12), for 'or the Service Civilian Court', substitute ', the Service Civilian Court or the Summary Appeal Court'.

(7) In paragraph (13), for 'or the Service Civilian Court', substitute ', the Service Civilian Court or the Summary Appeal Court'.

(8) In paragraph (16), for 'or the Service Civilian Court', substitute ', the Service Civilian Court or the Summary Appeal Court'.

6.—(1) Article 5 (application of protection of witnesses from cross-examination provisions) of the 2009 Order is amended as follows.

(2) After paragraph (1), insert –

> '(1A) The protection of witnesses from cross-examination provisions in paragraph (3) (b) to (e) shall apply in relation to proceedings before the Summary Appeal Court as they apply in relation to criminal proceedings, subject to the modifications specified in article 6.'.

7.—(1) Article 6 (modification of protection of witnesses from cross-examination provisions) of the 2009 Order is amended as follows.

(2) In paragraph (2), for 'or the Court Martial Appeal Court', substitute ', the Court Martial Appeal Court or the Summary Appeal Court'.

(3) In paragraph (3), for 'or the Service Civilian Court', substitute ', the Service Civilian Court or the Summary Appeal Court'.

(4) At the end of paragraph (4) (c) –

(a) substitute a semi-colon for the full stop;

(b) after that semi-colon, insert –

> '(d) in relation to proceedings before the Summary Appeal Court, as a reference to the appellant.'.

(5) In paragraph (6), after 'Court Martial Rules', insert 'or the SAC Rules'.

(6) In paragraph (8) , for 'or the Service Civilian Court', substitute ', the Service Civilian Court or the Summary Appeal Court'.

(7) In paragraph (9), for 'or the Service Civilian Court', substitute ', the Service Civilian Court or the Summary Appeal Court'.

(8) In paragraph (10) (a), for 'or the Service Civilian Court', substitute ', the Service Civilian Court or the Summary Appeal Court'.

(9) In paragraph (10) (b), for 'or the Court Martial Appeal Court', substitute ', the Court Martial Appeal Court or the Summary Appeal Court'.

(10) In paragraph (11), for 'or the Service Civilian Court', substitute ', the Service Civilian Court or the Summary Appeal Court'.

8.—(1) Article 9 (application of competence of witnesses and capacity to be sworn provisions) of the 2009 Order is amended as follows.

(2) At the end of paragraph (1) (b), delete 'and'.

(3) At the end of paragraph (1) (c), insert –

'and

(d) the Summary Appeal Court,'.

(4) At the end of paragraph (2) (a), delete 'and'.

(5) At the end of paragraph (2) (b), insert –

'and

(d) the Summary Appeal Court,'.

(6) In paragraph (3), after 'Court Martial', insert 'or the Summary Appeal Court'.

9.—(1) Article 10 (modification of competence of witness and capacity to be sworn provisions) of the 2009 Order is amended as follows.

(2) In paragraph (2), for 'or the Court Martial Appeal Court', substitute ', the Court Martial Appeal Court or the Summary Appeal Court'.

(3) In paragraph 3(a), for 'or the Service Civilian Court', substitute ', the Service Civilian Court or the Summary Appeal Court'.

(4) In paragraph (4), for 'or the Service Civilian Court', substitute ', the Service Civilian Court or the Summary Appeal Court'.

(5) In paragraph (5), after 'Court Martial', insert 'or the Summary Appeal Court'.

(6) In paragraph (6), for 'or the Service Civilian Court', substitute ', the Service Civilian Court or the Summary Appeal Court'.

(7) In paragraph (7), after 'Court Martial', insert 'or the Summary Appeal Court'.

(8) In paragraph (8), for 'or the Service Civilian Court', substitute ', the Service Civilian Court or the Summary Appeal Court'.

10.—(1) Article 11 (application of general and final provisions) of the 2009 Order is amended as follows.

(2) At the end of paragraph (1) (b), delete 'and'.

(3) At the end of paragraph (1) (c), insert –

'and

(d) the Summary Appeal Court,'.

11.—(1) Article 12 (modification of general and final provisions) of the 2009 Order is amended as follows.

(2) In paragraph (2), for 'or the Court Martial Appeal Court', substitute ', the Court Martial Appeal Court or the Summary Appeal Court'.

(3) In paragraph (5) (b), for 'or the CMAC Rules', substitute ', the CMAC Rules or the SAC Rules'.

(4) In paragraph (6) (a), for 'or the Service Civilian Court', substitute ', the Service Civilian Court or the Summary Appeal Court'.

(5) In paragraph (7), for 'or the Service Civilian Court', substitute ', the Service Civilian Court or the Summary Appeal Court'.

12.—(1) Article 13 (modification of reporting restrictions provisions) of the 2009 Order is amended as follows.

(2) In paragraph (2), for 'or the Court Martial Appeal Court', substitute ', the Court Martial Appeal Court or the Summary Appeal Court'.

(3) In paragraph (4), for 'or the Service Civilian Court', substitute ', the Service Civilian Court or the Summary Appeal Court'.

(4) At the end of paragraph (5) (c) –

(a) substitute a semi-colon for the full stop;

(b) after that semi-colon, insert –

'(d) in relation to proceedings before the Summary Appeal Court, "appellant"'.

(5) In paragraph (6), for 'or the Service Civilian Court', substitute ', the Service Civilian Court or the Summary Appeal Court'.

(6) In paragraph (7) (a), for 'or the Service Civilian Court', substitute ', the Service Civilian Court or the Summary Appeal Court'.

(7) In paragraph (8), for 'or the Service Civilian Court', substitute ', the Service Civilian Court or the Summary Appeal Court'.

INDEX

We hope that you have enjoyed this book.
Please consider posting a review for future readers.

Thank you.

www.ingramcontent.com/pod-product-compliance
Lightning Source LLC
LaVergne TN
LVHW052203200926
848481LV00001B/18

9781912440238